Medicinal Chemistry into the Millennium

Medicinal Chemistry into the Millennium

Edited by

M.M. Campbell
University of Bath, UK

I.S. Blagbrough
University of Bath, UK

ROYAL SOCIETY OF CHEMISTRY

Proceedings of the European Federation of Medicinal Chemistry Symposium held in Edinburgh on 6–11 September, 1998.

Special Publication No. 264

ISBN 0-85404-769-7

A catalogue record for this book is available from the British Library

Published by The Royal Society of Chemistry,
Thomas Graham House, Science Park, Milton Road,
Cambridge CB4 0WF, UK
Registered Charity No. 207890

For further information see our web site at www.rsc.org

Printed by Athanaeum Press Ltd, Gateshead, Tyne and Wear, UK

Preface

The XVth ISMC continued the now well established tradition of biennial medicinal chemistry symposia organised throughout Europe by the hosting Society under the auspices of the European Federation of Medicinal Chemistry. Judged by its multinational participation (over 1000 delegates from 48 different countries) and effective blend of industrialists, academics and students, Edinburgh undoubtedly enjoyed the success of preceding symposia and contributed to their international prestige. Comprising nine major themes, it is impossible to single out just one that is of particular importance, although supporting technologies, metabolism and pharmacokinetics studies underpin many of the advances made in specific therapeutic areas and are thus crucial to the successful exploitation of the plethora of newly discovered molecular targets.

The New Technologies in Drug Discovery session is particularly relevant to the pharmaceutical industry, and indeed all industries dependent on the discovery of biologically active molecules (*e.g.* pesticides, flavours and fragrances). With increasing numbers of molecular targets derived through genomic sequencing there has been a strong drive to enrich the quality and range of chemical libraries used to generate new leads, while at the same time enhancing the rate at which such libraries can be screened and hit molecules validated. Techniques such as library design, construction, analysis, quality control, robotisation, high throughput screening and informatics all feature strongly in modern lead identification.

Accompanying the New Technologies sessions were seven themes devoted to structurally distinct molecular targets. Such a selection can never be exhaustive, but those targets chosen cover areas perceived by the pharmaceutical industry to be therapeutically important. Undoubtedly a considerable amount of new data were presented during these and their supporting poster sessions, effectively contributing to the dissemination of current knowledge. While the Growth Factor and Glycochemistry and Glycobiology areas are less well established than the others, they are by no means less relevant to emerging therapeutic targets. It is perhaps in these areas that understanding at the molecular level is lacking, as opposed to the more established structural types where extensive research has paid significant dividends.

Having identified good leads on novel molecular targets one of the major problems confronting the pharmaceutical industry is the need to convert these into viable drugs. Drug metabolism and pharmacokinetics (DMPK) have always been a limitation in drug development, primarily because they lacked predictivity and were traditionally investigated late in the discovery programme. There is now a strong drive to develop predictive DMPK paradigms and to use these effectively early in the discovery process. The ninth session addressed some of the more pertinent problems and afforded insights into a number of the available solutions.

A large number of people contributed to the success of this symposium and it is not possible to thank them all. One person though, who through illness was forced to relinquish his chairmanship late in the organisational stage, does deserve a specific mention. Professor Malcolm Campbell worked hard to make Edinburgh a reality and its success is largely due to his insights and tremendous effort.

Derek Buckle
Chairman of the Organising Committee

Contents

Growth Factors

Intracellular Signalling

Protease Inhibition

Glycochemistry and Glycobiology

Nitric Oxide Synthase Inhibition

Predicting DMPK

New Technologies for Drug Discovery

DESIGN OF MACROCYCLIC PEPTIDASE INHIBITORS: THE RELATED ROLES OF STRUCTURE-BASED APPROACHES AND LIBRARY CHEMISTRY

Paul A. Bartlett,* Naeem Yusuff, Hyung-Jung Pyun, Alice C. Rico, J. Hoyt Meyer, Whitney W. Smith and Matthew T. Burger

Department of Chemistry, University of California, Berkeley, California 94720-1460

1 INTRODUCTION

New and improved approaches to the discovery and invention of biologically active compounds follow each other in rapid succession. Rational design approaches have at various times encompassed strategies that involve covalent and irreversible inactivation of receptors or enzymes with affinity labels, that exploit enzyme mechanism with suicide inhibitors and transition state analogs, or that rely on 3-dimensional characterization of the binding site or ligand for structure-based design. The currently fashionable combinatorial or library-based approaches would appear to represent a swing of the pendulum away from these "rational" methods. However, strategies that share a common goal are inevitably going to find common ground. There are elements of logic and serendipity in all of these approaches: mechanistic and structural insight can be used to advantage in the design of combinatorial libraries and the interpretation of screening results, and library screening can provide a key starting point in a structure-based optimization process. As we seek to demonstrate in this overview, one approach often sets the stage for another, and the newer strategies can provide insights and tools that enable the older methods to be applied more effectively.

2 MACROCYCLIC PEPTIDASE INHIBITORS

Our evolving interest in transition state analogs of peptidase inhibitors illustrates how these ideas can be applied to the design of potent inhibitors and used to understand some of the fundamental principles of protein-ligand binding. The underlying tenet of transition state analogy is embodied in Equations 1 and 2 (where TS and GS refer to transition and ground states, respectively).[1]

$$K_{TS} = (K_{GS}/k_{cat})k_{noncat} \qquad \text{Equation 1}$$

$$\log K_I \propto \log K_{TS} = \log(K_m/k_{cat}) + \text{constant} \qquad \text{Equation 2}$$

Linear relationships between inhibitor K_i and substrate K_m/k_{cat} values have enabled us to demonstrate transition state mimicry by phosphonamidate and phosphonate peptide inhibitors of zinc peptidases, such as thermolysin[2-4] and carboxypeptidase A,[5-7] and the aspartic peptidase pepsin[8,9] (Figure 1). The potency of these inhibitors, in one case reaching 11 fM,[6] and their conformational and electronic similarities to the transition state structures have made them valuable subjects for crystallographic analysis.[10-15] As a result, we have extensive structural information on how they bind to the target enzymes.

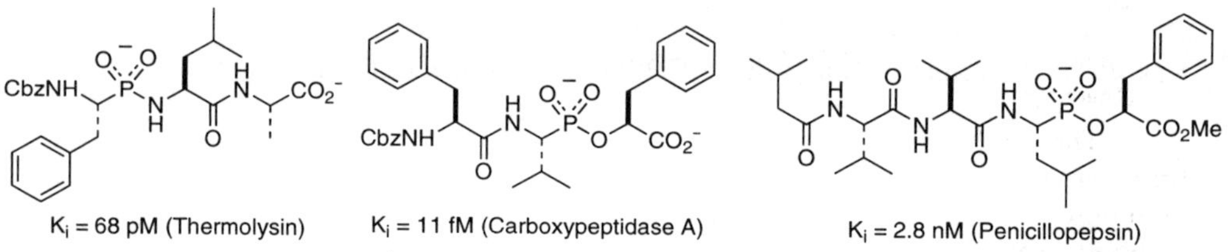

K$_i$ = 68 pM (Thermolysin)   K$_i$ = 11 fM (Carboxypeptidase A)   K$_i$ = 2.8 nM (Penicillopepsin)

Figure 1   *Representative phosphorus-containing peptidase inhibitors*

2.1 Macrocyclic Inhibitors of Thermolysin

In the bound conformations of phosphorus-containing inhibitors of thermolysin, represented by the complex with Cbz-Phe^P-Leu-Ala (K_i = 68 pM),[11] both the P1 side chain and the terminal carboxylate at P2' approach each other near the opening of the active site (Figure 2a). Their proximity stimulated us to design a rigidified analog in which the α-carbons at P1 and P2' are bridged by a bicyclic chroman unit.[16] This compound, **1**, is a potent inhibitor of thermolysin and binds as anticipated, with the aromatic ring spanning the narrow opening of the active site (Figure 2b). However, an attempt to determine how much of the binding enhancement for this analog (relative to an unbridged phosphonamidate **3**) arises from conformational constraint and how much is due to interactions between the enzyme and the bridging unit itself was undermined by the finding that the acyclic comparison compound **2** binds differently than **1**, with the bicyclic moiety rotated out of the active site (Figure 2c).

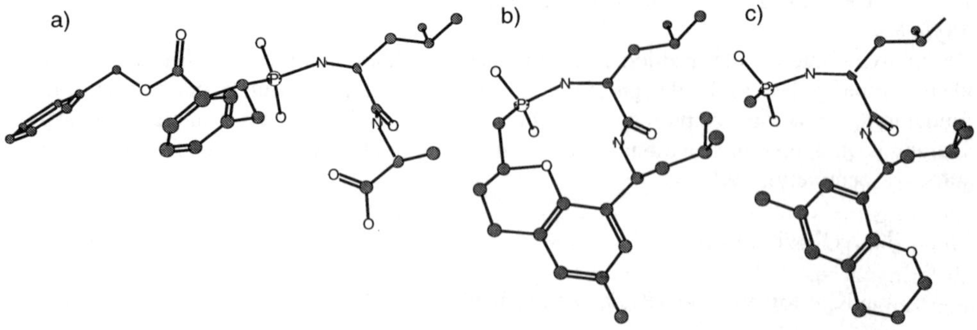

Figure 2   *Conformations of macrocyclic and acyclic inhibitors bound to thermolysin. a) Cbz-Phe^P-Leu-Ala; b) rigid tricycle **1**; c) seco-analog **2***

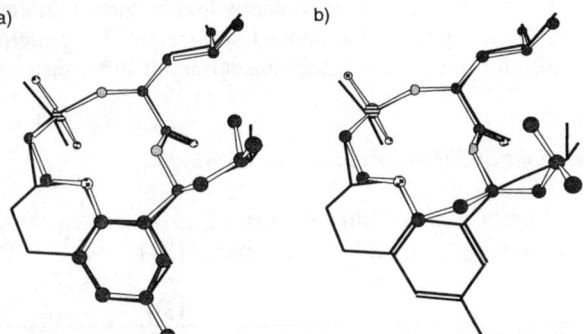

| **1** | **2** | **3** | **4** | **5** |
| 4 nM | 80 nM | 200 nM | 50 nM | 500 nM |

K_i (Thermolysin)

The complication of alternative binding modes has been minimized in the series of three macrocycles **1, 4,** and **5,** which span two orders of magnitude in binding affinity but differ only in the size and conformational flexibility of the bridging units. For this series of inhibitors, we have been able to distinguish the relative contributions that conformational mobility (or lack thereof) and hydrophobic interactions make to the differences in binding affinity. The bound conformations of the three analogs have been determined by Debbi Holland and Doug Juer in Brian Matthews' group in Oregon; **1** and **4** are essentially identical, except for the two-carbon segment that is missing from the latter (Figure 3a). The ring of **5** is slightly different because sp^3 hybridization and an additional degree of rotational freedom allow a more puckered conformation and a different rotamer of the isobutyl substituent corresponding to P2' (Figure 3b).

Figure 3   *Conformations of macrocyclic inhibitors bound to thermolysin; comparison of 1 (line) with a) 4 and b) 5 (ball-and-stick)*

The conformational properties of the three cyclic analogs were determined in solution, using a combined NMR and modeling method.[17] The findings are understandable, but nonetheless striking. The tricyclic analog **1** is essentially rigid, adopting the same conformation in solution that it does in the thermolysin active site (Figure 4a). The bicyclic derivative **4** has two comparable, low-energy conformations, differing at the O–C–C–P dihedral angle (Figure 4b); one corresponds to the bound conformation. Finally, the monocycle **5** is highly flexible in solution, particularly in the region of the bridging chain; a sample of low-energy conformers is shown in Figure 4c.

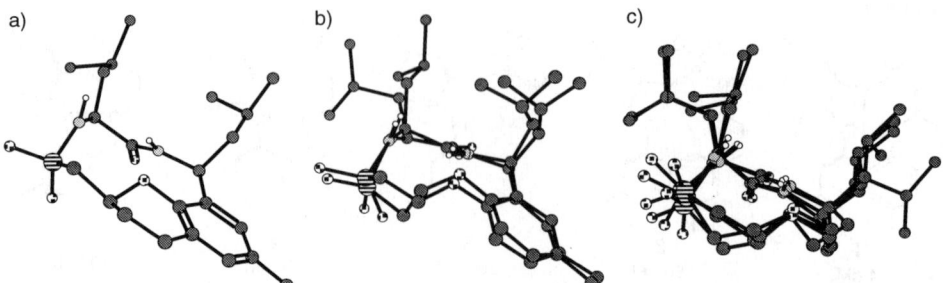

Figure 4 *Solution structures determined for macrocyclic thermolysin inhibitors from NMR and modeling: a) single conformation for 1, b) two conformations for 4; c) ensemble of low-energy conformations for 5*

The hydrophobic contributions to the differences in binding affinities of the three macrocycles were assessed through modeling as well as experimentally. The amount of hydrophobic surface area, on both enzyme and inhibitor, removed from contact with solvent on association was readily determined from the crystal structures of the native and inhibited enzymes (Table 1). The results are striking and unequivocal: while **1** and **4** differ slightly in size, there is almost the same reduction in exposed surface area when they bind to thermolysin. Since the extra ethylene unit of **1** does not contact the enzyme, it is exposed in both the bound and unbound states and thus does not contribute to the binding interaction. In contrast, monocycle **5** has significantly less contact with the enzyme than **1** or **4**; indeed, the reduction in hydrophobic contact surface for **1** represents a loss of 2.3 kcal/mole in binding affinity, much more than the observed difference of 1.4 kcal/mole versus **4** or **5**.

Table 1 *Calculations of Surface Area Buried on Binding*

Inhibitor Complex	Surface area of I (Å^2)	Surface area of E (Å^2)	Surface area of E·I complex (Å^2)	$\Delta \text{Å}^2$ (E+I $\rightarrow$ E·I)	$\Delta\Delta$ Å^2	$\Delta\Delta G°$ (kcal/mole)[a]
1	653.6	12703.8	12565.2	−792		
					2.5	<0.1
4	641.1	12703.8	12555.2	−790		
					97.1	2.3
5	544.4	12703.8	12555.5	−693		

[a] Calculated on the basis of 24 cal/mole/$\Delta\text{Å}^2$.[18]

Breslow's recent use of solvation effects to probe hydrophobic contact surfaces in organic reaction transition states[19] led us to attempt a similar approach to assess hydrophobic binding contributions experimentally. The inhibition constants for the three cyclic inhibitors, plus an acyclic phosphinate, Cbz-Gly^P-(C)Leu-Ala, were determined in the presence of increasing concentrations of ethanol, up to 12%. As the water content of the solvent decreases, the hydrophobic contributions to binding become less significant, and the binding becomes weaker. This effect is quite dramatic, leading to more than a 10-fold increase in K_i for the bi- and tricyclic inhibitors in the range 0-9% ethanol (Figure 5).

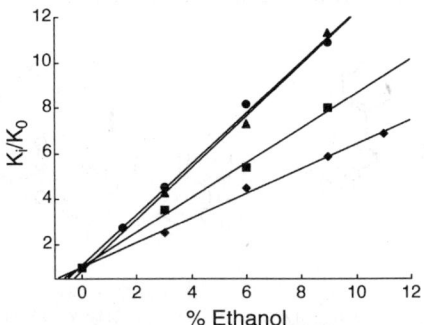

Figure 5   *Effect of solvent composition on inhibitor binding to thermolysin:* ● = *tricycle 1,* ▲ = *bicycle 4,* ■ = *monocycle 5,* ♦= *Cbz-Gly^P-(C)Leu-Ala*

Importantly, the effect is less prominent for the monocyclic and acyclic analogs. There is thus a remarkable correlation between the modeling results of Table 1 and the experimentally observed "anti-hydrophobic" solvent effect.

From these results, it is clear that there is no hydrophobic contribution to the difference in binding affinities between **1** and **4**; the factor of 10 therefore arises entirely from the increased conformational flexibility of **4**, and the possibility that the bound conformation is higher in energy than the other one observed. (However, no significant difference in energy between the two conformers of the ring systems of **4** is apparent from the modeling,) For the monocyclic inhibitor **5**, the reduced hydrophobic contribution accounts for *more* than the observed difference with **4**, even before the added conformational flexibility of **5** is taken into account. This conundrum is explained by the different conformations that the inhibitors adopt around the P2' units in the active site, which result in different interactions with protein. The orientation of the isobutyl side chain of the more flexible inhibitor **5** must lead to an interaction that is more favorable than those of **1** and **4**, largely compensating for the differences in size and flexibility.

2.2 Macrocyclic Inhibitors of Penicillopepsin

Although the zinc and aspartic peptidases differ in mechanism and active site configurations, they both catalyze direct addition of water to the peptide linkage and are potently inhibited by phosphonate analogs. As described above for thermolysin, the structure of the complex between penicillopepsin and the phosphonate inhibitor, isovaleryl-Val-Val-Leu^P-(O)Phe-OMe (K_i = 2.8 nM),[15] provided the foundation for design of more rigid derivatives and another opportunity to assess the quantitative value of this approach. Since the peptide analog adopts an extended conformation in the active site, alternate side chains adopt parallel orientations and thus can be linked in macrocyclic structures. The program CAVEAT[20] was used to identify molecular fragments that could bridge from the P3 to the P1 side chains, and from P2 to P1', without bumping into the protein. For the P2-P1' linked derivative, a variety of designs were evaluated to ensure that the desired conformer was favored in the low energy population. Two macrocycles were designed and synthesized: **6**, bridged between the P3 and P1 side chains,[21] and **9**, bridged between the P2 and P1' positions,[22] along with acyclic comparison compounds from which a methylene group had been removed. Relative to the macrocycles, these *seco* analogs have only slightly greater surface area and can more readily adopt similar conformations.

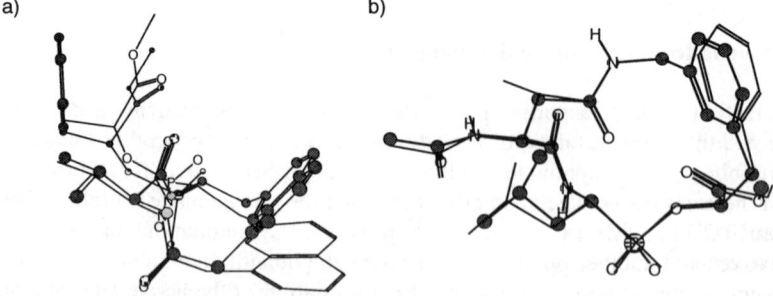

	K_i [a]
6: $Y_1 = Y_2 = CH_2$	800 nM
7: $Y_1 = CH_2$; $Y_2 = H,H$	7,600 nM
8: $Y_1 = H,H$; $Y_2 = CH_2$	110,000 nM

	K_i [b]
9: $Y_1 = Y_2 = CH_2$	0.1 nM
10: $Y_1 = CH_2$; $Y_2 = H,H$	42 nM
11: $Y_1 = H,H$; $Y_2 = CH_2$	1,300 nM

K_i (Penicillopepsin): [a] pH 3.5, [b] pH 4.5

Within each set of compounds, the observed binding affinities reflect the relative conformational flexibility expected: the macrocycles are the most potent, and among the acyclic analogs, the least-branched are the weakest. However, both the absolute affinity and the impact of the macrocyclic constraint are most pronounced for the P2-P1' linked analog **9**, which is much more potent than **6** and 3-4 orders of magnitude more tightly bound than the acyclic derivatives **10** and **11**.

The solution conformations of both **6** and **9** were determined (Figure 6); each appears to be well-defined, with the peptide backbone closely approximating the bound conformation of the original inhibitor (isovaleryl-Val-Val-Leu[P]-(O)Phe-OMe). However, while the conformation found for **9** is very similar to the design model, in **6**, the naphthalene ring is in a different orientation than originally envisaged, up and out of the plane of the macrocycle. In this conformation, the naphthalene ring of **6** could not fit into the P3-P1 binding pocket: either the macrocycle or the protein would have to deform in order for it to bind, which explains why **6** is such a weak inhibitor.

a)   b)

Figure 6   *Solution conformations of macrocycles 6 and 9; a) comparison of original model (line) of 6 bound to enzyme with solution conformation (ball-and-stick) determined from NMR and modeling; b) comparison of solution conformation of 9 (ball-and-stick) with bound conformation of isovaleryl-Val-Val-Leu[P]-(O)Phe-OMe (line)*

The structures of the penicillopepsin complexes of **6**, **7**, and **8**,[23] and of **9** and **10**,[24] reveal both exciting and unanticipated results, and enable us to make sense out of the binding data.   First, the bound conformation of the P3-P1-linked analog **6** is identical to that determined in solution: it is the *protein* that deforms to accommodate the unfavorable shape! The mobile flap that usually closes down over the active site when substrates and inhibitors bind is forced upward by the naphthalene ring, reducing the favorable contacts with the rest of the inhibitor and explaining the poor affinity of the analog.   In the acyclic derivative **7**, the naphthalene ring is not constrained in a macrocycle and can swing into a more favorable position in the binding pocket, thus allowing the protein flap to close into its usual position (Figure 7a).   As a result, the active site interactions for **7** are considerably improved and the loss of binding affinity from conformational flexibility is largely overcome.   The second surprise was the completely unexpected orientation found for the other acyclic analog, **8**, which binds out of register, with the phosphonate located where the P2-P1 peptide bond should be (Figure 7b).   Since every interaction between inhibitor and enzyme is different for **8** than for the other analogs, its binding affinity cannot be related to those of the others.   It is still unclear why **8** adopts this orientation; while it is hard to fathom why this abnormal position is more stable, it is unlikely that the normal orientation was blocked kinetically, since the complex was formed by co-crystallization.

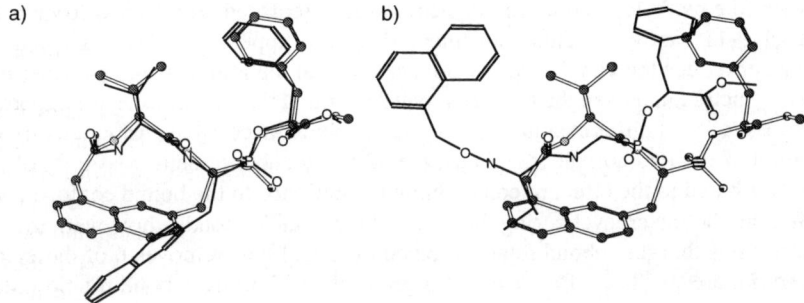

Figure 7   *Conformations of macrocycle 6 (ball and stick) and acyclic comparison compounds (line) a) 7 and b) 8 bound to penicillopepsin*

The recent results from structural analysis of the penicillopepsin complexes of macrocycle **9** and the acyclic comparison **10** are much less ambiguous (Figure 8).[24] The bound and unbound conformations of **9** are very similar, but in this case, it is the inhibitor that deforms on interaction with the enzyme.   The meta-substituted aromatic ring of the P1' residue is displaced slightly toward the plane of the macrocycle on binding, and the amide linkage between the P2 and P1' sidechains is rotated 1800 (Figure 8a); however, the peptide backbone and the P2 side chain are essentially unaffected.   The energy penalty for this deformation is modest; the two amide rotamers differ by less than one kcal/mole.

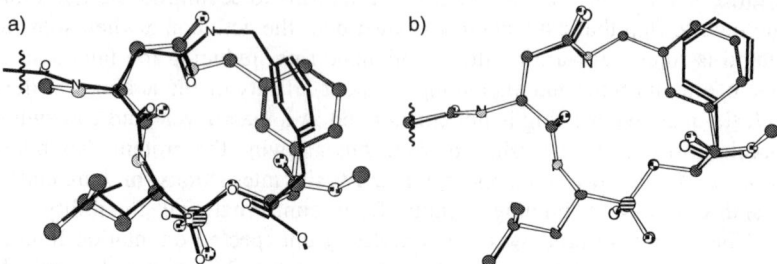

Figure 8   *a) Comparison of bound conformation of macrocycle **9** (ball-and-stick) with a) solution conformation of **9** (line) and b) bound conformation of acyclic control, **10** (line)*

Most striking is the similarity between the bound conformations of **9** and **10** (Figure 8b). Movement is seen around only two bonds in **10**, relative to **9**: rotations of 120 around Φ and 38° around χ^2 of the P1' residue allow a potential steric clash to be avoided in the region where the methylene group in the macrocycle is replaced with two hydrogens (Y_2 in structure of **9-11**, above). This movement does not appear to affect any active site interactions or reduce the van der Waals contact between the inhibitor and enzyme; indeed, the protein structure and even the hydration pattern around the binding site are unperturbed. Moreover, there is no difference in the amount of solvent-accessible surface that is buried on formation of the two complexes. Thus, the 420-fold enhancement in affinity of **9** over **10** can be attributed to the effect of constraining the inhibitor to the bound conformation.

Within the macrocyclic ring of **9** are 11 rotatable bonds; however, the cyclic structure requires that these bond rotations be coupled and thus removes 6 of these degrees of freedom (Figure 9). The difference of 4 degrees of freedom from bond rotation between **9** and **10** corresponds to an average contribution of 0.9 kcal/mole per rotatable bond, consistent with the values deduced from other approaches. A similar comparison between **9** and the other acyclic control, **11**, gives a value of 1.4 kcal/mole per rotatable bond, although the bound conformation of the latter compound has yet to be determined. Not all single bonds are as free in their rotations, however, and it would make sense that the more linear, less branched structure of **11** has more conformational mobility than **10**, even though they both have the same number of rotatable bonds.

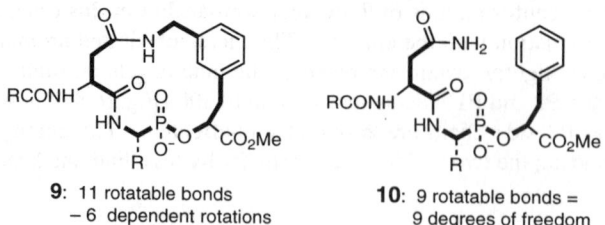

9: 11 rotatable bonds
− 6 dependent rotations
= 5 degrees of freedom

10: 9 rotatable bonds =
9 degrees of freedom

Figure 9   *Comparison of degrees of freedom from bond rotation: acyclic vs. macrocyclic inhibitors*

2.3 A Library-Screening Approach to the "Design" of Macrocyclic Inhibitors

The results above clearly show that macrocyclization can be an effective way to increase binding affinity without increasing molecular weight or adding hydrophobic groups. It is also risky (as shown by the relatively poor affinity of **6**), and relatively slow, since macrocyclic molecules are not always straightforward to synthesize. Indeed, this synthetic challenge makes it impractical to use combinatorial or parallel approaches to explore a greater variety of linking patterns and bridging units directly. We therefore devised a library-screening approach to reveal *transition state* binding, rather than to identify the structures of inhibitors themselves (Figure 10). The logic is as follows: Macrocyclic lactams that are most readily hydrolyzed contain the substituents and linking groups that would lead to the most potent inhibitor if incorporated in a transition state analog. Further simplification of the synthetic challenge comes from recognition that the tetrahedral transition state for cyclization of a bridged "product" analog is the same as that for cleavage of the macrocyclic lactam. Thus, the desirable structural features for an inhibitor will also be found in acyclic derivatives that are readily *cyclized* by the enzyme. This strategy has the attraction that acyclic, product-like analogs are readily assembled combinatorially on solid phase.

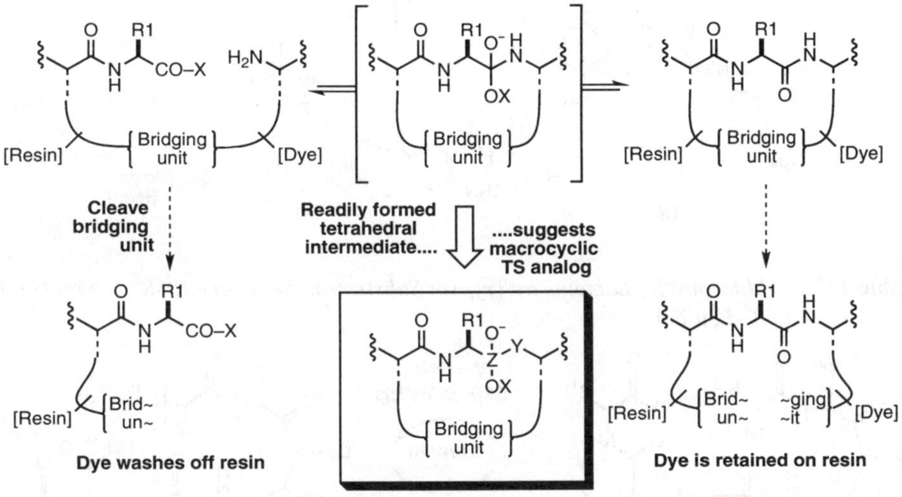

Figure 10 *Rationale that underlies the screen for enzymatic cyclization, and method for distinguishing cyclized from uncyclized analogs*

There is not a pure relationship between the transition state binding affinity and rate of enzymatic cyclization, since the inherent rate of cyclization (k_{noncat} of Equation 1 above) is not the same for all analogs. However, learning which structures are inherently easier to cyclize is not undesirable, considering that the macrocycles will eventually need to be synthesized. Crucial for the success of this scheme is to establish screening conditions under which peptide bond formation can occur.[25] For peptidases that act through an acyl-enzyme intermediate, the situation is relatively straightforward, since formation of this intermediate can be favored thermodynamically by starting with an activated ester. The process is more challenging for the aspartic and zinc peptidases, which require combination of ammonium and carboxylate groups in the reverse reaction; however, conditions involving low-pH or low-water content are known to favor peptide bond formation with representative peptidases from these classes.

Key to the success of the combinatorial screening strategy outlined in Figure 10 is a ready method to differentiate resin beads that contain cyclic product from those that don't, so the desired structures can be identified. If a dye substituent is incorporated in the substrate on the other side from the resin of a bond that can be cleaved in the presence of a peptide, cyclization can be detected readily (also illustrated in Figure 10). Cleavage of the susceptible bond in uncyclized material will result in release of the dye from the resin, while cleavage of a cyclic molecule leaves the dye connected to the resin via the peptide. Although a number of cleavable bonds can be envisaged, an ester linkage is perhaps the simplest to incorporate.

12

Table 2   *Macrocyclic Lactams as Trypsin Substrates. Sequence R^1-R^6 = Arg-Thr-R^3-R^4-Leu-Ser*

R^3-R^4	K_m (μM)	k_{cat} (s^{-1})	K_m/k_{cat} (10^5 M^{-1}s^{-1})
Val-Tyr	0.23	0.8	34
Gly-Asp	5	7	14
D-Ala-Asp	8	4	5

We first demonstrated this strategy by using trypsin to cyclize an amino ester, both in solution and on PEG-polyacrylamide resin, to form a cyanopeptolin[26,27] analog **12**, and demonstrating that beads with cyclized and uncyclized products can be readily distinguished after alkaline hydrolysis of the Val-Thr ester linkage.[28] We have since assembled a combinatorial library based on this motif to explore the effect of substituent variation, both conservative and radical (Figure 11). Preliminary screening results with the library demonstrate that this approach can identify a number of substituents that are readily cyclized in mixed-solvent systems. The lactams that they afford prove, in turn, to be effective trypsin substrates under hydrolytic conditions (Table 2).

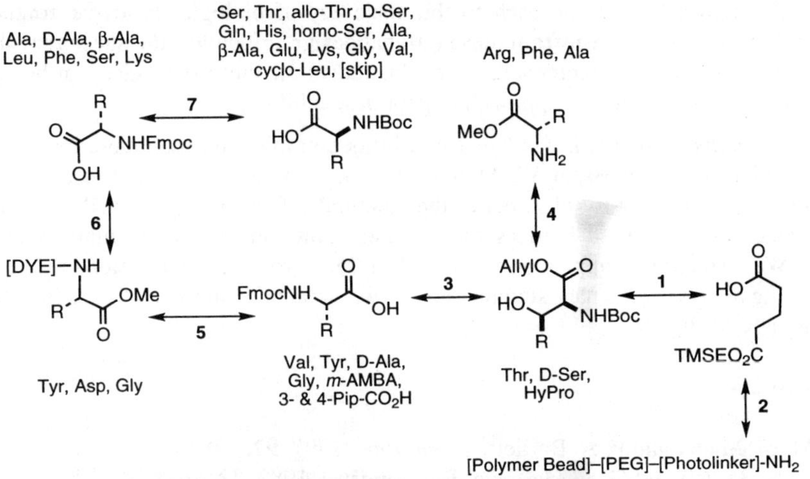

Figure 11 *Assembly sequence for a library of cyclization substrates*

Ultimately, of course, it will be interesting to see how these kinetic results correlate with the effectiveness of transition state analogs derived from the macrocyclic precursors. The initial application will entail preparation of the boronic acid analogs, which are capable of undergoing macrocyclization in the active site to give a mimic of the tetrahedral intermediate (e.g., **13** → **14**).[29,30]

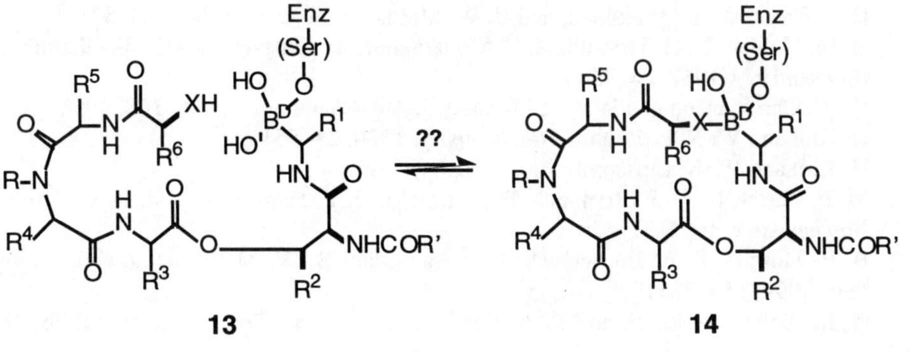

3 CONCLUSIONS

The potential advantage of conformational constraint through macrocyclization has long been recognized, but there are few examples in which the effect has been assessed quantitatively. Through structural evaluation of solution and active site forms of the cyclic inhibitors, and by comparison with appropriate acyclic analogs, we have been able to separate the influence of conformational control from binding effects that arise from direct interactions between the inhibitor and solvent or active site. The macrocyclic phosphonate analogs thus illustrate many of the concepts of mechanism- and structure-based, 'rational' design. Although a library approach to the discovery of biologically-active compounds may represent to some a retreat from these rational design principles, the success of the on-bead enzymatic screen for macrocyclization illustrates how these strategies can be applied to advantage in a combinatorial approach to peptidase inhibitors.

Acknowledgements. Our work has benefitted immeasurably from collaborations with the crystallography groups of Brian W. Matthews at the University of Oregon and Michael N.G. James at the University of Alberta; the stuctural information provided by them and their coworkers have served both as the starting points and the final analysis in these projects. We also express appreciation to Andrea Sefler for her contributions to the NMR and modeling analyses. Primary support for this work came from the National Institutes of Health, grant GM-30759.

REFERENCES

1.   M. M. Mader and P. A. Bartlett, *Chem. Rev.,* 1997, **97**, 1281.
2.   P. A. Bartlett and C. K. Marlowe, *Biochemistry,* 1983, **22**, 4618.
3.   P. A. Bartlett and C. K. Marlowe, *Biochemistry,* 1987, **26**, 8553.
4.   B. P. Morgan, J. M. Scholtz, M. Ballinger, I. Zipkin, and P. A. Bartlett, *J. Am. Chem. Soc.,* 1991, **113**, 297.
5.   J. E. Hanson, A. P. Kaplan, and P. A. Bartlett, *Biochemistry,* 1989, **28**, 6294.
6.   A. P. Kaplan and P. A. Bartlett, *Biochemistry,* 1991, **30**, 8165.
7.   M. A. Phillips, A. P. Kaplan, W. J. Rutter, and P. A. Bartlett, *Biochemistry,* 1992, **31**, 959.
8.   P. A. Bartlett, J. E. Hanson, and P. P. Giannousis, *J. Org. Chem.,* 1990, **55**, 6268.
9.   P. A. Bartlett and M. A. Giangiordano, *J. Org. Chem.,* 1996, **61**, 3433.
10.   D. E. Tronrud, H. M. Holden, and B. W. Matthews, *Science,* 1987, **235**, 571.
11.   H. M. Holden, D. E. Tronrud, A. F. Monzingo, L. H. Weaver, and B. W. Matthews, *Biochemistry,* 1987, **26**, 8542.
12.   D. W. Christianson and W. N. Lipscomb, *J. Am. Chem. Soc.,* 1986, **108**, 4998.
13.   H. Kim and W. N. Lipscomb, *Biochemistry,* 1990, **29**, 5546.
14.   H. Kim and W. N. Lipscomb, *Biochemistry,* 1991, **30**, 8171.
15.   M. E. Fraser, N. C. J. Strynadka, P. A. Bartlett, J. E. Hanson, and M. N. G. James, *Biochemistry,* 1992, **31**, 5201.
16.   B. P. Morgan, P. A. Bartlett, D. R. Holland, and B. W. Matthews, *J. Am. Chem. Soc.,* 1994, **116**, 3251.
17.   A. M. Sefler, G. Lauri, and P. A. Bartlett, *Int. J. Pept. Prot. Research,* 1996, **48**, 129.

18. C. Chothia and J. Janin, *Nature,* 1975, **256**, 705.
19. R. Breslow and R. Connors, *J. Am. Chem. Soc.,* 1996, **118**, 6323.
20. G. Lauri and P. A. Bartlett, *J. Comp. Aided Mol. Design,* 1994, **8**, 51.
21. J. H. Meyer and P. A. Bartlett, *J. Amer. Chem. Soc.,* 1998, **120**, 4600.
22. W. W. Smith and P. A. Bartlett, *J. Amer. Chem. Soc.,* 1998, **120**, 4622.
23. J. Ding, M. E. Fraser, J. H. Meyer, and P. A. Bartlett, *J. Amer. Chem. Soc.,* 1998, **120**, 4610.
24. A. R. Khan, J. C. Parrish, M. E. Fraser, W. W. Smith, P. A. Bartlett, and M. N. G. James, *Nature,* 1998, submitted.
25. J. Bongers and E. P. Heimer, *Peptides,* 1994, **15**, 183.
26. J. Weckesser, C. Martin, and C. Jakobi, *System. Appl. Microbiol.,* 1996, **19**, 133.
27. A. Y. Lee, T. A. Smitka, R. Bonjouklian, and J. Clardy, *Chem. Biol.,* 1994, **1**, 113.
28. M. T. Burger and P. A. Bartlett, *J. Amer. Chem. Soc.,* 1997, **119**, 12697.
29. Z.-Q. Tian, B. B. Brown, C. Hutton, D. P. Mack, and P. A. Bartlett, *J. Org. Chem.,* 1997, **62**, 514.
30. B. A. Katz, J. Finer-Moore, R. Mortezaei, D. H. Rich, and R. M. Stroud, *Biochemistry,* 1995, **34**, 8264.

MERGING RATIONAL DRUG DESIGN WITH COMBINATORIAL CHEMISTRY: REASONABLE AND UNREASONABLE EXPECTATIONS

Daniel H. Rich,[1,2] Natalie A. Dales,[2] Xiaodong Fan,[2] George Flentke, [1] Peter Glunz,[2] Stacy J. Keding,[1] Amy S Ripka,[1] and Kenneth Satyshur [1]

School of Pharmacy [1] and Department of Chemistry,[2] University of Wisconsin-Madison, 425 N. Charter St. Madison, WI 53706

1   INTRODUCTION

Medicinal chemists typically utilize three rational approaches for discovering protease inhibitors: design of enzyme inhibitors based on enzyme mechanism,[1] screening of natural products or corporate and combinatorial libraries,[2] and lead optimization by use of molecular modeling and computational chemistry.[3] In this paper we describe novel inhibitors of three proteases discovered by use of these strategies and discuss the prospects for accelerating drug discovery by interfacing combinatorial design with combinatorial synthesis.

2   DESIGN AND SYNTHESIS OF INHIBITORS OF METHIONINE AMINOPEPTIDASE-1

In eukaryotic cells protein synthesis is initiated by methionine,[4] which is removed co-translationally from the nascent polypeptide chain by Methionine Aminopeptidase (MetAP), a metalloprotease that requires divalent cobalt for activity. Two classes of Methionine Aminopeptidases, MetAP-1 and MetAP-2,[5,6] have been reported. MetAP-1 is found in prokaryotes and yeast and contains two zinc finger motifs at the amino terminus that are required for full function *in vivo*.[7] MetAP-2, which is found in yeast and humans, may play an important role in endothelial cell proliferation.[8,9] Knock-out of the yeast gene encoding for either MetAP-1 or MetAP-2 leads to slow growth, and disruption of both genes is lethal to the cell. Since deletion of the MetAP-1 gene in prokaryotes is lethal, inhibitors of MetAP-1 could have potential antibacterial activity, and inhibitors of either enzyme will be useful for studying mechanisms of cell replication and possibly controlling cell division. We have developed a series of mechanism based inhibitors of MetAP-1,[10] based on the transition-state isostere, (3R)-amino-(2S)-hydroxy heptanoic acid (2S, 3R AHHpA; **1**).

Inhibitors of bimetallic zinc aminopeptidases have been designed by incorporating a 2S-hydroxyl, 3R-amino functionality into substrate-derived sequences for the targeted APase.[11] Since MetAP-1 cleaves the N-terminal methionine residue from peptide chains, replacement of the methionine by the corresponding 2S-hydroxyl, 3R-amino acid was expected to produce an inhibitor. Our previous work with substrates showed that the methionyl sulfur was not essential for recognition by bacterial MetAP-1,[12] so inhibitors **2-**

4 contain a norleucine side chain. The 2S-hydroxyl, 3R-amino functionality needed to chelate the active site cobalt ion was efficiently synthesized regio- and stereoselectively by use of the Sharpless asymmetric aminohydroxylation (AA).[13] Inhibitors **2-4** inhibited MetAP-1 with K_i = 5-7 μM. A crystal structure of one inhibitor bound to MetAP-1 has been solved by Matthews and Lowther,[14] who found that the designed inhibitor binds in the active site of the enzyme as expected for a mechanism based inhibitor.

3 IDENTIFICATION OF ACTIVE CORE OF DIDEMNAKETALS FOR INHIBITION OF HIV PROTEASE

Screening of natural product inhibitors of proteases[15] is a powerful method for identifying potent and specific inhibitors of medicinally relevant enzymes.[16] In addition to providing novel peptidomimetic structural motifs, natural products often inhibit by new mechanisms, and reveal targets for further drug discovery. Study of the Didemnaketals A (**5**) and B, two HIV-1 protease inhibitors isolated from the *Ascidian Didemnum* sp. at Auluptagel Island, Palau, illustrates how such efforts can lead to unexpected mechanistic information. Didemnaketal A (**5**) inhibits HIV-1 protease with an IC_{50} of 2 μM,[17] but the absolute configuration of eight stereogenic centers were not determined in the original report. We synthesized all eight diastereomers of the simplified pentaester analogs **6** and found that several retained anti HIV protease activity; the activity of **6e** was identical to that of the parent natural product.[18] Since HIV protease is composed of two identical subunits which spontaneously dimerize to form active enzyme, and all clinically used protease inhibitors function by blocking the active site and not dimerization of the enzyme, we chose to analyze this system by using the kinetic method of Zhang *et al*,[19] which was developed to differentiate between pure competitive inhibition and non-competitive or dissociative inhibition for inhibitors of HIV protease. Kinetic data were consistent with the pentaester inhibiting dimerization of HIV-1 protease monomers.

Inhibitor **6e** represents a novel HIV-1 protease inhibitor that acts by an unusual mechanism. Compound **6e** is only the fourth non-nitrogen containing HIV-1 protease inhibitor reported to date,[20] and two previous examples[21,22] have been used to develop HIV protease inhibitors in clinical trials. Only a few dimerization inhibitors of HIV-1 PR are known[23] but these offer a new approach to inhibiting mutant strains of HIV-1 protease. It is clear that natural product screening can provide us with unexpected leads, even in this era of combinatorial synthesis.

5, Didemnaketal A   6e, (5S, 7R, 8S)

4   USE OF COMPUTER-GENERATED STRUCTURES TO DESIGN PROTEASE INHIBITORS

The development of inhibitors **2-4** and **6e** was achieved relatively quickly but neither type of inhibitor is likely to be a bioavailable compound; inhibitors **2-4** are too peptidic and inhibitor **6e** is a pentaester. Inhibitor discovery *per se* rarely is the rate-limiting step in drug development; the major road-block is finding a **selective, bioavailable inhibitor**. Although medicinal chemists can efficiently design and synthesize peptide-derived tight-binding enzyme inhibitors, it is the conversion of these peptide-derived inhibitors into orally active compounds that has been so difficult, primarily because we have no fundamental principles to guide optimization of bioavailability and selectivity. Empirical efforts that remove or replace amide bonds in an inhibitor (Type-I peptidomimetics) have proven successful, especially in the area of HIV protease inhibition but have required vast synthetic efforts. In some cases, rational design of inhibitors based on structural biology is ignored altogether, and replaced with screening of combinatorial libraries of synthetic compounds assayed against a variety of targets. The chasm between structural biology and combinatorial chemistry arises because there are no systematic pathways for converting the structure of an enzyme-bound peptide into completely non-peptide topographical "mimetics," i.e. Type-III peptidomimetics. In principle, it should be possible to discover principles for replacing peptide structure in the binding components of enzyme inhibitors with non-peptide "mimetics" in the same way that enzyme reaction pathway intermediates are replaced by transition state mimetics. If this could be achieved, then both structural biology and combinatorial synthesis could be used in concert to discover inhibitors. The hope is that the novel molecular scaffolds identified will possess the pharmacodynamic properties needed for efficient drug use.

Recently, we described our attempts to use computer programs to generate novel structures that would replace peptide structure in enzyme-bound inhibitors.[24] Our strategy for compound generation is based on the work of Bohacek and McMartin,[25] who created GrowMol and used this to generate useful inhibitors of metalloproteases.[26] We use GrowMol to help discover Type-III peptidomimetic inhibitors of aspartic proteases. Beginning with the X-ray crystal structure of the inhibitor A66702 (**7**) complexed to pepsin,[27] GrowMol generated 20,000-50,000 potential structures, which were then classified according to structural type and used to produce a manageable file of 200-400 distinct structures that were examined for synthetic feasibility. We have synthesized several of these compounds and determined which are inhibitors of the target aspartic protease. Most recently, we have been able to obtain X-ray crystal structures of several enzyme-inhibitor complexes to determine how the inhibitor binds to the enzyme.

We began by showing that GrowMol could successfully generate known pepsin inhibitors. We had synthesized a series of cyclic, bis-sulfide inhibitors of pepsin **8a-d**.[28] GrowMol was used to create a series of compounds that linked the P_1 with the P_3 side chain, and the backbones of known inhibitors **8a** and **8c** were re-generated. The X-ray crystal structure of **8c** bound to pepsin was obtained and is shown (Figure 1) superimposed on a closely related GrowMol generated structure. It is remarkable how closely the computer predicted the correct structure and mode of binding. Several additional inhibitors related to **8c** were synthesized and found by X-ray crystallography to bind to Rhizopus pepsin in the predicted manner (data not shown). These results established that our general protocol successfully generated realistic inhibitor structures.

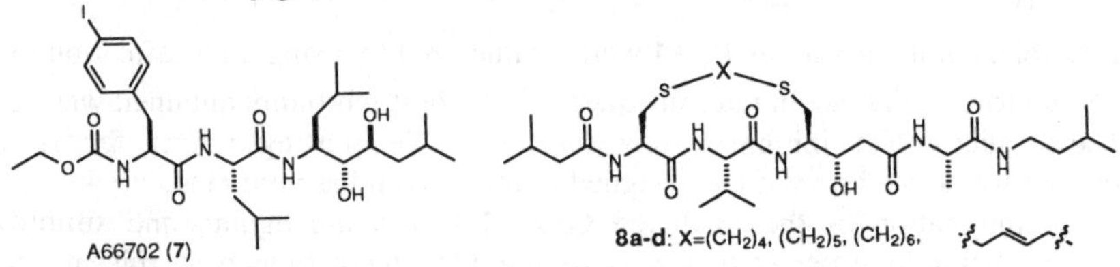

A66702 (7)

8a-d: X=$(CH_2)_4$, $(CH_2)_5$, $(CH_2)_6$,

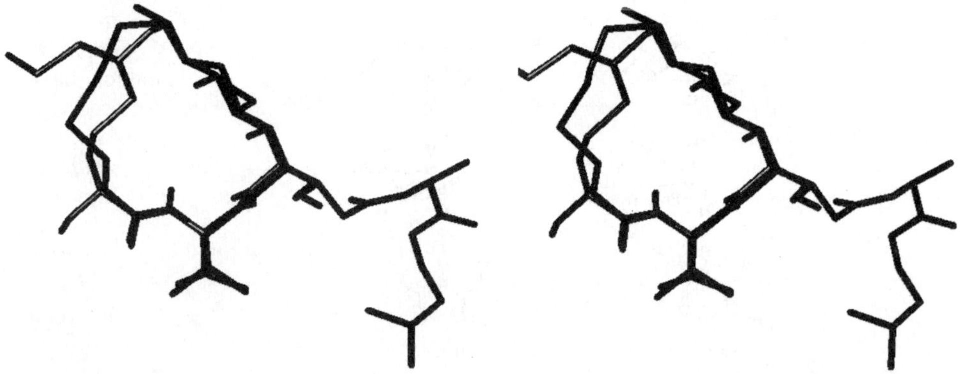

Figure 1 *Stereographic (crossed) comparison of crystal structure of 8c superimposed on GrowMol generated structure*

Another GrowMol calculation created a series of novel urea-derived inhibitors of pepsin in which a urea bond separated P_1 and P_2 in the inhibitor. At the time we began this work, this was an unknown structural feature for aspartic proteases. A variety of low molecular weight, micromolar urea alcohol derivatives, e.g. **9**, were synthesized to test this prediction. When **9** was elaborated into a more pepstatin-like structure by addition of P' substituents, several tight-binding inhibitors ($K_i < 1$ nM) such as **10** were formed, showing that the amide bond between P_1 and P_2 could be replaced by a urea to form a good inhibitor. X-ray crystal structures of **10** bound to Rhizopus pepsin were obtained and matched the predicted structure.

The success achieved with GrowMol calculations encouraged us to try more ambitious targets. The cyclohexanol-derived inhibitor **11** represented a completely non-peptide structure that lacks nitrogen. At the time we began this work, no non-nitrogen containing tight-binding inhibitors of pepsin were known, and we simplified **11** to produce a series of compounds designed to determine whether GrowMol had produced a realistic prediction.

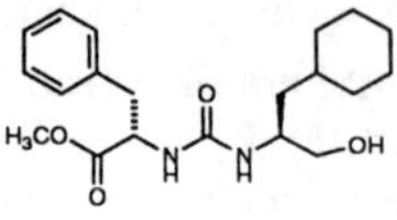

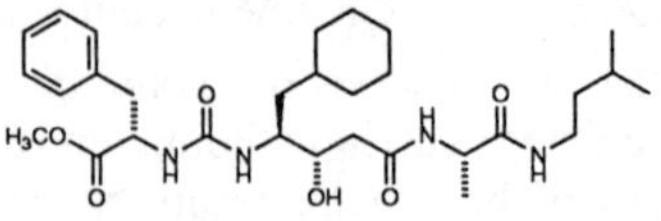

A series of cyclohexanol derivatives **12-14** were synthesized by using a modification of an eleven step route to a phylanthocin intermnediate.[29] The best inhibitors obtained were **13** and **14**. These micromolar inhibitors represent plausible starting points for further optimization, but we do not know if the designed inhibitors bind as predicted.

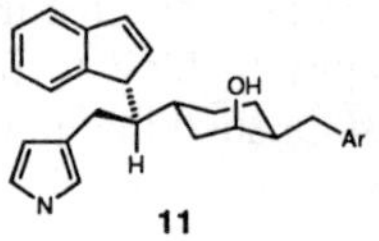

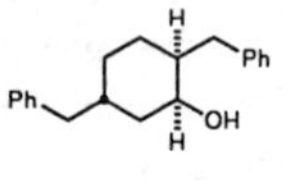

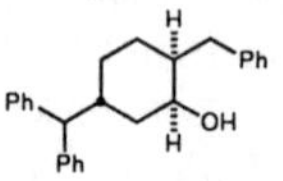

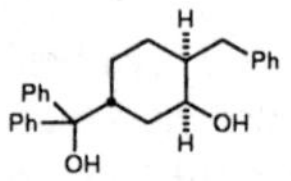

Careful re-examination of the predicted GrowMol structure against the simplified structure showed that at least two critical enzyme-inhibitor interactions were missing in **11** that are present in A66702 (**7**) and in **10** (Figure 2).

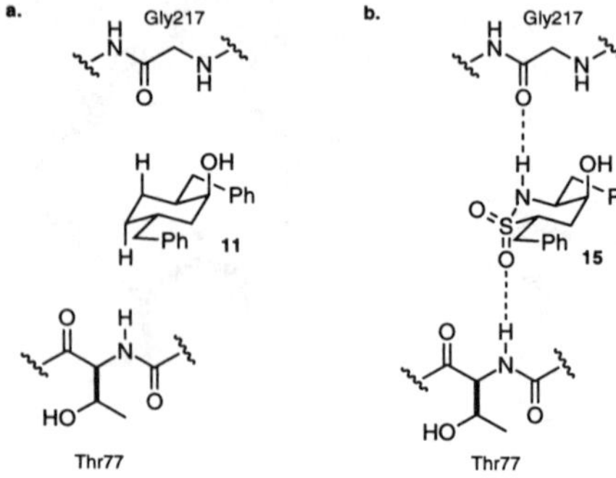

Figure 2 *A. Schematic representation of hydrogen bonding pattern if **11** binds in the active site of pepsin as predicted; B. Schematic representation of the hydrogen bonding pattern for sultam derivative **15** if it binds as predicted*

We synthesized analog **15** to test if restoring the extra hydrogen bonds might lead to better inhibitors. Sultam **15** binds better to pepsin than alcohol **11**. Compounds **16** (which contains the α-branching analogous to compound **13**) and **17** (which contains the extra hydroxyl group analogous to compound **14**) are also more potent than **11**. Interestingly, both structural changes increase potency but do not act in an additive fashion, which may suggest different binding modes for each compound. We are currently trying to crystallize these inhibitors in the active site of pepsin to determine if compounds **15-17** bind as predicted by GrowMol.

Structure-generating programs can help move the chemist out of old thought patterns, thereby stimulating new ideas for structures. A nice example of the unanticipated idea arising out of consideration of computer-generated structures was the realization that sultam **15** when opened is a new sulfonic acid analog **18** of the hydroyethylene

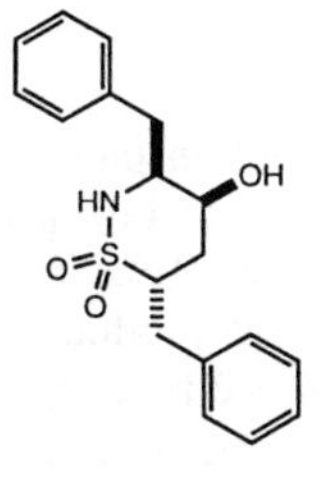

15	**16**	**17**
36 ± 3 µM	1.1 ± 0.08 µM	306 ± 18 nM

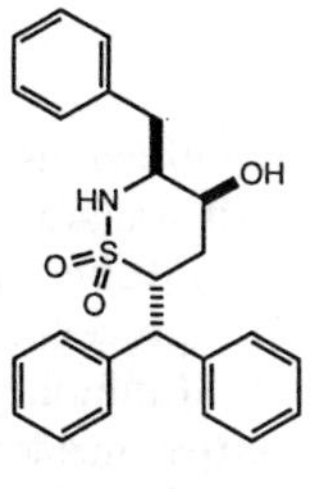

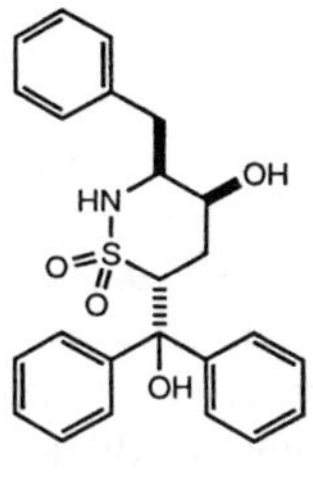

dipeptideisostere.The sultam intermediate was used to synthesize a carbon analog of the Vertex HIV protease inhibitor, VX-478.[30] The HIV protease inhibitor **19** showed reasonable activity *in vitro* and *in vivo*, but as yet we have made no attempt to optimize activity in this class of inhibitor.

In our early experiments, we grew structures starting from the transition state hydroxyl group so that the structure grew away from the catalytic groups toward protein and solvent. This strategy led to a vast variety of diverse structures, especially those that varied near the solvent interface. To restrict the structure-generation to space more constrained by the protein, we began some searches in a well-defined side-chain binding region and directed the computer to grow the structures into the catalytic groups, with instructions to terminate in the desired hydroxyl group. This strategy forced the program to grow structures into converging space rather than into expanding space. Compounds **20** and **21** illustrate two examples of potential inhibitors generated by this strategy. Amino alcohol **21**, a µM inhibitor of pepsin, was synthesized by Sharpless aminohydroxylation[14] in four steps, similar structures could be synthesized by combinatorial methods.

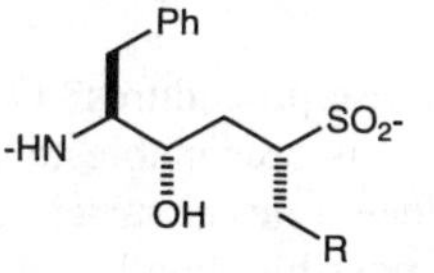

18 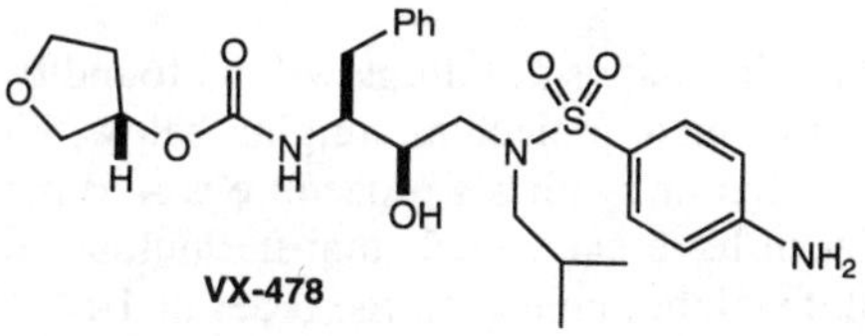   **VX-478**

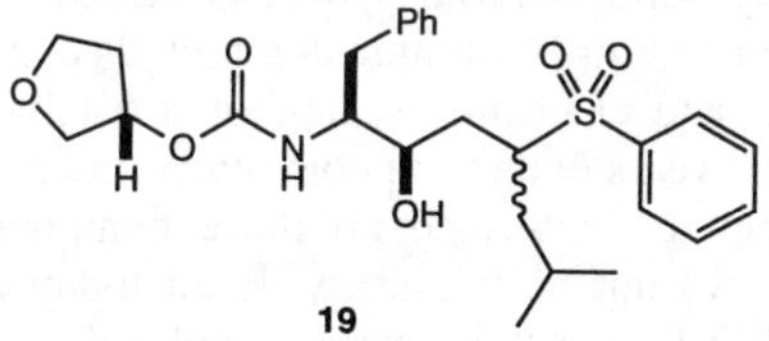

19

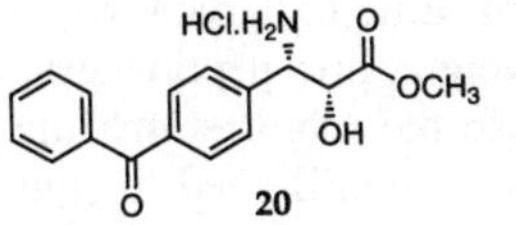

20 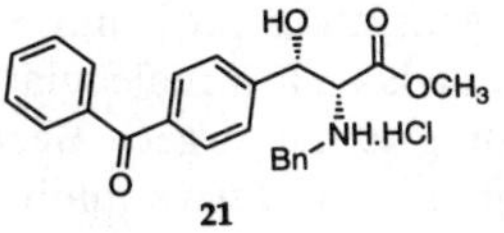   **21**

5 DISCUSSION

Protease inhibitors are often developed by mimicking the reaction pathway intermediates formed in the active site of an enzyme as it transforms substrate into product. Substrate analogs are synthesized in which the cleavable amide bond is replaced by non-hydrolyzable transition-state isosteres. This procedure usually produces an effective *in vitro* inhibitor. Ancillary portions of this lead inhibitor then are modified until inhibitors with the proper pharmacodynamic properties needed for drug use are obtained. The difficulty in this approach is that there is no theoretical basis for transforming the peptide-like structure into a non-peptide structure with good pharmacodynamic properties, e.g. oral activity. GrowMol offers a new approach to this problem by generating a diverse array of molecules from which medicinal chemists can extract suitable potential lead structures for further evaluation and synthesis. The work reported here demonstrates that GrowMol can generate known and novel inhibitors in the active site of porcine pepsin and *R. chinensis* pepsin, and complements the demonstrated success of GrowMol for generating thermolysin inhibitors. It is important to note that not all inhibitors reported here are established to bind to the enzyme in the computed fashion. The Type-I inhibitors **8-10** clearly bind as predicted but additional X-ray studies of the enzyme-inhibitor complexes are needed to determine the bound conformations for **11-18**.

The use of GrowMol to generate libraries of potential inhibitors for a target enzyme represents a combinatorial process of enormous power. Clearly, potent inhibitors have been obtained and when structure-generating programs are combined with powerful synthetic efforts it is reasonable to expect that optimization will lead to more potent inhibitors. Combinatorial synthesis of molecules is likely to be a particularly effective way to optimize lead structures to obtain tight-binding inhibitors. We believe that combinatorial design coupled with combinatorial synthesis will lead to new classes of enzyme inhibitors.

What is the probability that useful drugs will be found by these procedures? Given that most orally active drugs have molecular weights below 500, it is reasonable to ask how likely it is that any design or synthesis procedure would produce a novel structural motif. Bohacek and McMartin have calculated[5] that the number of possible small molecules in this general molecular weight range is on the order of 10^{62}! Diversity of this size cannot be explored exhaustively by synthetic or computational methods. For example random synthesis of all possible structures by combinatorial synthesis utilizing everyone on the planet would take greater than 10^{30} years; known organic compounds constitute only about $1/10^{52}$ of possible small compounds; and electronic search of a database with all those structures would take greater than 10^{30} years even with computers much faster than any in existence. Although it is impossible to synthesize all these compounds (not enough starting material in the universe), it will not be necessary. Even today about a thousand useful drugs exist in the known 10 billion small organic molecules. Clearly it is not necessary to make all organic structures to find better and new drugs.

Interfacing structure-generating programs with structural data will help focus the combinatorial synthetic methods on low molecular weight, non-peptide structures that have a high probability for binding to the target. Because the major stumbling block in drug discovery remains identifying orally active inhibitors, we still need to find ways to make improved structures that satisfy the necessary goal of inhibiting and reaching the target while by-passing non-productive interactions. This in turn requires discriminating high-throughput screening methods that will identify selective, bioavailable compounds in the libraries of active inhibitors. High through-put analoging will eventually produce the lead compound with the needed pharmacodynamic properties, but these may be found faster if the information obtained from structural biology can be used to help guide the design of the combinatorial libraries. Furthermore, if the estimates of small molecule structural diversity are anywhere near correct, then there are vast numbers of scaffolds that have not been evaluated in any known biological systems and it is possible that some of those might lead to future generations of "privileged" structures. It would seem there is much to be gained by merging combinatorial chemistry with structural biology to help accelerate the rate of discovery of more potent, sufficiently bioavailable and minimally toxic clinical candidates.

ACKNOWLEDGEMENTS

Support from the National Institutes of Health (GM50113) is appreciated. We thank Drs. Bohacek and McMartin for use of their software (GrowMol and Lazy Mouse) and for many helpful discussions.

REFERENCES

1. D. H. Rich, Peptidase Inhibitors. in "Comprehensive Medicinal Chemistry. The Rational Design, Mechanistic Study and Therapeutic Application of Chemical Compounds", C. Hansch, P.G. Sammes, and J.B. Taylor (Ed.), Pergamon Press 1989, pp. 391.
2. (a) E. M. Gordon, R. W. Barrett, W. J. Dower, S. P. A. Fodor, M. A. Gallop *J Med Chem* 1994, **37**, 1385. (b) M. A. Gallop, R. W. Barrett, W. J. Dower, S. P. A. Fodor, E. M. Gordon, *J Med Chem* 1994, **37**, 1233.
3. Recent Advances in Peptidomimetic Design. A.S. Ripka and D. H. Rich, <u>Current Opinions in Chemical Biology</u> 1998 in press.
4. C. Flinta, R. Persson, H. Jornvall, G.von Heijne *Eur. J. Biochem.* 1986, *154*, 193.
5. S. M. Arfin, R. L. Kendall, L. Hall, L. H. Weaver, A. E. Stewart, B. W. Matthews, R. A. Bradshaw. *Proc. Natl. Acad. USA* 1995, **92**, 7714.
6. X. Li, Y. Chang, *Proc. Natl. Acad. Sci. USA* 1995, **92**, 12357.
7. Y. Chang, U. Teichert, J. A. Smith, *J. Biol. Chem.* 1992, **267**, 8007.
8. E. C. Griffith, Z. Su, B. E. Turk, S. Chen, Y. Chang, Z. Wu, K. Biemann, J. O. Liu, *Chemistry & Biology* 1997, **4**, 461.
9. N. Sin, L. Meng, M. Q. W. Wang, J. J. Wen, W. G. Bornmann, C. M. Crews, *Proc. Natl. Acad. Sci. USA* 1997, **94**, 6099.
10. S. J. Keding, N. A. Dales, S. Lim, D. Beaulieu, and D. H. Rich, *Synthetic Communications*, in press.
11. (a) D. H. Rich,B. J. Moon, A. Boparai, A. *J. Org. Chem.* 1980, **45**, 2288; (b) D. H. Rich, B. J. Moon, S. Harbeson, S. *J. Med. Chem.* 1984, **27**, 417.
12. S. Lim and D.H. Rich, in "Peptides: Chemistry and Biology (Proceedings of the Thirteenth American Peptide Symposium)", 1994, R.S. Hodges and J.A. Smith, Ed. ESCOM, 625-627.
13. (a) G. Li, H. H. Angert, K. B. Sharpless,*Angew. Chem. Int. Ed. Engl.* 1996, **35**, 2913. (b) G. Li, H. Chang, K. B. Sharpless, *Angew. Chem. Int. Ed. Engl.* 1996, **35**, 451.
14. W. T. Λοωτηερ and B. Matthews, B. Personal communication.
15. R. A. Wiley and D. H. Rich, *Med. Research Review*, 1993, **13**, 327.
16. R. E. Babine and S. L. Bender, *Chem. Rev.* 1997, **97**, 1359.
17. B. C. M. Potts, D. J. Faulkner, J. A. Chan, G. C. Simolike, P. Offen, M. E. Hemling, T. A. Francis, *J. Am. Chem. Soc.* 1991, **113**, 6321.

18. X. Fan, G.R. Flentke, and D.H. Rich, *J. Amer. Chem. Soc.*, 1998, in press.
19. Z. Zhang, R. A. Poorman, L. L. Maggiora, R. L. Heinrikson, F. J. Kezdy,*J. Biol. Chem.* 1991, **266**, 15591.
20. K. R. Romines, R. A. Chrusciel, *Current Med. Chem.* 1995, **2**, 825.
21. (a) P. J. Tummino, J. V. N. Vara Prasad, D. Ferguson, C. Nouhan, N. Graham, N. J. M. Domagala, E. Ellsworth, C. Gajda, S. E. Hagen, E. A. Lunney, K. S. Para, B. D. Tait, A. Pavlovsky, J. W. Erickson, S. Gracheck, T. J. McQuade, D. J. Hupe, *Bioorg. and Med. Chem.* 1996, **4**, 1401. (b) S. E. Hagen, J. V. N. Vara Prasad, F. E. Boyer, J. M. Domagala,E. Ellsworth, C. Gajda, Hamilton, L. J. Markoski, B. A. Steinbough, B. D. Tait, E. A. Lunney, P. J. Tummino, D. Ferguson, D. J. Hupe, C. Nouhan, S. J. Gracheck, J. M. Saunders, and S. VanderRoest, *J. Med. Chem.* 1997, **40**, 3707.
22. Schwartz, T. M.; Bundy, G. L.; Strohbach, J. W.; Thaisrivongs, S.; Johnson, P. D.; Skulnick, H. I.; Tomich, P. K.; Lynn, J. C.; Chong, K. T.; Hinshaw, R. R.; Raub, T. J.; Padbury, G. E.; Toth, L. N.; *Bioorganic. Med. Chem. Lett.* **1997**, 7, 399.
23. (a) Zutshi, R.; Franciskovich, J.; Shultz, M.; Schweitzer, B.; Bishop, P.; Wilson, M.; Chmielewski, J. *J. Am. Chem. Soc.* **1997**, *119*, 4841. (b) Schramm, H.J.; Nakashima, H.; Schramm, W.; Wakayama, H.; Yamamoto, N. *Biochem. Biophys. Res. Commun.* **1991**, *179*, 847. (c) Franciskovich, J.; Houseman, K.; Mueller, R.; Chmielewski, J. *Bioorganic Med. Chem. Lett.* **1993**, *3*, 765. (d) Schramm, H.; Billich, A.; Jaeger, E.; Rucknagel, K.; Arnold, G.; Schramm, W. *Biochem. Biophys. Res. Commun.* **1993**, *194*, 595. (e) Babe, L. M.; Rose, J.; Craik, C. S.; *Protein Sci.* **1992**, *1*, 1244.
24. D.H. Rich, R.S. Bohacek, N.A. Dales, P. Glunz, A.S. Ripka in "Actualités de chimie Thérapeutique-22e série", Elsevier 1996, pp 101.
25. R. S. Bohacek, and C. McMartin *J Amer Chem Soc* 1994, **116,** 5560.
26. R. S. Bohacek, C. McMartin, and W. C. Guida, *Med. Res. Rev.* 1996, **16**, 3.
27. L. Chen, J. W. Erickson, T. J. Rydel, C. H. Park, D. Neidhart, J. Luly and C. Abad-Zapatero, *Acta Cryst.*, 1992, **B48**, 476.
28. Z. Szweczuk, K. Rebholz and D. H. Rich, *Int. J. Pept. Protein Res.*, 1992, **40**, 233.
29. A. B. Smith, M. Fukui, H. A. Vaccaro, J. R. Empfield, *J. Am. Chem. Soc.*, 1991, **113**, 2071.
30. E. E. Kim, C. T. Baker, M. D. Dwyer, M. A. Murko, B. G. Rao, R. D. Tung, M. A. Navia, *J. Am. Chem. Soc.*, 1995, **117**, 1181.

THE USE OF STRUCTURAL GENOMICS AND PROTEIN SUPERFAMILES IN DRUG DISCOVERY: PREDICTION THAT THE BINDING SITE OF THE INSULIN RECEPTOR ECTODOMAIN LIES AT THE CONCAVE FACE OF A β-HELIX

Mercedes Martin-Martinez[+], Ken Siddle[*] and Tom L. Blundell[+ X]

[+] Department of Biochemistry, 80 Tennis Court Rd, Cambridge CB2 1GA
[*] Department of Clinical Biochemistry, Addenbrookes hospital level 4, Hills Road, Cambridge CB2 2QR
[X] To whom correspondence should be addressed

1 INTRODUCTION

Analyses of protein sequences derived from genome analyses indicate that about 50% of the 'new' clustered sequences can be recognised as members of homologuos families of known function.[1,2] However, proteins with sequence identities in the twilight zone or even with statistically insignificant sequence similarities, can also have similar topologies[3,4] and similar functions[5,6]; these clusters probably represent distant homologues and are known as superfamilies. Recognition of topology from sequence and association of a new protein with a known superfamily may therefore give rise to testable hypotheses about function; this is structural genomics.

The identification of new members of superfamilies is greatly facilitated by the existence of databases of aligned protein structures and sequences.[7] The alignments of protein families described here are based on the conservation of structural features and relationships using the program COMPARER.[8,9] They include a database of structurally aligned homologous proteins (HOMSTRAD: HOMologous STRucture Alignment Database)[7,10] and a database of protein superfamilies (CAMPASS: CAMbridge database of Protein Alignments organized as Structural Superfamilies).[11] These are available on the World Wide Web at http://www-cryst.bioc.cam.ac.uk/~homstrad for HOMSTRAD and http://www-cryst.bioc.cam.ac.uk/~campass for CAMPASS). HOMSTRAD and CAMPASS are distinct from but complementary to other databases. SCOP,[12] CATH[13] and FSSP.[14]

We have used our databases to cluser a number of proteins into existing families. Here we illustrate this with the use of our data bases, to identify a protein with a similar fold to that of the repeats in the insulin receptor ectodomain and to obtain clues about the likely ligand binding region. The insulin receptor (IR) is an integral membrane glycoprotein synthesised as a single polypeptide chain and processed to an α-chain and a β-chain linked by disulphide bonds to give an $\alpha_2\beta_2$ active receptor.[16] The α-chain is extracellular and contains the insulin-binding site. Analysis of the sequences of IR α-chains available in 1987 suggested that the extracellular region of the insulin receptor is comprised of two large, homologous domains (L1 and L2) with repeating motifs containing leucines[17]. The first domain, L1, is followed by several smaller cystine-rich domains, and the second, L2, by an unrelated region which contains a few cystines. Examination of the structural databases HOMSTRAD and CAMPASS showed that X-ray crystal structures of several leucine-rich structures have been determined in the past decade; these new structures include the leucine-rich repeat structures,[18] the β-roll and the β-helix,[19,20,21] which are composed mainly of β-sheets. A preliminary comparison of the sequences with these in databases showed that the receptor was probable similar to the known β-helix structures, so allowing us to construct a model.

We compare our results with those defined more recently by X-ray analysis of the ectodomain of the homologous type 1 insulin-like growth factor receptor (IGF1R).[15] We show that structural genomics approaches usefully identify functional regions of proteins in the absence of experimental data and suggest that this may be useful in target identification in drug discovery.

2 METHODS

The alignment of the sequences of the different motifs of IR and other homologous receptors was generated with CLUSTALX[22] and manually adjusted (Figure 1). Comparison of the sequences of the motifs reveals only weak similarities except for the conservation of four hydrophobic residues and some asparagines and glycines. We investigated the CAMPASS and HOMSTRAD databases to identify motifs in which similar patterns of motifs might be conserved for structural purposes. Similar patterns were identified in leucine rich repeats,[18] in the β-roll and in the β-helix,[19,20,21] with the latter appearing to be more similar.

In order to investigate whether these common features, between IR and the β-helix structure, imply a similar tertiary structure, we constructed a three-dimensional model of the IR L1 using pectin lyase A (IDK)[20] and tailspike protein (TSP)[21] as templates. The percentage identity between the IR L1 domain and TSP is 8.3% and 9.3% when compared with IDK; the percentage of similarity is much higher at 49.9% and 52.4%, respectively. We built the 3D model of IR using the comparative modelling restraint-based approach implemented in MODELLER.[23] The final models were checked with the programs PROCHECK[24] and Verify3D.[25]

3 RESULTS & DISCUSSION

3.1 A helical model for the repeated sequence

Figure 2 shows the structural alignment of the repeated motifs of IDK and TSP generated with COMPARER.[8,9] The parallel β-helix is composed of parallel β-strands wound into a large right-handed coil. Each turn of the β-helix has three short parallel β-sheets (ranging from two to five amino acids in length), two of which pack against each other to form a parallel β-sandwich with the β-strands of one sheet oriented antiparallel to the β-strands of the other. The third β-sheet lies nearly perpendicular to one end of the parallel β-sandwich. A striking feature of the β-helix is the stacking interactions[26] among the side chains in the interior of the central helix. Four different types of side chain stacks can be observed: aliphatic stacks, asparagine ladders, serine stacks and aromatic stacks. Although their 3D structures are remarkably similar, there is little sequence conservation among the motifs. There are four hydrophobics conserved within the β-strands.

Additionally, in several β-strands the amino acids immediately before and after the β-strand have backbone dihedral angles in an α_L conformation and are quite often Asn and Gly. It is evident that the conserved residues in the alignment of IDK and TSP are similar to the conserved residues in those IR homologues; moreover, the number of residues per turn is also similar.

```
IR_hu.L11       -RLHELE -CSVIE- HLQILLMFKTRPEDFR
IGF1R_hu.L11    -QLKRLE -CTVIE- YLHILLISKAEDYR
IRR_hu.L11      -ELRQLE -CSVVE- HLQILLMFTATGEDFR
IRR_ra.L11      -ELRRLE -CSVVE- HLQILLMFAATGEDFR
IR_hu.L12       --DLSFPK--LIMIT-DY--LLLFRVYGLES
IGF1R_hu.L12    --SYRFPK--LTVIT-EY--LLLFRVAGLES
IRR_hu.L12      --GLSFPR--LTQVT-DY--LLLFRVYGLES
IRR_ra.L12      --GLSFPR--LTQVT-DY--LLLFRVYGLE
IR_hu.L13       -LKDLFP -LTVIR- S-RLFFNYALVIFEMVHLKE
IGF1R_hu.L13    -LGDLFP -LTVIR- W-KLFYNYALVIFEMTNLKD
IRR_hu.L13      -LRDLFP -LAVIR- T-RLFLGYALVIFEMPHLRD
IRR_ra.L13      -LRDLFP -LAVIR- A-RLFLGYALIIFEMPHLRD
IR_hu.L14       ---LGLY -LMNITR --SVRIEKNNEL
IGF1R_hu.L14    ---IGLY -LRNITR --AIRIEKNADL
IRR_hu.L14      ---VALPA-LGAVLR --AVRVEKNQELC
IRR_ra.L14      ---IGLPS-LGAVLR --AVRVEKNQEL
IR_hu.L21       TSAQELR -CTVIN- --SLIINIRGGNNLAAEL
IGF1R_hu.L21    TSAQMLQ -CTIFK- --NLLINIRRGNNIASEL
IRR_hu.L21      QAAQDLV -CTHVE- --SLILNLRQGYNLEPQL
IRR_ra.L21      QATQDLV -CTHVE- --SLILNLRQGYNLEPEL
IR_hu.L22       --EANLGL--IEEIS- --YLKIRRSYALVS
IGF1R_hu.L22    --ENFMGL--IEVVT- --YVKIRHSHALVS
IRR_hu.L22      --QHSLGL--VETIT- --FLKIKHSFALVS
IRR_ra.L22      --QRNLGL--VETIT- --FLKIKHSFALVT
IR_hu.L23       --LSFFRK--LRLIR- -ETLEIGNYSFYALDNQN
IGF1R_hu.L23    --LSFLK -LRLIL- EEQLEGNYSFYVLDNQN
IRR_hu.L23      --LGFFK -LKLIR- DAMVDGNYTLYVLDNQN
IRR_ra.L23      --LGFFK -LKLIR- DSMVDGNYTLYVLDNQN
IR_hu.L24       --LRQLWDWSKHNLTITQ KLFFHYNPKL
IGF1R_hu.L24    --LQQLWDWDHRNLTIKA KMYFAFNPKL
IRR_hu.L24      --LQQLGSWVVAAGLTIPV KIYFAFNPRL
IRR_ra.L24      --LQQMGSWVVAAGLTIPV KIYFAFNPRL
```

Figure 1 *Sequence alignment of four motifs from each of the L domains of the human IR (IR_hu), the human IGF1R (IG1_hu) and the human and rat insulin receptor-related protein (IRR_hu and IRR_ra). L1x: motifs of the L1 domain, L2x: motifs of the L2 domain, x: the number of the motif*

```
              A B C    300   A B C D E F G H I         310      A B C D
TSP1 (291)  k m v d a n - - - ñ p s g g k - - - - - - - - - - d G i l ĩ f e ñ l s g d - - - - w g k̃ g ñ
TSP2 (320) - ỹ V i g - - - - G r̃ T̃ s̃ y G s - - - - - - - - - - v S S̃ A q̃ F l r̃ Ñ ñ G g - - - f ẽ r̃ d̃ G G
TSP3 (348) - G V i g - - - - F t S̃ y r A g - - - - - - - - - - ẽ S G V k̃ I w̃ q̃ g t v g s t t S̃ Ĩ N y ñ L
TSP4 (380) - q̃ F r̃ d - - - - S v V i y p v - - - - - - - - - w d g f̃ d l g A D̃ t   q y p l h q l p l a
IDK1 (110) k s l i g e g s s̲ g a i k g k - - - - - - - - - G - - - - l r Ĩ v s̲ - - - - - - - - g a - e n
IDK2 (135) i i I q̲ n̲ - - - - I a V t d̃ I n̲ - p k y v w g g - - d̃ A I t L d - - - - - - - - - d̃ C - d̃ l
IDK3 (164) V w̃ I D̲ h - - - - V t T A r̃ I G - - - - - - - - r̃ Q H Y v l g T̃ - - - - - - - - s̃ a d n̲ r̃
IDK4 (189) v s l i ñ - - - - N̲ y i d g v S d y s a i̲ c d̃ g y H̲ y w a i ỹ l d̃ G d a d̃
              β β β        β β β                          β β β
```

Figure 2 *Structural alignment of the motifs of TSP and IDK used in the modelling. The formatting convention of program JOY[27] is applied*

The model of the L1 domain of the insulin receptor (Figure 3) includes four complete turns, each of which has three β-strands, with the exception of the first where one strand is replaced by a less regular region. Although we have modelled four complete turns, there is at least a further similar motif before these. As in tailspike protein and pectin lyase A, the β-strands of consecutive turns line up to form three parallel β-sheets, two of them form an antiparallel β-sandwich, while the other lies approximately perpendicular to them. In addition the conserved aliphatic side chains are aligned and oriented towards the interior of the β-helix and aligned.

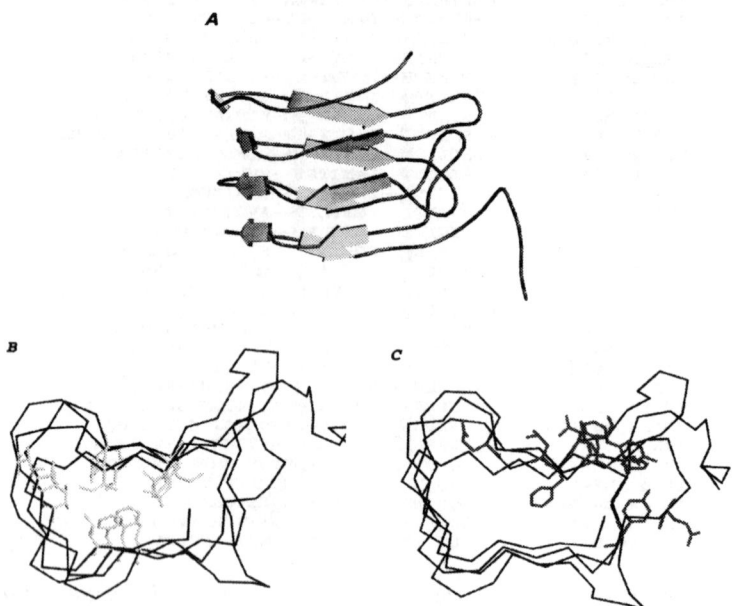

Figure 3   *A. Ribbon diagram of the model of L1 domain of IR viewed perpendicular to the β-helix, arrows represent β-strands. B. L1 domain of IR viewed along the axis of the helix from the N-terminus with the conserved hydrophobics in black. C. Residues that appear to play a role in high affinity insulin binding in black, and the probably glycosylated asparagines in grey*

3.2 Functional residues in the insulin receptor model

In tailspike protein and pectin lyase A, the binding sites lie in the concave face of the β-helix. Is this a useful predictor of the binding site of insulin on the β-helix?

It is known that in IR there are important contributions to the ligand-binding site located between amino acids 1 and 120 (L1 domain),[28,29,30] particularly from 14 amino acids arranged in four discontinuous peptide segments. Segment 1: Asp-12, Ile-13, Arg-14 and Asn-15, Segment 2:

Gln-34, Leu-36, Met-38, Phe-39 and Glu-44, Segment 3: Phe-64 and Tyr-67, and Segment 4: Phe-89, Asn-90 and Tyr-91. Segments two to four in our model are located in the same zone of the molecule. Moreover, all the lateral chains of the key amino acids, with the exception of Phe-89, are on a concave face on the exterior of the central helix. The insulin receptor is highly glycosylated. Several studies[31] have shown that most, if not all, of the potential sites are glycosylated. In our model of the L1 domain there are three possible glycosylation sites, and our model places the sidechain of these asparagines pointing towards the solvent.

Mutational studies have shown that mutation of Val 28[32] caused a defect in oligomerization; this residue is on the exterior of the β-helix in our model. Thus, it is likely that this residue is implied in the association of two L-domains. It is also interesting to note that three β-helices form a trimer in the P22 tailspike protein[21].

In summary, the model of the L1 domain shown in Figure 3 presents some of the characteristic interactions of the core of a β-helix, as well as grouping the residues that have been shown to interact with insulin on a concave face of the molecule. In addition there are some hydrophobic residues that may be involved in forming an interface between the two L domains. Thus, we conclude that the L1 and L2 domains of the IR are likely to adopt a 3D structure quite similar to that of the β-helix. The L1 and L2 domains of IR, IG1 and IRR follow the same pattern and it is likely that they also adopt a similar fold to the L1 domain of IR, proposed here.

3.3 A comparison with the crystal structure of the IGF receptor ectodomain

Recently the X-ray structure of the IGF1R has been solved by Garrett et al,[15] showing that it is a new type of right-handed super-helical structure. The authors found that it was similar to both the Leucine-rich repeats of the ribonuclease inhibitor and the β-helix, but argued that it is more similar to the former as the cross sections are rectangular and the large β-sheets untwisted in both. The structure of IGF1R is comprised of three β-sheets, like that of our model. However the position of one of the sheets differs. The binding site was correctly localised on the large β-sheet, that is the proposal binding site of the L1 domain suggested by Garret et al.,[15] after building a model of the IR on the basis of IGF1R.

4 CONCLUSIONS

This analysis shows that such approaches, using data bases of aligned protein superfamilies, can give an indication of the 3D structure of a protein, even when the identity between the sequences is below 10%. It demonstrates that not only can the protein tertiary structure be identified and modelled in a useful way, but also the binding site can be identified by consideration of binding sites of other members of a superfamily. Such models may give helpful ideas both in target identification and lead optimisation in drug discovery.

ACKNOWLEDGEMENTS

We thank Dr. Mark Williams for the help in the use of modelling programs. We thank Drs R. Sowdhamini, K. Mizuguchi, N. Srinivasan and D. Burke for making available the CAMPASS database. Dr. Martin-Martinez is a European Community fellow. The research is supported by a grant from the Wellcome Trust to TLB.

REFERENCES

1. P. Bork, C. Ouzounis and C. Sander, *Curr. Opin. Strc. Biol.*, 1994, **4**, 393.
2. E.V. Koonin, P. Bork and C. Sander, *EMBO J.*, 1994, **13**, 493.
3. F.C. Bernstein, T.F. Koetzle, G.J.B. Williams, E.F. Meyer, M.D. Brice, J.R. Rodgers, O. Kennard, T. Shimanouchi and M. Tasumi, *J. Mol. Biol.*, 1977, **112**, 535.
4. S.E. Brenner, C. Chothia, and T.J.P. Hubbard, *Curr. Opin. Strc. Biol.*, 1997, **7**, 369.
5. T.L. Blundell, and R.E. Humbel, *Nature*, 1980, **287**, 781.
6. A.G. Murzin, and C. Chothia, *Curr. Opin. Struc. Biol.*, 1992, **2**, 895.
7. J.P. Overington, M.S. Johnson, A. Sali, and T.L. Blundell, *Proc. R. Soc. (London)*, 1990,. **B241**, 132.
8. A. Sali, and T.Blundel, *J. Mol. Bio.*, 1990, **212**, 403.
9. Z. Y. Zhu, A. Sali, and T. L. Blundell, *Protein Engineering*, 1992, **5**, 43.
10. K. Mizuguchi, C.M. Deane, T. L. Blundell and J.P. Overington, *Protein Science*, 1998, in press.
11. R. Sowdhamini, D.F. Burke, J.F. Huang, K. Mizuguchi, H.A. Nagarajaram, N. Srinivasan, R.E. Steward and T.L. Blundell, *Structure*, 1998, **6**, 1087.
12. A.G. Murzin, S.E. Brenner, T. Hubbard, and C. Chothia, *J. Mol. Biol.*, 1995, **247**, 536.
13. C.A. Orengo, D.T. Jones, and J.M. Thornton, *Nature*, 1994, **372**, 631.
14. L. Holm, and C. Sander, *Nucleic Acid Res.*, 1994, **22**, 3600.
15. T.P.J. Garrett, N.M. Mckern, M.Lou, M.J. Frenkel, J.D. Bentley, G.O. Lourecz, T.C. Elleman, L.J. Cosgrove and C.W. Ward, *Nature*, 1998, **394**, 395.
16. A. Ullrich, J. R. Bell, E. Y. Chen, R. Herrera, L. M. Petruzzelli, T. J. Dull, A. Gray, L. Cussens, Y. C. Liao, M. Tsubokawa, A. Mason, P. H. Seeburg, C. Grunfield, O. M. Rosen and J. Ramachandran, *Nature*, 1985, **313**, 756.
17. M. Bajaj, D. M. Waterfield, J. Schlessinger, W. R. Taylor and T. L. Blundell, *Biochimica et Biophysica Acta*, 1987, **916**, 220-226.
18. S. G. ST. C. Buchanan and N. J. Gay, *Prog. Biophys. Molec. Biol.*, 1996, **65**, 1.
19. M. D. Yoder, and F. Jurnak, *FASEB J.*, 1995, **9**, 335-342.
20. O. Mayans, M. Scott, J. Connerton, T. Gravesen, J. Benen, J. Visser, R. Pickersgill and J. Jenkins, *Structure*, 1997, **5**, 677.
21. S. Steinbacher, R. Seckler, S. Miller, B. Steipe, R. Huber and P. Reinemer, *Science*, 1994, **265**, 383.
22. D. G. Higgins, J. S. Thompson and T. J. Gibson, *Methods in Enzymology*, 1996, **266**, 383.
23. A. Sali, and T. L. Blundell, *J. Mol. Biol.*, 1993, **234**, 779.
24. R. A. Laskowski, M. W. MacArthur, D. S. Moss and J. M. Thornton, *J. Applied Crystallogr.*, 1993, **26**, 283.
25. R. Lüthy, J. U. Bowie and D. Eisenberg, *Nature*, 1992, **356**, 83.
26. M. D. Yoder, S. E. Lietzke, and F. Jurnak, *Structure*, 1993, **1**, 241.
27. K. Mizuguchi, C. M. Deane, T. L. Blundell, M. S. Johnson and J. P. Overington, *Bioinformatics*, 1998, **14**, 617.
28. P. F. Williams, D. C. Mynarcik, G. Q. Yu and J. Whittaker, *J. Biol. Chem.*, 1995, **270**, 3012.
29. P. DeMeyts, J. L. Gu, R. M. Shymko, B. E. Kaplan, G. I. Bell and J. Wittaker, *Mol Endocrinol*, 1990, **4**, 409.

30. T.Kjeldsen, F. C. Wiberg and A. S. Andersen, *J. Biol. Chem.*, 1994, **269**, 32942.
31. C. Collier, J. L. Carpentier, L. Beitz, L. H. P. Caro, S. I. Taylor and P. Gorden, *Biochemistry*, 1993, **32**, 7818.
32. E. Wertheimer, F. Barbertti, M. Muggeo, J. Roth and S. I. Taylor, *J. Biol. Chem.*, 1994, **269**, 7587.

Ion Channels

LIGAND-GATED ION CHANNELS AS TARGETS: PROBLEMS AND POSSIBILITIES

Paul D Leeson

Department of Medicinal Chemistry, AstraZeneca R&D Charnwood, Bakewell Road, Loughborough, Leicestershire, UK, LE11 5RH

1 INTRODUCTION

Ion channels, which function as key modulators of cellular excitability, are important drug discovery targets in almost all therapeutic areas. There are two broad classes of membrane-bound ion channels: voltage-gated and ligand-gated. In this review we focus on the ligand-gated ion channels,[1] which are commonly termed as the receptors for their specific agonists, or 'ionotropic' receptors. The emphasis here is on the special issues associated with drug discovery at these targets. The intensive efforts to identify useful modulators of the N-methyl-*D*-aspartate (NMDA) subtype of glutamate channel/receptor, for central nervous system (CNS) disorders, are surveyed. Many of the drug discovery challenges encountered in the NMDA receptor area are likely to be relevant to ligand-gated ion channels in general.

2 OVERVIEW OF LIGAND-GATED ION CHANNELS

Those ion channel/receptors of interest are postsynaptic targets which respond to natural ligands of varying structure, from protons acting at acid sensing ion channels (ASIC)[2] to adenosine triphosphate acting at purinergic P2X channels[3] (Table 1). Excitatory channels on activation gate non-selective trans-membrane transport of sodium, potassium and calcium ions. Inhibitory channels transmit extracellular chloride ions. The purinergic P2X$_7$ receptor appears unique in that low molecular weight substances can also pass through the channel pore.[3] The glutamate,[4] 5HT$_3$,[5] nicotinic acetylcholine (nACh),[6] γ-aminobutyric acid$_A$ (GABA$_A$)[7] and glycine[8] receptors have so far been the most extensively investigated in drug discovery programmes. The ligand-gated ion channels share a number of common characteristics, briefly summarised below.

2.1 Protein subunits

Over the past decade, the multiple protein subunits which compose the ligand-gated ion channels have been identified and cloned (Table 1).[1]

Table 1   *Ligand-gated Ion Channels/Receptors[1]*

Agonist Ligand	Ion Channel Receptor	Protein Subunits	Effector Ions	Ref.
Glutamate	NMDA	NR1, NR2A-D, NR3A	Na^+, K^+, Ca^{2+}	4, 31
	AMPA	GluR1-4	Na^+, K^+ (Ca^{2+})	
	Kainate	GluR5-7, KA1-2	Na^+, K^+ (Ca^{2+})	
ATP	P2X	$P2X_{1-7}$	Na^+, K^+ (Ca^{2+}, low M.Wt. solutes)	3
5HT	$5HT_3$	$5HT_3$	Na^+, K^+, Ca^{2+}	5
Acetylcholine	Nicotinic ACh	α1-9, β2-4, γ, δ, ϵ	Na^+, K^+, Ca^{2+}	6
H^+	ASIC	α, β, γ, δ	Na^+	2
GABA	$GABA_A$	α1-6, β1-4, γ1-3, δ, ρ1-3, ϵ, π	Cl^-	7, 13
Glycine	Glycine	α1-4, β	Cl^-	8

These major advances have allowed generation of recombinant channels to become commonplace, and have hugely advanced drug discovery programmes with provision of novel targets and selective *in vitro* assay systems. The membrane topologies of the protein subunits suggest the existence of three distinct superfamilies (Figure 1):   a) four transmembrane units (TM) with both C- and N-termini extracellular, for example GABA and nACh receptors;[6,7] b) three TM plus an insert loop with N terminus extracellular and C-terminus intracellular, for example the glutamate receptors;[4] and c) two TM with C- and N-termini intracellular, for example the ASICs and P2X receptors.[2,3]   A large number of studies suggest that agonist binding sites are present on extracellular domains of $GABA_A$, nACh, and glutamate receptors.

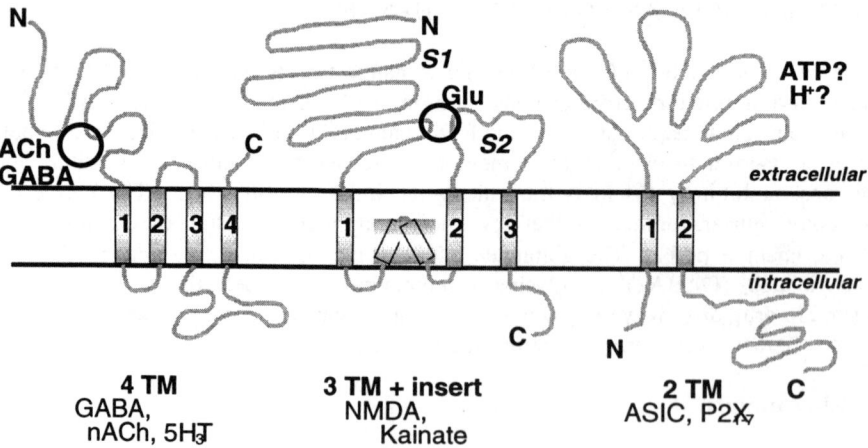

Figure 1   *Proposed membrane topologies and transmitter binding sites*

2.2 Channel architecture

Knowledge of structures of the ligand-gated ion channels at atomic resolution, as with other membrane bound receptors, is currently lacking. However, images of the nACh receptor complex, determined in the open state at 9Å resolution by electron microscopy, provide overall details of the channel architecture.[9] There are five protein subunits forming the channel, with each subunit contributing TM2 to the inner lining of the channel. The transmitter binding site is well defined, within a large extracellular domain, approximately 50Å from the membrane. Mechanisms of channel opening involving movement of TM2 have been proposed. As well as the nACh channel, there is evidence that the glycine, GABA$_A$, and 5HT$_3$-gated channels also exist as pentamers. Since the subunits in these channels have similar membrane topologies, they may be structurally homologous.[10] In contrast, the balance of evidence suggests that the glutamate receptors are tetrameric[11] whilst the P2X channels may be trimers.[12]

2.3 Channel heterogeneity

The existence of multiple protein subunits for each of the channels except 5HT$_3$ (Table 1), together with various possible alignments around the channel (Figure 2), suggests that considerable heterogeneity is theoretically possible. The uncovering of the physiologically relevant subunit compositions of the channel/receptor complexes has provided novel drug discovery targets at GABA$_A$,[13] NMDA[4] and nACh[5] receptors. The best characterised subtype targets to date are the GABA$_A$ receptors, where it has been shown that one each of the α, β and γ subunits is necessary for channel function. Although the 13 available GABA$_A$ subunits provide >10,000 possible pentameric combinations, it has been shown that there are only *four* major benzodiazepine-sensitive classes in rat brain, defined by receptors containing the α1, α2, α3 and α5 subunits.[13] From the repertoire of existing benzodiazepine-site GABA$_A$ ligands, some α-subunit selective compounds have emerged. Selection for or against α5-containing receptors has been shown, and the imidazobenzodiazepines RY-80[14] and L-655,708[15] show ~50 fold binding selectivity for the α 5-containing receptor. Certain β-carbolines and hypnotics such as zolpidem and CL 218,872 have modest (~10 fold) selectivity in binding to the α1-containing receptor,[16] but there appear to be no compounds published to date with significant selectivity for α2- or α3-containing receptors. Preliminary definition of benzodiazepine-site pharmacophores at the new GABA$_A$ receptor subtypes has been reported,[16] and these new findings may begin to map the way forward for discovery of selective, non-sedating anxiolytics, cognition enhancers and other novel CNS agents.

2.4 Ligand-binding domains

The recent X-ray determination of the structure of kainic acid bound to the extracellular S1/S2 ligand-binding domain, excised from the rat GluR2 subunit,[17] is a major advance for drug discovery at glutamate receptors. The binding site is a deep, hinged gorge, created by the two extracellular regions of the protein (see Figure 2, centre) and the amino acids which make contact with kainate are identical or conserved amongst the four glutamate receptor subfamilies, despite sequence identities of <10%.[4]

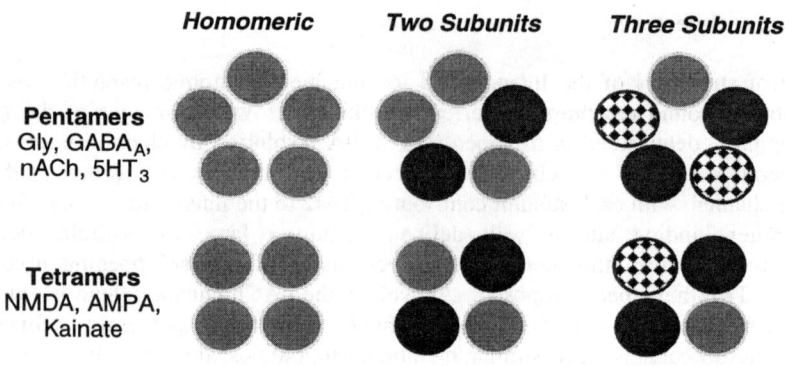

Figure 2 *Possible arrangements of protein subunits forming central channels*

Aspects of the structure-activity relationships of kainate derivatives can be explained, as well as the allosteric effects of cyclothiazide, and redox behaviour.[17] These exciting findings open the way for homologous models to be generated of the ligand binding sites in other glutamate receptors,[18] including the glycine site of the NR1 NMDA receptor subunit, where the corresponding soluble S1/S2 construct has been characterised.[19] This information might also permit more speculative molecular models of the binding of competitive antagonists to be built. Since agonist and antagonist interactions must differentially affect channel opening and overall receptor conformation, it is experimentally determined protein structures with the antagonists bound that will provide the most value to drug discovery efforts. Such structures of ligand-gated ion channel domains, if not of the entire complexes, can now be considered a realistic prospect. Amongst the non-glutamate receptor/channels, it has been suggested that the extracellular ligand-binding domains of the pentameric glycine and nACh complexes, which differ from the glutamate receptors (Figure 1), are formed by SH2 and SH3-like domains.[20]

Identifying the amino acid residues and channel domains which can influence small molecule binding sites has relied predominantly upon a wealth of site-directed mutagenesis and other experiments, especially with the glutamate, GABA$_A$ and nACh channels. Although site-directed mutagenesis has successfully predicted the important amino acid residues for binding of glutamate agonists,[4, 17] it is often difficult to use this information to derive meaningful models of the actual ligand-protein binding interactions, unless there is accompanying structure-binding data for different ligands. For example, it appears that homologous amino acids can influence the binding of dissimilar molecules, γ-aminobutyric acid and benzodiazepines, to the GABA$_A$ receptor, at apparently comparable sites on the interfaces of the α-β and α-γ subunits respectively.[21] However, the observation that aromatic residues are important for the binding of glycine to the NMDA receptor[22] and acetylcholine (ACh) to the nACh receptor[23] suggests the involvement of π-cation interactions in the binding of the ammonium groups of the ligands. This interaction has been validated in host-guest chemistry, where it is energetically favoured in aqueous media, and it is of significance in a number of biological systems.[23]

2.5 General properties relevant to drug discovery

The ligand-gated ion channels can exist in open, closed and desensitized states, each of which can in principle be targeted by drug molecules. The best characterised channel/receptors (NMDA, GABA$_A$, nACh) all show multiple binding sites for small molecules. Typically there are sites for agonists, various modulatory molecules, channel blockers (open and closed state blockers), and divalent cations, which provide several options for drug discovery approaches at each target channel. In addition, all of the ligand-gated ion channels show characteristic, and often complex, allosteric interactions between the different agonist, modulatory and channel binding sites. The interactions between binding sites are an intriguing design feature and probably exist to permit operation of multiple control mechanisms for channel opening under various physiological conditions. Generally, it has been shown that molecules acting at different sites of the same channel complex usually show differing *in vivo* pharmacologies and side-effect profiles, an observation which is of primary significance to drug design efforts.

3 CHALLENGES IN DRUG DISCOVERY AT LIGAND-GATED ION CHANNELS

3.1 Problems

The likely occurrence of receptor/channel diversity means that obtaining selectivity, and overcoming side-effects, are both potentially serious issues for drug discovery. Lack of selective activity *in vivo* has hampered discovery of clinically useful NMDA receptor antagonists (see below) and the lack of subtype specificity may underlie the absence of non-sedating and non-addictive GABA$_A$ benzodiazepine-site drugs. Unfortunately, identifying the relevant *native* subtype targets and establishing assays for drug discovery are not trivial. For example, recombinant receptors which are functional *in vitro* may not have significance *in vivo*, and more than one receptor/channel complex may be relevant to particular disease processes. In addition, all the relevant subunits may not have been characterised. The generation of useful assays further requires functional, rather than binding screens, especially when targeting non-competitive or allosteric sites. Current evidence also suggests that subunit selective agents may be difficult to identify. The majority of molecules acting at ligand-gated ion channels, which were initially identified using non-subtype selective assays, have been subsequently found to lack significant subunit selectivity. Also, it is by no means a certainty that subunit selective compounds will in fact lack side-effects. Finally, the current paucity of useful 3-D structural information on antagonist-receptor binding interactions is seriously hampering structural approaches to drug design at ligand-gated ion channels. Medicinal chemists, in identifying lead molecules, have had to rely largely on fortuitous observations and synthetic manipulation of agonist molecules, such as α-amino acids, which have undesirable physicochemical properties from a "drug-like" perspective.

3.2 Possibilities

Given the issues described above, and considering the structural complexity of ligand-gated ion channels, one might feel pessimistic about the prospects for application of modern techniques of drug discovery in this area. However, much has been learned so far, largely from approaches which have not targeted individual receptor/channel subtypes.

Exceptionally, the ability to 'fine-tune' pharmacological profiles of drug molecules has been of major significance. This has been achieved by targeting distinct sites on the various channels for antagonist action, and by identifying agonists and inverse agonists of varying efficacy. The phenomenon of open channel or 'use-dependent' blockade additionally provides a unique opportunity for uncompetitive antagonism to occur, only under conditions where the channel is activated. Clinically useful drugs have been discovered in the absence of knowledge of the channel subtypes, including the GABA$_A$ benzodiazepines and agonists, nACh channel blockers and 5HT$_3$ antagonists. NMDA receptor antagonists are now in the later stages of drug development, including mechanistically novel NR2B subunit-selective modulators, and progress is being made towards GABA$_A$ and nACh subtype-selective compounds. The newer technologies of high throughput functional screening, combined with high speed chemistry and targeted library synthesis, have yet to be applied fully in this area, but as with other targets, will be expected to provide improved 'drug-like' lead compounds.

4 THE NMDA RECEPTOR AS A DRUG DISCOVERY TARGET FOR CNS DISORDERS

There has been a massive effort over the past decade within the Pharmaceutical Industry to discover useful antagonists of the NMDA receptor.[4,24] The major physiological role of the NMDA receptor is in mediating synaptic plasticity, such as that underlying learning and development. However, the predominant therapeutic aim has been the acute treatment of stroke and CNS trauma, based on the "Excitotoxic Hypothesis", which states that the increased levels of extracellular glutamate in these diseases increases cellular Ca^{2+} uptake *via* NMDA receptors, resulting in delayed neuronal death.[25] Each of the major classes of NMDA receptor antagonists (reviewed below) do indeed demonstrate consistent and effective neuroprotection in a variety of animal models of focal cerebral ischemia. In addition, further evidence exists for use of NMDA receptor antagonists and partial agonists in epilepsy, analgesia, schizophrenia and various neurodegenerative disorders. Although several NMDA receptor antagonists have reached the clinic, neuroprotection in stroke or head trauma patients has not yet been demonstrated, largely as a result of a high attrition rate due mainly to side-effect problems.[26]

4.1 NMDA receptor subtypes

The NMDA receptor is activated by agonists only after postsynaptic membrane depolarisation relieves voltage-dependent channel block by Mg^{2+}, and is additionally unique amongst the ligand-gated ion channels in its requirement for both glutamate and glycine (or *D*-serine) for activation.[27] These amino acids are often termed "co-agonists"[28] and two molecules each of glutamate and glycine appear to be required for channel opening.[29] The glycine site, which is probably not saturated *in vivo*, has been the focus for intensive drug discovery efforts.[30,31] The open channel of the NMDA receptor gates passage of extracellular Na$^+$ and Ca^{2+} and intracellular K$^+$. Receptor subunits identified so far are NR1 (8 alternatively spliced isoforms), NR2A-D and NR3A.[4,32] Only *heteromeric* recombinant receptors are functional and the presence of the NR1 subunit is obligatory, for example, NR1 + one of NR2A-D and three-subunit complexes such as NR1 + NR2A + NR2B and NR1 + NR2A + NR2C. The glycine binding site is present only on NR1[22] whilst glutamate binding needs an NR2 subunit.[33,34] Recent evidence suggests the NMDA receptor is tetrameric.[35] Given the stoichiometry of amino acid binding,[29] it seems probable that two each of the NR1 and NR2 subunits make up the receptor. If one type of NR2 subunit is employed, only two possible alignments of the subunits around the channel will be possible in a tetrameric complex (Figure 2). Although the receptor subunits show regional distribution in rat brain, suggesting receptor heterogeneity, little is currently known about the composition of the native NMDA receptors.[4]

4.2 **Drug binding sites on the NMDA receptor**

In addition to the glutamate and glycine sites, at which competitive antagonists bind, the NMDA receptor channel complex is activated by low concentrations of polyamines such as spermine. High concentrations of polyamines block the receptor channel. Sites exist within the open channel for voltage-dependent block by Mg^{2+} and use-dependent block by synthetic channel blockers such as MK-801 (dizocilpine) and AR-R15896 (Figure 3). The glutamate, glycine, 'polyamine' and channel sites have received the most attention in drug discovery efforts[4,24,26] and the profiles of antagonists targeted at each of them will be reviewed below. Several other binding sites on the NMDA receptor are known, including sites for cannabinoids and peptides such as conontoxins, and Zn^{2+}. The receptor is subject to additional physiological control by H^+ block, oxidation and reduction, and phosphorylation.

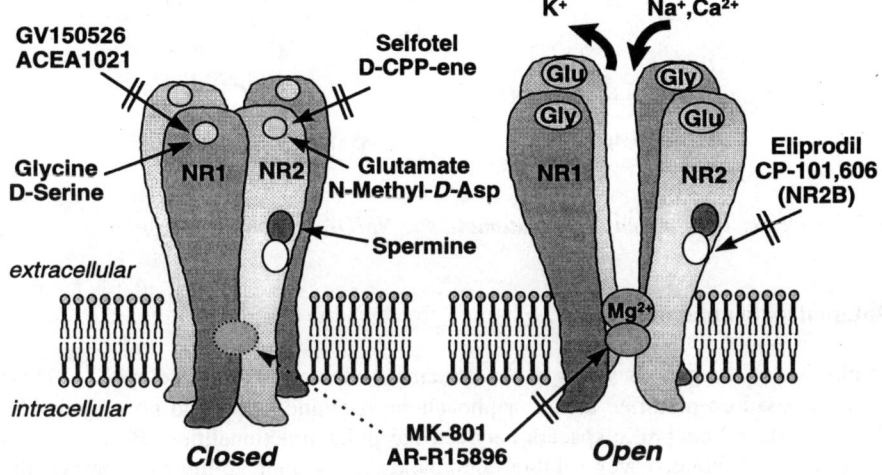

Figure 3 *Model of the NMDA receptor showing putative binding sites for drug molecules*

4.3 **Side-effect profiles of NMDA receptor antagonists**

The *in vivo* outcome of blockade of the NMDA receptor by various antagonists is a cascade of effects, including, as well as beneficial neuroprotective and antiseizure activity, several unwanted side-effects. The major side-effects observed in aminal studies are ataxia, motor stimulation, 'phencyclidine-like' psychotomimetic behavioural effects, neuronal vacuolation, activation of cerebral glucose metabolism, cardiovascular changes and blockade of long term potentiation (LTP). Potent open channel blockers, such as MK-801, display all these effects at comparable doses, whereas degrees of separation of neuroprotection from these side-effects are evident with antagonism mediated *via* the glycine, glutamate and polyamine sites. For example, under the neuroprotective infusion regime employed in the rat middle cerebral artery model,[36] the glycine-site partial agonist L-687,414[37] was free from the typical side-effects[38,39,40] exhibited by MK-801. Similarly, the glycine-site antagonist L-701,324 also lacks the side-effects[41] and has a different *in vivo* profile from MK-801, exhibiting atypical neuroleptic activity.[42]

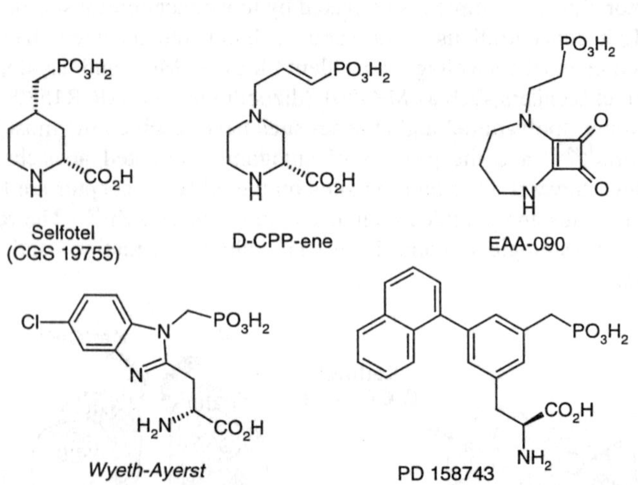

Figure 4 *Structures of representative glutamate-site NMDA receptor antagonists*

4.4 Glutamate-site antagonists

Many high-affinity antagonists acting at the glutamate site have been developed.[43] Their structures are based on modified acidic or phosphono α-amino acids, and potency can be increased by certain liphophilic spacers between the polar functionalities (Figure 4). In general these highly polar, water soluble compounds show poor ability to penetrate the blood-brain barrier. It has not proven easy to replace the α-amino acid with more "drug-like" bioisosteres, with the notable exception of the diaminobutanedione EAA-090.[44] Competition with high glutamate levels *in vivo* could also contribute to the generally modest *in vivo* activity shown by these compounds. In principle, as the glutamate site requires the NR2 subunit, glutamate antagonists should be capable of distinguishing NR2A-D subunit containing receptors; however convincing evidence of such specificity is lacking at present. Although improved side-effects were found in animal studies, in comparison with high affinity channel blockers, the clinical trials with Selfotel have been halted.[26]

4.5 NR2B-selective antagonists

Compounds initially reported to antagonise the effect of polyamine action at the NMDA receptor have been shown to possess potent neuroprotective activity, for example the prototypical compound, eliprodil (Figure 5).[45] 'Polyamine' antagonists are all basic, lipophilic amines and may be expected to show good CNS penetration. Notably, these compounds are devoid of the typical *in vivo* side effects of the other classes of NMDA antagonists. This has not been easy to explain, until two recent observations. Firstly, 'polyamine' antagonists appear to act selectively at a novel site, which does not bind polyamines, on receptor/channel complexes containing the NR2B subunit.[46,47] Secondly, they exhibit an unusual 'activity-dependent' block of NMDA receptor responses, by appearing to stabilise an agonist-bound state of the receptor that has 'low open probability'.[46] This receptor state has been suggested to be the inactivated, protonated receptor, and eliprodil and related compounds can increase the potency of proton blockade.[48] The intriguing mechanism of antagonism, combined with NR2B subunit specificity, may explain the absence of expected NMDA-related side effects.

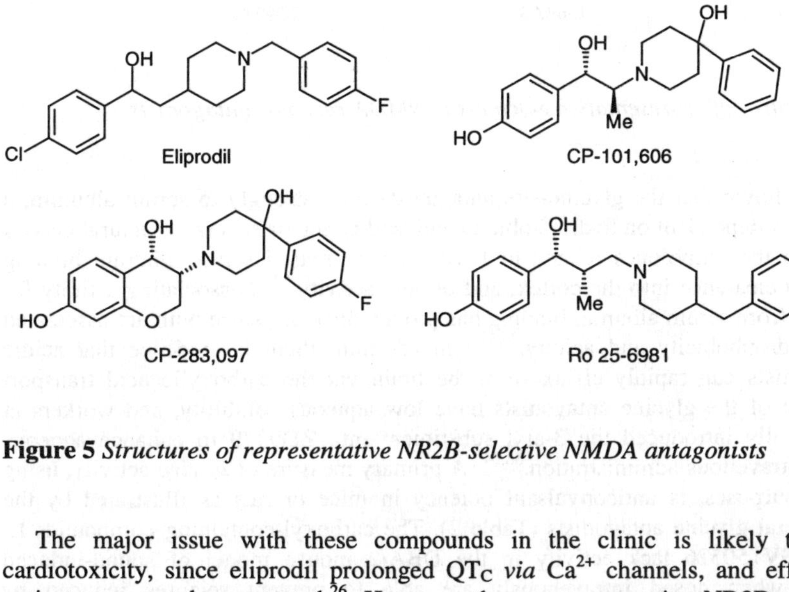

Figure 5 *Structures of representative NR2B-selective NMDA antagonists*

The major issue with these compounds in the clinic is likely to be potential cardiotoxicity, since eliprodil prolonged QT_C *via* Ca^{2+} channels, and efficacy in stroke patients was not demonstrated.[26] However, the newer generation NR2B-compounds, CP-101,606[49] and Ro25-6981[50] (Figure 5) are likely to have been designed to overcome this problem and if they prove to be safe, present exciting opportunities to show neuroprotective efficacy of NMDA receptor antagonists in man.

4.6 Glycine-site antagonists

Following the discovery of the unique "co-agonist" action of glycine in 1987,[27] there has been intensive medicinal chemical activity leading to several classes of high affinity NMDA receptor antagonists specific for the glycine site (Figure 6).[30,31] Structural manipulation of the initial leads, kynurenic acid (2-carboxy-4-quinolone) derivatives,[51] has provided the most potent glycine antagonists. However, the physicochemical and pharmacokinetic properties of the glycine antagonists generally are not ideal for access to the CNS, and *in vivo* activity in animal models is consequently limited.

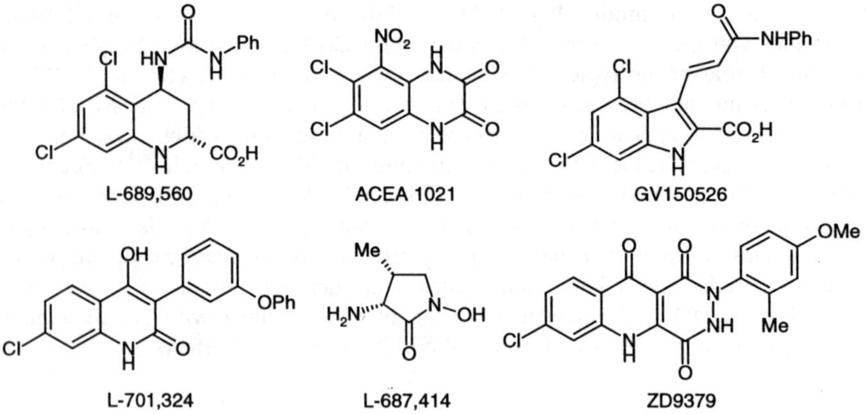

Figure 6 *Structures of representative glycine-site NMDA receptor antagonists*

It has been shown that the glycine-site antagonists bind strongly to serum albumin, a property which is dependent on hydrophobicity and acidity across several structural classes (Table 2).[52] In the quinolones related to L-701,324, increased serum albumin binding directly reduces clearance into the cortex, and *in vivo* activity.[52] Dissociating affinity for the glycine site from serum albumin binding has proven difficult, since both are associated with ligand hydrophobicity and acidity.[30,52] In addition, there is evidence that acidic glycine antagonists can rapidly efflux from the brain *via* the carboxylic acid transport system.[31] Most of the glycine antagonists have low aqueous solubility, and workers at Zeneca specifically introduced the 3-aryl substituent into ZD9379 to enhance aqueous solubility for intravenous administration.[53,54] A primary measure of *in vivo* activity, using for screening purposes, is anticonvulsant potency in mice or rats as illustrated by the profiles of several glycine antagonists (Table 2). The carboxyl-containing compounds L-689,560 and GV150526 lack activity in the DBA/2 mouse model of sound-induced seizure,[52] but when dosed intravenously are able to prevent seizures induced by intracerebrovascular (i.c.v.) dosing of NMDA to rats.[55] In contrast, replacement of the carboxyl by acidic bioisosteres, as in the hydroxyquinolone L-701,324[56] and the quinoxalinedione ACEA 1021,[57] leads to activity in both models, possibly as a result of improved brain levels. However, it has been pointed out that the DBA/2 mouse anticonvulsant test is particularly sensitive and compounds active in this assay may not possess neuroprotective efficacy.[31] The data in Table 2 suggest that the rat i.c.v. NMDA anticonvulsant assay[55] is even more sensitive to glycine-site antagonism.

The glycine antagonists have improved side-effect profiles relative to all other classes of NMDA antagonists,[31,40] with the exception of the polyamine antagonists. Their activity is not dependent on glutamate concentrations, and the availability of orally active compounds[56] has helped to establish therapeutic potential beyond neuroprotection,[30,31] for example in schizophrenia [42] and pain.[58] The glycine binding site is on the NR1 subunit, which is a prerequisite for receptor activity, but the affinity of glycine itself can vary by 10-fold depending on which NR2 subunit is present. In addition, many studies show NR1-NR2A-D allosteric interactions between the glycine site and glutamate sites (reviewed in ref. 31) showing that there are differences dependent on antagonist structure, and recently modest selectivity (~10 fold) has been found in derivatives of L-689,560.[59] ACEA 1021[57] (despite problems with solubility[26]), GV150526[55,60] and ZD9379[53,54] (Figure 6) are currently in clinical trials for stroke treatment.[26,31]

Table 2 *Profiles of Glycine-site NMDA Receptor Antagonists* [52,55]

Compound Log P	IC_{50} (nM) vs. [³H]L-689,560	ED_{50} (mg/kg, i.p.) DBA/2 mouse	ED_{50} (mg/kg, i.v.) i.c.v. NMDA, rat	HSA binding HPLC time vs. L-701,324
L-689,560 *1.1*	4.0	>100	0.21	0.18
GV150526 *2.2*	1.0	>10	0.06	2.0
L-701,324 *2.9*	2.0	0.88	0.78	1.0 (5 x 10^5 L/mol)
ACEA 1021 *0.21*	2.8	3.3	0.21	0.28

4.7 Channel blockers

The NMDA receptor antagonists which have been most widely used to characterise receptor function and effects, are the prototypical open channel blockers dizocilpine (MK-801) and phencyclidine (PCP, Figure 7).[61] A large number of structural classes of such channel blockers have been developed from PCP and MK-801-like molecules and in general these compounds are lipophilic amines which rapidly cross the blood-brain barrier. They demonstrate "use-dependent" channel blockade, becoming more potent as glutamate and glycine concentrations are increased. This class of NMDA receptor antagonist displays the greatest *in vivo* potency and neuroprotective activity, but there is no significant separation from unwanted side-effects with *potent* channel blockers, which appear to display steep dose-response relationships and "all or nothing" receptor blockade. Interestingly, and perhaps counter-intuitively, it is the *less potent* channel blockers, including AR-R15896 and remacemide, that consistently show improved therapeutic ratios and reduced propensity for 'psychotomimetic' PCP-like effects (see Table 3).[61,62]

Rationales for the improved profiles of the low versus high affinity channel blockers have been proposed.[61,62] The arguments are based on experimentally determined differences in kinetics of channel block[63,64] and in forebrain/cerebellar and receptor subtype specificity [65] (Table 3). However in these cases, and in general, only marginal receptor subtype specificity has been achieved so far with channel blockers. A further possibility for reduced side-effects has been put forward based on the improved *in vivo* profile of the quaternary compound WIN 63840, which is poorly soluble in lipid (Log D = -4.1) and can only access the open channel from the extracelluar aqueous medium.[66] In contrast, MK-801 has greater lipid solubility (Log D = 1.8) and the protonated form blocks the open channel, but the uncharged neutral form can also block the closed channel, possibly gaining access to the binding site *via* the lipid membrane (see: Figure 3).

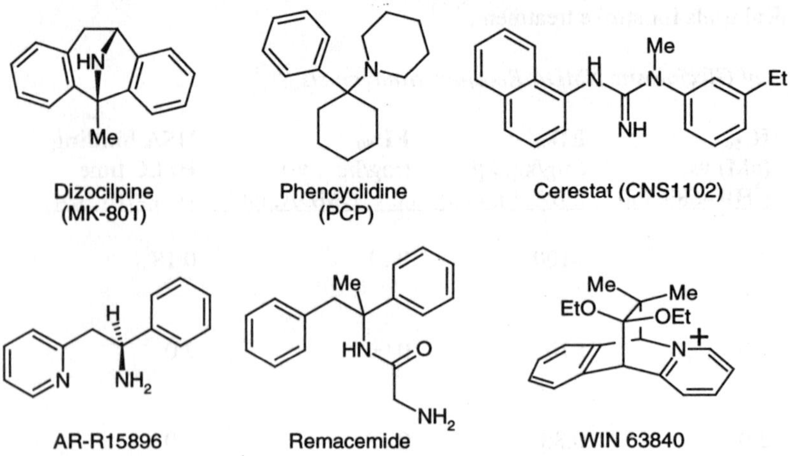

Dizocilpine   Phencyclidine   Cerestat (CNS1102)
(MK-801)   (PCP)

AR-R15896   Remacemide   WIN 63840

Figure 7 *Structures of representative NMDA receptor channel blockers*

The lack of closed channel block with WIN 63840 was suggested to result in reduced overall antagonism and side effects,[67] although the compound might also be expected to show poor CNS penetration. Closed channel block can also occur by the phenomenon of 'trapping' of the channel blocker, where, following agonist dissociation, an open channel blocker remains bound to the closed channel. Differing degrees of trapping amongst channel blockers reflect differences in rates of dissociation from the closed channel binding site. Lowered levels of trapping correlated with improved therapeutic ratio between three different channel blockers.[68] Unfortunately for drug design, the molecular properties of channel blockers which control blocking kinetics and the degree of trapping do not appear to be well understood.

In agreement with the results from animal studies, the low affinity NMDA receptor channel blockers appear to be safe for administration to humans.[26] Remacemide[69] is currently in Phase III clinical trials for epilepsy, where it is efficacious at a dose of 300 mg b.i.d. AR-R15896[70] is in Phase II clinical trials for stroke. In Phase I studies, single and multiple doses of AR-R15896 of up to 160 mg i.v. were tolerated, and animal neuroprotective plasma levels were achieved.

4.8 Conclusions

Drug discovery at the NMDA receptor has been a frustrating process, compounded by serious side-effects as well as poor pharmacokinetic and physical properties of synthetic ligands. A simplistic view of the outcome of antagonising the NMDA receptor, suggesting dependence on the overall degree of receptor block at all subtypes achieved *in vivo*, is illustrated in Figure 8. The chemical challenges, in producing suitable 'drug-like' physical properties when starting CNS medicinal chemistry programmes with highly polar amino acids as leads, have been considerable.

Table 3 *Comparison of Properties of Low and High Affinity NMDA Receptor Channel Blockers, AR-R15896 and MK-801* [61,63,65,68]

	Low affinity AR-R15896	High affinity MK-801
IC_{50} (μM, vs. [^{3}H]MK-801)	1.3	0.010
ED_{50} (rat MES, mg/kg, i.p.)	13	0.43
PCP score (x ED_{50}; max = 25)	1 @ 50x ED_{50}	22 @ 6x ED_{50}
Therapeutic index ED_{50} MES/TD_{50} inv. Screen	10	0.66
Kinetics [Ca^{2+}]$_{int.}$, cortical neuron T_{on} at IC_{50}	fast, partial block instant	slow, full block 118s
Degree of trapping	54%	complete
Specificity Forebrain/cerebellum Recombinant NMDA-R	0.38 NR2C>NR2A	2.4 NR2A>NR2C

Low affinity channel blockers and glycine antagonists are now progressing in the clinic, but it is the discovery of the NR2B subunit-selective agents, with a novel mechanism of antagonism, that may provide the best hope so far of establishing clinical efficacy of NMDA receptor antagonists in CNS disorders. The prospects of discovering other subunit selective agents seems to depend also on identifying novel allosteric binding sites, since the glutamate, glycine and channel-site NMDA receptor ligands so far studied have shown insufficient selectivity to be useful enough for drug discovery. The subtype selectivities reported so far for these agents appear to be no greater than ~10-fold,[4,31] which is not sufficient for adequate characterisation of pharmacological effects. The identification of new NMDA receptor modulators and antagonists with meaningful subtype specificity is a significant and highly attractive challenge for CNS drug discovery.

5 OUTLOOK: TARGETING ALLOSTERIC SITES

The definition of novel ligand-gated ion channels at the molecular level will undoubtedly lead to increasing numbers of validated drug discovery targets. The traditional approach employed by medicinal chemists, of synthetic manipulation of the natural competitive agonists, is unlikely to be the best means of obtaining subunit specificity between multiple, related ion channels which employ the same agonist. In the cases of the new and little-studied P2X and ASIC channels, where ATP and protons respectively are the agonists, this approach is in any event chemically unattractive. Targeting sites within the open or closed channel has allowed opportunities to optimise pharmacological profiles, in the absence of significant subunit specificity, at NMDA receptors. Although it has not proven simple to discover open channel blockers at the non-NMDA glutamate receptors, searching for subunit-selective channel blockers should be considered as part of a screening strategy on any new ligand-gated ion channel.

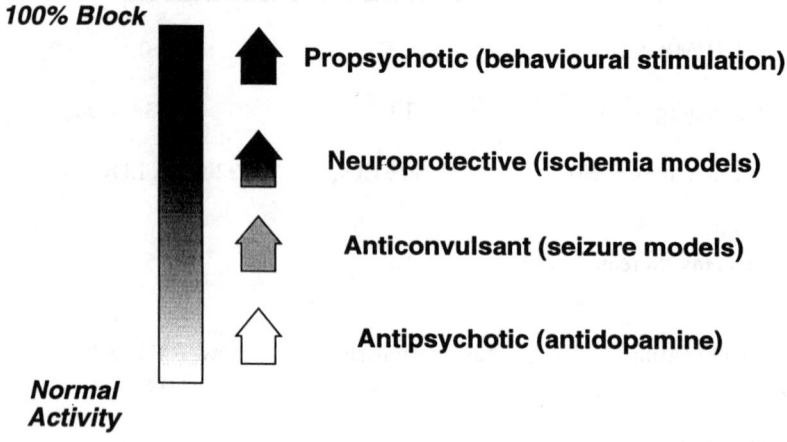

Figure 8   *A Proposal for NMDA Receptor Blockade and Responses*

The identification of novel allosteric ligands, which act non-competitively or uncompetitively with respect to the agonist ligand, as exemplified above by the NMDA NR2B subunit selective agents, is likely to be a better prospect for subunit specificity than competitive antagonists or channel blockers. To search for such novel sites, and identify lead compounds acting on them, requires high throughput functional screens using the relevant specific human receptor subtypes, the availability of "drug-like" compound libraries, and ideally 3-D structural information on ligand binding domains. These tools are now becoming available and are expected to play a key part in the continuing challenge of uncovering new and specific modulators for the ligand-gated ion channels.

6 ACKNOWLEDGEMENTS

We thank colleagues at Astra: Ed Harris, Margareta Lönngren, Dennis McCarthy, Bob Murray, Gene Palmer, George Smith and John Suschitzky, for helpful information on remacemide and AR-R15896, and for stimulating discussions.

References

1. N. Le Novere and J. P. Changeux, *Nucleic Acids Research*, 1999, **27**, 340. The address of a useful Ligand-Gated Ion Channel Database is: http://www.pasteur.fr/units/neubiomol/LGIC.html
2. R. Waldmann and M. Lazdunski, *Curr. Opinion Neurobiol.*, 1998, **8**, 418.
3. R. A. North and E. A. Barnard, *Curr. Opinion Neurobiol.*, 1997, **7**, 346.
4. R. Dingledine, K. Borges, D. Bowie and S. F. Traynelis, *Pharmacol. Revs.*, 1999, **51**, 7.
5. R. M. Eglen and D. W. Bonhaus, *Curr. Pharm. Design*, 1996, **2**, 367.
6. M. W. Holladay, M. J. Dart and J. K. Lynch, *J. Med. Chem.*, 1997, **40**, 4169.
7. W. Hevers and H. Luddens, *Mol. Neurobiol.*, 1998, **18**, 35.
8. H.-G. Breitinger and C.-M. Becker, *Curr. Pharm. Design*, 1998, **4**, 315.
9. N. Unwin, *Nature*, 1995, **373**, 37.
10. N. Nayeem, T. P. Green, I. L. Martin and E. A. Barnard, *J. Neurochem.*, 1994, **62**, 815.
11. C. Rosenmund, Y. Stern-Bach and C.F. Stevens, *Science*, 1998, **289**, 1596.
12. A. Nicke, H. G. Baumert, J. Rettinger, A. Eichele, G. Lambrecht, E. Mutschler and G. Schmalzing, *EMBO J.*, 1998, **17**, 3016.
13. R. M. McKernan and P. J. Whiting, *TiNS.*, 1996, **19**, 139.
14. P. Skolnick, R. J. Hu, C. M. Cook, S. D. Hurt, J. D. Trometer, R. Liu, Q. Huang and J. M. Cook, *J. Pharm. Exp. Ther.*, 1997, **283**, 488.
15. K. Quirk, P. Blurton, S. Fletcher, P. Leeson, F. Tang, D. Mellilo, C. I. Ragan and R. M. McKernan, *Neuropharmacology*, 1996, **35**, 1331.
16. E. D. Cox, H. Diaz-Arauzo, Q. Huang, M. S. Reddy, C. Ma, B. Harris, R. McKernan, P. Skolnick and J. M. Cook, *J. Med. Chem.*, 1998, **41**, 2537.
17. N. Armstrong, Y. Sun, G.-Q. Chen and E. Gouaux, *Nature*, 1998, **395**, 913.
18. Y. Paas, *TiNS.*, 1998, **21**, 117.
19. A. Ivanovic, H. Reilander, B. Laube and J. Kuhse, *J. Biol. Chem.*, 1998, **273**, 19933.
20. J. E. Gready, S. Ranganathan, P. R. Schofield, Y. Matsuo and K. Nishikawa, *Protein Science*, 1997, **6**, 983.
21. E. Sigel and A. Buhr, *TiPS.*, 1997, **18**, 425.
22. H. Hirai, J. Kirsch, B. Laube, H. Betz and J. Kuhse, *Proc. Nat. Acad. Sci. U.S.A.*, 1996, **93**, 6031.
23. J. C. Ma and D.A. Dougherty, *Chem. Revs.*, 1997, **97**, 1303.
24. G. L. Collingridge and J. C. Watkins (Eds.), The NMDA Receptor, 2nd Ed., IRL Press, Oxford, 1994.
25. D. W. Choi and S. M Rothman, *Ann. Rev. Pharmacol. Toxicol.*, 1991, **31**, 171.
26. P. L. Wood and J. E. Hawkinson. *Exp. Opin. Invest. Drugs*, 1997, **6**, 389.
27. J. W. Johnson and P. Ascher, *Nature*, 1987, **325**, 529.
28. N. W. Kleckner and R. Dingledine, *Science*, 1988, **241**, 835.
29. M. Benveniste and M. L. Mayer, *Biophys. J.*, 1991, **59**, 560.
30. P. D. Leeson and L. L. Iversen, *J. Med. Chem.*, 1994, **37**, 4053.
31. W. Danysz and C. G. Parsons, *Pharmacol. Revs.*, 1998, **50**, 597.
32. N. J. Sucher, M. Awobuluyi, Y. B. Choi and S. A. Lipton, *TiPS.*, 1996, **17**, 348.

33. L. C. Anson, P. E. Chen, D. J. A. Wylie, D. Colquhoun and R. Schoepfer, *J. Neurosci.*, 1998, **18**, 581.
34. B. Laube, H, Hirai, M. Sturgess, H. Betz and J. Kuhse, *Neuron*, 1997, **18**, 493.
35. B. Laube, J. Kuhse and H. Betz, *J. Neurosci.*, 1998, **18**, 2954.
36. R. Gill, R. J. Hargreaves and J. A. Kemp, *J. Cereb. Blood Flow Metab.*, 1995, **15**, 197.
37. P. D. Leeson, B. J. Williams, M. Rowley, K. W. Moore, R. Baker, J. A. Kemp, T. Priestley, A. C. Foster and A. E. Donald, *Bioorg. Med. Chem. Lett.*, 1993, **3**, 71.
38. M. D. Tricklebank, L .J. Bristow, P. H. Hutson, P. D. Leeson, M. Rowley, K. Saywell, L. Singh, F. D. Tattersall, L. Thorn, and B. J. Williams, *Br. J. Pharmacol.*, 1994, **113**, 729.
39. R. J. Hargreaves, M. Rigby, D. Smith, and R. G. Hill, *Br. J. Pharmacol.* 1993, **110**, 36.
40. T. Priestley, G. R. Marshall, R. G. Hill and J. A. Kemp, *Br. J. Pharmacol.*, 1998, **124**, 1767.
41. L. J. Bristow, P. H. Hutson, J. J. Kulagowski, P. D. Leeson, S. Matheson, F. Murray, D. Rathbone, K. L Saywell, L. Thorn, A. P. Watt and M. D. Tricklebank, *J. Pharm. Exp. Ther.*, 1996, **279**, 492.
42. L. J. Bristow, K. L. Flatman, P. H. Hutson, J. J. Kulagowski, P. D. Leeson, L. Young and M. D. Tricklebank, *J. Pharm. Exp. Ther.*, 1996, **277**, 578.
43. G. Johnson and P. L. Ornstein, *Curr. Pharm. Design*, 1996, **2**, 331.
44. W. A. Kinney, M. Abou-Gharbia, D. T. Garrison, J. Schmid, D. M. Kowal, D. R. Bramlett, T. L. Miller, R. P. Tasse, M. M. Zaleska and J. A. Moyer, *J. Med. Chem.*, 1998, **41**, 236.
45. C. J. Carter, K. G. Lloyd, B. Zivkovic and B. Scatton, *J. Pharm. Exp. Ther.*, 1990, **253**, 475.
46. M. J. Gallagher, H. Huang, D. B. Pritchett and D. R. Lynch, *J. Biol. Chem.*, 1996, **19**, 9603.
47. J. N. C. Kew, G. Trube and J. A. Kemp, *J. Physiol.*, 1996, **497.3**, 761.
48. D. D. Mott, J. A. Doherty, Z. Zhang, M. S. Washburn, M. J. Fendley, P. Lyuboslavsky, S. F. Traynelis and R. Dingledine, *Nat. Neurosci.*, 1998, **1**, 659.
49. F. Menniti, B. Chenard, M. Collins, M. Ducat, I. Shalaby and F. White, *Eur. J. Pharmacol.*, 1997, **331**, 117.
50. G. Fischer, V. Mutel, G. Trube, P. Malherbe, J. N. C. Tew, E. Mohacsi, M. P. Heitz and J. A. Kemp, *J. Pharm. Exp. Ther.*, 1997, **283**, 1285.
51. P. D. Leeson, R. Baker, R. W. Carling, N. R. Curtis, K. W. Moore, B. J. Williams, A. C. Foster, A. E. Donald, J. A. Kemp and G. R. Marshall, *J. Med. Chem.*, 1991, **34**, 1243.
52. M. Rowley, J. J. Kulagowski, A. P. Watt, D. Rathbone, G. I. Stevenson, R. W. Carling, R. Baker, G. R. Marshall, J. A. Kemp, A. C. Foster, S. Grimwood, R. Hargreaves, C. Hurley, K. L. Saywell, M. D. Tricklebank and P. D. Leeson, *J. Med. Chem.*, 1997, **40**, 4053.
53. T. M. Bare, *J. Heterocycl. Chem.*, 1998, **35**, 1171.
54. T. Tatlisumak, K. Takano, M. R. Meiler and M Fisher, *Stroke*, 1998, **29**, 190.
55. R. Di Fabio, A.M. Capelli, N. Conti, A. Cugola, D. Donati, A. Feriani, P. Gastaldi, G. Gaviraghi, C.T Hewkin, F. Micheli, A. Missio, M. Mugnaini, A. Pecunioso, A.M.

Quaglia, E. Ratti, L. Rossi, G. Tedesco, D. G. Trist and A. Reggiani, *J. Med. Chem.*, 1997, **40**, 841.

56. J. J. Kulagowski, R. Baker, N. R. Curtis, P. D. Leeson, I. M. Mawer, A. M. Moseley, M. P. Ridgill, M. Rowley, I. Stansfield, A. C. Foster, S. Grimwood, R. G. Hill, J. A. Kemp, G. R. Marshall, K. L. Saywell and M. D. Tricklebank, *J. Med. Chem.*, 1994, **37**, 1402.

57. J. F. W. Keana, S. M. Kher, S. X. Cai, C. M. Dinsmore, A. G. Glenn, J. Guastella, J-C Huang, V. Ilyin, Y. Lu, P. L. Mouser, R. M. Woodward and E. Weber, *J. Med. Chem.*, 1995, **38**, 4367.

58. J. M. A. Laird, G. S. Mason, J. Webb, R. G. Hill and R. J. Hargreaves, *Br. J. Pharmacol.*, 1996, **117**, 1487.

59. M. Honer, D. Benke, B. Laube, J. Kuhse, R. Heckendorn, H. Allgeiers, C. Angst, H. Monyer, P. H. Seeburg, H. Betz and H. Mohler. *J. Biol. Chem.*, 1998, **273**, 11158.

60. R. Di Fabio, A. Cugola, D. Donati, A. Feriani, G. Gaviraghi, E. Ratti, D. G. Trist and A. Reggiani, *Drugs. Fut.*, 1998, **23**, 61.

61. E. W. Harris and R. J. Murray, *Curr. Pharm. Des.*, 1996, **2**, 429.

62. M. A. Rogawski, *Trends Pharm. Sci.*, 1993, **14**, 325.

63. M. Black, T. Lanthorn, D. Small, G. Mealing, V. Lam and P. Morley, *Eur. J. Pharmacol.*, 1996, **317**, 377.

64. G. A. R. Mealing, T. H. Lanthorn, D. L. Small, M. A. Black, N.B. Laferriere and P. Morley, *J. Pharm. Exp. Ther.*, 1997, **281**, 376.

65. D. T. Monaghan and H. Larsen, *J. Pharm. Exp. Ther.*, 1997, **280**, 614.

66. W. G. Earley, V. Kumar, J. P. Mallamo, C. Subramanyam, J. A. Dority, M. S. Miller, D. L. DeHaven-Hudkins, L. D. Aimone, M. D. Kelly and B. Ault, *J. Med. Chem.*, 1995, **38**, 3586.

67. B. Ault, M. S. Miller, M. D. Kelly, L. M. Hildebrand, W. G. Earley, D Luttinger, J. P. Mallamo and S. J. Ward, *Neuropharmacology*, 1995, **34**, 1597.

68. G. A. R. Mealing, T. H. Lanthorn, C. L. Murray, D. L. Small and P. Morley, *J. Pharm. Exp. Ther.*, 1999, **288**, 204.

69. G. C. Palmer and J. B. Hutchinson, 'Excitatory Amino Acids - Clinical Results with Antagonists,' Academic Press, 1997, Chapter 10, p 109.

70. G. C. Palmer, J. A. Miller, E. F. Cregan, P. Gendron and J. Peeling, *Ann. N.Y. Acad. Sci.*, 1997, **825**, 220.

STRUCTURAL AND STEREOCHEMICAL REQUIREMENTS FOR ACTIVATION AND BLOCKADE OF EXCITATORY AMINO ACID RECEPTORS

Frank A. Sløk, Hans Bräuner-Osborne, Tine B. Stensbøl, Tommy N. Johansen, *Bjarke Ebert, Martin Mortensen, Birgitte Nielsen, Ulf Madsen, Lotte Brehm, Erik Falch and Povl Krogsgaard-Larsen

Departments of Medicinal Chemistry and *Pharmacology, The Royal Danish School of Pharmacy, 2 Universitetsparken, DK-2100 Copenhagen, Denmark

1 MULTIPLICITY OF EXCITATORY AMINO ACID RECEPTORS

(S)-Glutamic acid ((S)-Glu), which is the major excitatory neurotransmitter in the central nervous system (CNS), and other excitatory amino acids (EAAs) operate through four different classes of receptors. In addition to the three heterogeneous classes of ionotropic EAA receptors (iGluRs), named N-methyl-D-aspartic acid (NMDA), (RS)-2-amino-3-(3-hydroxy-5-methyl-4-isoxazolyl)propionic acid (AMPA), and kainic acid receptors,[1-3] a heterogeneous class of metabotropic EAA receptors (mGluRs) (Figure 1) has been shown to have important functions in the central excitatory neurotransmission processes.[4] It is now generally agreed that iGluRs as well as mGluRs play important roles in the healthy as well as the diseased CNS, and that all subtypes of these receptors are potential targets for therapeutic intervention in a number of neurologic and psychiatric disorders.[5,6]

EAA receptors are involved in the mechanisms of long-term potentiation, which is believed to play an important role in learning and memory functions, and the deficits of these functions in, for example, Alzheimer patients may, to some extent, be caused by hypoactivity at CNS synapses primarily operated by iGluRs and mGluRs.[7-10] There is also growing evidence of an implication of EAA receptors in schizophrenia.[11-13] As in Alzheimer's disease, the role of these receptors in the etiology and in the clinical manifestations of schizophrenia is still incompletely understood, but there is evidence to suggest that hypoactivity at EAA receptors is also a factor of importance in the latter CNS disorder.[13-15]

2 DESIGN OF SELECTIVE EAA RECEPTOR LIGANDS USING NATURALLY OCCURRING AMINO ACIDS AS LEADS

A large number of plants ranging from microorganisms to flowering plants biosynthesize amino acids structurally related to (S)-Glu. A number of these naturally occurring (S)-Glu analogues are recognized by EAA receptors as agonists, which in most cases explains their excitotoxic effects in vitro and in vivo.[16,17] A few of these amino acids, notably ibotenic acid and kainic acid (Figure 2), have been extensively used as neurotoxins in experimental neurobiology, although the lack of receptor specificity has frequently made interpretation of the data in terms of degenerative mechanisms difficult. Some of these amino acids, in particular ibotenic acid[16] and willardiine,[18] have been used as lead structures for the design of subtype-specific or -selective receptor ligands.

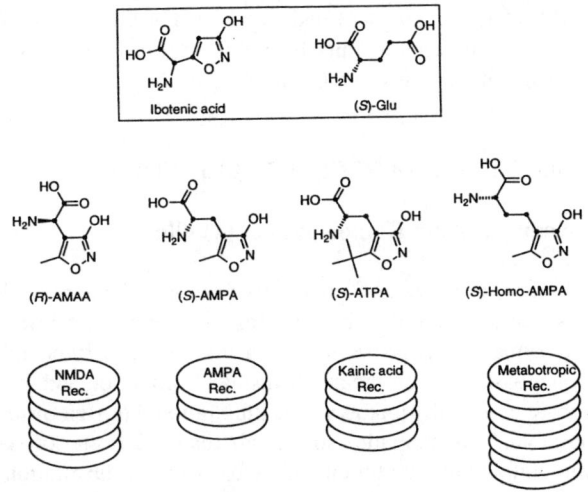

Figure 1 *Schematic illustration of the multiplicity of EAA receptors and the structures of some key agonists. The number of symbols for the iGluRs (NMDA, AMPA, and kainic acid receptors) indicate the number of receptor subunits known, whereas the number of symbols for the mGluRs (metabotropic receptors) indicate the number of receptor subtypes known*

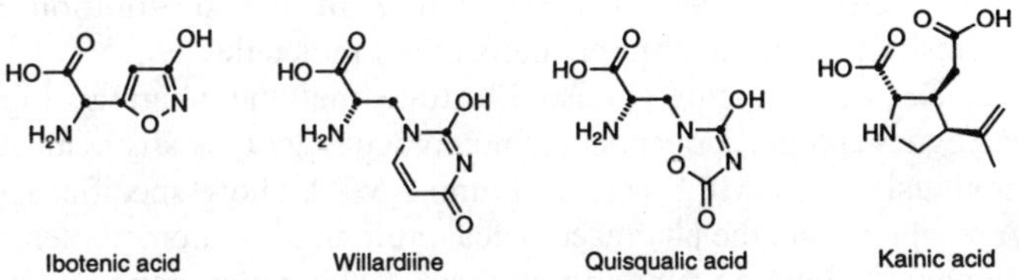

Figure 2 *Structures of some naturally occurring heterocyclic analogues of (S)-Glu, which have been used as leads for the design of specific/selective iGluR ligands*

Ibotenic acid, in which the terminal propionic acid part is constituted by a bioisosteric 3-isoxazolol group, is a potent NMDA receptor agonist. This structurally unique, but chemically and stereochemically unstable, natural amino acid also interacts with other EAA receptor subtypes.[16] Systematic variations of the structure of ibotenic acid have, however, led to the development of a variety of EAA receptor subtype-selective agonists, as exemplified in Figure 1. Thus, AMPA specifically activates AMPA receptors,[19] and whereas (S)-AMPA is the active enantiomer,[20,21] it is the (R)-form of (RS)-2-amino-2-(3-hydroxy-5-methyl-4-isoxazolyl)acetic acid (AMAA), which selectively activates NMDA receptors.[22] (RS)-2-Amino-3-(5-tert-butyl-3-hydroxy-4-isoxazolyl)propionic acid (ATPA) was designed as a lipophilic analogue of AMPA and as a tool for studies of the topography of the AMPA receptor recognition site.[23] ATPA turned out to be a relatively weak AMPA receptor agonist,[24] which is active after systemic administration,[25] but ATPA has recently been shown to be a potent agonist at the kainic acid-preferring iGluR5 receptor, being some three orders of magnitude more potent at this receptor than at AMPA receptors.[26] It has been demonstrated that the iGluR5 agonist effect of ATPA resides in the (S)-enantiomer (T.B. Stensbøl *et al.*, unpublished). The higher homologue of AMPA, (RS)-2-amino-4-(3-hydroxy-5-methyl-4-isoxazolyl)butyric acid (homo-AMPA) does not show significant affinity for iGluRs,[27] but Homo-AMPA,[28] or rather (S)-homo-AMPA,[29] is a specific agonist at mGluR6 (Figure 1).

3 PHARMACOLOGY OF EAA RECEPTOR LIGANDS

3.1 Bioisosteric Analogues and Homologues of (S)-Glu

A prerequisite for rational drug design,[30] in the strict sense of the term, is detailed information about the mechanism of action of the pharmacological target. Such information is available for a number of enzymes, and, consequently, a variety of enzyme inhibitors have been designed on a rational basis. So far, the mechanisms of activation and inactivation of iGluRs as well as mGluRs are essentially unknown, making rational design of receptor agonists and antagonists impossible. Systematic structural manipulations of lead structures by conformational immobilization, enantioseparation, bioisosteric substitution, and homologation have been successfully used, separately or in combination, to design and develop a broad spectrum of specific/highly selective EAA receptor ligands.[1,2]

The relationships between structure and EAA receptor pharmacology of a selection of acidic amino acids designed along these lines are summarized in Figure 3. Within the series of analogues of (S)-Glu, the pharmacology of the compounds is strongly dependent on the nature of the terminal carboxyl bioisosteric group. Although both ibotenic acid and (S)-AMPA are 3-isoxazolol bioisosteric analogues of (S)-Glu, the distinctly different pharmacology of these two amino acids illustrates the importance of the substitution and degree of conformational restriction of the terminal parts of the molecules.

The examples shown in Figure 3 also illustrate that the pharmacological effects of bioisosteric analogues and the corresponding homologues are typically distinctly different. As mentioned previously, (S)-AMPA and (S)-homo-AMPA show specific agonist effects at different EAA receptors, and the pharmacological profiles of (S)-homoibotenic acid[31] and (S)-homoquisqualic acid[32] show no similarity to those of the parent amino acids. Interestingly, neither (S)-2-amino-4-phosphonobutyric acid ((S)-AP4) nor the homologous amino acid, (S)-AP5 interact with iGluRs, but each of these compounds interact quite selectively with mGluR subtypes as an agonist and antagonist, respectively.[32]

3.2 Alkyl Analogues of AMPA: An Unsuccessful Approach to the Design of Partial AMPA Receptor Agonists

Growing evidence suggests an involvement of central EAA receptors in Alzheimer's disease[7-10] and schizophrenia,[11-13] and although the role of EAA receptors in these disorders is far from being fully elucidated, hypoactivity at synapses primarily operated by such receptors appears to be an important factor.[13-15] In principle, EAA receptor agonists might have therapeutic potential in these disorders, although long-term treatment with such compounds may cause excitotoxic neuronal damage. Partial agonists, on the other hand, may

Compound	Structure	Effects at	
		iGluRs	mGluRs
(S)-Glu	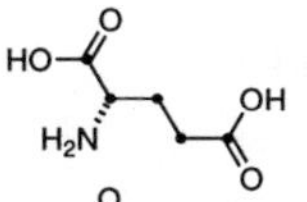	Nonselective ago	Nonselective ago
(S)-α-AA	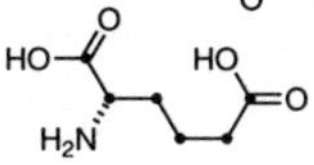	Nonselective effects	mGluR$_2$ ago mGluR$_6$ ago
Ibotenic acid	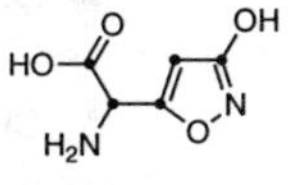	NMDA ago Kainic acid ago	Nonselective ago
(S)-Homoibotenic acid	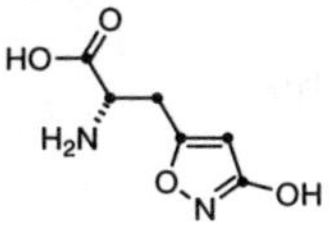	AMPA ago	mGluR$_1$ ant mGluR$_5$ ant
(S)-AMPA	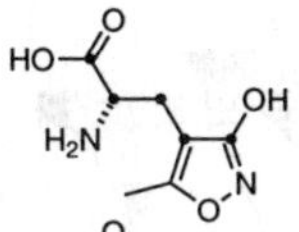	AMPA ago	Inactive
(S)-Homo-AMPA	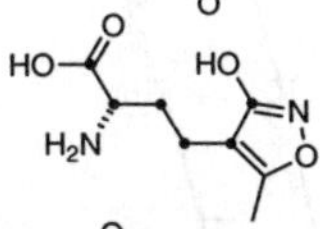	Inactive	mGluR$_6$ ago
(S)-Quisqualic acid	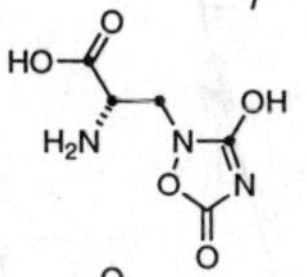	AMPA ago Kainic acid ago	mGluR$_1$ ago mGluR$_5$ ago
(S)-Homoquisqualic acid	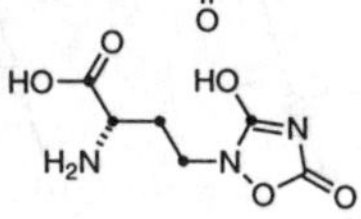	Inactive	mGluR$_1$ ant mGluR$_2$ ago mGluR$_5$ ago
(S)-AP4	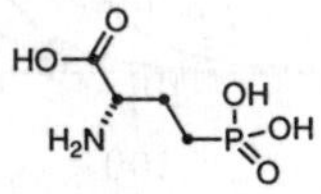	Inactive	mGluR$_4$ ago mGluR$_6$ ago mGluR$_7$ ago mGluR$_8$ ago
(S)-AP5	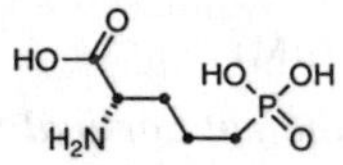	Inactive	mGluR$_2$ ant

Figure 3 *A comparison of the effects at iGluRs and mGluRs of a number of key EAAs and the corresponding homologues listed in pairs*

have therapeutic interest in Alzheimer's disease as well as schizophrenia. Such compounds, showing appropriately balanced agonist/antagonist profiles, may be capable of restoring EAA receptor activity in a nontoxic manner in brain areas suffering from EAA transmitter hypoactivity.[13]

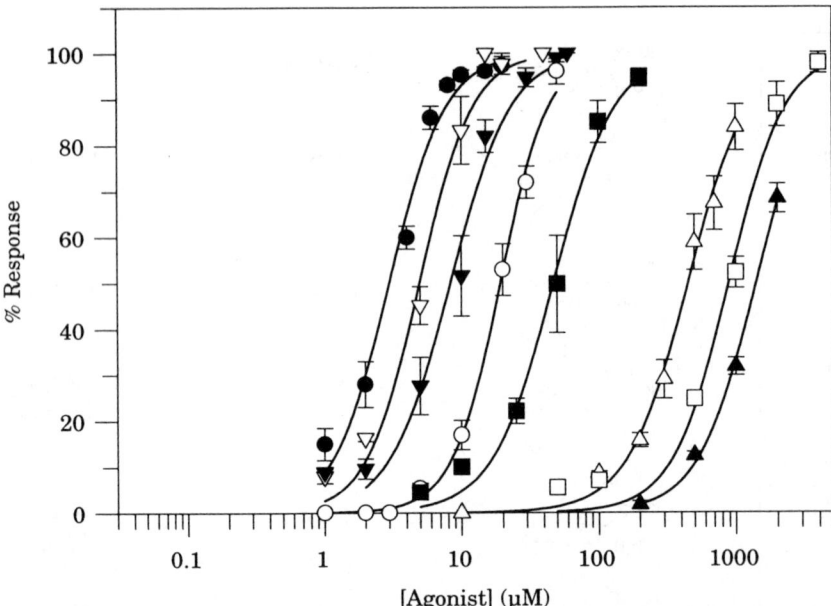

Figure 4 *Dose-response curves as determined in the rat cortical wedge preparation for AMPA and analogues. Values are mean values ± SEM relative to the maximal AMPA response. Data were fitted to the equation: % Response = MAX χ [Ago]^n (EC$_{50}$^{n} + [Ago]^n), where MAX is the maximal response relative to the AMPA plateau response, [Ago] is the agonist concentration in µM, and n is the Hill slope, determined to be close to 2 for all compounds. A 100% reponse is determined as the maximal response for AMPA (for details see ref. 34)*

As an attempt to develop AMPA receptor partial agonists, we have synthesized a series of AMPA analogues in which the methyl group has been replaced by alkyl groups of different steric volumes (Figure 4).[23,27] On the basis of previous structure-activity studies on AMPA analogues, we have proposed the existence of a lipophilic cavity at the AMPA recognition site capable of accomodating alkyl substituents of limited size.[27,33] The extension of this line of research was based on the expectation that AMPA analogues carrying alkyl substituents just accomodatable by the proposed cavity would possess limited capacity to activate the receptor and thus show partial agonism.[34] This simple concept proved to be incorrect. With the exception of Et-AMPA, containing an ethyl group in the 5-position of the 3-isoxazolol ring, which is slightly more potent than AMPA as an AMPA agonist,[27] increases of the size of the alkyl groups led to analogues showing decreasing agonist potency (Figure 4).[34] However, all of the active compounds within this series of AMPA analogues showed relative efficacies not significantly different from that of AMPA, and further increases of the size of the alkyl group resulted in inactive compounds.[34]

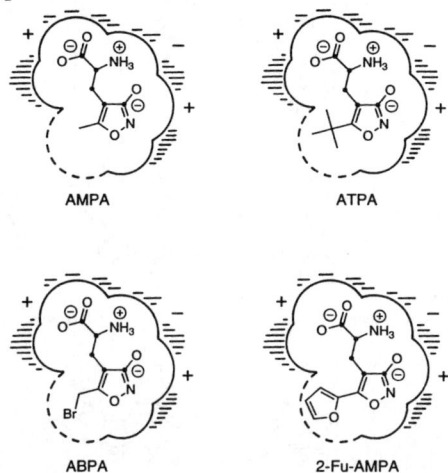

Figure 5 *A hypothetical model of the recognition site of the AMPA receptor indicating the presence of a cavity capable of accommodating certain substituents in the 5-position of the 3-isoxazolol ring of AMPA analogues, as illustrated for AMPA, ATPA, ABPA and 2-Fu-AMPA*

None of these alkyl-substituted AMPA analogues showed AMPA antagonist effects suggesting that this proposed cavity is present in the agonist rather than the antagonist conformation of the AMPA receptor. Since demethyl-AMPA (R=H in Figure 4), which has rather high affinity for the AMPA receptor site, shows low, but still full, agonist potency, we hypothesize that occupancy of this proposed cavity by an appropriately sized alkyl group (Figure 5) contributes to the stabilization of the agonist conformation of the AMPA receptor. In this context it is noteworthy that structure-activity studies on structurally related AMPA receptor antagonists have indicated that larger alkyl groups in the same position of the 3-isoxazolol ring of these compounds are tolerated at the recognition site of the antagonist conformation of the AMPA receptor.[35]

3.3   (R)- and (S)-APPA as Key Tools for the Development of the Pharmacological Principle, Functional Partial Agonism

Whereas all alkyl-substituted AMPA analogues, so far synthesized, show full agonism or inactivity at AMPA receptors (see preceding section), (RS)-2-amino-3-(3-hydroxy-5-phenyl-4-isoxazolyl)propionic acid (APPA) (Figure 6) showed the characteristics of a weak partial AMPA receptor agonist with an efficacy of approximately 60% relative to that of AMPA (Figure 7).[36] In order to shed further light on this observation, APPA was resolved, and, quite surprisingly, (S)-APPA was shown to be a full AMPA agonist, whereas (R)-APPA turned out to be a competitive AMPA antagonist.[37] Further experiments revealed that co-administration of (S)- and (R)-APPA at different fixed concentration ratios gave dose-response curves showing different levels of partial AMPA receptor agonism.[38]

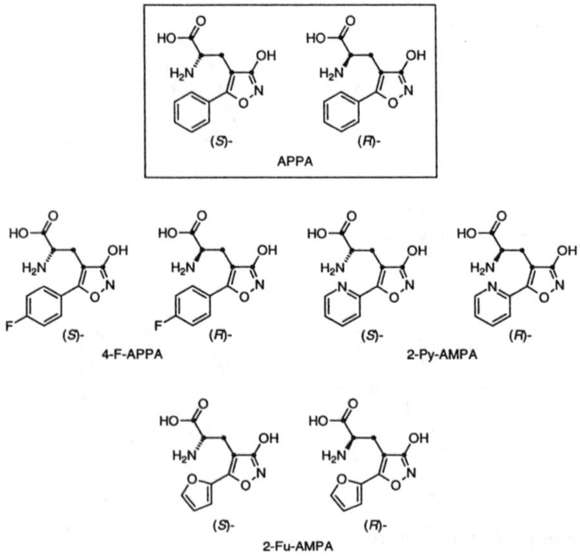

Figure 6   *Structures of the enantiomeric pairs of the AMPA analogues APPA, 4-F-APPA, 2 Py-AMPA, and 2-Fu-AMPA.   The (S)-forms of all of these analogues are AMPA receptor agonists, whereas the (R)-forms, with the exception of (R)-2-Fu-AMPA, are competitive AMPA antagonists*

When compared to the full agonist curve of (S)-APPA, these dose-response curves illustrate the principle of functional partial agonism (Figure 7). The pharmacology of the (RS)-, (S)-, and (R)-forms of 4-F-APPA (Figure 6) is almost identical with that of the racemic and enantiomeric forms of APPA.[39] A theoretical analysis indicates that this principle can be applied to any pair of agonist and competitive antagonist at all types of receptors to produce any desired level of relative efficacy (Figure 8).[13]

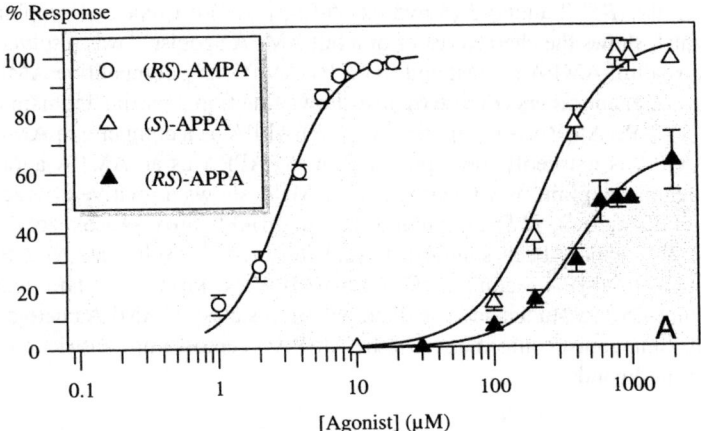

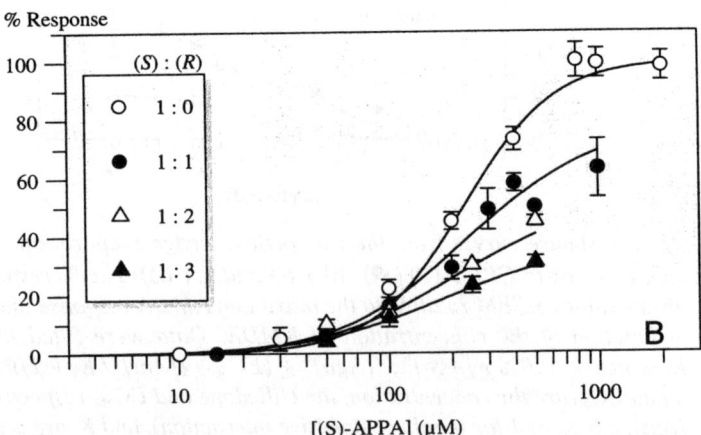

Figure 7 *Dose-response curves from the rat cortical wedge preparation. A: AMPA (°), (S) APPA (Δ) and (RS)-APPA (). Values are mean values ± SEM relative to the maximum AMPA response. Data were fitted to the equation: % Response = MAXχ [Ago]^n / (EC$_{50}$^{n} + [Ago]^n), where MAX is the maximum response relative to the AMPA plateau response, [Ago] is the agonist concentration in μM and n is the Hill slope (close to 2 for all compounds). 100% response is determined as the maximum response for AMPA. B: (S)-APPA at fixed molar ratios to (R)-APPA: 1:0 (°), 1:1(●) (racemate), 1:2 (Δ) and 1:3 (). The % response values are mean values ± SEM relative to the maximum AMPA response and are plotted as a function of the concentration of(S)-APPA. Data were fitted to the equation: Response = 100% χ [Ago]^n / ([Ago]^n + (EC$_{50}$ χ ([Ant]^s / K$_i$ + 1))^n), where [Ago], n and EC$_{50}$ are the concentration, the Hill slope and EC$_{50}$, respectively for (S)-APPA, [Ant], s (equals 1 for purely competitive interaction), and K$_i$ are the concentration, the Hill slope and potency for (R)-APPA, respectively. Responses are plotted relative to the response to 30 μM AMPA, which gives the maximum response*

Subsequently, (*RS*)-2-amino-3-(3-hydroxy-5-(2-pyridyl)-4-isoxazolyl)propionic acid (2-Py-AMPA), which shows the characteristics of a full AMPA agonist,[40] was resolved to give (*S*)-2-Py-AMPA as a full AMPA agonist and (*R*)-2-Py-AMPA as a competitive AMPA antagonist, and this pair of enantiomers (Figure 6) also shows functional partial agonism as predicted.[41] Whereas (*R*)-2-Py-AMPA is equipotent with (*R*)-APPA as a competitive AMPA antagonist, (*S*)-2-Py-AMPA is markedly more potent than (*S*)-APPA as an AMPA agonist, and these relative potencies explain why racemic 2-Py-AMPA shows a relative efficacy close to full agonism (not illustrated). (*RS*)-2-Amino-3-(5-(2-furyl)-3-hydroxy-4-isoxazolyl)propionic acid (2-Fu-AMPA) has now been synthesized and resolved.[40] Whereas (*S*)-2-Fu-AMPA is a potent and full AMPA agonist, (*R*)-2-Fu-AMPA is inactive. The results of these stereostructure-activity studies indicate that, within this class of AMPA analogues, the agonist as well as the antagonist conformation of the AMPA receptors impose strict constraints on the structure of the ligands.

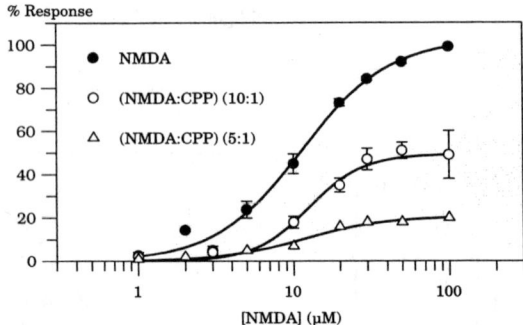

Figure 8 *Dose-response curves from the rat cortical wedge preparation. NMDA at fixed molar ratios to CPP : 1:0 (●), 10:1 (°) and 5:1 (Δ). The % response values are mean values ± SEM relative to the maximum NMDA response and are plotted as a function of the concentration of NMDA. Data were fitted to the equation: Response = 100% χ [Ago]^n / ([Ago]^n + (EC$_{50}$ χ ([Ant]^s / K$_i$ + 1))^n), where [Ago], n and EC$_{50}$ are the concentration, the Hill slope and EC$_{50}$, respectively for NMDA, [Ant], s (equals 1 for purely competitive interaction), and K$_i$ are the concentration, the Hill slope and potency for CPP, respectively. Responses are plotted relative to the response to 100 μM NMDA, which gives the maximum response*

3.4 Attempts to Design Alkylating and Photoactive Labels for AMPA Receptors

Quite extensive structure-activity studies on different series of AMPA agonists and antagonists have shed light on the structural requirements for interaction with the agonist and antagonist conformations of the AMPA receptor. However, essentially nothing is known about the precise location and topography of the AMPA recognition site(s). In order to facilitate such studies, we have made attempts to design ligands capable of interacting irreversibly with this receptor recognition site. Such ligands, in radiolabeled forms, would be useful for the mapping of the amino acid residues, which form the recognition site, and for the elucidation of the atomic contacts between ligands and receptor protein. Furthermore, covalently bound specific ligands might facilitate crystallization and X-ray crystallographic studies of AMPA receptors.

(*RS*)-2-Amino-3-(5-bromomethyl-3-hydroxy-4-isoxazolyl)propionic acid (ABPA) (Figure 9) was designed as an AMPA agonist capable of alkylating the AMPA receptor recognition site, but although ABPA is a high-affinity and potent AMPA agonist,[42] attempts to demonstrate covalent receptor binding under a variety of experimental conditions were unsuccessful.[43] There are several possible explanations for these negative results. As the bromomethyl group of ABPA is thought to occupy space in the proposed cavity during the receptor binding of ABPA (Figure 5), the lack of covalent binding may indicate that this cavity does not contain an appropriately located and/or sufficiently reactive nucleophile.

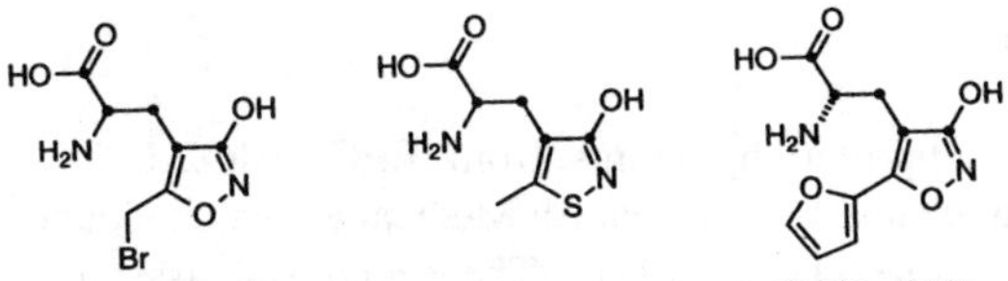

Figure 9 *The structures of three specific AMPA receptor agonists designed as alkylating (ABPA) or photolabeling (thio-AMPA and (S)-2-Fu-AMPA) ligands. None of these compounds show detectable irreversible AMPA receptor interactions*

The heterocyclic 4-aminobutyric acid (GABA) bioisostere, 5-aminomethylisothiazol-3-ol (thiomuscimol), is a specific and potent $GABA_A$ receptor agonist.[44] Radiolabelled thiomuscimol[45] has been shown to be an effective photolabel for $GABA_A$ receptors,[46] and this observation prompted us to synthesize the 3-isothiazol analogue of AMPA, (*RS*)-2-amino-3-(3-hydroxy-5-methyl-4-isothiazolyl)propionic acid (thio-AMPA) (Figure 9) as a potential AMPA receptor photolabel.[47] Thio-AMPA binds effectively to AMPA receptor sites, but in spite of extensive studies, these receptor sites have, so far, escaped photolabeling by thio-AMPA.

(*S*)-2-Fu-AMPA, which is a high-affinity AMPA receptor agonist, is sensitive to UV light-induced decomposition *in vitro*,[40] but photo-labelling studies using crude membrane preparations, recombinant AMPA receptor subunits, or peptide fragments comprising the AMPA binding site[48] have, so far, failed to detect irreversible receptor binding of (*S*)-2-Fu-AMPA (M. Mortensen *et al.*, unpublished). The mechanisms underlying the photodecomposition of (*S*)-2-Fu-AMPA have not been elucidated yet, but evidently decomposition products formed at the AMPA recognition site are unable to interact covalently with the receptor protein(s). The AMPA recognition site may contain water molecules, and this possibility is taken into consideration in the current attempts to design effective AMPA receptor photolabels.

4 CONCLUSIONS

A prerequisite for rational drug design is detailed information about the reaction or activation/inactivation mechanisms of the pharmacological target. Whereas there are numerous examples of rational design of enzyme inhibitors, based on knowledge of enzyme reaction mechanisms, precise information about activation/inactivation mechanisms for ionotropic or metabotropic receptors, including iGluRs and mGluRs, is not available yet.

Using naturally occurring (*S*)-Glu analogues as leads, a wide variety of agonist and antagonist ligands for iGluRs and mGluRs have been designed. Based on structure-activity studies of such ligands, it has been possible to gain some indirect, and superficial, insight into the topography of the recognition sites of these receptors. Some aspects of the design of AMPA receptor ligands along these lines have been summarized, and it may be concluded that in this area, ligand design projects have now reached the level of semirational drug design.

These approaches combined with ongoing attempts to develop computer-based pharmacophores relevant to iGluRs and mGluRs, and to determine the structures of receptor proteins or key protein fragment using X-ray crystallographic techniques may smooth the path for rational drug design in this research area.

Acknowledgements

This work was supported by grants from EEC (BIO2-CT93-0243), the Lundbeck Foundation and H. Lundbeck A/S, the Danish Medical Research Council and the Danish State Biotechnology Programme (1991-1995). The secretarial assistance of Anne Nordly is gratefully acknowledged.

References

1. P. Krogsgaard-Larsen and J. J. Hansen (eds.). *Excitatory Amino Acid Receptors: Design of Agonists and Antagonists*, Ellis Horwood, Chichester, 1992.
2. G. L. Collingridge and J. C. Watkins (eds.). *The NMDA Receptor*, Oxford University Press, Oxford, 1994.
3. H. V. Wheal and A. M. Thomson (eds.). *Excitatory Amino Acids and Synaptic Transmission*, Academic Press, London, 1995.
4. P. J. Conn and J. Patel (eds.). *The Metabotropic Glutamate Receptors*, Humana Press, New Jersey, 1994.
5. W. Danysz, C. G. Parsons, I. Bresink and G. Quack, *Drug News Perspect.*, 1995, **8**, 261.
6. T. Knöpfel, R. Kuhn and H. Allgeier, *J. Med. Chem.*, 1995, **38**, 1417.
7. S. I. Deutsch and J. M. Morihisa, *Clin. Neuropharmacol.*, 1988, **11**, 18.
8. J. T. Greenamyre and A. B. Young, *Neurobiol. Ag.*, 1989, **10**, 593.
9. D. M. Bowen, *Br. J. Psychiat.*, 1990, **157**, 327.
10. U. Madsen, B. Ebert and P. Krogsgaard-Larsen, *Biomed. Pharmacother.*, 1994, **48**, 305.
11. M. Carlsson and A. Carlsson, *TiNS.*, 1990, **13**, 272.
12. J. Ulas and C. W. Cotman, *Schizophrenia Bull.*, 1993, **19**, 105.
13. B. Ebert, K. K. Søby, U. Madsen and P. Krogsgaard-Larsen. *Schizophrenia - An Integrated View*, R. Fog, J. Gerlach and R. Hemmingsen (eds.), Munksgaard, Copenhagen, 1995, p. 379.
14. S. I. Deutsch, J. Mastropaolo, B. L. Schwartz, R. B. Rosse and J. M. Morihisa, *Clin. Neuropharmacol.* 1989, **12**, 1.
15. J. Drejer, *Excitatory Amino Acid Receptors: Design of Agonists and Antagonists*, P. Krogsgaard-Larsen and J. J. Hansen (eds.), Ellis Horwood, Chichester, 1992, p. 352.
16. P. Krogsgaard-Larsen, B. Ebert, T. M. Lund, H. Bräuner-Osborne, F. A. Sløk, T. N. Johansen, L. Brehm and U. Madsen, *Eur. J. Med. Chem.*, 1996, **31**, 515.
17. P. Krogsgaard-Larsen and J. J. Hansen, *Toxicol. Lett.*, 1992, **64/65**, 409.
18. L. M. Hawkins, K. M. Beaver, D. E. Jane, P. M. Taylor, D. C. Sunter and P. J. Roberts, *Neuropharmacol.*, 1995, **34**, 405.
19. P. Krogsgaard-Larsen, T. Honore, J. J. Hansen, D. R. Curtis and D. Lodge, *Nature*,

1980, **284**, 64.

20. J. J. Hansen, J. Lauridsen, E. Nielsen and P. Krogsgaard-Larsen, *J. Med. Chem.*, 1983, **26**, 901.

21. J. J. Hansen, *Current Topics in Med. Chem.*, 1993, **1**, 377.

22. U. Madsen, K. Frydenvang, B. Ebert, T. N. Johansen, L. Brehm and P. Krogsgaard-Larsen, *J. Med. Chem.*, 1996, **39**, 183.

23. J. Lauridsen, T. Honoré and P. Krogsgaard-Larsen, *J. Med. Chem.*, 1985, **28**, 668.

24. P. Krogsgaard-Larsen, J. J. Hansen, J. Lauridsen, M. J. Peet, J. D. Leah and D. R. Curtis, *Neurosci. Lett.*, 1982, **31**, 313.

25. L. Turski, P. Jacobsen, T. Honoré and D. N. Stephens, *J. Pharmacol. Exp. Ther.*, 1992, **260**, 742.

26. V. R. J. Clarke, B. A. Ballyk, K. H. Hoo, A. Mandelzys, A. Pellizzari, C. P. Bath, J. Thomas, E. F. Sharpe, C. H. Davies, P. L. Ornstein, D. D. Schoepp, R. K. Kamboj, G. L. Collingridge, D. Lodge and D. Bleakman, *Nature*, 1997, **389**, 599.

27. U. Madsen, B. Frølund, T. M. Lund, B. Ebert and P. Krogsgaard-Larsen, *Eur. J. Med. Chem.* 1993, **28**, 791.

28. H. Bräuner-Osborne, F. A. Sløk, N. Skjærbæk, B. Ebert, N. Sekiyama, S. Nakanishi and P. Krogsgaard-Larsen, *J. Med. Chem.*, 1996, **39**, 3188.

29. H. Ahmadian, B. Nielsen, H. Bräuner-Osborne, T. N. Johansen, T. B. Stensbøl, F. A. Sløk, N. Sekiyama, S. Nakanishi, P. Krogsgaard-Larsen and U. Madsen, *J. Med. Chem.*, 1997, **40**, 3700.

30. T. Liljefors, F. S. Jørgensen and P. Krogsgaard-Larsen (eds.). *Rational Molecular Design in Drug Research*, Munksgaard, Copenhagen, 1998.

31. H. Bräuner-Osborne, B. Nielsen and P. Krogsgaard-Larsen, *Eur. J. Pharmacol.*, 1998, **350**, 311.

32. H. Bräuner-Osborne and P. Krogsgaard-Larsen, *Br. J. Pharmacol.*, 1998, **123**, 269.

33. L. Brehm, F. S. Jørgensen, J. J. Hansen and P. Krogsgaard-Larsen, *Drug News Perspect.*, 1988, **1**, 138.

34. F. A. Sløk, B. Ebert, Y. Lang, P. Krogsgaard-Larsen, S. M. Lenz and U. Madsen, *Eur. J. Med. Chem.*, 1997, **32**, 329.

35. U. Madsen, B. Bang-Andersen, L. Brehm, I. T. Christensen, B. Ebert, I. T. S. Kristoffersen, Y. Lang and P. Krogsgaard-Larsen, *J. Med. Chem.*, 1996, **39**, 1682.

36. I. T. Christensen, A. Reinhardt, B. Nielsen, B. Ebert, U. Madsen, E. Ø. Nielsen, L. Brehm and P. Krogsgaard-Larsen, *Drug Des. Del.*, 1989, **5**, 57.

37. B. Ebert, S. M. Lenz, L. Brehm, P. Bregnedal, J. J. Hansen, K. Frederiksen, K. P. Bøgesø and P. Krogsgaard-Larsen, *J. Med. Chem.*, 1994, **37**, 878.

38. B. Ebert, U. Madsen, T. M. Lund, S. M. Lenz and P. Krogsgaard-Larsen, *Neurochem. Int.*, 1994, **24**, 507.

39. N. Skjærbæk, L. Brehm, T. N. Johansen, L. M. Hansen, B. Nielsen, B. Ebert, K. K. Søby, T. B. Stensbøl, E. Falch and P. Krogsgaard-Larsen, *Bioorg. Med. Chem.*, 1998, **6**, 119.

40. E. Falch, L. Brehm, I. Mikkelsen, T. N. Johansen, N. Skjærbæk, B. Nielsen, T. B. Stensbøl, B. Ebert and P. Krogsgaard-Larsen, *J. Med. Chem.*, 1998, **41**, 2513.

41. T. N. Johansen, B. Ebert, E. Falch and P. Krogsgaard-Larsen, *Chirality*, 1997, **9**, 274.

42. P. Krogsgaard-Larsen, L. Brehm, J. S. Johansen, P. Vinzents, J. Lauridsen and D. R. Curtis, *J. Med. Chem.*, 1985, **28**, 673.

43. E. Ø. Nielsen, U. Madsen, K. Schaumburg, L. Brehm and P. Krogsgaard-Larsen, *Eur.*

J. Med. Chem. Chim. Ther., 1986, **21**, 433.

44. P. Krogsgaard-Larsen, H. Hjeds, D. R. Curtis, D. Lodge and G. A. R. Johnston, *J. Neurochem.*, 1979, **32**, 1717.
45. B. Frølund, B. Ebert, L. W. Lawrence, S. D. Hurt and P. Krogsgaard-Larsen, *J. Labelled Compd. Radiopharm.*, 1995, **36**, 877.
46. M. Nielsen, M.-R. Witt, B. Ebert and P. Krogsgaard-Larsen, *Eur. J. Pharmacol. Mol. Pharmacol. Sect.*, 1995, **289**, 109.
47. L. Matzen, A. Engesgaard, B. Ebert, M. Didriksen, B. Frølund, P. Krogsgaard-Larsen and J. W. Jaroszewski, *J. Med. Chem.*, 1997, **40**, 520.
48. A. Kuusinen, M. Arvola and K. Keinänen, *EMBO J.*, 1995, **14**, 6327.

Glycine Antagonists

ANTAGONISTS ACTING AT THE NMDA RECEPTOR COMPLEX: POTENTIAL FOR THERAPEUTIC APPLICATIONS

David E. Jane

Department of Pharmacology, School of Medical Sciences
University of Bristol, Bristol BS8 1TD, UK

1 INTRODUCTION

The acidic amino acid (S)-glutamic acid is thought to be the predominant excitatory transmitter in the central nervous system (CNS) acting at a range of excitatory amino acid (EAA) receptors.[1] The first evidence for heterogeneity amongst these receptors came from the differential activity of a range of agonists based on the glutamate structure in different regions of the CNS.[2,3a,3b] More definitive evidence came from the observation that (R)-α-aminoadipate (D-αAA) blocked N-methyl-(R)-aspartic acid (NMDA)-induced and synaptic excitation in spinal motoneurones, but not depolarisations due to either of the two natural products kainate or quisqualate (Figure 1).[4] The case for EAA receptor subtypes was further strengthened by the discovery of the selective blockade of NMDA-induced responses on spinal neurones by magnesium ions.[5] This initial work led to the classification of glutamate receptors into NMDA, quisqualate and kainate receptor subtypes.[1]

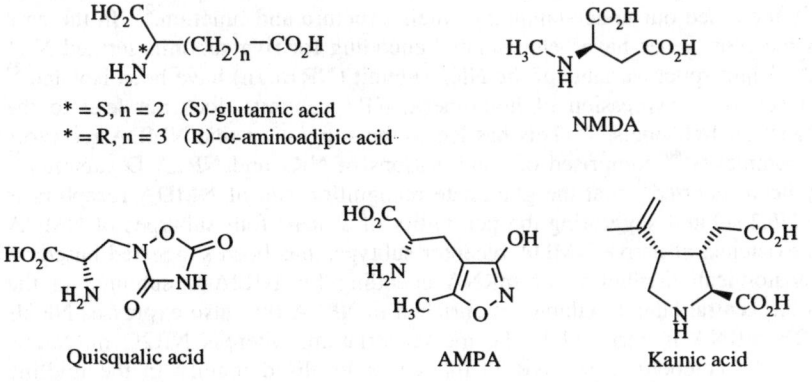

* = S, n = 2   (S)-glutamic acid
* = R, n = 3   (R)-α-aminoadipic acid

NMDA

Quisqualic acid   AMPA   Kainic acid

Figure 1 *Structures of ligands used to define EAA receptors*

It was later discovered that (*RS*)-2-amino-3-(3-hydroxy-5-methyl-4-isoxazolyl)propionic acid (AMPA) was more potent and selective than quisqualate and this led to the quisqualate receptor being renamed the AMPA receptor.[6,7,8] These three groups of receptors termed ionotropic glutamate (iGlu) receptors, make up a class of ligand-gated ion channels, which are thought to mediate fast synaptic transmission in the CNS. More recently, genes encoding for eight subtypes of G-protein coupled metabotropic glutamate receptors (termed $mGlu_1$–$mGlu_8$ receptors), thought to play a modulatory role in synaptic transmission, have been isolated.[9] It is now known that quisqualate, as well as being able to activate AMPA receptors, is also an agonist at several mGlu receptor subtypes including $mGlu_1$, $mGlu_3$ and $mGlu_5$.[9]

1.1 Properties of the NMDA Receptor Complex

A number of features of the NMDA receptor complex distinguish it from other ligand-gated ion channels (Figure 2). As mentioned earlier Mg^{2+} is known to specifically antagonise NMDA receptors.[5] At resting membrane potential extracellular Mg^{2+} blocks the NMDA receptor channel by binding to a site within it and opening of these channels can only occur when depolarisation occurs upon agonist binding.[10,11] In addition to the influx of Na^+ ions and efflux of K^+ ions seen upon activation of iGlu receptors NMDA receptors are also highly permeable to Ca^{2+} ions.[12] It is this property that is thought to be responsible for the role of NMDA receptors in various neuropathologies such as neurotoxicity[13] and in many forms of synaptic plasticity such as long-term potentiation (LTP)[14] which has been implicated in learning and memory.[15] Another important finding involves the necessity of glycine to bind as a co-agonist[16] to a site of the NMDA receptor distinct from the glutamate binding site[17] in order to achieve optimal activation of NMDA receptors. Inside the channel there lies another binding site at which phencyclidine (PCP) and a number of other non-competitive antagonists bind leading to prevention of ion flux.[18] Modulatory sites on the NMDA receptor complex for Zn^{2+} and polyamines have also been proposed.[19,20]

1.2 Molecular Biology of NMDA Receptors

The recent isolation of genes encoding for subunits for each subgroup of iGlu receptors has enormously increased our understanding of their structure and function.[21] In the case of the NMDA receptor, genes have been isolated encoding for five subunits termed NR1 and NR2A-D.[22] Eight splice variants of the NR1 subunit (NR1a-1h) have been isolated.[23] In mammalian cell lines expression of homomeric NR1 subunits does not lead to the formation of functional channels.[24a] This has led to the suggestion that NMDA receptors are tetrameric complexes[24b] comprised of combinations of NR1 and NR2A-D subunits.[22] It has recently been reported[25] that the glutamate recognition site of NMDA receptors is located on the NR2 subunit suggesting the possibility of at least four subtypes of NMDA receptor. The existence of native NMDA receptor subtypes has been suggested based on the different anatomical distribution of mRNA encoding for NR2A-D subunits in the CNS.[26] Thus the ventral lateral thalamus is enriched in NR2A (but also expresses NR2B subunits), NR2B mRNA is enriched in the medial striatum, whereas NR2C mRNA is found mainly in the cerebellum and NR2D mRNA is localised mainly in the midline thalamus.[26] The co-agonist glycine binding site is thought to be located on the NR1 subunit[27] while it would appear that one of the polyamine binding sites is on the NR2B subunit, based on the preferential activity of polyamine antagonists on cells expressing NMDA receptors containing NR2B subunits.[28]

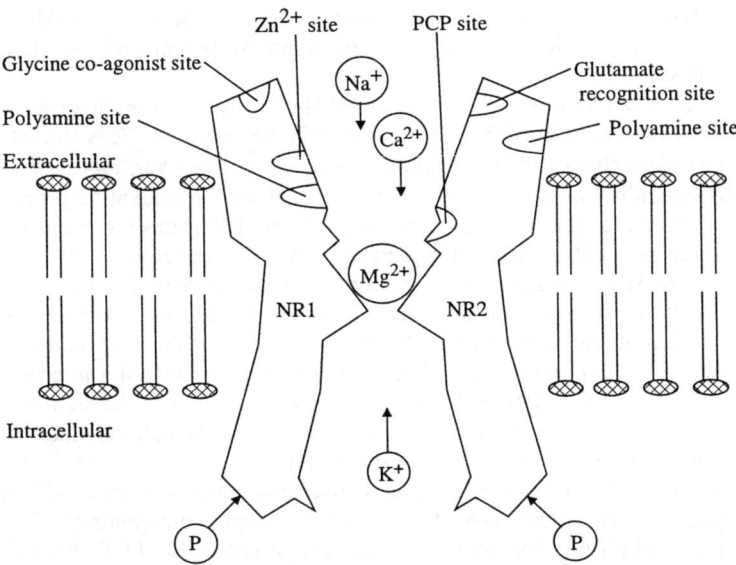

Figure 2   *Schematic diagram of the NMDA receptor complex showing modulatory sites and ion fluxes. Phosphorylation (P) of the NMDA receptor may play a role in the regulation of receptor function*

2 COMPETITIVE NMDA RECEPTOR ANTAGONISTS

As outlined above D-αAA, an analogue of glutamate with an extra carbon atom in the inter-acidic group chain, was shown to selectively antagonise NMDA-induced responses. However, D-αAA was not a particularly potent antagonist.[1] The discovery that ѡ-phosphono analogues of D-αAA such as (R)-2-amino-5-phosphonopentanoic acid ((R)-AP5) and (R)-2-amino-7-phosphonoheptanoic acid ((R)-AP7) (Figure 3) had greatly increased potency and selectivity was a milestone in the design of NMDA receptor antagonists.[29,30]   In all cases the active isomer had the R absolute stereochemistry at the α-amino acid stereogenic centre; however, as will be detailed later, there are now known to be compounds with NMDA receptor antagonist properties in which the α-amino acid grouping has the S absolute stereochemistry.

Further enhancement of potency was achieved on synthesising conformationally restricted analogues of (R)-AP5 and (R)-AP7. One way of achieving conformational restriction was by including some of the flexible backbone of AP5 or AP7 in a heterocyclic ring.   This led to the discovery of some of the most potent competitive NMDA receptor antagonists known to date such as (2RS,4SR)-4-phosphonomethyl-piperidine-2-carboxylic acid (CGS19755),[31] (R)-(E)-4-3-(3-phosphonoprop-2-enyl)-piperazine-2-carboxylic acid ((R)-CPP-ene)[32]   and   (3S,4aR,6S,8aR)-6-phosphono-methyl-1,2,3,4,4a,5,6,7,8,8a-decahydroisoquinoline-3-carboxylic acid (LY 235959) (see Figure 3 for structures).[33] Another way of achieving conformational restriction was to include a double bond in the carbon backbone of AP5 to give compounds such as (RS)-(E)-2-amino-4-methyl-5-phosphono-3-pentenoic acid (CGP 37849, active isomer CGP 40116)[34] and (RS)-(E)-2-amino-4-propyl-5-phosphono-3-pentenoic acid (CGP 39653)[34] or similarly to add a keto group to give (R)-2-amino-4-oxo-5-phosphonopentanoic acid (MDL 100,453) (see Figure 3 for structures).[35] All of these compounds proved to be potent and selective NMDA receptor antagonists.

The optimal inter-acidic group chain length for NMDA receptor antagonists was found to be a chain of 4 (as found in AP5) or 6 (as found in AP7) atoms. This rule of thumb is obeyed for most open chain and heterocyclic analogues.[36] However, there are exceptions including (RS)-α-amino-6,7-dichloro-3-(phosphonomethyl)-2-quinoxaline-propanoic acid (2, Figure 3) which has a chain of 5 atoms between the acidic groups but is one of the most potent NMDA receptor antagonists yet reported.[37] Another exception is the cyclobutane analogue (1, Figure 3) which again has a chain of 5 atoms between the two acidic groups; however, molecular modelling indicated that this analogue (and not the analogues with a carbon more or less in the chain) gave the best fit to the CGS19755 template.[38] The quinoxaline analogue (2, Figure 3) also had excellent overlap with the minimum energy conformation of (2R,4S)-CGS19755.[37] Thus, in some cases, due to the shorter C-C bond lengths imposed by incorporating the carbon chain into rings, the optimal chain length may be different than predicted from the general rule of thumb.

A range of phenylalanines synthesised as conformationally restricted AP7 analogues have been reported to be potent competitive NMDA receptor antagonists.[39,40] None of these compounds had affinity for the glycine co-agonist site or the PCP channel-blocking site of the NMDA receptor nor did they have any effect on AMPA- or kainate-induced responses in rat cortical slices.[40] Substitution of 3-phosphonomethylphenylalanine with a phenyl group at the meta position to give α-amino-5-(phosphonomethyl)[1,1'-biphenyl]-3-propanoic acid (SDZ-EAB-515, Figure 3) led to a marked increase in antagonist potency at NMDA receptors.[39] Unlike the parent compound, (R)-AP7, the potent NMDA receptor antagonist activity of these compounds resides in the S enantiomer.[39,40]

Structure-activity studies on this series of compounds showed that hydroxy-substitution at R^1 and substitution of the second phenyl ring in the para- (see SDZ-215-439, Figure 3) and especially the ortho-position (optimally with a chloro group, see SDZ-220-040, Figure 3) led to increased NMDA receptor antagonist potency.[40] Substitution of the second phenyl ring with a group in the ortho position is thought to increase the torsion angle between the two phenyl rings and this is thought to be the optimal arrangement for NMDA receptor antagonist activity.[40] The hydroxy group in SDZ-220-040 may be critical in producing the correct conformation by promoting intermolecular hydrogen bonds between this group and the phosphonomethyl and alanine side chains.

The only other group of compounds where the S enantiomer has the potent NMDA receptor antagonist activity is the series of decahydroisoquinolines, which can be looked upon as conformationally restricted analogues of CPP-ene.[33,41] The most potent of these are the AP7 analogues with the 3S,4aR,6S,8aR configuration and either a phosphonomethyl (LY 235959) or tetrazolylmethyl (LY 202157) substituent at the 6-position of the decahydroisoquinoline ring (see Figure 3).[33]

By comparison of a range of compounds differing only in the nature of the terminal acidic group a rank order of potency can be established for a number of possible terminal acidic groups: PO_3H_2 > tetrazole > CO_2H >> SO_3H.[36] The methylphosphinate group is also an effective terminal acidic group.[42]

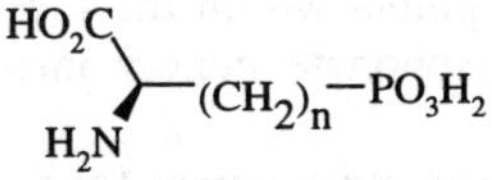

n = 3 (R)-AP5 (0.62, [³H]D-AP5)
n = 5 (R)-AP7 (1.7, [³H]D-AP5)

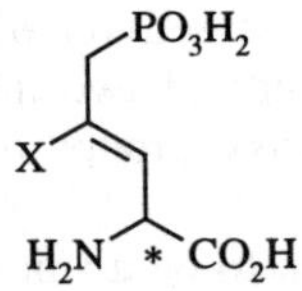

X = CH₃ CGP 37849 (0.035)
* = R CGP 40116 (0.019)
X = n-C₃H₇ CGP 39653
(0.007, [³H]CGS 39653)

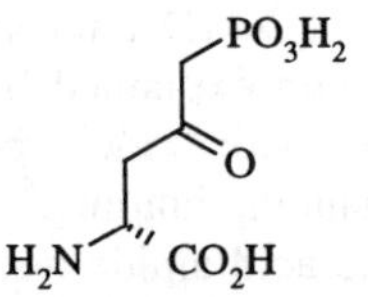

MDL 100,453 (0.109)

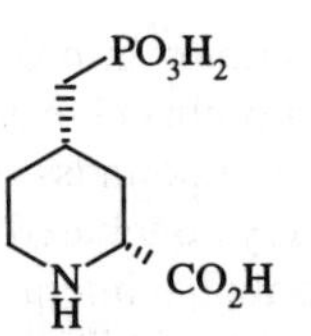

CGS19755 (0.183)

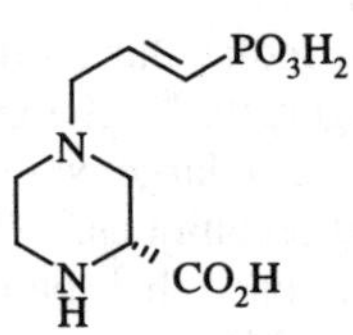

(R)-CPP-ene (0.044)

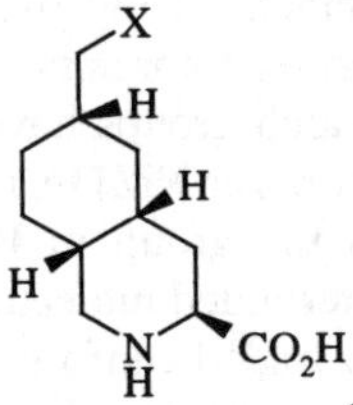

LY 235959 X = PO₃H₂ (0.025, [³H]CGS19755)
LY 202157 X = tetrazole (0.42, [³H]CGS19755)

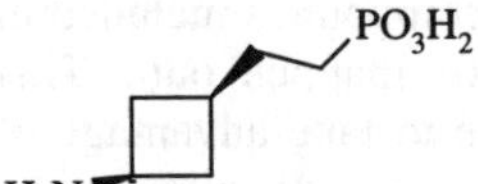

1 (No binding data, equipotent with (RS)-CPP in an electrophysiological assay)

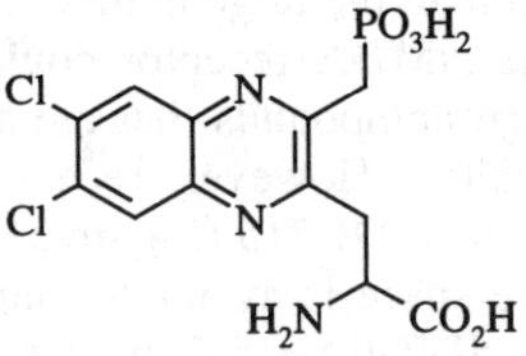

2 (0.0034)

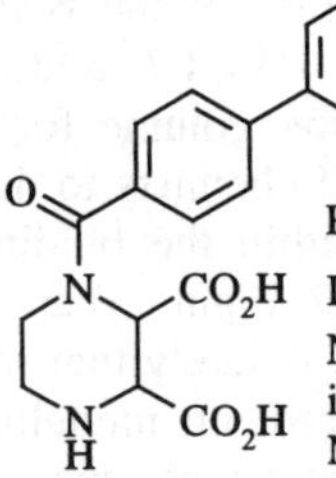

PBPD

IC₅₀ (µM) for antag. of NMDA-induced depols in cells expressing:
NR1/NR2A 15.79
NR1/NR2B 5.01
NR1/NR2C 8.98
NR1/NR2D 4.29

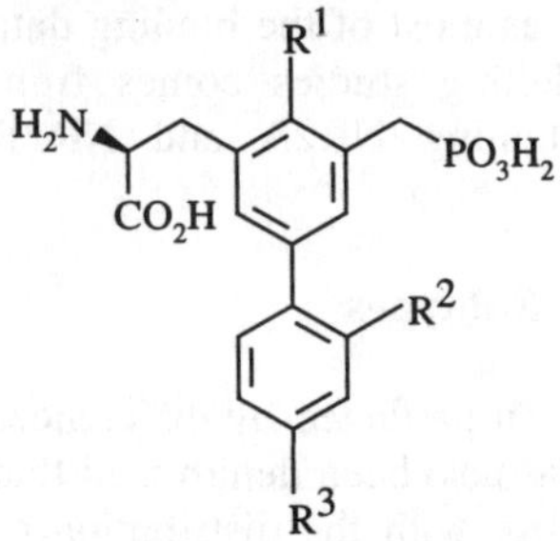

3 (0.47)

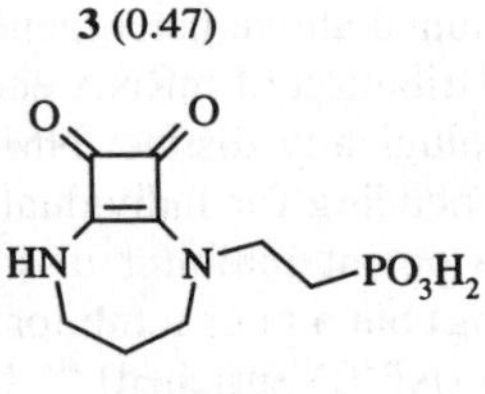

4 (0.030)

	pK$_i$ [³H]CGP-39653
R¹ = R² = R³ = H SDZ-EAB-515	6.6
R¹ = H, R² = H, R³ = Ph SDZ-215-439	7.8
R¹ = OH, R² = R³ = Cl SDZ-220-040	8.5
R¹ = H, R² = Cl, R³ = H SDZ-220-581	7.7

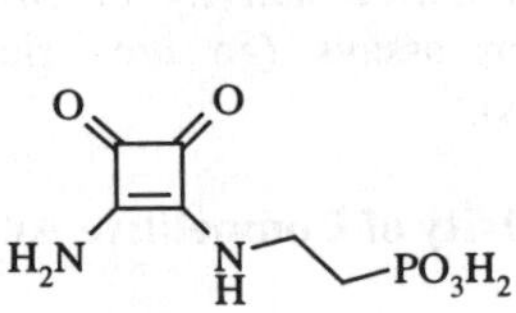

Figure 3 *Selective competitive antagonists for NMDA receptors. IC₅₀ (µM) values for displacement of [³H]-CPP binding to rat brain membranes given in parenthesis (unless otherwise stated)*

The potent NMDA receptor antagonist activity observed with a terminal phosphono group has been explained by the existence of two interaction points within the receptor binding site, with which the two hydroxy groups of the phosphonate moiety interact, thereby enhancing binding.[43]

A squaric acid amide has been shown to be a suitable isosteric replacement for the α-amino acid moiety found in most competitive NMDA receptor antagonists. The N-phosphonoethyl analogue (3, Figure 3) was found to have almost the same affinity as (RS)-AP7.[44] A significant increase in potency was obtained when the two nitrogen atoms of the squaric acid amide moiety were linked by a trimethylene chain (4, Figure 3).[45,46] Interestingly, replacement of the α-carboxyl of AP5 with an α-phosphonate moiety was detrimental to NMDA receptor antagonist activity.[47]

A number of research groups have undertaken molecular modelling studies in order to define a pharmacophore for NMDA receptor antagonists.[36] The most comprehensive study to date comes from the group in Parke-Davis on a large variety of compounds using sophisticated computer-aided molecular modelling techniques.[43] This study concluded that antagonists bind in a folded conformation and proposed that there is a primary interaction point on the receptor, proximal to the distal NMDA receptor agonist binding site, positioned so that a receptor moiety can simultaneously hydrogen bond to both the ϖ-phosphonate and α-carboxylic acid groups. This model also provided an explanation for the periodicity in affinity for the NMDA receptor observed when the inter-acidic chain length is varied in the AP4-AP7 series. Due to the large number of compounds included in this study, the exclusion volume for the NMDA receptor could be mapped out. This should allow medicinal chemists to design compounds that are able to take advantage of the space available within the binding site. However, recent compounds such as the quinoxaline analogue (2, Figure 3) and SDZ-EAB-515 (Figure 3) were not included in this model and therefore it is likely that more space is available than was apparent from this modelling study. Indeed, on modelling SDZ-EAB-515 it was shown that the biphenyl moiety of this compound was able to occupy space previously thought to be excluded volume.[20,48] In all these models, it has been assumed that the molecules are interacting with the same type of NMDA receptor. This may be the case as most of the binding data used to define activity at NMDA receptors in these modelling studies comes from forebrain assays (an area rich in NMDA receptors containing NR2A and NR2B subunits).[26]

2.1 Activity of Competitive Antagonists at NMDA Receptor Subtypes

As mentioned above heterogeneity of NMDA receptors has been predicted by differences in the distribution of mRNA encoding NR2A-D subunits. It has also been determined that pharmacologically distinct NMDA receptor subtypes co-localise with the distribution of mRNA encoding for individual NMDA receptor subunits.[26,49] Thus (R)-CPP-ene (Figure 3) was a potent inhibitor of [³H]glutamate binding in the forebrain (NR2A and NR2B containing) but a poor inhibitor of binding in the cerebellum (NR2C enriched) and midline thalamus (NR2D enriched).[26] However, both SDZ-EAB-515 and PBPD (Figure 3) inhibit NMDA receptors in the forebrain and midline thalamus with similar affinities.[49] These results have been confirmed in studies using NR1/NR2A, NR1/NR2B, NR1/NR2C or NR1/NR2D receptors expressed in *Xenopus* oocytes.[26,49] Thus, (R)-CPP-ene had high affinity for NR1/NR2A or NR1/NR2B receptors, but low affinity for NR1/NR2C or NR1/NR2D receptors while both SDZ-EAB-515 and PBPD had higher affinity for NR1/NR2D and NR1/NR2B receptors than for NR1/NR2A or NR1/NR2C receptors.[49] In the same study it was determined that co-expression of NR1a, NR2B and NR2D subunits in *Xenopus* oocytes led to heterooligomeric receptors with a unique pharmacological profile suggesting that the two NR2 subunits co-exist in the same NMDA receptor complex. In the case of the biphenyl-AP7 analogue, SDZ 220-581 (Figure 3), a slightly higher affinity for NR1/NR2A or NR1/NR2B-containing receptors over NR1/NR2C or NR1/NR2D-containing receptors was observed.[40]

2.2 Antagonists Binding to the Glycine Co-agonist Site

During the course of early structure activity studies it was noted that 3-amino-1-hydroxypyrrolidin-2-one (HA-966) (Figure 4) had NMDA receptor antagonist properties.[50] However, unlike other antagonists HA-966 did not displace labelled competitive antagonists from rat brain membranes and was therefore thought to be acting non-competitively.[51] It was later discovered that HA-966 was an antagonist of the agonist potentiating action of glycine and was therefore acting at the glycine binding site of the NMDA receptor.[52] It is important to note that the glycine binding site on the NMDA receptor is distinct from the strychnine-sensitive glycine binding site of inhibitory synaptic receptors found mainly in the spinal cord and brain stem. It was concluded that HA-966 was acting as a partial agonist on observation that HA-966 can potentiate the action of NMDA-induced responses when care is taken to exclude glycine contamination.[17,53] The action of HA-966 at the glycine site of the NMDA receptor complex has been attributed to the R-(+) enantiomer, the S-(-) form behaving as a sedative by an as yet undetermined mechanism.[54] A structure-activity study on HA-966 analogues revealed that the 4-methyl derivative, (+)-*cis*-3-amino-1-hydroxy-4-methylpyrrolidin-2-one (L-687,414, Figure 4) was a potent glycine site antagonist.[55,56,57]

A number of kynurenic acid analogues have been reported to be glycine site antagonists.[58] One of the most potent of these, 5,7-dichloro-4-hydroxyquinoline-2-carboxylic acid (5,7-dichlorokynurenic acid, 5,7-DCKA, Figure 4),[59] has the optimal size of lipophilic 5,7-substituents for high affinity at the glycine recognition site.[58] Further development of kynurenic acid analogues led to the (2R,4S)-4-substituted tetrahydro-quinoline-2-carboxylic acid, L-689,560 (Figure 4), which has high affinity for the glycine site.[60] Based on detailed structure-activity analysis a pharmacophore has been proposed for L-689,560 in which regions of bulk tolerance and intolerance have been identified.[61]

As well as providing a range of AMPA receptor selective antagonists,[62] quinoxalinediones have provided a number of selective antagonists for the glycine site of the NMDA receptor,[61] and amongst these, 5-nitro-6,7-dichloro-1,4-dihydro-2,3-quinoxaline-dione (ACEA-1021, Figure 4) is a potent and selective glycine site antagonist, over 250-fold more potent at this site than at AMPA receptors.[63,64] A further development in the structure-activity studies in this series involved modification of the quinoxaline ring to give 6,7,8-trichloro-2,3-dihydroxypyrido[2,3-b]pyrazine[65] (1, Figure 4) and 6,7-dichloro-2,3-dihydroxy-pyrido[2,3-b]pyrazine-5-oxide[66] (2, Figure 4), which have been reported as potent and selective glycine site antagonists with *in vivo* activity. A tricyclic quinoxalinedione (3, Figure 4) has been reported to be a glycine site antagonist with nanomolar affinity.[67] It was shown that only the S-isomer had the glycine site antagonist activity.[67]

The indole-2-carboxylates represent successful templates for the identification of potent glycine site antagonists. The most successful have been based around the 4,6-dichloroindole-2-carboxylate nucleus[68] (4, Figure 4) bearing substituents such as a 2-carboxyethyl group at C-3 of the indole ring (5, Figure 4).[69] Further increases in potency were obtained upon adding an α,β-unsaturated amide at C-3 of the indole leading to 3-[2'-[(phenylamino)carbonyl]ethenyl]-4,6-dichloroindole-2-carboxylic acid (GV150526A, Figure 4) which has nanomolar potency at the glycine site.[70] It has been recently reported that the phenyl ring of the indole nucleus is not a requirement for potent antagonist activity.[71] Thus, pyrrole analogues of GV150526A (6, Figure 4) retained potent antagonist activity as long as substituents were present on the 4- and 5-positions of the pyrrole ring. The compound with highest affinity for the glycine site was the 4,5-dibromo substituted analogue (6, Figure 4). A quantitative structure-activity study pointed to a correlation between activity and electron-withdrawing ability, bulk and lipophilicity of the 4,5-substituents. These pyrrole analogues are therefore a new lead for the development of even more potent antagonists in the future.

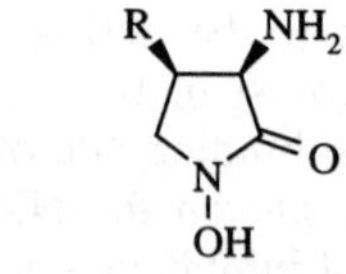

R = H = (R)-HA-966 (12.5)
R = Me = L-687,414 (1.4)

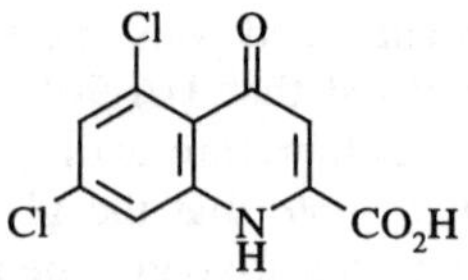

5,7-DCKA (0.2)

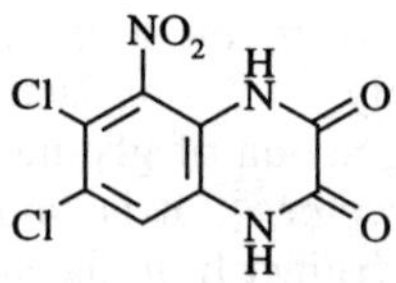

ACEA-1021 (0.0059, [^{3}H]5,7-DCKA)

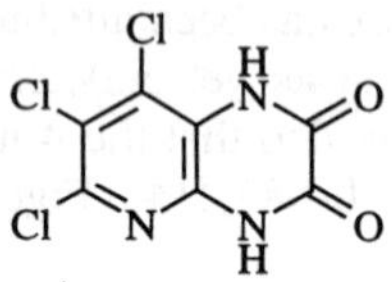

1 (pK$_i$ value 6.96)

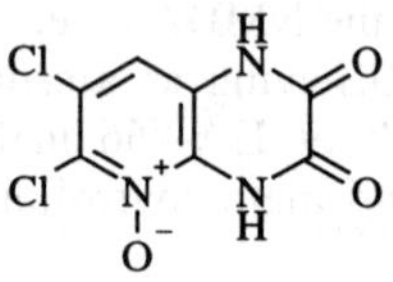

2 (pK$_i$ value 7.0)

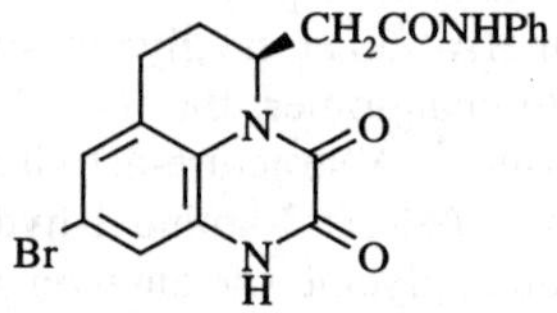

3 (0.0026)

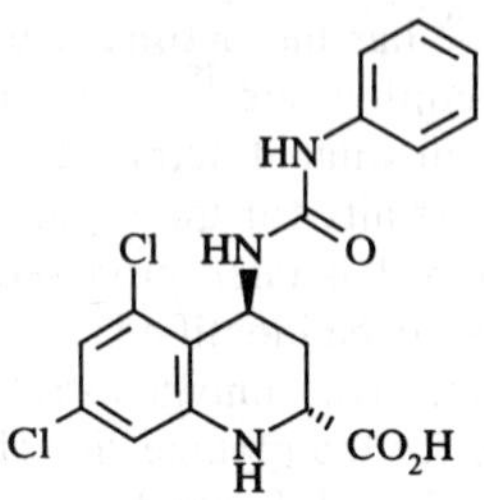

L-689,560 (0.0078)

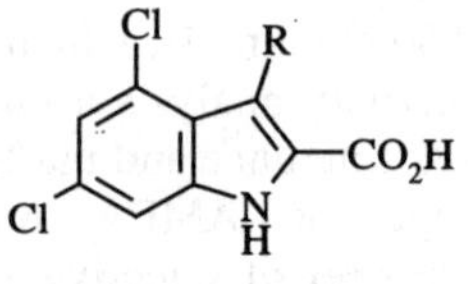

4 R = H (pK$_i$ value 5.7)
5 R = -(CH$_2$)$_2$CO$_2$H (0.14)
GV1500526A R = -CH=CHC(O)NHPh
(pK$_i$ value 8.5)

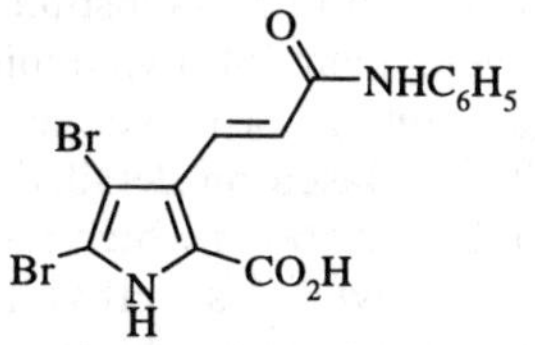

6 (pK$_i$ value 7.95)

Figure 4 *Structures of ligands acting at the glycine binding site of the NMDA receptor. IC$_{50}$ (μM) values for displacement of [^{3}H]glycine from rat brain membranes given in parenthesis (unless otherwise stated)*

2.3 Channel Blocking NMDA Receptor Antagonists

The dissociative anaesthetic phencyclidine (PCP, Figure 5) once used in clinical practice was withdrawn from use due to psychotomimetic effects. It was subsequently found that

PCP and a similarly acting drug with a shorter duration of action, ketamine (Figure 5), blocked responses evoked by NMDA.[72] A highly potent antagonist with PCP-like action, MK801 (Figure 5) has been developed.[73] The actions of these PCP-like antagonists on the NMDA receptor complex are thought to be of a non-competitive nature and are both use dependent (require initial activation of the receptor by an agonist) and voltage sensitive.[18,74] Binding studies have established that binding of PCP type ligands is enhanced by NMDA receptor agonists, and reduced by competitive NMDA receptor antagonists, but once binding of the PCP like ligand has taken place it is not easily displaced by NMDA receptor agonists. This evidence has led to the conclusion that PCP-like ligands bind within the ion channel of the NMDA receptor complex.[18,74] It has been demonstrated that the NR2 but not the NR1 subunit contributes to the heterogeneity of action of channel blocking antagonists in different brain regions.[75] In addition, MK801 has been shown to have slower kinetics at NMDA receptors containing NR1/NR2C subunits.[75]

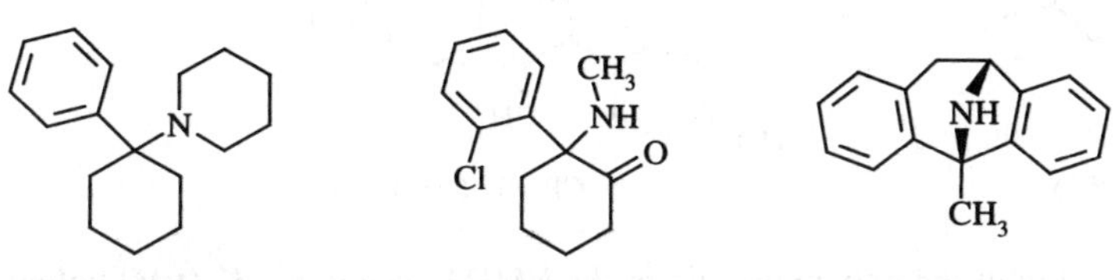

Phencyclidine (PCP)   Ketamine   MK801

Figure 5 *Non-competitive NMDA receptor antagonists*

2.4 Antagonists of the Polyamine Site of the NMDA Receptor

The polyamine, spermine is a known neurotoxin acting via modulatory sites on the NMDA receptor complex.[76,77] The precise mechanism of action of polyamines is currently a matter of debate but it would appear that it is more complex than a direct modulation of ion channel opening. Evidence is gathering[78] that it may involve complex allosteric interactions between the polyamine, glutamate recognition[79] and glycine[80,81] binding sites on the NMDA receptor complex. Ifenprodil (Figure 6), originally developed as an α_1 adrenoreceptor antagonist,[82] has been shown to be a potent non-competitive NMDA receptor antagonist acting via a polyamine modulatory or a distinct but overlapping site. There is evidence for the presence of an ifenprodil binding site on the NR1 subunit which overlaps with one of the polyamine sites.[78] It is assumed that the glycine-independent effects of ifenprodil are mediated via an NR1 subunit. However, the potent glycine-dependent high affinity effects of ifenprodil are confined to NMDA receptors containing the NR2B subunit.[83] There is evidence that ifenprodil is a state-dependent blocker of NMDA receptors containing the NR2B subunit displaying higher affinity for the agonist bound activated and desensitized states in comparison to the resting state (no agonist bound) of the NMDA receptor.[84] A number of 1,4-disubstituted piperidine analogues of ifenprodil such as eliprodil,[85] Ro 25-6981[86] and CP-101,606[87] (for structures see Figure 6) have also been reported to display NR2B selectivity. Ifenprodil as well as having the effects already described is known to be a potent σ ligand[88] and a moderately potent blocker of L-, N- and P-type neuronal calcium channels.[89] In contrast, Ro 25-6981 does not bind to σ receptors, α_1-adrenergic, or serotonergic binding sites but binds to NR2B-containing NMDA receptors with high affinity.[90] A radiolabelled form of Ro 25-6981 has recently been used to probe the distribution of NR2B containing NMDA receptors in the CNS and to characterise the binding of a range of ifenprodil analogues to rat brain membranes.[90]

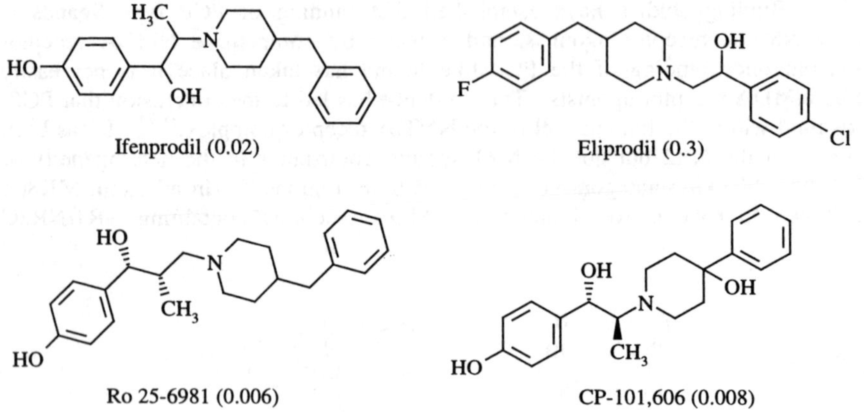

Ifenprodil (0.02)   Eliprodil (0.3)

Ro 25-6981 (0.006)   CP-101,606 (0.008)

Figure 6 *Antagonists acting at the polyamine site of the NMDA receptor. K_i (μM) values for displacement of [^3H]Ro 25-6981 binding to rat membranes are given in parenthesis*

2.5 Therapeutic Potential Of Nmda Receptor Antagonists

The field of EAA research had its origin in the observation that both (S)-glutamate and (S)-aspartate cause convulsions in mammalian brain.[91] Since this discovery NMDA receptors have been implicated in neurodegeneration following cerebral ischaemia,[92] Alzheimer's disease[93] and Huntingdon's disease,[94,95] and a range of CNS dysfunctional conditions such as epilepsy,[96,97] Parkinson's disease,[98,99] schizophrenia,[100] and anxiety,[101] and in chronic pain.[102,103,104]

The therapeutic potential of NMDA receptor antagonists will be illustrated by a discussion of their use in the prevention of excitotoxicity, as anticonvulsants and in the treatment of chronic pain.[105] Over-activation of NMDA receptors by excessive release of glutamate following ischaemic injury is thought to cause an increase in intracellular calcium[106] leading to neuronal cell death.[13,107] NMDA[108,109] and the endogenous NMDA receptor agonist, quinolinic acid[110] have been shown to induce excitotoxicity which can be reversed by competitive NMDA receptor antagonists.[111,112] This has led to an investigation of the application of NMDA receptor antagonists to prevent brain damage following stroke. In a model of focal ischaemia (R)-CPP-ene (Figure 3) (at a dose of 15 mg/kg) was reported to give 75% cortical protection when administered 15 min prior to occlusion of the cat middle cerebral artery (MCA).[113] In order to obtain maximal protection the drug should be given shortly before the injury. However, the precise therapeutic time window for humans is not easily established from animal models, but it would appear that the earlier the drug is administered the greater the likelihood of obtaining neuroprotection. In a similar study CGS 19755 (10 mg/kg) (Figure 3) was shown to give 82% cortical protection when administered 5 min prior to occlusion of the rat MCA.[114] Both CGS19755

In a recent study, the biphenyl-AP7 analogue, SDZ-220-581 (dose range 3-15 mg/kg i.p. or 10-50 mg/kg p.o.) (Figure 3) protected against quinolinic acid-induced striatal lesions in rats with a potency comparable to that of (*R*)-CPP-ene.[115] In addition, SDZ-220-581 (1.25 mg/kg) reduced the infarct size by 40% when administered i.v. 15 min. prior to rat MCA occlusion. Although the neuroprotective effects of SDZ-220-581 were not superior to (*R*)-CPP-ene when given i.v., upon oral administration the former drug was more effective. This can be explained by the more effective brain penetration of SDZ-220-581 probably aided by the increased lipophilicity of this class of compound and the possibility that it utilises carriers for active transport of amino acids from the intestine and through the blood brain barrier.[48,115]

Competitive NMDA receptor antagonists have been reported to be anticonvulsants when tested in a variety of animal models. Thus, (R)-CPP-ene protects against sound-induced seizures in DBA/2 mice,[97,116] electroshock-induced convulsions in rats[117] and chemically induced siezures.[118] The AP7 analogue, SDZ-220-581 also demonstrated protection against maximal electroshock seizures at oral doses of 10 mg/kg in rats and mice displaying a fast onset and long duration ($\geq$ 24 hr) of action.[115] A good separation between anticonvulsant action and undesirable ataxic side effects (~ 10-fold in mice) was observed in this study.[115] In the case of seizures induced by repetitive daily stimulation of the amygdala with a subthreshold electrical shock (a process known as kindling) competitive NMDA receptor antagonists are less effective, requiring high doses that also elicit side-effects.[119] Preliminary results from a clinical trial conducted with (R)-CPP-ene on patients with complex partial seizures suggest that low potency on kindled animals translates to poor efficacy on partial seizures in humans.[120]

It has been noted that NMDA receptors are involved in the transmission and modulation of nociceptive information in the spinal cord.[102,103,104] A close analogue of (R)-CPP-ene, (RS)-4-(3-phosphonopropyl)piperazine-2-carboxylic acid ((RS)-CPP) has been used clinically to treat a patient suffering from severe neuropathic pain resulting from nerve injury.[103,104] It was observed that the afterdischarge (the pronounced increase in pain level lasting from minutes to hours after the termination of stimulation) and the spread of pain beyond the area of the injured nerve, but not the continuous deep pain, was abolished upon intrathecal injection of (RS)-CPP (200 nmol). Normal sensory and motor function was not affected by blockade of NMDA receptors. This observation is important to the clinical application of NMDA receptor antagonists as use of other spinal analgesics results in immobilisation and reduced sensibility. In a separate study, SDZ-220-581 (Figure 3) was shown to be an effective orally active analgesic when tested in an animal model of neuropathic pain free from side effects at the low doses required for effective activity.[115] SDZ-220-581 was also effective in a model of persistent inflammatory pain, but not in models of thermal hyperalgesia or acute nociception. If the experiences with (*RS*)-CPP and SDZ-220-581 are confirmed in clinical trials then NMDA receptor antagonists may represent a novel treatment of pain in humans.

A number of adverse side effects such as motor impairment (ataxia),[105,117] impairment of spatial learning[121] and psychotomimetic effects[105] have been noted for competitive NMDA receptor antagonists in animal models. These side effects have also been observed in trials on the effectiveness of (R)-CPP-ene in patients with treatment resistant epilepsy,[105] in patients with severe head injury[105] and in a trial of the effectiveness of (RS)-CPP in a patient suffering from neuropathic pain.[104] In addition, CGS19755 (Figure 3) has been demonstrated to produce vacuoles in rat cortical neurones leading to concern that the use of NMDA receptor antagonists may lead to neuronal damage.[122] These side effects, coupled with the low oral bioavailability of competitive antagonists, may limit the clinical usefulness of such compounds to conditions such as stroke and head trauma where treatment is given for short periods of time and under medical supervision. In contrast, the recently developed antagonist, SDZ-270-581 displays long lasting oral activity coupled with a reduction in the severity of side effects.[115]   Compounds with this profile offer the chance of increasing the scope of therapeutic applications to conditions such as chronic pain and chronic neurodegenerative diseases.[115] The development of subtype selective competitive NMDA receptor antagonists may also lead to a more accurate targeting to the particular subset of NMDA receptors responsible for the abnormal function thereby reducing side effects.

Non-competitive antagonists such as MK801 (Figure 5) have also proved to be effective neuroprotectants, anticonvulsants and blockers of dependence on drugs of abuse.[123] However, due to the poor side effect profile of channel blocking non-competitive antagonists (such as the production of psychotomimetic effects, ataxia and vacuolation of neurones) the possibilities of therapeutic application are likely to be limited.[123]

Non-competitive antagonists acting at the polyamine site of the NMDA receptor complex are effective neuroprotectants,[78,87,124] block ethanol withdrawal seizures[125] and are antinociceptive and enhance the effect of morphine in mice.[126] The neuroprotective effect of ifenprodil and eliprodil (for structures see Figure 6) may in part be due to their calcium channel blocking activity.[78]   In studies with eliprodil it has been demonstrated that at neuroprotective doses none of the side effects associated with competitive and channel blocking NMDA receptor antagonists are observed.[78]   In phase II safety studies with eliprodil on acute stroke patients, no adverse pyschotomimetic effects or any effect on blood pressure or heart rate was reported.[78] The low side effect profile observed with these drugs may in part be explained by their effect at σ sites but until the function of such sites has been established this will remain only a possibility.   Another possible explanation for the low side effect profile of these antagonists is their selective action on NMDA receptors containing NR2B subunits and as such they have a role in indicating the importance of NR2B-containing NMDA receptors in normal and abnormal CNS function.   If NMDA receptor subtype selectivity is indeed the explanation for the low side effect profile then this offers hope for therapeutic applications of subtype selective competitive NMDA receptor antagonists.

In common with polyamine site antagonists, antagonists acting at the strychnine-insensitive glycine site of the NMDA receptor complex show promising therapeutic utility coupled with a wider therapeutic ratio.[61,63,65,66,70] First generation glycine site antagonists such as 5,7-DCKA and L-689,560 (Figure 4), although potent and selective *in vitro*, do not readily cross the blood brain barrier and therefore demonstrate poor *in vivo* profiles.[61] Nonetheless, early studies have shown that glycine site antagonists show neuroprotective, anticonvulsant, anxiolytic and antipsychotic effects in animal models.[61]   There are also indications that glycine site antagonists may be useful in the treatment of persistent pain and hyperalgesia.[61]   More recently discovered antagonists such as ACEA-1021,[127] GV150526A[70] and the substituted pyrido[2,3-b]pyrazine analogues (1 and 2)[65,66] (for structures see Figure 4) show improved *in vivo* activity.   Thus GV150526A inhibited NMDA-induced convulsions in mice when administered either i.v. or p.o. (ED_{50} values 0.06 and 6 mg/kg respectively) and showed excellent neuroprotective activity in a model of focal ischaemia in rats (ED_{50} value 0.76 mg/kg).[70]   No ataxic effects or impairment of performance was observed in mice treated with GV150526A up to a dose of 30 mg/kg (500-fold higher than the ED_{50} for inhibition of NMDA-induced convulsions).[70]   These results indicate that glycine site antagonists are promising candidates for the development of therapeutically useful agents as they have good systemic activity and low side effect profiles. If results from animal studies can be repeated in clinical trials with humans then glycine site antagonists may be useful to treat not only ischaemia, but also a range of chronic neurodegenerative disorders and also epilepsy and chronic pain.

3 CONCLUSIONS

Much progress has been made in the design of potent and selective systemically active antagonists acting at either the glutamate recognition site or one of the various modulatory sites of the NMDA receptor complex. In addition, inroads have been made into the identification of subtype selective competitive NMDA receptor antagonists; however, further work is required to improve potency and selectivity. As far as therapeutic indications are concerned, the main indication is for ischaemic brain damage following stroke or head injury. Once antagonists have been identified with an improved side effect profile then the scope may widen to include chronic neurodegenerative disorders, epilepsy and chronic pain. Subtype selective competitive NMDA receptor antagonists and, in particular, antagonists acting at either the glycine or polyamine binding sites would seem to have the greatest potential as therapeutically useful agents.

References
1. J. C. Watkins and R. H. Evans, *Ann. Rev. Pharmacol. And Toxicol.*, 1981, **21**, 165.
2. H. Mclennan, R. D. Huffman and K. C. Marshall, *Nature*, 1968, **219**, 387.
3a. A. W. Duggan, *Exp. Brain Res.*, 1974, **19**, 522.
3b. G. A. R. Johnston, D. R. Curtis, J. Davies and R. M. McCulloch, *Nature*, 1974, **248**, 804.
4. T. J. Biscoe, R. H. Evans, A. A. Francis, M. R. Martin, J. C. Watkins, J. Davies and A. Dray, *Nature*, 1977, **270**, 743.
5. R. H. Evans, A. A. Francis and J. C. Watkins, *Experientia*, 1977, **33**, 489.
6. P. Krogsgaard-Larsen, T. Honoré, J. J. Hansen, D. R. Curtis and D. Lodge, *Nature*, 1980, **284**, 64.
7. P. Krogsgaard-Larsen, J. J. Hansen, J. Lauridsen, M. J. Peet, J. D. Leah and D. R. Curtis, *Neurosci. Lett.*, 1982, **31**, 313.
8. D. T. Monaghan, R. J. Bridges and C. W. Cotman. (1989) *Annu. Rev. Pharmacol. Toxicol.*, 1989, **29**, 365.
9. P. J. Conn and J-P. Pin, *Annu. Rev. Pharmacol. Toxicol.*, 1997, **37**, 205.
10. L. Nowak, P. Bregestovski, P. Ascher, A. Herbet and A. Prochiantz, *Nature*, 1984, **307**, 462.
11. M. L. Mayer and G. L. Westbrook, *J. Physiol.*, 1985, **361**, 65.
12. M. L. Mayer and G. L. Westbrook, *J. Physiol.*, 1987, **394**, 501.
13. C. F. Bigge and P. A. Boxer, *Ann. Rep. Med. Chem.*, 1994, **29**, 13.
14. T. V. P. Bliss and G. L. Collingridge, *Nature*, 1993, **361**, 31.
15. R. G. M. Morris and M. Davis (1994) The role of NMDA receptors in learning and memory. In: The NMDA receptor. Eds. G. L. Collingridge and J. C. Watkins Oxford University Press, Oxford, UK p. 340.
16. J. W. Johnson and P. Ascher, *Nature*, 1987, **325**, 529.
17. G. Henderson, J. W. Johnson and P. Ascher, *J. Physiol.*, 1990, **430**, 189.
18. D. Lodge and K. M. Johnson, *Trends Pharm. Sci*, 1990, **11**, 81.
19. E. H. F. Wong and J. A. Kemp, *Ann. Rev. Pharmacol. Toxicol.*, 1991, **31**, 401.
20. C. F. Bigge, *Biochemical Pharmacology*, 1993, **45**, 1547.
21. M. Hollmann and S. Heinemann, *Ann. Rev. Neurosci.*, 1994, **17**, 31.

22.   P. H. Seeburg, H. Monyer, R. Sprengel and N. Burnashev, (1994) Molecular biology
       of NMDA receptors. In: The NMDA receptor. Eds. G. L. Collingridge and J. C.
       Watkins Oxford University Press, Oxford, UK p. 147.
23.   G. M. Durand, M. V. Bennett and R. S. Zurkin, *Proc. Natl. Acad. Sci. USA*, 1993,
       90, 6731.
24a.   R. A. J. Mcllhinney, E. Molnár, J. R. Atack and P. J. Whiting, *Neuroscience*, 1996,
       70, 989.
24b.   B. Laube, J. Kuhse and H. Betz, *J. Neurosci.*, 1998, **18**, 2954.
25.   B. Laube, H. Hirai, M. Sturgess, H. Betz and J. Kuhse, *Neuron*, 1997, **18**, 493.
26.   A. L. Buller, H. C. Larson, B. E. Schneider, J. A. Beaton, R. A. Morrisett and D. T.
       Monaghan, *J. Neurosci.*, 1994, **14**, 5471.
27.   H. Hirai, J. Kirsch, B. Laube, H. Betz and J. Kuhse, *Proc. Natl. Acad. Sci. USA*,
       1996, **93**, 6031.
28.   M. J. Gallagher, H. Huang and D. R. Lynch, *J. Neurochem.*, 1998, **70**, 2120.
29.   J. Davies, A. A. Francis, A. W. Jones and J. C. Watkins, *Neuroscience Letters*, 1981,
       21, 77.
30.   R. H. Evans, A. A. Francis, A. W. Jones, D. A. S. Smith and J. C. Watkins, *Br. J.
       Pharmacol.*, 1982, **75**, 65.
31.   J. Lehmann, A. J. Hutchison, S. E. McPherson, C. Mondadori, M. Schmutz, C. M.
       Sinton, C. Tsai, D. E. Murphy, D. J. Steel, M. Williams, D. L. Cheney and P. L.
       Wood, *J. Pharmacol. Exp. Ther.*, 1988, **246**, 65.
32.   B. Aebischer, P. Frey, H. P. Haerter, P. L. Herrling, W. A. Mueller, H. J. Olverman
       and J. C. Watkins, *Helv. Chim. Acta*, 1989, **72**, 1043.
33.   P. L. Ornstein and V. J. Klimkowski, (1992). Competitive NMDA receptor
       antagonists. In Excitatory amino acid receptors. Design of agonists and antagonists,
       (ed. P. Krogsgaard-Larsen and J. J. Hansen) Ellis Horwood Ltd, Chichester, UK p.
       183.
34.   G. E. Fagg, H.-R. Olpe, M. F. Pozza, J. Baud, M. Steinmann, M. Schmutz, C. Portet,
       P. Baumann, K. Thedinga, H. Bittiger, H. Allgeier, R. Heckendorn, C. Angst, D.
       Brundish and J. G. Dingwall, *Br. J. Pharmacol.*, 1990, **99**, 791.
35.   J. P. Whitten, B. M. Baron, D. M. F. Miller, H. S. White and I. A. McDonald, *J.
       Med. Chem.*, 1990, **33**, 2961.
36.   D. E. Jane, H. J. Olverman, and J. C. Watkins, (1994) Agonists and competitive
       antagonists: structure-activity and molecular modelling studies. In: The NMDA
       receptor. Eds. G. L. Collingridge and J. C. Watkins Oxford University Press,
       Oxford, UK p. 31.
37.   R. B. Baudy, L. P. Greenblatt, I. L. Jirkovsky, M. Conklin, R. J. Russo, D. R.
       Bramlett, T. A. Emrey, J. T. Simmonds, D. M. Kowal, R. P. Stein and R. P. Tasse, *J.
       Med. Chem.*, 1993, **36**, 331.
38.   Y. Gaoni, A. G. Chapman, N. Parvez, P. C-K. Pook, D. E. Jane and J. C. Watkins, *J.
       Med. Chem.*, 1994, **37**, 4288.
39.   W. Müller, D. A. Lowe, H. Neijt, S. Urwyler, P. L. Herrling, D. Blaser and D.
       Seebach, *Helv. Chim. Acta*, 1992, **75**, 855.
40.   S. Urwyler, D. Laurie, D. A. Lowe, C. L. Meier and W. Müller,
       Neuropharmacology, 1996, **35**, 643.
41.   P. L. Ornstein, D. D. Schoepp, M. B. Arnold, N. K. Augenstein, D. Lodge, J. D.
       Millar, J. Chambers, J. Campbell, J. W. Paschal, D. M. Zimmerman and J. D.
       Leander, *J. Med. Chem.*, 1992, **35**, 3547.

42. S. J. Hays, C. F. Bigge, P. M. Novak, J. T. Drummond, T. P. Bobovski, M. J. Rice, G. Johnson, L. J. Brahce and L. L. Coughenour, *J. Med. Chem.*, 1990, **33**, 2916.

43. D. F. Ortwine, T. C. Malone, C. F. Bigge, J. T. Drummond, C. Humblet, G. Johnson and G. W. Pinter, *J. Med. Chem.*, 1992, **35**, 1345.

44. W. A. Kinney, N. E. Lee, D. T. Garrison, E. J. Podlesney Jr., J. T. Simmonds, D. Bramlett, R. R. Notvest, D. M. Kowal and R. P. Tasse, *J. Med. Chem.*, 1992, **35**, 4720.

45. G. Johnson and P. L. Ornstein, *Current Pharmaceutical Design*, 1996, **2**, 331.

46. W. A. Kinney and D. C. Garrison, European Patent Application, 1992, 496,561 A2.

47. J. B. Monahan and J. Michel, *J. Neurochem.*, 1987, **48**, 1699.

48. J-H. Li, C. F. Bigge, R. M. Williamson, S. A. Borosky, M. G. Vartanian and D. F. Ortwine, *J. Med. Chem.*, 1995, **38**, 1955.

49. A. L. Buller and D. T. Monaghan, *Eur. J. Pharmacol.*, 1997, **320**, 87.

50. R. H. Evans, A. A Francis, K. Hunt, D. J. Oakes and J. C. Watkins, *Br. J. Pharmacol.*, 1979, **67**, 591.

51. H. J. Olverman, A. W. Jones and J. C. Watkins, *Neuroscience*, 1988, **26**, 1.

52. E. J. Fletcher and D. Lodge, *Eur. J. Pharmacol.*, 1988, **151**, 161.

53. A. C. Foster and J. A. Kemp, *J. Neurosci.*, 1989, **9**, 2191.

54. L. Singh, A. E. Donald, A. C. Foster, P. H. Hutson, L. L. Iversen, S. D. Iversen, J. A. Kemp, P. D. Leeson, G. R. Marshall, R. J. Oles, T. Priestley, L. Thorn, M. D. Tricklebank, C. A. Vass and B. J. Williams, *Proc. Nat. Acad. Sci. U.S.A.*, 1990, **87**, 347.

55. J. A. Kemp and P. D. Leeson, *Trends Pharmacol. Sci.*, 1993, **14**, 20.

56. P. D. Leeson, B. J. Williams, M. Rowley, K. W. Moore, R. Baker, J. A. Kemp, T. Priestley, A. C. Foster, E. A. Donald, *Bioorg. Med. Chem. Lett.*, 1993, **3**, 71.

57. P. D. Leeson, B. J. Williams, R. Baker, T. Ladduwahetty, K. W. Moore and M. Rowley, *J. Chem. Soc. Chem. Commun.*, 1990, 1578.

58. P. D. Leeson, R. Baker, R. W. Carling, N. R. Curtis, K. W. Moore, B. J. Williams, A. C. Foster, A. E. Donald, J. A. Kemp and G. R. Marshall, *J. Med. Chem.*, 1991, **34**, 1243.

59. B. M. Baron, B. L. Harrison, F. P. Miller, I. A. McDonald, F. G. Salituro, C. J. Schmidt, S. M. Sorensen, H. S. White and M. G. Palfreyman, *Mol. Pharmacol.*, 1990, **38**, 554.

60. P. D. Leeson, R. W. Carling, K. W. Moore, A. M. Moseley, J. D. Smith, G. Stevenson, T. Chan, R. Baker, A. C. Foster, S. Grimwood, J. A. Kemp, G. R. Marshall and K. Hoogsteen, *J. Med. Chem.*, 1992, **35**, 1954.

61. P. D. Leeson and L. L. Iversen, *J. Med. Chem.*, 1994, **37**, 4053.

62. C. F. Bigge, P. A. Boxer and D. F. Ortwine, *Current Pharmaceutical Design*, 1996, **2**, 397.

63. R. M. Woodward, J. E. Huettner, J. Guastella, J. F. W. Keana and E. Weber, *Mol. Pharmacol.*, 1995, **47**, 568.

64. S. X. Cai, S. M. Kehr, Z-L. Zhou, V. Ilyin, S. A. Espitia, M. Tran, J. E. Hawkinson, R. M. Woodward, E. Weber and J. F. W. Keana, *J. Med. Chem.*, 1997, **40**, 730.

65. A. Cugola, D. Donati, M. Guarneri, F. Micheli, A. Missio, A. Pecunioso, A. Reggiani, G. Tarzia and V. Zanirato, *Bioorg. & Med. Chem. Lett.*, 1996, **6**, 2749.

66. F. Micheli, A. Cugola, D. Donati, A. Missio, A. Pecunioso, A. Reggiani and G. Tarzia, *Bioorg. & Med. Chem.*, 1997, **5**, 2129.

67. R. Nagata, N. Tanno, T. Kodo, N. Ae, H. Yamaguchi, T. Nishimura, F. Antoku, T. Tatsuno, T. Kato, Y. Tanaka and M. Nakamura, *J. Med. Chem.*, 1994, **37**, 3956.

68.   F. G. Salituro, B. L. Harrison, B. M. Baron, P. L. Nyce, K. T. Stewart and I. A. McDonald, *J. Med. Chem.*, 1990, **33**, 2944.

69.   F. G. Salituro, B. L. Harrison, B. M. Baron, P. L. Nyce, K. T. Stewart, J. H. Kehne, H. S. White and I. A. McDonald, *J. Med. Chem.*, 1992, **35**, 1791.

70.   R. Di Fabio, A. M. Capelli, N. Conti, A. Cugola, D. Donati, A. Feriani, P. Gastaldi, G. Gaviraghi, C. T. Hewkin, F. Micheli, A. Missio, M. Mugnaini, A. Pecunioso, A. M. Quaglia, E. Ratti, L. Rossí, G. Tedesco, D. G. Trist and A. Reggiani, *J. Med. Chem.*, 1997, **40**, 841.

71.   C. Balsamini, A. Bedini, G. Diamantini, G. Spadoni, A. Tontini, G. Tarzia, R. Di Fabio, A. Feriani, A Reggiani, G. Tedesco and R. Valigi, *J. Med. Chem.*, 1998, **41**, 808.

72.   N. A. Anis, S. C. Berry, N. R. Burton and D. Lodge, *Br. J. Pharmacol.*, 1983, **79**, 565.

73.   E. H. F. Wong, J. A. Kemp, T. Priestley, A. R. Knight, G. N. Woodruff and L. L. Iversen, *Proc. Nat. Acad. Sci. U.S.A.*, 1986, **83**, 7104.

74.   D. Lodge, M. Jones and E. Fletcher, (1994) Non-competitive antagonists of *N*-methyl-D-aspartate. In: The NMDA receptor. Eds. G. L. Collingridge and J. C. Watkins Oxford University Press, Oxford, UK p. 105.

75.   D. T. Monaghan and H. Larsen, *J. Pharmacol. Exp. Ther.*, 1997, **280**, 614.

76.   R. W. Ransom and N. L. Stec, *J. Neurochem.*, 1988, **51**, 830.

77.   J.-C. Marvizon and M. Baudry, *J. Neurochem.*, 1994, **63**, 963.

78.   C. Carter, P. Avenet, J. Benavides, F. Besnard, B. Biton, A. Cudennec, D. Duverger, J. Frost, C. Giroux, D. Graham, S. Z. Langer, J. P. Nowicki, A. Oblin, G. Perrault, S. Pigasse, P. Rosen, D. Sanger, H. Schoemaker, J. P. Thénot and B. Scatton, (1997) Ifenprodil and eliprodil: neuroprotective NMDA receptor antagonists and calcium channel blockers In: Excitatory amino acids - clinical results with antagonists. Ed. G. P. L. Herrling Academic Press, London, UK p. 57.

79.   L. Pullan and R. J. Powell, *Eur. J. Pharmacol.*, 1991, **207**, 173.

80.   C. Carter, C. Minisclou and J. P. Rivy, *Br. J. Pharmacol.*, 1992, **105**, 18P.

81.   C. Voltz, D. Fage and C. Carter, *Eur. J. Pharmacol.*, 1994, **255**, 197.

82.   C. Carron, A. Jullien and B. Bucher, *Arzneimittelforschung. Drug Res.*, 1971, **21**, 1992.

83.   K. Williams, *Mol. Pharmacol.*, 1993, **44**, 851.

84.   J. N. C. Kew, G. Trube and J. A. Kemp, *J. Physiol.*, 1996, **497**, 761.

85.   P. Avenet, J. Léonardon, F. Besnard, D. Graham, H. Depoortere and B. Scatton, *Neurosci. Letts.*, 1997, **223**, 133.

86.   G. Fischer, V. Mutel, G. Trube, P. Malherbe, J. N. C. Kew, E. Mohacsi, M. P. Heitz and J. A. Kemp, *J. Pharmacol. Exp. Ther.*, 1997, **283**, 1285.

87.   B. L. Chenard, J. Bordner, T. W. Butler, L. K. Chambers, M. A. Collins, D. L. Decosta, M. F. Ducat, M. L. Dumont, C. B. Fox, E. E. Mena, F. S. Menniti, J. Nielsen, M. J. Pagnozzi, K. E. G. Richter, R. T. Ronau, I. A. Shalaby, J. Z. Stemple and W. F. White, *J. Med. Chem.*, 1995, **38**, 3138.

88.   H. Schoemaker, J. Allen and S. Z. Langer, *Eur. J. Pharmacol.*, 1990, **176**, 249.

89.   J. Church, E. J. Fletcher, K. Baxter and J. F. Macdonald, *Br. J. Pharmacol.*, 1994, **113**, 499.

90.   V. Mutel, D. Buchy, A. Klingelschmidt, J. Messer, Z. Bleuel, J. A. Kemp and J. G. Richards, *J. Neurochem.*, 1998, **70**, 2147.

91.   T. Hayashi, *Keio Journal of Medicine*, 1954, **3**, 183.

92.   B. S. Meldrum, *Clinical Science*, 1985, **68**, 113.

93.　R. J. Bridges, J. W. Geddes, D. T. Monaghan and C. W. Cotman, (1988) Excitatory amino acid receptors in Alzheimer's disease. In Excitatory amino acids in health and disease, Ed. D. Lodge, John Wiley, London, UK p. 321.

94.　M. F. Beal, *Current Opin. Neurobiol.*, 1992, **2**, 657.

95.　F. Moroni, G. Lombardi, V. Carla, D. Pellegrini, G. L. Carassale and C. Cortesini, *J. Neurochem.*, 1986, **46**, 869.

96.　M. J. Croucher, J. F. Collins and B. S. Meldrum, *Science*, 1982, **216**, 899.

97.　A. G. Chapman, J. Graham, and B. S. Meldrum, *Eur. J. Pharm.*, 1990, **178**, 97.

98.　L. Turski, M. Schwarz, W. A. Turski, T. Klockgether, K. -H. Sontag and J. F. Collins, *Neurosci. Letts.*, 1985, **53**, 321.

99.　J. T. Greenamyre and C. F. O'Brien, *Archives of Neurology*, 1991, **48**, 977.

100.　B. S. Meldrum and R. W. Kerwin, *J. Psychopharmacol.*, 1987, **1**, 217.

101.　D. N. Stephens, B. S. Meldrum, R. Weidmann, C. Schneider and M. Grutzner, *Psychopharmacol.*, 1986, **90**, 166.

102.　S. N. Davies and D. Lodge, *Brain Research*, 1987, **424**, 402.

103.　J. D. Kristensen, R. Karlsten, T. Gordh and O. -G. Berge, *Pain*, 1994, **56**, 59.

104.　J. D. Kristensen (1997) Intrathecal administration of a competitive NMDA receptor antagonist for pain treatment. In: Excitatory amino acids - clinical results with antagonists. Ed. G. P. L. Herrling Academic Press, London, UK p. 23.

105.　P. L. Herrling, M. Emre and J. C. Watkins (1997) D-CPP-ene (SDZ EAA-494)-A competitive NMDA antagonist: pharmacology and results in humans. In: Excitatory amino acids - clinical results with antagonists. Ed. G. P. L. Herrling Academic Press, London, UK p. 7.

106.　A. B. MacDermott, M. L. Mayer, G. L. Westbrook, S. J. Smith and J. L. Barker, *Nature*, 1986, **321**, 519.

107.　B. S. Meldrum and J. Garthwaite, *Trends Pharm. Sci.*, 1990, **11**, 379.

108.　J. W. Olney, O. L. Ho and V. Rhee, *Experimental Brain Research*, 1971, **14**, 61.

109.　D. W. Choi (1991) Excitotoxicity. In Excitatory amino acids antagonists, Ed. B. S. Meldrum, Blackwell, Oxford, UK p. 216.

110.　T. W. Stone, *Pharmacol. Rev.*, 1993, **45**, 309.

111.　L. Massieu, K. H. Thedinga, M. McVey and G. E. Fagg, *Neuroscience*, 1993, **5**, 883.

112.　E. Aizenman and K. A. Hartnett, *Brain Research*, 1992, **585**, 28.

113.　M. Chen, R. Bullock, D. I. Graham, P. Frey, D. Lowe and J. McCulloch, *Annals of Neurology*, 1991, **30**, 62.

114.　R. P. Simon and K. Shiraishi, *Annals of Neurology*, 1990, **27**, 606.

115.　S. Urwyler, E. Campbell, G. Fricker, P. Jenner, M. Lemaire, K. H. McAllister, H. C. Neijt, C. K. Park, M. Perkins, M. Rudin, A. Sauter, L. Smith, K-H, Wiederhold and W. Müller, *Neuropharmacology*, 1996, **35**, 655.

116.　S. Patel, A. G. Chapman, J. L. Graham, B. S. Meldrum and P. Frey, *Epilepsy Research*, 1990, **7**, 3.

117.　D. A. Lowe, M. Emre, P. Frey, P. H. Kelly, J. Malanowski, K. H. McAllister, H. C. Neijt, C. Ruedeberg, S. Urwyler, T. G. White and P. L. Herrling, *Neurochem. Int.*, 1994, **25**, 583.

118.　K. McAllister, *Eur. J. Pharmacol.*, 1992, **231**, 309.

119.　W. Loescher and D. Honack, *J. Pharmacol. Exp. Ther.*, 1991, **256**, 432.

120.　S. Sveinbjornsdottir, J. W. A. S. Sander, D. Upton, P. J. Thompson, P. N. Patsalos, D. Hirt, M. Emre, D. Lowe and J. S. Duncan, *Epilepsy Research*, 1993, **16**, 165.

121.　R. G. M. Morris, E. Anderson, G. S. Lynch and M. Baudry, *Nature*, 1986, **319**, 774.

122. M. Schmutz, A. Arthur, H. Faleck, G. Karlsson, A. Kotake, L. Lantwicki, L. LaRue, S. Markabi, D. Murphy, M. Powell and D. Sauer (1997) Selfotel (CGS19755). In: Excitatory amino acids - clinical results with antagonists. Ed. G. P. L. Herrling Academic Press, London, UK p. 1.
123. L. L. Iversen and J. A. Kemp (1994) Non-competitive NMDA antagonists as drugs. In: The NMDA receptor. Eds. G. L. Collingridge and J. C. Watkins Oxford University Press, Oxford, UK p. 469.
124. M. Reyes, A. Reyes, T. Opitz, M. A. Kapin and P. K. Stanton, *Brain Research*, 1998, **782**, 212.
125. J. Kotlinska and S. Liljequist, *Psychopharmacology*, 1996, **127**, 238.
126. M. Bernardi, A. Bertolini, K. Szczawinska and S. Genedani, *Eur. J. Pharmacol.*, 1996, **298**, 51.
127. D. S. Warner, H. Martin, P. Ludwig, A. McAllister, J. F. W. Keana and E. Weber, *J. Cereb. Blood Flow Metab.*, 1995, **15**, 188.

RECENT ADVANCES IN GLYCINE ANTAGONISTS AS NEUROPROTECTIVE AGENTS

G. Gaviraghi*, R. Di Fabio, D. Donati, A. Feriani, E. Ratti, A. Reggiani and D. G. Trist

GlaxoWellcome S.p.A., Medicines Research Centre,
Via Fleming 4, 37135 Verona, Italy

1 INTRODUCTION

Stroke is a devastating disease caused by a sudden reduction of cerebral blood flow due to an ischaemic occlusion or an haemorragic episode, which leads to an irreversible neurological impairment. No effective neuroprotective therapies are currently available. Thus, stroke is a therapeutic area with a large unmet medical need. Since the discovery of the neurotoxic properties of glutamate, substantial observations have linked the glutamatergic hypothesis of acute neurodegeneration with stroke.[1] This research has generated both an advancement in the understanding of the pathophysiology of the disease and a promising approach to identify novel compounds to treat the disease.

At the CNS level, glutamate exerts its neurotransmitter action by activating two broad classes of receptors: ionotropic and metabotropic receptors respectively. The main difference between them is that the intracellular signalling mechanism is through an ion channel in the former case and through an enzymatic response (adenylate cyclase and PLC) in the latter case. Among the different types of glutamate receptors it is now widely accepted that the NMDA sub-type could play a major role in the excitotoxic cascade, according to the scheme depicted in Figure 1.[1,2] Once the reduction of blood flow takes place (i.e. after ischaemia), the subsequent loss of oxygen and nutrients to the brain causes an extensive depolarization, which disrupts the function of several mechanisms including glutamate reuptake mechanisms. Thus, an excess of glutamate occurs transynaptically causing an excess of Ca^{2+} entry mainly through the overactivation of NMDA receptors.[3,4] The abnormal Ca^{2+} overload cannot be buffered by intracellular mechanisms, thus over-activation of intracellular enzymes occurs (like nNOs and peroxidase, leading to free radicals production, proteases and nucleases).

If blood flow is not restored within a few minutes, neurones undergo death creating the so-called "*core*" of the infarct. The *core* is surrounded by a peri-infarct zone called "*penumbra*" where cells are suffering due to ischaemia, but are still alive. However, these cells will inevitably be permanently damaged if not protected until blood flow is restored. Therefore, the "*penumbra*" is the potential salvageable tissue and the blockade of NMDA receptors could be the biological target to counter the mechanism through which glutamate sustains the *"penumbra"*.

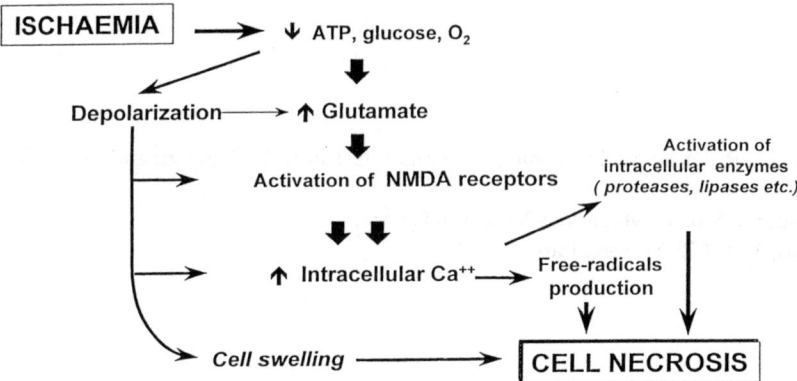

Figure 1 *Involvement of NMDA receptors in the excitotoxic cascade*

The NMDA receptor is a receptor associated ion channel selectively permeable to Ca^{2+}. The molecular structure of NMDA receptor is not completely elucidated, and by analogy with other receptor gated ion channels (i.e. nicotinic receptors) it is believed that NMDA receptor is constituted by five different subunits called NR1 and NR2$_{A-B-C-D}$ respectively.[5-7] Each subunit is encoded by a separate gene and splice variants for each gene have been described (more than 15 considering all the variants) (for a review, see: Danysz and Parsons).[8] As NMDA receptors are probably heteromeric assembly of at least four subunits, the possible combinations to achieve the final structure are multiple. Thus, it can be hypothesized that more than one subtype of NMDA receptor should exist. This conclusion is indeed supported by morphological studies showing that the NR1 element is always present where the NMDA is localized, but NR2 might vary geographically thus suggesting different receptor composition depending on the brain area.[8] Similarly, electrophysiological and pharmacological studies showing that the pharmacology of ligands is strongly influenced by the receptor composition support the heterogeneity of the NMDA receptor. The pathophysiological implication of these differences is still unknown, but this remains a fascinating area of research interest.

In addition to the molecular/structural complexity described above, the NMDA receptor is also characterized by a further complexity which is due to the different regulatory sites which control its function.[9] Among them a special interest has been attracted by the strichnine-insensitive glycine site (glycine receptor) as several studies have clearly shown that NMDA receptor function requires the simultaneous presence of both glutamate and glycine; thus glycine can be considered the glutamate co-agonist at the same receptor complex.[8,10]

Following the original description by Johnson and Ascher, in 1987, of the peculiar role of glycine on NMDA receptor function, several laboratories in Universities and in private Institutions have been involved in glycine receptor research.[11] These studies have built up a strong rationale for glycine antagonists in stroke which are summarized as follows:
1. Glycine levels are increased during stroke (increase is smaller compared to glutamate but persists longer).

2. Blockade of glycine sites with selective glycine antagonists allows complete blockade of NMDA receptor overactivation, as with truly competitive glutamate antagonists.
3. Glycine antagonists appear to be devoid of the typical side-effects associated to NMDA receptor blockade such as ataxia, memory impairments, psychotomimetic effects, PCP like effects.
4. Glycine antagonists are powerful neuroprotectives in animal models of stroke.

Based on the above rationale some years ago we became interested in the glycine antagonists field and this activity led to the identification of a novel series of indole-2-carboxylates which showed encouraging preliminary pharmacological properties.[12] Among them we identified GV150526 as a potential lead candidate for its unique profile of action.[13,14] Herein we describe the follow-on of this research, which led us to understand more about the chemistry of this class as well as more on the structural requirements to target the NMDA receptor. Taking advantage of this increased knowledge, the identification of more potent and more soluble indole-derivatives was achieved.

1.1 Pharmacological Profile of GV150526

GV150526, the indole-2-carboxylate shown in Figure 2, was the first compound belonging to a new series of indole derivatives and allowed the construction of the pharmacological profile of glycine antagonists as well as a pharmacophore for the glycine binding site.

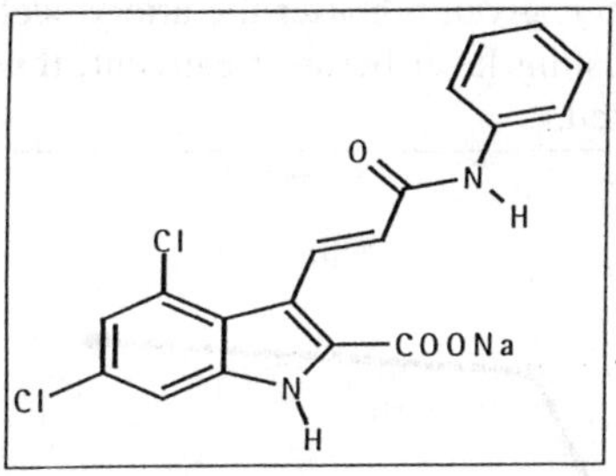

Figure 2 *GV150526*

Compounds synthesized in the optimization phases were evaluated according to the following screening sequence:
a) Binding assay to evaluate the affinity for the glycine site;
b) Selectivity for the glutamate receptors (NMDA / AMPA / KA);
c) In vivo anticonvulsant activity in the NMDA induced convulsions model in mice;
d) Neuroprotective activity in the MCAo model in rats both in pre ischaemia and in post ischaemia paradigm.

In Table 1, a brief summary of the overall profile of GV150526 is reported. GV150526 is a potent (nanomolar) and selective competitive antagonist at the NMDA associated glycine site. The high affinity value found on rat cortical receptors has been confirmed on human cortical receptors, thus affinity at the receptor level will not be a critical factor for its biological action in man. GV150526 was also found to have a potent in vivo activity (μg/kg range by the i.v. route) on NMDA induced convulsions in mice, we concluded that GV150526 crosses the blood brain barrier to such an extent that in vivo blockade of central NMDA associated glycine receptors can be achieved. This latter property results in a substantial neuroprotective effect in a model of Middle Cerebral Artery occlusion (MCAo) in rat both when given pre-ischaemia (70% of maximal protection at 3 mg/kg i.v.) and when it was given post-ischaemia up to 6 h from the induction of the damage.

Table 1 *Summary of the pharmacological profile of GV150526*

Affinity	Ki	3nM (rat cortex) 4nM (human cortex)
Antagonism of glycine induced potentiation of ^{3}H-TCP binding	Ki	3nM (rat cortex) Competitive
Receptor Selectivity	IC_{50}	>10uM (70 receptors)
NMDA-induced convulsions	ED_{50}	0.06 mg/kg iv (mice)
MCAo model Pre-ischaemia	ED_{50} max. protection	0.8 mg/kg iv (rat) 70% at 3 mg/k/iv
Post-ischaemia	Fully effective dose (at 6hrs post-ischaemia)	3 mg/kg iv (rat)

In this respect, it is important to underline that the neuroprotective effect of GV150526 is not due to a simple delay of damage progression, but to an effective neuronal salvage. As shown in Figure 3, when GV150526 is given 6 h after the artery occlusion, the progression of damage is stopped at the pre-existing level before treatment, thus suggesting that *core* expansion towards *penumbra* is halted.

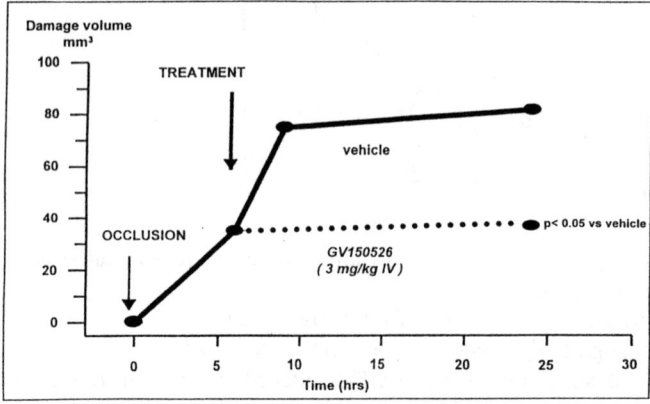

Figure 3 *Neuroprotective effect of GV150526 after post-ischaemia administration (6 h)*

These results support the hypothesis that the blockade of the NMDA receptor can reduce the infarct size in animal models of ischaemia and, in this respect, glycine antagonists appears to be extremely effective. These results also suggest that the indole 2-carboxylates, to which GV150526 belongs, can be used as a molecular template amenable to be optimized by chemical manipulation.

1.2 Identification of Novel Indole-2-Carboxylates

After the identification of GV150526, the further exploration of the "North-Eastern" region of the glycine binding site, allowed a better definition of the 3D-pharmacophore model, giving additional information on the electronic and steric requirements necessary to design novel classes of glycine antagonists within the indole-2-carboxylate series.

From previous explorations it was hypothesized that the terminal phenyl ring belonging to the C-3 side chain of GV150526 should lay in a non-hydrophobic pocket of limited size.[13] According to these studies the region of the receptor surrounding the *para* position of the terminal phenyl ring belonging to the C-3 side chain, could be able to accept hydrophilic substituents endowed with limited steric bulk. To prove this hypothesis, the series of urea derivatives shown in Figure 4A was synthesized. Also, a series of conformationally restricted analogue of GV150526 (Figure 4B) were studied with the aim of further reducing the conformational entropy of the C-3 side-chain.

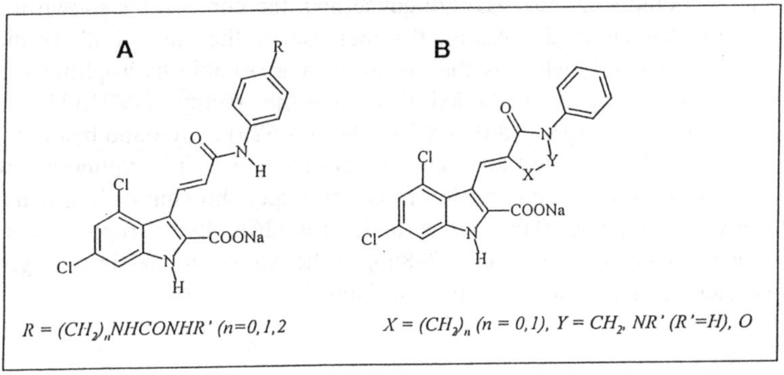

A

B

$R = (CH_2)_n NHCONHR'$ $(n=0,1,2)$

$X = (CH_2)_n$ $(n = 0,1)$, $Y = CH_2$, NR' $(R'=H)$, O

Figure 4 *Novel classes of indole-2-carboxylates*

Table 2 *In vitro affinity of a new class of substituted analogues of GV150526*

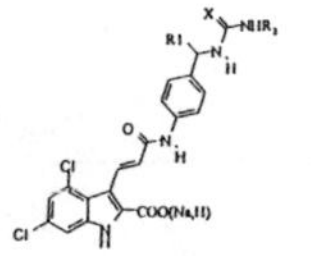

R1	R2	X	pKi[a]
GV150526	-		8.5
H	H	O	8.7
H	CH₂CH₃	O	8.6
H	CH₂COOH	O	8.6
H	c-C₃H₇	O	8.2
H	4-THP	O	8.1
H	C₆H₅	O	7.7
H	4-OCH₃-C₆H₅	O	7.7
H	3-C₅H₃N	O	7.8

[a] Displacement of [³H]-glycine

As far as class A of new indole-2-carboxylates is concerned, the compounds reported in Table 2 were synthesized and characterized in vitro for their affinity at the glycine binding site. As expected, substituent R_2 of limited size and/or increased hydrophilicity improved the in vitro affinity. Moreover, the introduction of bulky aryl groups caused a ten times reduction of the receptor affinity, further supporting the notion that there is limited space available within this region of the receptor.

The most in vitro potent urea derivative, GV228869 (R_2 = H, X = O, pK_i = 8.7), was also the most potent compound in vivo in the NMDA-induced convulsion model in mice, showing similar activity with respect to GV150526 (ED_{50}=0.07 mg/kg *vs* 0.06 mg/kg for GV150526). GV228869 was further progressed throughout the screening cascade and evaluated in the MCAo model in rats.

As far as the restricted analogues (B) are concerned, the compounds shown in Table 3 were prepared and characterized. Again, the increase of the steric bulk resulted in a reduction of in vitro potency, whereas the presence of a suitable hydrophilic substituent within the five membered ring allowed maximization of the affinity. GV213237 (X = CH_2, Y = NH) was the most potent (pK_i = 8.0 *vs* 8.5 for GV150526) compound belonging to this class of conformationally restricted analogues of GV150526. This compound was then evaluated in the NMDA-induced convulsion model in mice, showing a significant in vivo activity (ED_{50} = 0.07 mg/kg *vs.* 0.06 mg/kg for GV150526). Its neuroprotective activity was then evaluated, as in the case of GV228869, in the MCAo model in rats, given both pre and post-ischaemia. These data are given in Table 3.

Table 3 *In vitro affinity of new conformationally restricted analogues of GV150526*

X	Y	pK_i^a
GV150526	-	8.5
CH_2	CH_2	7.5
$(CH_2)_2$	CH_2	7.3
CH_2	O	7.1
CH_2	NH	8.0
CH_2	$N-CH_3$	7.5

[a] Displacement of [^{3}H]-glycine

Table 4 *Summary of the pharmacological profile of GV228869A and GV213327A compared to GV150526*

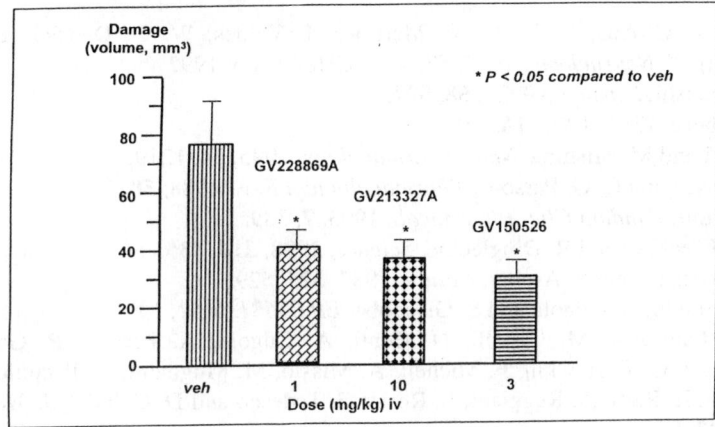

	GV150526	GV228869A	GV213327A
Affinity Ki	3 nM	2nM	10nM
Receptor Selectivity IC$_{50}$	>10 uM *(70 receptors)*	>10 uM *(70 receptors*	>10 uM *(70 receptors)*
NMDA-induced convulsions ED$_{50}$	0.06 mg/kg *(mice, iv)*	0.07 mg/kg *(mice, iv)*	0.1 mg/kg *(mice, iv)*
MCAo model pre-ischaemia ED$_{50}$ max protection	0.8 mg/kg iv *(rat)* 70% at 3 mg/kg iv	0.2 mg/kg iv *(rat)* 70% at 3 mg/k iv	1 mg/kg iv *(rat)* 70% at 3 mg/k iv

1.4 Pharmacological Profile of GV228869 and GV213237

GV228869A or GV213237A (Table 4) displayed a nanomolar affinity on glycine receptors associated to high receptor selectivity. Similarly, both compounds were very effective in blocking NMDA induced convulsions in mice. When tested in the MCAo model in rats both compounds showed a neuroprotective activity comparable to that of GV150526. In addition, both compounds were endowed with a post-ischaemia activity up to 6 h with a comparable efficacy as GV150526 (Figure 5).

Figure 5 *Post-ischaemic effect (6 h) of GV228869A and GV213327A compared to GV150526 in the MCAo model in rats*

2 CONCLUSIONS

The discovery of GV150526, a selective and potent glycine antagonist, enabled us to validate the glutamate induced neuronal toxicity hypothesis in animal model of brain ischaemia.

GV150526 is being tested in the stroke patients to confirm that the same neurotoxic pathway is working in man and consequently, GV150526 could become an effective stroke therapy.

To identify potential back-up compounds to GV150526, the indole-2-carboxylate template was explored in detail. According to the pharmacophoric model for glycine modulatory site of the NMDA receptor, the terminal phenyl ring belonging to the C-3 side chain of GV150526 was suitably substituted at the *para* position with hydrophilic groups of limited size. In this study, we have identified GV228869 as a novel glycine antagonist endowed with nanomolar antagonist affinity. At the same time, a new class of conformationally restricted analogues of GV150526 was designed and tested to gather additional information on the pharmacophoric requirements useful to design new glycine antagonists. Among the different compounds prepared from each series of indole-2-carboxylates, two compounds were identified, namely GV228869 and GV213237, which were selected and fully characterized according to a pharmacological screening protocol. On the basis of the observation of their outstanding neuroprotective activity observed, both pre- and post-ischaemia, these new indole-2-carboxylate derivatives can be considered as back-ups to GV150526.

References

1.   D. W. Choi, *Neuron,* 1988, **1**, 623.
2.   S. A. Lipton and P. A. Rosemberg, *N. Engl. J. Med.,* 1994, **330**, 613.
3.   H. Benveniste, J. Drejer, A. Schusboe and N. H. Dierner, *J. Neurochem.,* 1984, **43**, 1369.
4.   M. Y. T. Globus, R. Busto, E. Martinez, I. Valdes, W. D. Dietrich and M. D. Ginsberg, *J. Neurochem.,* 1992, **57**, 470. CHECK not 1992
5.   S. Nakanishi, *Science,* 1992, **258**, 597.
6.   H. Seeburg, *TiNS,* 1993, **16**, 359.
7.   H. Mori and M. Mishina, *Neuropharmacology,* 1995, **34**, 1219.
8.   W. Danysz and C. G. Parsons, *Pharmacological Rev.,* 1988, **50**, 1.
9.   B. Scatton, *Fundam Clin. Pharmacol.,* 1993, **7**, 389.
10.   N. W. Kleckner and R. Dingledine, *Science,* 1988, **214**, 835.
11.   J. W. Johnson and P. Ascher, *Nature,* 1987, **325**, 529.
12.   G. Gaviraghi, A. Cugola and S. Giacobbe, *EP 0568136 A1,* 1993.
13.   R. Di Fabio, A. M. Capelli, N. Conti, A. Cugola, A. Feriani, P. Gastaldi, G. Gaviraghi, C. T. Hewkin, F. Micheli, A. Missio, M. Mugnaini, A. Pecunioso, A. M. Quaglia, E. Ratti, A. Reggiani, L. Rossi, G. Tedesco and D. G. Trist, *J. Med. Chem.,* 1997, **40**, 841.
14.   R. Di Fabio, A. Cugola, D. Donati, A. Feriani, G. Gaviraghi, E. Ratti, A. Reggiani and D. G. Trist, *Drugs Fut.,* 1998, **23**, 61.

7TM Receptors

BOMBESIN, TACHYKININS AND MIMETICS OF PROTEIN-PROTEIN INTERACTIONS

David C. Horwell*, Julia A. H. Lainton, Jacqueline A. O'Neill, Martyn C. Pritchard, and Jennifer Raphy

Parke-Davis Neuroscience Research Centre, Cambridge University Forvie Site, Robinson Way, Cambridge CB2 2QB, UK

1 INTRODUCTION

Ligand-based design has a proven track record in drug design strategies. The chemical modification of histamine, for example, to give the anti-ulcer drugs cimetidine and ranitidine testifies *inter alia* to the validity of this approach in drug design[1]. We elected to utilize the ligand-design strategy to modify neuropeptides in order to give therapeutically useful non-peptide small molecules. The first example we investigated led to the design of CI-988, a non-peptide tryptophan moiety derived from the neuropeptide cholecystokinin (CCK)[2].

The successful development of such monomeric small molecules that we termed "peptoids" in the CCK area, led us to examine the generality of this approach to other neuropeptides[2]. We describe here further examples with the neuropeptides bombesin[3] and the tachykinin substance-P[4]. We also propose developing the chemistry of a new motif termed "dendroids" (Greek: *dendron* - a tree) which are self-organizing small molecules that are proposed as non-peptide mimetics of protein-protein interactions[5].

2 BOMBESIN

Bombesin is a 14-mer *C*-terminal amidated neuropeptide isolated from amphibian skin. Its mammalian counterparts are Gastrin Releasing Peptide (GRP) (Neuromedin C) and Neuromedin B (NMB). There are at least 5 sub-classes of bombesin receptors. The human BB-1 and BB-2 subtypes have been cloned and used for screening in our drug discovery programme. *N*-Terminal amino acid deletion studies of bombesin showed that the octapeptide Ac-Gln Trp Ala Val Gly His Leu Met-NH$_2$ was the minimal fragment to bind with nanomolar affinity for both BB-1 (NMB preferring) and BB-2 (GRP) receptors. An alanine scan on this fragment revealed the non-continuous Trp^7, Leu^2 (Phe) amino acid residues were essential for high binding affinity. Modelling studies indicated the peptide to be folded in a series of γ-turns from which it was concluded that the Trp-Leu(Phe) motif was close in 3-D space. A search of the Parke-Davis collection of Trp-X compounds gave **1** (Figure 1) as a micromolar hit, which was modified at both the *N*- and *C*-termini, and conformationally constrained by backbone methylation to give PD 168368 (Figure 1) with nanomolar affinity for the human BB-1 receptor (Ki = 0.15 nM)[3]. This compound was shown to be a functional antagonist *in vitro* by inhibition of calcium mobilization (K_B = 1 nM) and attenuation of acidification rates (K_D = 2 nM) by microphysiometry. The *in vivo* pharmacology of this novel non-peptide antagonist is under investigation in our laboratory.

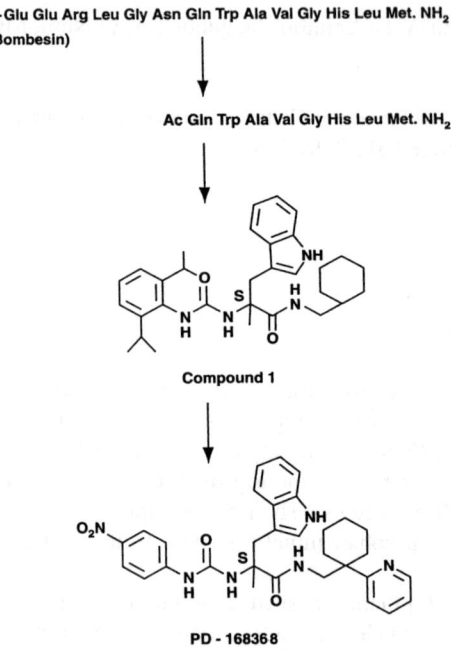

p-Glu Glu Arg Leu Gly Asn Gln Trp Ala Val Gly His Leu Met. NH₂
(Bombesin)

Ac Gln Trp Ala Val Gly His Leu Met. NH₂

Compound 1

PD - 168368

Figure 1 *Development of the Non-Peptide BB-1 Antagonist PD 168368 from the Chemical Structure of the Neuropeptide Bombesin*

3 TACHYKININS (SUBSTANCE-P)

The prototypic tachykinin, substance-P, is a *C*-terminal amidated undecapeptide and has been implicated in many physiological roles including the facilitation of pain transmission at the level of the dorsal horn in the spinal cord. Hence, the development of a substance-P antagonist may have utility in the treatment of pain, among other disorders of the central nervous system.

Our ligand-design approach was similar to that for cholecystokinin and bombesin, except that a continuous motif Phe-Phe- was found to be essential for binding to the substance-P preferring human NK-1 receptor. An alanine scan of substance-P revealed that the continuous Phe[5]-Phe[4] dipeptide motif was essential for high binding. This was modified to a Trp-Phe dipeptide which bound with micromolar affinity to both NK-1 and NK-2 (neurokinin-A) receptor subtypes. *N*- and *C*-terminal modifications and introduction of the semi-rigid backbone constraint of an α-methyl group, together with chiral optimisation at the two chiral centres, led to PD 154075 (CI-1021) (Figure 2) as a non-peptide functional antagonist at the human NK-1 receptor (Ki = 0.55 nM)[4]. This compound, in common with other NK-1 antagonists derived from different chemical classes, has been shown to have therapeutic potential both as a novel type of anti-emetic agent and in the treatment of neuropathic pain (pain associated with nerve damage). Figure 3 shows the effects of PD 154075 (CI-1021) in blocking acute and chronic emesis induced by administration of *cis*-platin in ferrets. This compound is not as potent as the 5-HT$_3$ antagonist *ondansetron* in the acute study, but is superior on chronic administration over several days.

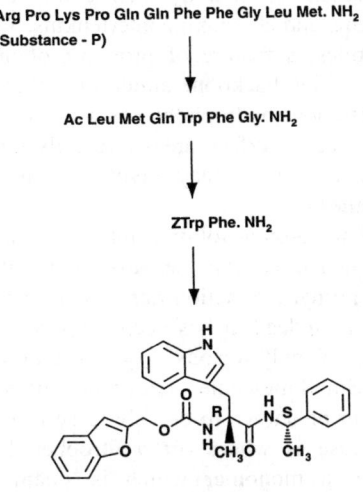

Figure 2 *Development of the Non-Peptide Tachykinin NK-1 Antagonist PD 154075 from the Chemical Structure of the Neuropeptide Substance-P*

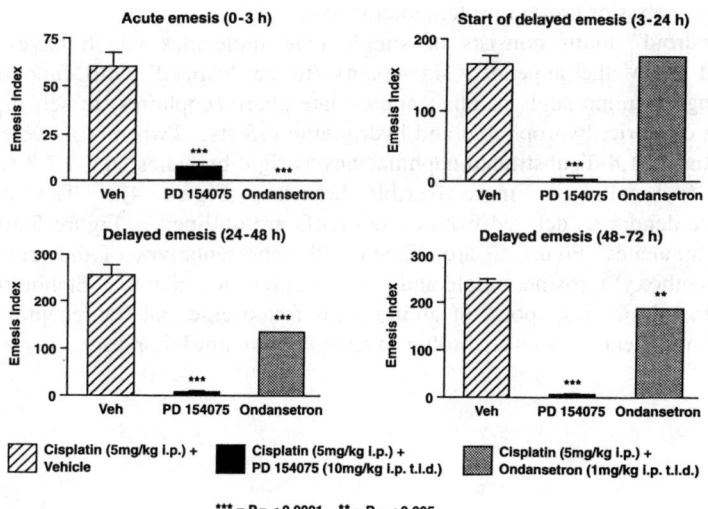

Figure 3 *Effect of the Tachykinin NK-1 Antagonist PD 154075 on Cisplatin-induced Acute and Delayed Emesis*

3.1 Dendroids: A Non-Peptide Motif as Mimetics of Protein-Protein Interactions

An emerging challenge to medicinal chemists is the design or discovery of small drug-like molecules that mimic protein-protein interactions. A wealth of structural (e.g. X-ray, NMR) and site-directed mutagenesis data has emerged that highlights "hot spots" of amino acid residues that are responsible for protein-protein (receptor) affinity and/or receptor activation. Examples include the Ras/Raf protein interaction as a target for treatment of cancer, conotoxins in the treatment of pain by their ability to block neuronal N-type calcium channels, and the cytokine and chemokine interleukins.

In common with neuropeptides, a feature of protein-protein interactions is that the structural and functional epitopes (i.e. backbone amide or amino acid side-chains) consist of both continuous and non-continuous hydrophobic, as well as hydrophilic, moieties. The main difference, however, is the large surface areas (typically 600-1000 sq. Å) associated with protein-protein interactions.   These may involve weak interactions from many partners, including water molecules.

Considering the requirement for protein folding (intra-molecular self-organization) and protein-protein interactions (inter-molecular self-assembly) it is the global energy minimum that determines the optimal interaction rather than local minima adopted by any particular amino acid side-chain. Indeed, it has been proposed that evolution of optimal interactions involving the use of poly-ligands requires that each step is at or near equilibrium[6], i.e. is reversible. Small molecules, rather than the bio-oligomers proteins and polysaccharides, were the first to be shown to be able to self-organize and self-assemble. For example, a very simple case is where *ortho*-nitrophenol self-organizes by intra-molecular H-bonding, to form a monomer which is steam volatile, whereas *para*-nitrophenol forms inter-molecular H-bonds with water and is not steam volatile. Hence, we were attracted to the notion that small molecules could be examined to show a variety of self-organizing properties using non-covalent forces that would allow them to search for wide areas of 3-D space and access conformations that correspond to the global minimum required as mimetics of protein-protein interactions.

The "dendroid" motif consists of simple core molecules which have substitution patterns that allow the appended side-chains (to be 'capped' with functional groups corresponding to amino acids or other appropriate pharmacophores) to self-organize by a combination of steric, hydrophobic and hydrophilic effects. Two motifs were chosen for this initial study: 1,8-disubstituted naphthalenes as rigid branches, and 1,2,3-trisubstituted aryl ethers (pyrogallol) as more flexible branches (Figure 4).   The syntheses of representative dendroids derived from these motifs are outlined in Figure 5 and Figure 6. The 1,8-naphthalenes (Figure 5) are capped with representatives of the aromatic amino acids, (paramethoxy) tyrosine, indole and phenylalanine side-chains[5]. Elaboration of these derivatives to mimic "hot spots" of amino acids found essential for receptor binding by site-directed mutagenesis is under further investigation in our laboratory.

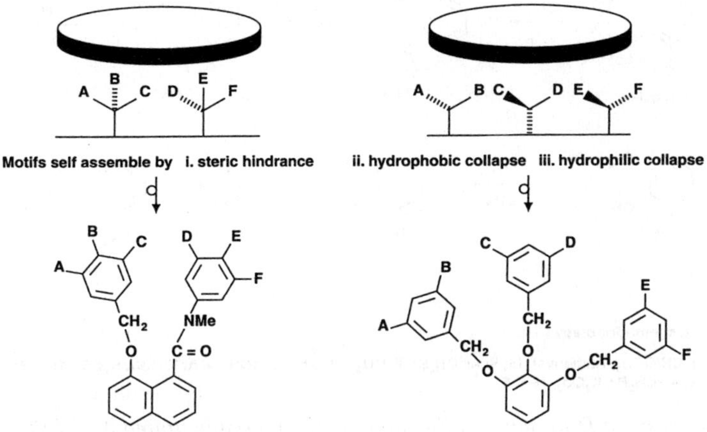

Figure 4 *1,8-Naphthalene and 1,2,3-Trisubstituted Aryl Ether Derivatives (Pyrogallols) as examples of "Dendroids"*

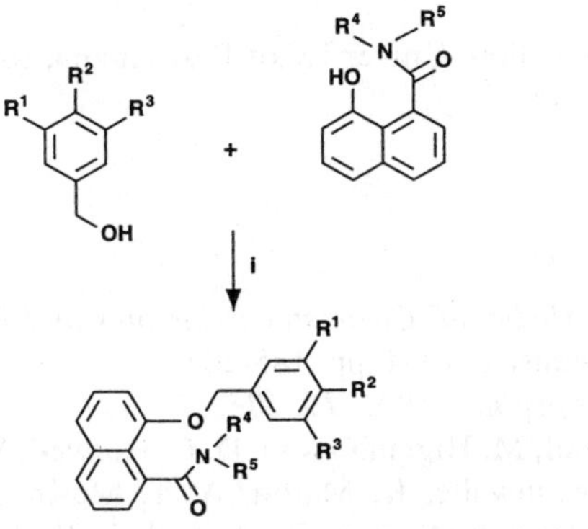

$R^1 = R^3 = H, R^2 = OBn, R^4 = Pr^i, R^5 = (R,S)CHMe-(indol-3-yl)$

$R^1 = R^3 = H, R^2 = OBn, R^4 = Bn, R^5 = CH_2C_6H_4OMe-p$

$R^1 = R^3 = H, R^2 = OBn, R^4 = Bn, R^5 = CH_2CH_2-(indol-3-yl)$

$R^1 = R^3 = H, R^2 = OBn, R^4 = CH_2C_6H_4OMe-p, R^5 = CH_2CH_2-(indol-3-yl)$

$R^1 = R^3 = H, R^2 = OBn, R^4 = CH_2CH_2C_6H_4OMe-p, R^5 = CH_2-(indol-3-yl)$

Reagents and conditions: i, PPh₃ (1 equiv.), diisopropyl azodicarboxylate (1 equiv.), THF, 0°C to room temp., 4 h.

Figure 5 *Synthesis of 1,8-Naphthalene "Dendroids"*

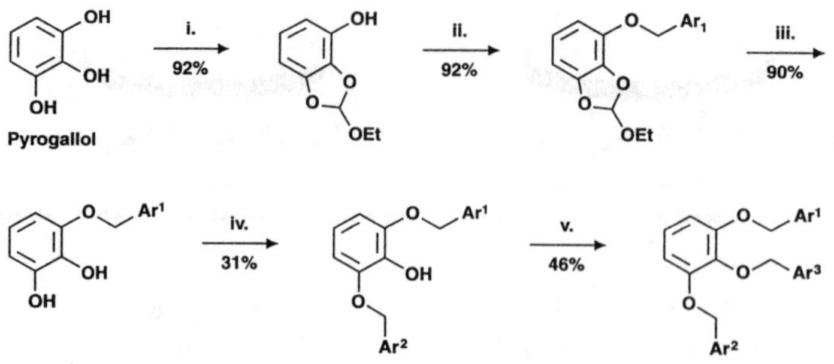

Reagents for steps i – v.

i. CH(OEt)$_3$, Amberlyst-15; ii. Ar^{1}CH$_2$Br, K$_2$CO$_3$, 18-C-6; iii. TsOH; iv. Ar^{2}CH$_2$Br, K$_2$CO$_3$, 18-C-6; v. Ar^{3}CH$_2$Br, K$_2$CO$_3$, 18-C-6.

Figure 6 *Synthesis of Differentially Substituted 1,2,3-Trisubstituted Aryl Ether Derivatives*

Acknowledgements

We thank Professor A. McKillop, University of East Anglia, for many helpful comments on the approaches to this work.

References

1.   C. R. Ganellin, In: *Medicinal Chemistry Principles and Practice*; King F. D. (Ed); Royal Society of Chemistry, 1994, pp. 185-205.
2.   D. C. Horwell, *Neuropeptides* 1991, *19*, 57.
3.   J. M. Eden, M. D. Hall, M. Higginbottom, D. C. Horwell, W. Howson, J. Hughes, R. E. Jordan, R. A. Lewthwaite, K. Martin, A. T. McKnight, J. C. O'Toole, R. D. Pinnock, M. C. Pritchard, N. Suman-Chauhan, S. C. Williams, *Bioorg. Med. Chem. Lett.* 1996, *6*, 2617.
4.   D. C. Horwell, *Bioorg. Med. Chem.* 1996, *10*, 1573.
5.   J. V. Allen, D. C. Horwell, J. A. H. Lainton, J. A. O'Neill, G. S. Ratcliffe, *Chem. Commun.* 1997, 2121.
6.   G. M. Whitesides, J. P. Mathias, C. T. Seto, *Science* 1990, *254*, 1312.

POTENT AND SELECTIVE 5-HT$_6$ RECEPTOR ANTAGONISTS

Steven M. Bromidge

Assistant Director, Department of Medicinal Chemistry,
SmithKline Beecham Pharmaceuticals, New Frontiers Science Park,
Third Avenue, Harlow, Essex CM19 5AW

1 INTRODUCTION

This chapter begins with a brief introduction to the 5-HT$_6$ receptor covering the cloning, localisation and some possible biological functions. It then goes on to describe the development and initial SAR of a novel series of antagonists starting from a high-throughput screening (HTS) lead and some of the *in vivo* effects of these compounds. This work has culminated in the identification of SB-271046 which is currently being investigated for its clinical potential. A compound from this series was also developed as a specific 5-HT$_6$ radioligand and used to map the distribution of the receptor binding protein in rat and human brains.

5-Hydroxytryptamine (5-HT, serotonin) is an endogenous neurotransmitter involved in a wide variety of physiological functions and disease states. Its diverse actions are mediated through a multiplicity of receptors and so far 7 classes (5-HT$_1$ - 5-HT$_7$) have been identified that embrace 14 human subclasses.[1] Apart from the 5-HT$_3$ family whose unique member is a ligand-gated cation channel, the other serotonergic receptors are all currently classified as belonging to the G-protein-coupled superfamily of receptors. The 5-HT$_1$ family are negatively coupled to adenylate cyclase, 5-HT$_2$ are positively linked to phospholipase C, and 5-HT$_4$, 5-HT$_6$ and 5-HT$_7$ families are positively coupled to adenylate cyclase. At the time of writing the coupling mechanism of 5-HT$_5$ receptors remains to be definitively determined. The 5-HT$_6$ receptor is one of the most recent additions to the family of serotonergic receptors and was identified using molecular biological techniques without prior knowledge of function and pharmacology. The rat 5-HT$_6$ receptor was first cloned in 1993 from striatal mRNA and contains a 438 amino acid sequence.[2,3] The human equivalent cloned later by Kohen *et al.*[4] is a 440 amino acid polypeptide with 7-transmembrane (7-TM) spanning domains typical of G-protein coupled receptors. The human receptor has 89% overall identity to its rat homologue, which increases to 96% within the more conserved TM region. In contrast to this close species homology, within the TM region there is only 30-40% identity between the 5-HT$_6$ receptor and other 5-HT receptors in both rat and human.

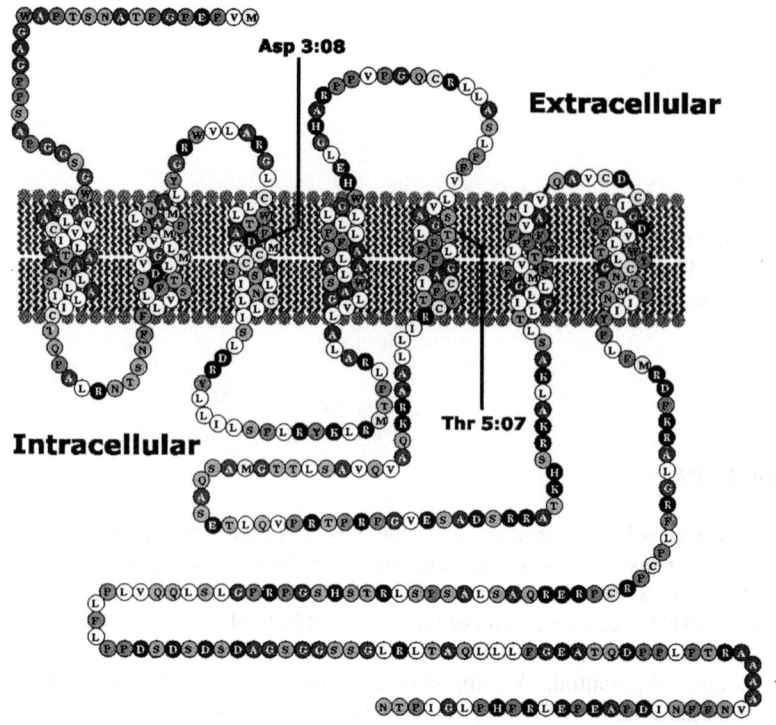

Figure 1 *Schematic representation of the 5-HT$_6$ receptor*

2 THE 5-HT$_6$ RECEPTOR: STRUCTURE, DISTRIBUTION AND FUNCTION

A schematic representation of the human 5-HT$_6$ receptor showing the putative 7-TM spanning domains as defined by hydropathy analysis is shown in Figure 1. There are 3 extracellular and 3 intracellular loops. The third cytoplasmic loop, of the intracellular regions of the receptor, is relatively short (50 amino acids) while the C-terminal tail is long (120 amino acids). The N-terminus contains a number of potential sites for glycosylation or phosphorylation by cAMP dependent protein kinase and protein kinase C which are likely to play a role in the regulation of this receptor. Looking in more detail at the amino acid sequence, the aspartic acid (Asp-308) in helix 3 is conserved throughout all G-protein coupled 5-HT receptor subtypes cloned to date and is thought to be responsible for their activation *via* an ionic interaction with the primary amino functional group of 5-HT. Site-directed mutagenesis studies by the Hoffman La Roche research group has provided strong evidence that this is the case for the 5-HT$_6$ recombinant receptor.[5] Another residue which on the basis of mutation studies is thought to be important for agonist binding and is unique to the 5-HT$_6$ receptor is a threonine in helix 5 (Thr-196) which has been proposed to hydrogen-bond to the indole NH of 5-HT. The corresponding residue in most other mammalian 5-HT receptors is alanine and this is one of the two characteristic positions of TM region 5 important for ligand binding in several biogenic amine receptors.

Phenylalanine (Phe-290) in helix 6, which is also characteristic of all biogenic amine receptors, is also conserved in the 5-HT$_6$ receptor.

The distribution of 5-HT$_6$ receptor mRNA in rat and human is broadly similar. *In situ* hybridisation studies and Northern blots found highest levels of receptor message in the olfactory tubercle, nucleus accumbens, striatum, cerebral cortex, and hippocampus.[2-4,6] In rat, studies using polyclonal antibodies have confirmed that, based on immunoreactivity, the distribution of the receptor protein is in good agreement with that of its message.[6] Importantly, only low levels of mRNA have been detected in the periphery (rat stomach and adrenal gland). *In situ* hybridisation studies in rat indicate that the 5-HT$_6$ receptor mRNA is found in 5-HT projection fields, but not in the raphe nuclei; suggesting there is a post-synaptic rather than an autoreceptor role for the 5-HT$_6$ receptor. As mentioned above, the recombinant 5-HT$_6$ receptor is positively coupled to adenylyl cyclase and confirmation of function in native tissue has been obtained in both rat and pig.[7]

2.1 Unique Pharmacological Profile of the 5-Ht$_6$ Receptor

In addition to its lack of sequence similarity, the 5-HT$_6$ receptor is distinguished from other 5-HT receptors by its unique pharmacological profile (Figure 2). It has high affinity for tryptamines, lysergic acid diethylamide (LSD) (which can be used to radiolabel the receptor) and methiothepin.[2] Several tricyclic antidepressants such as amitryptilin and mianserin, in addition to a large number of typical and atypical antipsychotic agents such as clozapine, also bind with high affinity.[8,9] Compounds which are selective for other 5-HT receptors such as mesulergine, ketanserin, 8-OH-DPAT etc. have low affinity for the 5-HT$_6$ receptor. Whereas 5-HT, 5-CT, 5-methoxytryptamine and LSD were shown to be agonists or partial agonists in a functional assay of 5-HT$_6$ receptor activation, methiothepin, clozapine, mianserin and ritanserin were found to be antagonists.[9] At present, no major species differences in the pharmacology of the rat and human receptors have been reported which is perhaps not surprising given close cross-species homology.

Drug	5-HT$_6$ Affinity (pKi)	Function
5-HT	7.3	agonist
5-CT	6.0	"
5-Methoxytryptamine	7.7	"
LSD	8.9	"
Clozapine	7.9	antagonist
Methiothepin	8.3	"
Amitryptiline	6.9	"
Mianserin	7.2	"
Ritanserin	7.5	"

Figure 2 *The unique pharmacological profile of the human 5-HT$_6$ receptor*

2.2 The Biological Functions of the 5-Ht$_6$ Receptor

The biological functions of the 5-HT$_6$ receptor are currently poorly understood. The presence of human receptor message in the hippocampus striatum and nucleus accumbens, in addition to the high affinity of the 5-HT$_6$ receptor for several therapeutically important antipsychotic and antidepressant agents, leads to possible roles for this receptor in the treatment of schizophrenia and depression.[2,8] Most atypical antipsychotic drugs, which lack extrapyrimidal side effects, bind with high affinity to the 5-HT$_6$ receptor. In fact, the prototypic atypical antipsychotic agent, clozapine, exhibits greater affinity for the 5-HT$_6$ receptor than for any other receptor subtype. There is also emerging evidence of a possible role for 5-HT$_6$ antagonists in the treatment of memory dysfunction. Antisense oligonucleotides produced a behavioural syndrome in rats consisting of yawning, stretching and chewing which could be dose-dependently blocked by the muscarinic antagonist atropine, but not by haloperidol.[10] This suggests that 5-HT$_6$ receptors may modulate cholinergic neurotransmission and hence that 5-HT$_6$ receptor antagonists may be useful for the treatment of memory dysfunction.

2.3 First Selective Ligands for the 5-Ht$_6$ Receptor

Pharmacological evaluation of the function of the receptor has been hampered by the lack of selective ligands. However, recently the first selective 5-HT$_6$ antagonists, Ro 04-6790 and Ro 63-0563, were reported by workers at Hoffman-La Roche (Figure 3).[11] These compounds had reasonably good affinity for both the rat and human 5-HT$_6$ receptor with no appreciable affinity for 23 other receptor subtypes. Although these compounds had rather low CNS penetration (<1%), when Ro 04-6790 was administered intraperitoneally to rats, after several days of habituation, sufficient brain levels were achieved to evoke a significant effect on stretching similar to that seen following treatment with antisense oligonucleotides. This work provides additional evidence that 5-HT$_6$ receptors modulate cholinergic neurotransmission.

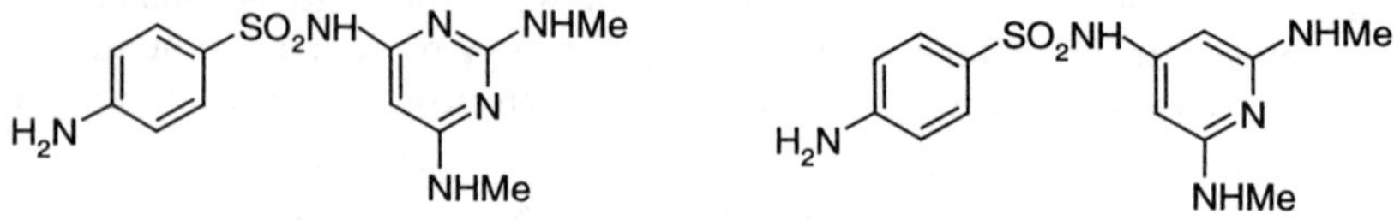

Ro 04-6790

Human 5-HT$_6$ p*K*i 7.3

Ro 63-0563

Human 5-HT$_6$ p*K*i 7.9

Figure 3 *The first reported selective 5-HT$_6$ receptor antagonists*

2.4   High   Throughput   Screening   for   5-Ht$_6$ Activity: Receptor   Binding   Profile Of Sb-214111

We became active in the area several years ago when high-throughput screening (HTS) of the SmithKline Beecham Compound Bank was carried out against the human 5-HT$_6$ receptor, expressed in membranes from HeLa cells, using [^{3}H]-LSD as radioligand. This identified a number of hits including the *bis*-aryl sulfonamide (SB-214111), containing the methoxyphenyl *N*-methylpiperazine right-hand side. SB-214111 demonstrated excellent affinity for the human 5-HT$_6$ receptor (p*K*i 8.3) with similar affinity at the rat 5-HT$_6$ receptor. On cross-screening, SB-214111 showed greater than 50-fold selectivity over a number of other key receptors including 10 other 5-HT receptor subtypes (Figure 4). After 5-HT$_6$ it showed greatest affinity for 5-HT$_1$ receptor subtypes. In addition, SB-214111 showed no appreciable affinity for over 30 other receptor, enzyme or ion-channel binding sites.

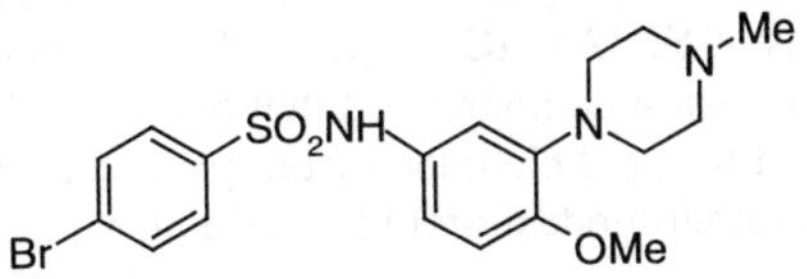

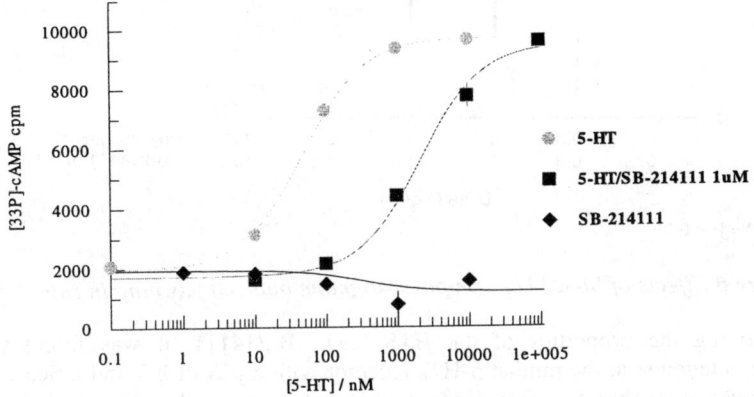

Receptor	5-HT$_{1A}$	5-HT$_{1B}$	5-HT$_{1D}$	5-HT$_{1E}$	5-HT$_{1F}$	5-HT$_{2A}$	5-HT$_{2B}$	5-HT$_{2C}$	5-HT$_4$
Affinity (pKi)	6.6	6.4	6.6	5.8	6.5	5.9	6.2	6.0	5.6

Receptor	5-HT$_6$	5-HT$_7$	adrenergic α_{1B}	dopaminergic D$_2$	dopaminergic D$_3$
Affinity (pKi)	**8.3 ± 0.2** (n > 400)	5.6	5.6	5.4	6.1

Figure 4 *Receptor binding profile of high-throughput screening hit SB-214111*

2.5 Sb-214111 in a Functional Model Of 5-Ht$_6$ Receptor Activation

SB-214111 was evaluated in a functional model of 5-HT$_6$ receptor activation. In this model, 5-HT stimulated adenylyl cyclase activity is determined by measuring the conversion of [□-^{33}P]-ATP into [^{33}P]-cAMP in HeLa cells expressing the human cloned 5-HT$_6$ receptor.[12] 5-HT elicited a dose-dependent 3-5-fold increase over basal cAMP levels.

Figure 5 *Activity of SB-214111 in a functional model of 5-HT$_6$ receptor activation*

In the presence of SB-214111, the 5-HT concentration-response curve had the same maximal response, but was shifted rightwards in a parallel manner (Figure 5). Thus, it had the profile of a competitive antagonist with a calculated pK_b at the 5-HT$_6$ receptor of 7.8 ± 0.2 (*n* = 3) which is in reasonable agreement with the binding affinity. In addition, SB-214111 showed no evidence of intrinsic activity in this system as demonstrated by the lack of effect on basal formation of cAMP with increasing concentrations of compound alone. The cytochrome P450 inhibitory potential of SB-214111 was determined using isoform selective assays to access the potential liability for drug interactions.[13] In general, SB-214111 demonstrated little propensity to inhibit the major human P450 enzymes with highest activity against CYP3A4 (IC_{50} 6 µM). Rat pharmacokinetic studies at steady state following a 16 h *i.v.* infusion demonstrated that SB-214111 was moderately brain penetrant (25%), but was subject to rapid clearance from plasma (~60 ml/min/kg), thus it has a short half-life (~0.5 h) and rather modest oral bioavailability (12%).

2.6 In Vivo Effects Of Sb-214111 on Physostigmine Induced Yawning

SB-214111 was found to have *in vivo* effects in a number of animal models, Figure 6 shows the effects in a physostigmine induced yawning model. In our studies, SB-214111 had no effect on yawning or stretching in rats following administration of the compound alone. However, the acetylcholine esterase inhibitor, physostigmine was found to elicit yawning in rats. *Subcutaneous* administration of SB-214111 dose-dependently potentiated this effect and this was significant at 3 and 10 mg/kg using a pair-wise statistical analysis. These results suggest that this compound enhances cholinergic neurotransmission through blockade of $5\text{-}HT_6$ receptors.

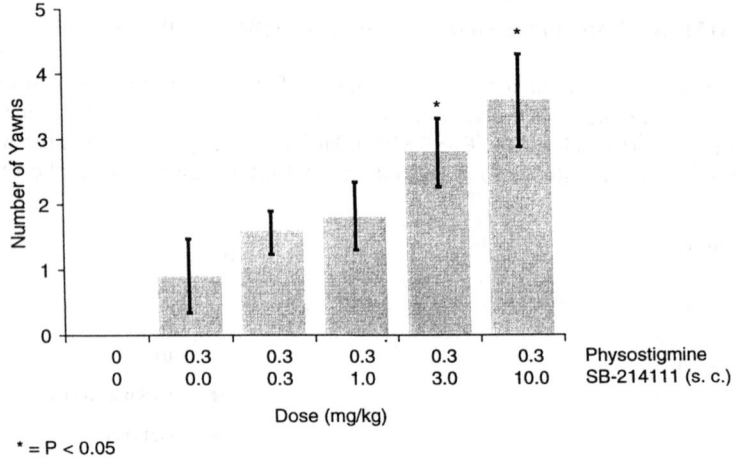

Figure 6 *Effects of SB-214111 on physostigmine induced yawning in rats*

Summarizing the properties of the HTS lead SB-214111, it was found to be a competitive antagonist at the human $5\text{-}HT_6$ receptor with a pKi of 8.3 and at least 50-fold selectivity over more than 50 other binding sites so far tested. In addition, it had a low propensity to inhibit human P450 enzymes, a reasonably promising pharmacokinetic profile, and *in vivo* activity in the physostigmine induced yawning model in rats. Finally, the structure was deemed to be readily amenable to exploitation by rapid parallel synthesis.

SB-214111 therefore represented an excellent starting point for a chemical programme and preliminary SAR investigations were made to all parts of the molecule. Initially, variation of the left-hand side aromatic group was investigated by parallel synthesis and over 100 compounds were rapidly prepared in this manner. The preparation of the key 5-amino-2-methoxyphenylpiperazine intermediate is shown (Figure 7). Thus, commercially available 1-(2-methoxyphenyl)piperazine was nitrated in good yield using potassium nitrate in conc. sulfuric acid. Reductive amination, using formic acid and formaldehyde, introduced the *N*-methyl group and catalytic reduction gave the required aniline. This was coupled with a range of sulfonyl chlorides containing a wide variety of aromatic nuclei [14,15]

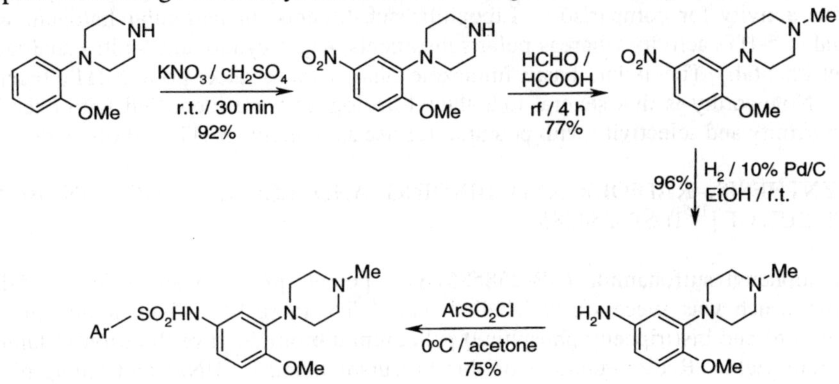

Figure 7 *Parallel synthesis of analogues of SB-214111: variation of LHS aromatic*

2.7 5-Ht$_6$ Activity of *Bis*-Aryl Sulfonamides

The resultant sulfonamides had a range of 5-HT$_6$ affinities and pleasingly many compounds were found to have an improved binding profile relative to the lead compound SB-214111 (Figure 8). A wide range of aryl groups including monocyclic and bicyclic aromatics gave excellent 5-HT$_6$ affinity.

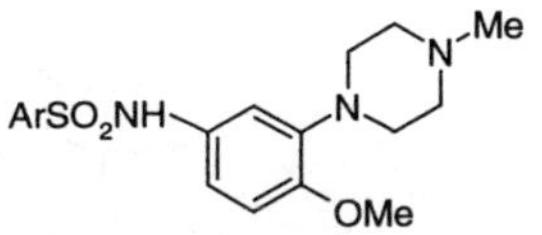

Ar	5-HT$_6$ pKi	Selectivity[a]	Ar	5-HT$_6$ pKi	Selectivity[a]
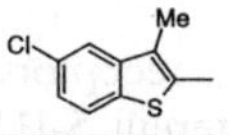	9.2	>300	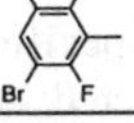	8.5	>300
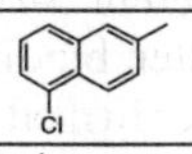	9.1	>250	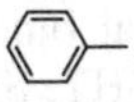	8.0	>80
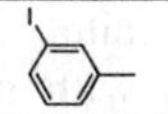	9.1	>200	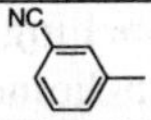	7.2	-
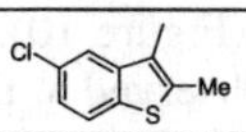	8.9	>300	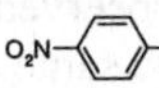	7.1	-
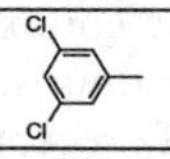	8.7	>100	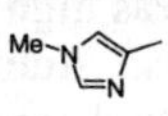	6.1	-
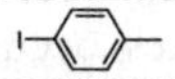	8.6	>160			

[a]Selectivity *vs* 13 receptor subtypes

Figure 8 *The 5-HT$_6$ receptor binding affinity and selectivity of bis-aryl sulfonamides*

The 2-substituted 5-chloro-3-methylbenzothiophene (SB-258510) was optimal in this study demonstrating sub-nanomolar 5-HT$_6$ receptor affinity and greater than 300-fold selectivity against a range of other receptors. The unsubstituted phenyl, pKi 8.0, provided a baseline activity for comparison. Lipophilic substituents, in particular halogen, were beneficial to 5-HT$_6$ activity whereas polar substituents, e.g. 3-cyano and 4-nitro analogues, were detrimental. The polar, basic imidazole also demonstrated poor 5-HT$_6$ receptor affinity. Noteworthy is that several iodophenyl analogues were identified with excellent receptor affinity and selectivity with potential for use as selective 5-HT$_6$ radioligands.

3 SYNTHESIS, RADIOLIGAND BINDING AND LOCALISATION IN BRAIN TISSUE OF [^{125}I]-SB-258585

The 4-iodophenyl sulfonamide (SB-258585) is a potent (pK_i 8.6) and selective 5-HT$_6$ antagonist which was successfully labelled with ^{125}I (Figure 9). Thus, treatment with bis(tributyltin) and bis(triphenylphosphine) palladium dibromide gave the tributylstannane in moderate yield. Radioiodination of this precursor using [^{125}I]NaI and chloramine-T afforded [^{125}I]-SB-258585 in >95% radiochemical purity following purification in 41% radiochemical yield (specific activity ~2000 Ci/mmol).

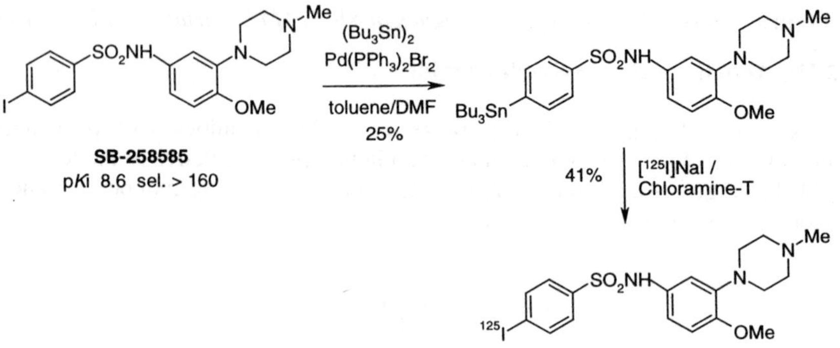

Figure 9 *Synthesis of the radioligand [^{125}I]-SB-258585*

This ligand bound with >90% specificity to human cloned 5-HT$_6$ receptors in HeLa cells. No specific binding was detected in a cell line expressing a different 5-HT receptor (5-HT$_{1B}$) and less than 10% of total binding was associated with filter binding (Figure 10).[16] Displacement studies with 5-HT, SB-214111 and SB-258585 confirmed that [^{125}I]-SB-258585 and [^{3}H]-LSD were labelling the same receptor binding site. Binding to membranes prepared from rat brain homogenates was 40% specific, with non-specific binding somewhat higher than that observed in recombinant systems (Figure 10). Binding was displaced by SB-214111 with the same affinity (pK_i 8.2) as that observed at the human and rat cloned receptors. Binding was high in the striatum, modest in the cortex and below detection in cerebellar homogenates. Autoradiographic radioligand binding studies were carried out using [^{125}I]-SB-258585 to determine the distribution of 5-HT$_6$ receptors in both rat and human brain. In the rat, the highest densities of binding sites occurred in the cerebral cortex, nucleus accumbens, striatum, hippocampus (CA1 and dentate gyrus) and the molecular layer of the cerebellum. Moderate densities were seen in the thalamus and substantia nigra. These data are in agreement with previous studies using polyclonal antibodies raised against the *C*-terminal domain of the 5-HT$_6$ receptor to study 5-HT$_6$ receptor immuno-localisation in rat brain.[6]

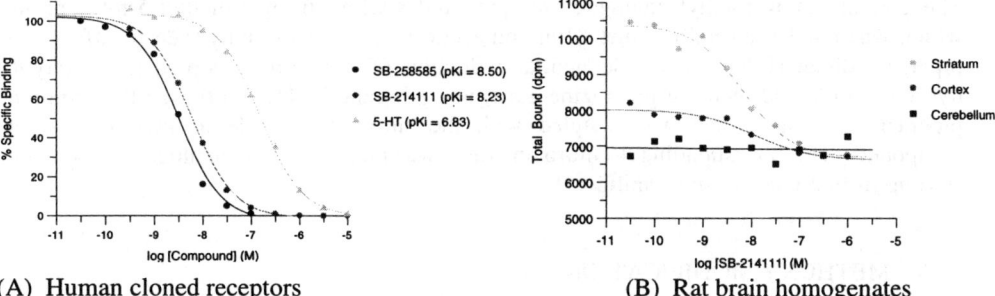

(A) Human cloned receptors (B) Rat brain homogenates

Figure 10 *Radioligand binding of [^{125}I]-SB-258585 to human cloned receptors (A) and rat brain homogenates (B)*

To establish whether the distribution of [^{125}I]-SB-258585 was similar in human brain, autoradiographic localisation was carried out on post-mortem brain tissue from a single 88-year old female with no history of neurological or psychiatric disease. The highest density of binding was seen in the frontal cortex, striatum, Purkinje cell layer of the cerebellum, and hippocampus (CA fields and dentate gyrus). Moderate densities were also seen in other layers of the cerebellar cortex. These data represent the first definitive evidence that 5-HT$_6$ receptors are present in human brain.

4 SYNTHESIS OF 4-ALKOXY-3-PIPERAZINYLPHENYL SULFONAMIDES

Returning to the discussion of SAR, the effect of modifying the methoxy substituent was investigated in a series of compounds containing the optimal 5-chloro-3-methylbenzothiophene left-hand side. *O*-Demethylation of the methoxy analogue, using boron tribromide-dimethyl sulfide complex, gave the phenol in reasonable yield (Figure 11). Alkylation of this phenol with the appropriate alkyl halide using KH and 18-crown-6 gave the required alkoxy sulfonamides.

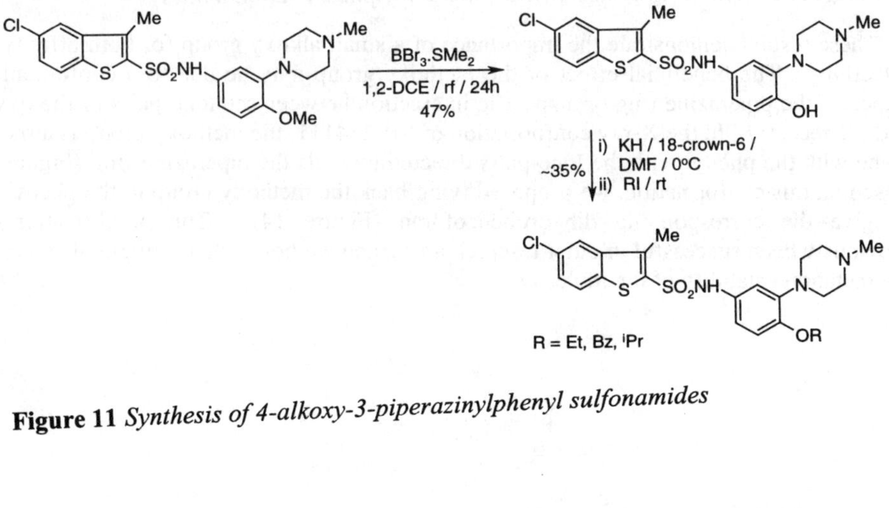

Figure 11 *Synthesis of 4-alkoxy-3-piperazinylphenyl sulfonamides*

4.1 Synthesis of 4-Methyl-3-piperazinylphenyl Sulfonamides

The corresponding methyl analogue was prepared starting from 2-methyl-5-nitroaniline which was condensed with *N*-methyliminodiacetic acid in acetic anhydride to afford the piperazinedione (Figure 12). Sequential reduction of the nitro group, using catalytic hydrogenation, and then the piperazinedione, using borane in THF, afforded the required piperazine aniline which was coupled with the sulfonyl chloride to give the target compound. The corresponding 4-chloro analogue was prepared by an analogous procedure starting from 2-chloro-5-nitroaniline.

5 4-METHOXY MODIFICATIONS

Considering the binding results, the ethoxy analogue had a very similar *in vitro* binding profile to the methoxy compound (Figure 13). However, increasing the size of the alkoxy group further to isopropyloxy and benzyloxy resulted in analogues with an order of magnitude reduction in 5-HT$_6$ affinity and loss of selectivity. The methyl analogue had similar moderate 5-HT$_6$ affinity and also lacked selectivity. Replacement with an electron withdrawing chloro group produced a larger decrease in potency.

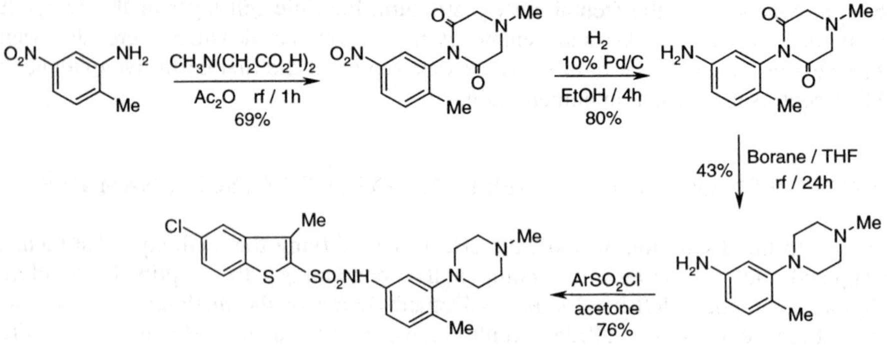

Figure 12 *Synthesis of 4-methyl-3-piperazinylphenyl sulfonamides*

These results demonstrate the importance of a small alkoxy group for both affinity and selectivity. The beneficial effect of the methoxy group may be due to a conformational effect on the piperazine ring or a specific interaction between the lone-pairs of the oxygen and the receptor. In the X-ray conformation of SB-214111, the methoxy group is almost in plane with the phenyl with the lone-pairs directed towards the piperazine ring (Figure 14). Based on this conformation, we proposed tying-back the methoxy group to the phenyl ring to give the corresponding dihydrobenzofuran (Figure 14). This modification had previously been successful in other research areas, and we hoped that it might also increase the metabolic stability of the molecule.

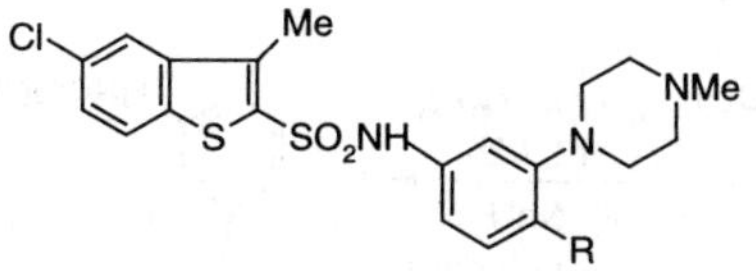

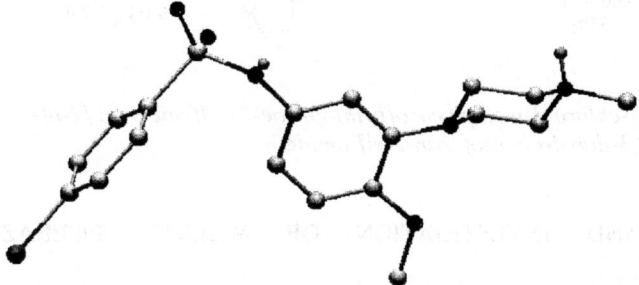

R	5-HT$_6$ pKi	Sel. *vs* 13 receptor subtypes
OMe	9.3	>300
OEt	9.2	400
O^{i}Pr	8.4	50
OBz	8.1	20
Me	8.1	<20
Cl	7.8	-

Figure 13 *Activity of 4-substituted bis-aryl sulfonamides*

Figure 14 *Single crystal X-ray structure of SB-214111*

6 SYNTHESIS OF 2,3-DIHYDROBENZOFURANS

The 2,3-dihydrobenzofuran was prepared by the sequence of reactions shown (Figure 15). Formation of the anion of 2,3-dihydrobenzofuran, using *n*-BuLi and TMEDA in hexane at 25 °C, followed by reaction with carbon dioxide gave exclusively the 7-carboxylic acid. This was converted into the acid azide, *via* the mixed anhydride, which underwent Curtius rearrangement to the trifluroacetanilide. Base hydrolysis afforded 7-amino-2,3-dihydrobenzofuran in reasonable overall yield. The 4-methylpiperazine ring was then constructed by reaction with mechlorethamine in chlorobenzene, heated under reflux in basic conditions. The nitro group was introduced selectively into the 5-position under standard conditions using KNO$_3$ in conc. H$_2$SO$_4$. Catalytic hydrogenation of this group followed by coupling with the benzothiophene sulfonyl chloride afforded the target compound. This compound was found to maintain good 5-HT$_6$ receptor affinity, but showed reduced selectivity *vs.* 5-HT$_{1B/D}$.

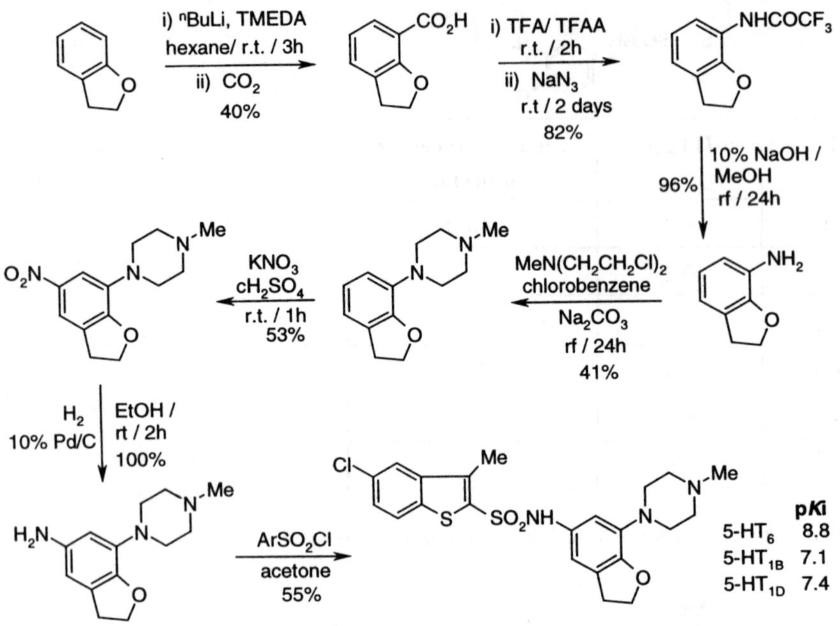

Figure 15 *Synthesis of 5-chloro-3-methylbenzo[b]thiophene-2-sulfonic acid [7-(4-methylpiperazin-1-yl)-2,3-dihydrobenzofuran-5-yl] amide*

7   SYNTHESIS   AND   INVESTIGATION   OF   *N*-ALKYL   PIPERAZINE   ANALOGUES

A number of *N*-piperazine substituted analogues were prepared to explore the effect of increasing the size of the methyl group. These compounds were prepared *via* acylation of 1-(2-methoxy-5-nitrophenyl)piperazine. Sequential reductions, first by hydrogenation of the nitro group followed by lithium aluminium hydride reduction of the acetyl moiety afforded the required anilines for the coupling reaction (Figure 16). Increasing the size of the piperazine substituent from methyl to ethyl, in combination with the 4-bromophenyl of the original lead structure (SB-214111), caused an order of magnitude reduction in affinity (Figure 17). Similarly, the *N*-cyclopropylmethylpiperazinyl analogue and *N*-benzylpiperazinyl showed approximately 50-fold reduced 5-HT$_6$ affinity relative to the *N*-methyl analogue in combination with the optimal benzothiophene. These results indicate either quite severe steric constraints or a hydrophilic pocket in the receptor binding site.

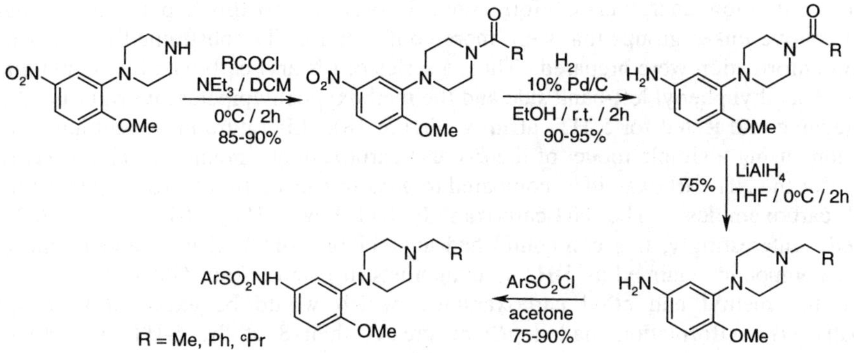

Figure 16 *Synthesis of N-alkylpiperazine analogues*

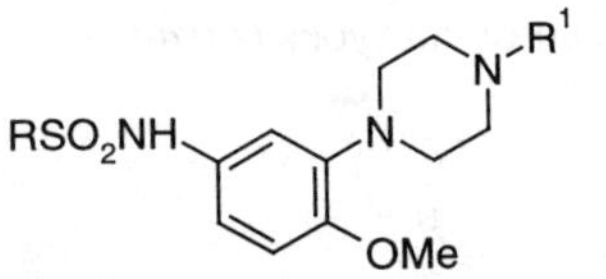

R	R¹	5-HT$_6$ pKi	selectivity
Br–⟨⟩–	Me	8.3	50
Br–⟨⟩–	Et	7.3	-
Cl–(benzothiophene, Me)	Me	9.3	>300
Cl–(benzothiophene, Me)	CH$_2$^{c}Pr	7.7	-
Cl–(benzothiophene, Me)	Bz	7.6	-

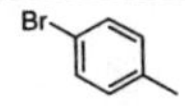

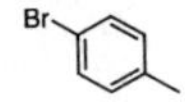

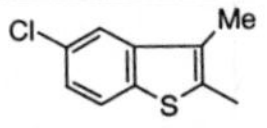

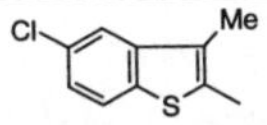

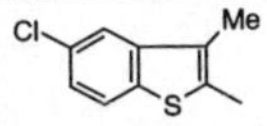

Figure 17 *Activity of N-alkylpiperazine analogues*

8 INVESTIGATION OF THE SULFONAMIDE LINKER: *BIS*-ARYL CARBOXAMIDES

The final area of SAR to be covered involves investigations of the sulfonamide linker group. Molecular modelling of the *bis*-aryl sulfonamides indicated considerable conformational freedom about this sulfonamide linker, but revealed that in the lowest energy conformations, the two aromatic groups were *gauche* with respect to each other. This is in agreement with the X-ray structure of SB-214111 where the torsion angle is 74° compared to 64° in the lowest energy conformation from modelling. The solution conformation based on NMR nOe experiments in D$_2$O indicates a similar conformation and, taken together with the high 5-HT$_6$ affinity, we proposed that this compound binds to the receptor in a low energy *cis* conformation. In order to test this hypothesis, analogues with alternative linker groups that were more conformationally constrained in either a *cis* or *trans* conformation were prepared. Thus, a series of *bis*-aryl carboxamides containing a 4-bromo-3-methyl phenyl left-hand side and the methoxyphenylpiperazinyl right-hand side were prepared and tested for 5-HT$_6$ affinity (Figure 18). Literature precedent and energy calculations using a simple model of the *bis*-aryl carboxamides predict a *trans* to *cis* ratio of 5 to 1 for the NH-carboxamides, compared to a *cis* to *trans* ratio of over 800 to 1 for the *N*-alkyl carboxamides. The NH-carboxamide had low 5-HT$_6$ affinity (pKi 6.3) as predicted. Interestingly, this compound had appreciable 5-HT$_1$ affinity and is one of a series of compounds claimed as 5-HT$_{1D}$ antagonists in patents from Glaxo workers.[17] In contrast, the methyl and ethyl carboxamides, which would be expected to adopt a principally *cis* conformation, had a pKi of greater than 8 at the 5-HT$_6$ receptor and approximately 100-fold selectivity over 5-HT$_1$ receptor subtypes. The *N*-benzyl carboxamide was considerably less potent suggesting a limit to the size of this substituent. These findings support our hypothesis that the optimum conformation for 5-HT$_6$ receptor binding is *cis*, whereas, that for the 5-HT$_1$ receptor subtypes is *trans*.

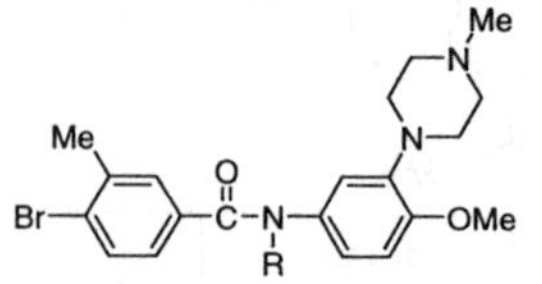

R	pKi 5-HT				carboxamide
	6	1A	1B	1D	stereochemistry
H	6.3	7.2	8.1	7.7	*trans*
Me	8.0	6.0	6.2	5.9	*cis*
Et	8.1	<5.2	5.9	5.7	*cis*
Bz	6.5	<5.5	6.0	6.0	*cis*

Figure 18 *Activity of bis-aryl carboxamides*

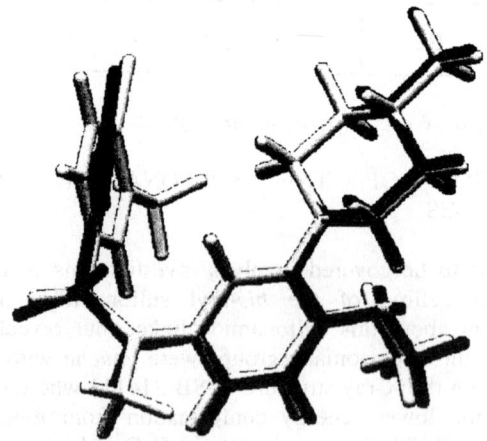

Figure 19 *Overlap of the lowest energy conformations of the sulfonamide SB-214111 (dark) and the NMe-carboxamide (light)*

Variation of the left-hand side aromatic in the *N*-methyl carboxamide series indicates that SAR mirror those seen in the sulfonamide series, but with a relative drop in affinity. Although there is a good overlap between the lowest energy conformations of the sulfonamide SB-214111 and the *N*-methyl carboxamide (Figure 19), the observed drop in affinity may be due to subtle differences between this lowest energy conformation and the actual binding conformation which the carboxamide is less able to accommodate due to reduced conformational freedom. For instance, conformations with the carbonyl out of plane with the aromatic ring will be less favoured and there are likely to be steric interactions between the *N*-methyl group and the phenyl protons. The effect of introducing methyl substitution into the sulfonamide linker was to reduce 5-HT$_6$ affinity by an order of magnitude compared to SB-214111. Modelling studies revealed that the *N*-methyl sulfonamide can adopt many alternative conformations not accessible to the NH analogue, suggesting that entropy factors may account for the differences of affinity observed in this case. This is supported by solution NMR nOe experiments that indicate that the major conformation of this compound is *trans*. The 5-HT$_1$ activity of the *N*-methyl sulfonamide was also low. In concluding the preliminary SAR around SB-214111, the benzothiophene sulfonamide SB-258510 emerged from this series as the compound with the best overall profile. This analogue demonstrated sub-nanomolar affinity for the 5-HT$_6$ receptor and over 300-fold selectivity against all other receptor subtypes tested (Figure 20). In the functional cyclase assay, SB-258510 was found to be a potent competitive antagonist with an apparent pKb of 8.5 ± 0.2 ($n = 3$) which again is in reasonable agreement with the binding affinity.

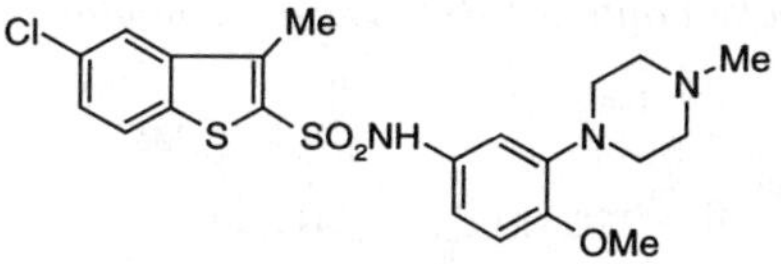

Receptor	5-HT$_{1A}$	5-HT$_{1B}$	5-HT$_{1D}$	5-HT$_{1E}$	5-HT$_{1F}$	5-HT$_{2A}$	5-HT$_{2B}$	5-HT$_{2C}$	5-HT$_4$
Affinity (pKi)	6.3	6.1	6.7	5.6	6.6	6.0	6.0	6.0	6.3

Receptor	5-HT$_6$	5-HT$_7$	adrenergic α_{1B}	dopaminergic D$_2$	dopaminergic D$_3$
Affinity (pKi)	9.2 ± 0.1 (n = 3)	5.5	5.7	6.1	6.7

Figure 20 *Receptor binding profile of SB-258510*

9 SB-258510 DOCKED INTO A MODEL OF THE 5-HT$_6$ RECEPTOR

SB-258510 is shown in the low energy *cis*-conformation docked into our model of the 5-HT$_6$ receptor (Figure 21). This model was constructed by homology modelling based on the electron microscopy alpha-carbon template for the transmembrane helices of frog rhodopsin published by Baldwin.[18] In the proposed binding mode, the protonated basic piperazine nitrogen forms an ionic interaction with the aspartic acid (3:08) on helix 3. The other major interactions are aromatic π-stacking interactions with phenylalanine (6:12) and tryptophan (6:16).

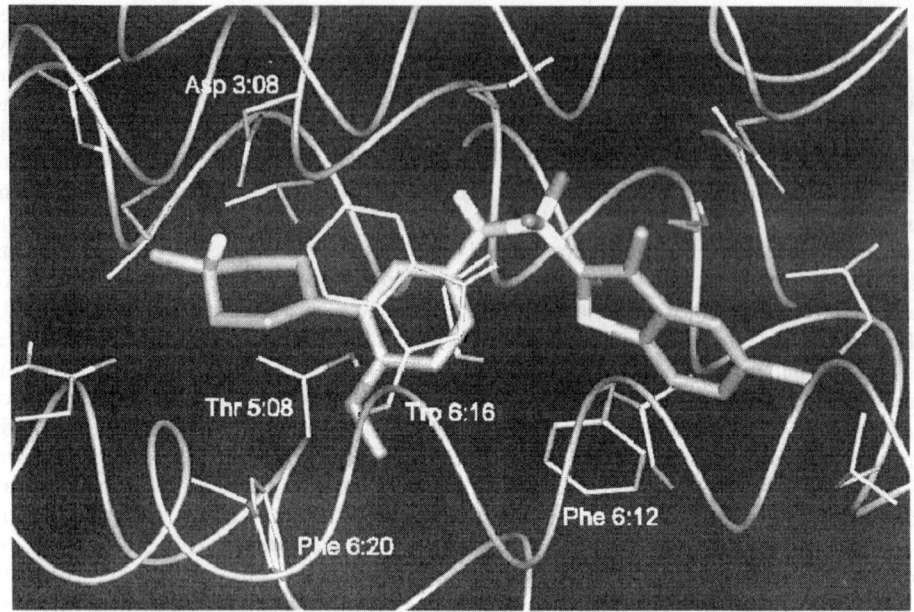

Figure 21 *SB-258510 Docked onto a 5-HT$_6$ receptor model*

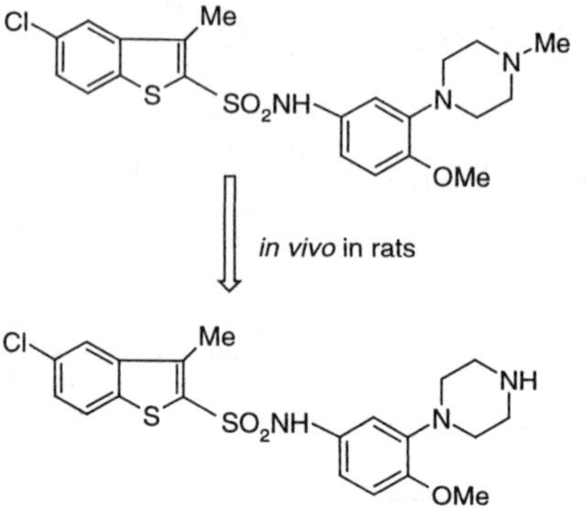

Figure 22 *Metabolic demethylation of SB-258510*

9.1 P450 Metabolic and Pk Profile of Sb-258510

In addition to an excellent binding profile, SB-258510 demonstrated an improved P450 profile relative to SB-214111. There was no significant inhibitory activity at the major human P450 enzymes predicting that the compound will have no major drug interactions. Pharmacokinetic studies at steady state in rats demonstrated it to have similar brain penetration compared to the lead SB-214111, but in contrast, it was subject to low blood clearance (12.5 ml/min/kg). However, the major metabolic fate of SB-258510 was piperazine *N*-demethylation and high levels of the NH-piperazine were found to be present in rat plasma (Figure 22). It was therefore imperative to prepare and test this metabolite for biological activity. This compound (SB-271046) was prepared via the *N*-BOC derivative of 2-methoxy-5-nitrophenylpiperazine which was reduced and coupled with the benzothiophene sulfonyl chloride. Removal of the BOC protecting group with HCl in THF gave SB-271046 as the hydrochloride salt in about 70% overall yield (Figure 23). On screening, this compound was also found to have excellent 5-HT$_6$ affinity (pKi 8.9) and greater than 200-fold selectivity over more than 50 other binding sites (Figure 24). It was a competitive antagonist with a pA_2 of 8.7. SB-271046 also demonstrated low propensity to inhibit cytochrome P450 enzymes. Although the brain penetration (10%) was somewhat reduced relative to the *N*-methyl piperazine, the clearance was even lower (7.7 ml/min/kg) which is reflected in a half-life of almost 5 h in rat. In addition, it showed excellent bioavailability (Fpo>80%) (Figure 24). SB-271046 has shown excellent oral activity in a number of animal models and is currently being further evaluated for its therapeutic potential.[15,19]

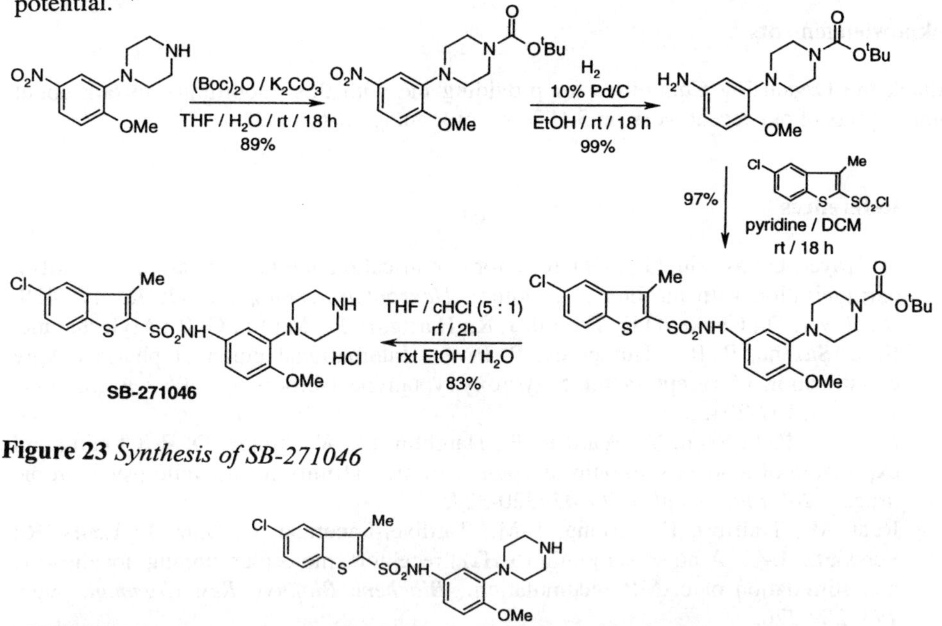

Figure 23 *Synthesis of SB-271046*

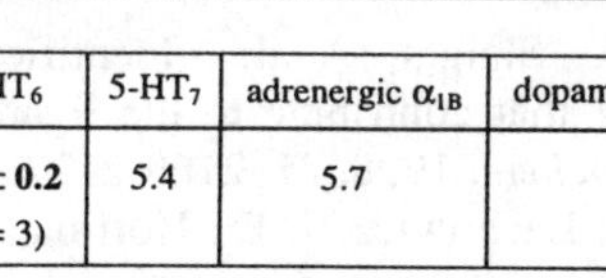

Receptor	5-HT$_{1A}$	5-HT$_{1B}$	5-HT$_{1D}$	5-HT$_{1E}$	5-HT$_{1F}$	5-HT$_{2A}$	5-HT$_{2B}$	5-HT$_{2C}$	5-HT$_4$
Affinity (pKi)	6.4	6.1	6.6	5.6	<5.0	<6.0	<5.6	5.7	5.4

Receptor	5-HT$_6$	5-HT$_7$	adrenergic α_{1B}	dopaminergic D$_2$	dopaminergic D$_3$
Affinity (pKi)	8.9 ± 0.2 (n = 3)	5.4	5.7	5.6	6.3

- Clearance (CLb) - 7.7 mL/min/kg
- Half-life - 4.8 ± 0.1 h
- Brain : blood ratio - 0.10
- Oral Bioavailability > 80%

Figure 24 *Receptor binding profile and pharmacodynamic profile in rats of SB-271046*

10 CONCLUSIONS

From a HTS lead a novel series of *bis*-aryl sulfonamides which are selective and potent 5-HT_6 receptor ligands have been developed. The benzothiophene SB-271046 was identified with nanomolar affinity and >200-fold selectivity over a range of other receptor subtypes. This compound is metabolically stable and has oral activity in animal models.   The first reported highly specific radioligand has also been identified from this series. Autoradiography studies in rat and human brain slices using this ligand provided the first compelling evidence that 5-HT_6 receptors are present in human brain.   SB-271046 is currently being used to elucidate further the therapeutic potential of 5-HT_6 receptor ligands.

Acknowledgements

I thank the Organising Committee for providing me with this opportunity to talk about some aspects of our recent work on 5-HT_6 receptor antagonists.

References

1.   (a) Hoyer, D.; Martin, G.   5-HT receptor classification and nomenclature: towards a harmonisation with the human genome.   *Neuropharmacology.* 1997, *36*, 419-428. (b) Hoyer, D.; Clarke, D.E.; Fozard, J. R.; Hartig, P. R.; Martin, G. R.; Mylecharane, E. J.; Saxena, P. R.;   Humphrey, P. P. A.   International union of pharmacology classification of receptors for 5-hydroxytryptamine (serotonin).   *Pharmacol. Rev.* 1994, *46*, 157-204.

2.   Monsma, F. J.; Shen, Y.; Ward R. P.; Hamblin, M. W.; Sibley. D. R. Cloning and expression of a novel serotonin receptor with high affinity for tricyclic psychotropic drugs. *Mol. Pharmacol.* 1993, *43*, 320-327.

3.   Ruat, M.; Traiffort, E.; Arrang, J.-M.; Tardivel-Lacombe, J.; Diaz, J.; Leurs, R.; Schwartz, J.-C.   A novel serotonin (5-HT_6) receptor: molecular cloning, localisation and stimulation of cAMP accumulation.   *Biochem. Biophys. Res. Commun.* 1993, *193*, 269-276.

4.   Kohen, R.; Metcalf, M. A.; Druck, T.; Huebner, K.; Sibley, D. R.; Hamblin, M. W. Cloning and chromosomal localization of a human 5-HT_6 serotonin receptor. *Soc. Neurosci. Abstracts.* 1994, *20*, 476.8.

5.   Boess, F. G.; Monsma, F. J. J., Sleight, A. J.   Identification of residues in transmembrane regions III and IV that contribute to the ligand binding site of the serotonin 5-HT_6 receptor. *J. Neurochem.* 1998, *71*, 2169-2177.

6.   (a) Ward, R. P.; Hamblin, M. W.; Lachowicz, J. E.; Hoffman, B. J.; Sibley, D. R.; Dorsa, D. M. Localization of serotonin subtype 6 receptor messenger RNA in the rat brain by *in situ* hybridization histochemistry. *Neuroscience.* 1995, *64*, 1105-1111. (b) Gerard, C.; El Mestikawy, S.; Lebrand, C.; Adrien, J.; Ruat, M.; Traiffort, E.; Hamon, M.; Martres, M-P.   Quantitative RT-PCR distribution of serotonin 5-HT_6 receptor mRNA in the central nervous system of control or 5,7-dihydroxytryptamine-treated rats. *Synapse.* 1996, *23*, 164-173.

7.   Schoeffter, P.; Waeder, C. 5-Hydroxytryptamine receptors with a 5-HT$_6$ receptor like profile stimulating adenylyl cyclase activity in pig caudate membranes. *Naunyn-Schmiedebergs Arch Pharmacol.* 1994, *350*, 356-360.

8.   Roth, B. L.; Craig, S.C.; Choudhary, M. S.; Uluer, A.; Monsma, F. J.; Shen, Y.; Meltzer, H. Y.; Sibley, D. R. Binding of typical and atypical antipsychotic agents to 5-hydroxytryptamine-6 and 5-hydroxytryptamine-7 receptors. *J. Pharmacol. Exp. Ther.* 1994, *268*, 1403-1410.

9.   Boess, F. G.; Monsma, F. J. J.; Carolo, C.; Meyer, V.; Rudler, A. Functional and radioligand binding characterization of rat 5-HT$_6$ receptors stably expressed in HEK293 cells. *Neuropharmacology*, 1997, *36*, 713-720.

10. (a) Bourson, A.; Borroni, E.; Austin, R. H.; Monsma, F. J.; Sleight, A. J. Determination of the role of the 5-HT$_6$ receptor in the rat brain: a study using antisense oligonucleotides. *J. Pharmacol. Exp. Ther.* 1995, *274*, 173-180. (b) Sleight, A. J.; Monsma, F. J., Borroni, E.; Austin, R. H.; Bourson, A. Effects of altered 5-HT$_6$ expression in the rat: functional studies using antisense oligonucleotides. *Behav. Brain Res.* 1996, *73*, 245-248.

11. Sleight, A. J.; Boess, F. G.; Bös, M.; Levet-Trafit, B.; Riemer C.; Bourson, A. Characterization of Ro 04-6790 and Ro 63-0563: potent and selective antagonists at human and rat 5-HT$_6$ receptors. *Br. J. Pharmacol.* 1998, *124*, 556-562.

12. Salomon, Y. Adenylate Cyclase Assay. *Adv. Cyclic Nucleotide Res.* 1979, *10*, 35-55.

13. Lewis, D. F. V. *Cytochromes P450: Structure, Function and Mechanism;* Taylor and Francis Publishing, 1996.

14. Bromidge, S. M.; King, F. D.; Wyman, P. A. WO Patent 98/27081.

15. Bromidge, S. M.; Brown, A. M.; Clarke, S. E.; Dodgson, K.; Gager, T.; Grassam, H. L.; Jeffrey, P. M.; Joiner, G. F.; King, F. D.; Middlemiss, D. N.; Moss, S. F.; Newman, H.; Riley, G.; Routledge, C.; Wyman, P. 5-Chloro-N-(4-methoxy-3-piperazin-1-ylphenyl)-3-methyl-2-benzothiophenesulfonamide (SB-271046): a potent, selective and orally bioavailable 5-HT$_6$ receptor antagonist. *J. Med. Chem.* 1999, *42*, 202-205.

16. Hirst, W. D.; Minton, J. A. L.; Bromidge, S. M.; Moss, S. F.; Latter, A. J.; Riley, G.; Routledge, C.; Middlemiss, D. N.; Price, G. W. Characterisation of [^{125}I]SB-258585 binding to human recombinant and native 5-HT$_6$ receptors in rat, pig and human brain tissue. *Br. J. Pharmacol.*, in press.

17. Carter, M. WO Patent 94/15920.

18. Baldwin, J. M.; Schertler, G. F. X.; Unger, V. M. An alpha-carbon template for the transmembrane helices in the rhodopsin family of G-protein-coupled receptors. *J. Mol. Biol.* 1997, *272*, 144-164.

19. Routledge, C.; Price, G. W.; Bromidge, S. M.; Moss, S. F.; Hirst, W. D.; Newman, H.; Riley, G.; Gager, T.; Brown, A. M.; Middlemiss, D. N. Characterisation of SB-271046: a potent and selective 5-HT$_6$ receptor antagonist. *Br. J. Pharmacol.*, in press.

DUAL D_2-RECEPTOR AND β_2-ADRENOCEPTOR AGONISTS FOR THE TREATMENT OF AIRWAYS DISEASES

F. Ince

Department of Medicinal Chemistry, Astra Charnwood, Bakewell Road, Loughborough, Leics., LE11 5RH, UK

1 INTRODUCTION

Airways hyperreactivity is a characteristic of asthma and, to a lesser extent, chronic obstructive airways disease (COPD).[1] This hyperreactivity can lead to an exaggerated reflex response to a variety of stimuli and can give rise to many of the symptoms of both COPD and asthma (bronchoconstriction, dyspnoea, cough and mucus production).[2] Whilst topically administered β_2-adrenoceptor agonists[3] and topically applied corticosteroids[4] are well established treatments for asthma, COPD is still a poorly treated disease. Neither the β_2-adrenoceptor agonists nor the corticosteroids were designed to reduce airways hyperreactivity and relatively little attention has been paid to the provision of compounds that could specifically reduce sensory nerve activity in the lung.

Hyperreactivity of the airways is intimately associated with neural reflex pathways which contain both afferent and efferent nerves.[5] The receptors that modulate the activity of afferent and/or efferent nerves in the lung are not well characterised, although it has been recognised that efferent nerve activity is not solely mediated by acetylcholine and that neuropeptides are also probably involved in this process.[5] An efficient approach to the control of reflex nerve activity in the airways would be to modulate the activity of afferent (sensory) nerves. Our objective has been to discover compounds that are receptor agonists (or antagonists) which would suppress afferent nerve activity in the lung. We also required that these compounds should possess anti-bronchoconstrictor activity.

2 DOPAMINE RECEPTORS AND SENSORY NERVE ACTIVITY

A number of studies have shown that dopamine can act as an inhibitor of a variety of neural systems, e.g., sympathetic,[6] chemoreceptor,[7] *via* the stimulation of D_2-receptors.[8] Whilst the presence of D_2-receptors on sensory nerve endings in the lung has yet to be demonstrated, D_2-receptor mRNA has been shown to be present in rat vagal afferent neurones[9] and dorsal root ganglia.[10] There is also an anecdotal report[11] that L-DOPA has produced some beneficial effects in a patient with chronic bronchitis (in addition to Parkinson's Disease) and it was presumed that these effects were produced by a metabolite of L-DOPA. The mechanism of action was not discussed but the involvement of dopamine could not be ruled out. In the β-blocked anaesthetised dog it has also been possible to demonstrate that D_2-receptor agonists are capable of reducing the histamine-induced increase in sensory nerve traffic[12] (Figure 1).

Our working hypothesis has been that the stimulation of D_2-receptors on afferent nerves in the lung would lead to the suppression of sensory nerve traffic and the associated reflex nerve activity. Given the involvement of direct-acting bronchoconstrictor agents in both COPD and asthma, it seemed unlikely that D_2-receptor agonist activity alone would produce the desired anti-bronchoconstrictor activity, consequently we set out to discover

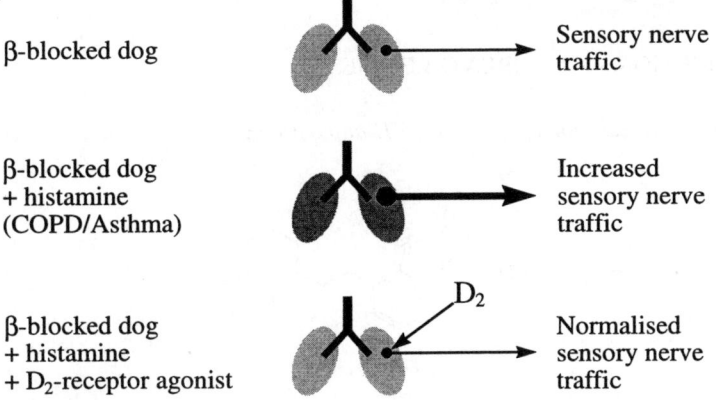

Figure 1 *Sensory Nerve Traffic in the Dog Lung (In Vivo)*

compounds that were dual D_2-receptor and β_2-adrenoceptor agonists. This dual activity should provide an effective symptomatic treatment for both COPD and asthma and has the added advantage that, if neurogenic inflammation is a significant component of human airways disease, this approach would beneficially affect some aspects of the underlying inflammation in the lung (Figure 2).

Our objective has been to discover compounds that are potent, dual D_2-receptor and β_2-adrenoceptor agonists where the β_2-adrenoceptor agonist activity was at least as potent as salbutamol and salmeterol. Our initial aim was to find compounds that possessed $D_2 = \beta_2$ activity (*in vivo* animal studies) and select the most promising compound for further development as a potential treatment for airways diseases such as COPD and asthma. We proposed to administer the compounds topically to the lung (nebuliser, pMDI or DPI), thus minimising the systemic side effects usually associated with D_2-receptor stimulation (nausea and emesis) or β_2-adrenoceptor agonist activity (tachycardia or tremor).

The compounds to be described were evaluated in the following *in vitro* systems: D_2-receptor activity: D_2LB - pK_H (ligand binding to bovine pituitary membranes[13]); D_2REA - $p[A]_{50}$, intrinsic activity (IA) vs 2-amino-6,7-dihydroxy-1,2,3,4-tetrahydronaphthalene = 1.0 (field stimulated rabbit ear artery[14]). β_2-adrenoceptor activity: β_2GPT - $p[A]_{50}$, IA vs isoprenaline = 1.0 (guinea pig trachea[15]). α_1-adrenoceptor activity: α_1REA - $p[A]_{50}$, IA vs phenylephrine = 1.0 (rabbit ear artery[16]).

INFLAMMATION
(D2)
HYPERREACTIVITY
D2 (β2)
COUGH
D2
MUCUS
β2 (D2)
D2
BRONCHOCONSTRICTION
BREATHLESSNESS

Figure 2 *D_2-receptors and β_2-adrenoceptors in COPD and asthma*

HO
HO
H
N
N
H
Ph
1
O
S
HN
HO
N
2
O
S
HN
HO
H
N
X
Ph
3
3a X = NH
3b X = S
3c X = O

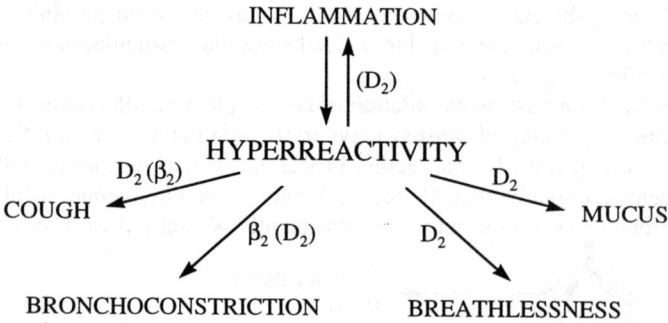

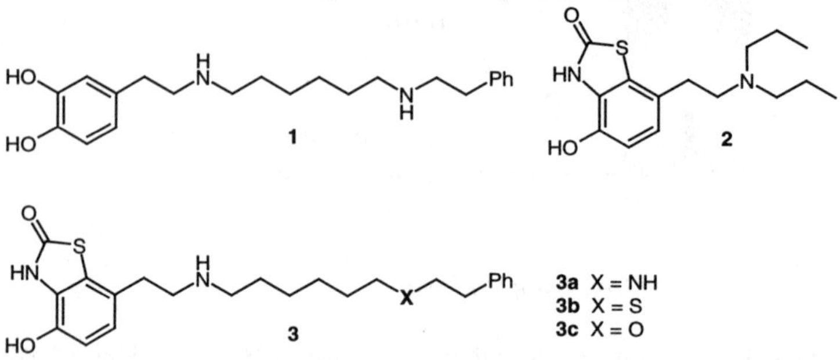

Figure 3 *Structure of Compounds 1-3*

3 DUAL D_2-RECEPTOR AND β_2-ADRENOCEPTOR AGONISTS?

The wisdom of attempting to combine D_2-receptor and β_2-adrenoceptor agonist activity in the same molecule can be questioned, however, a number of factors gave us confidence that this should be possible:

a) dopaminergic and adrenergic receptors belong to the same 7 transmembrane helix, G-protein coupled receptor family,[9]

b) dopamine and adrenaline are very similar in structure,

c) dopamine is an agonist at dopamine receptors and β_1-adrenoceptors (and α_1-adrenoceptors),[17]

d) our studies in the D_1-receptor area[18] had already established that Dopacard® (**1**) possesses D_2-receptor and β_2-adrenoceptor agonist activity as well as being a D_1-receptor agonist.

During our search for compounds that were potent D_2-receptor agonists we prepared hybrids of **1**[18] and the potent D_2-receptor agonist **2**[19] (Figure 3 and Table 1). An example of these compounds (**3a**) is a potent D_2-receptor ligand (D_2LB: pK_H 8.99). For a number of the compounds described their α_1-adrenoceptor agonist activity (vasoconstriction) does not allow a reliable estimate of D_2-receptor agonist activity to be made using the field stimulated REA. **3a** is also a β_2-adrenoceptor partial agonist with low intrinsic activity

(β_2GPT: p[A]$_{50}$ 7.19, IA 0.18) and an α_1-adrenoceptor agonist (α_1REA: p[A]$_{50}$ 6.81, IA 0.82). At this stage a number of analogues of **3a** were prepared in which the second amino group had been modified.

The sulphide (**3b**) retains useful D$_2$-receptor binding activity but is a weak β_2-adrenoceptor agonist; the ether (**3c**) possesses a much more encouraging profile of activity in that it is a potent D$_2$-receptor agonist (D$_2$LB: pK$_H$ 9.02) and β_2-adrenoceptor partial agonist with an increased intrinsic activity (β_2GPT: p[A]$_{50}$ 7.89, IA 0.40), however, **3c** is still a potent α_1-adrenoceptor agonist (α_1REA: p[A]$_{50}$ 7.56, IA 0.96).

4 CAN α_1-ADRENOCEPTOR AGONIST ACTIVITY BE REDUCED ?

In the context of the treatment of respiratory diseases, α-adrenoceptor agonist activity is

Table 1 *In Vitro Biological Activities of 1, 2 and 3a-c*

Entry	D$_2$LB pK$_H$	D$_2$REA p[A]$_{50}$ (IA)	β_2GPT p[A]$_{50}$ (IA)	α_1REA p[A]$_{50}$ (IA)
1	6.67	5.90 (0.16)	5.81 (0.16)	-
2	9.34	9.60 (0.81)	< 5.0	6.03 (0.63)
3a	8.99	~ 8.0[a]	7.19 (0.18)	6.81 (0.82)
3b	9.12	-	6.14 (0.08)	6.00 (0.77)
3c	9.02	~ 7.8[a]	7.89 (0.40)	7.56 (0.96)

a α_1-adrenoceptor agonist activity does not allow a reliable
    estimate of D$_2$-receptor agonist activity to be made

undesirable, since it could produce bronchoconstriction[20] and/or pulmonary vasoconstriction.[21] It was, therefore, important that this inappropriate activity be reduced (by reducing affinity and/or intrinsic activity) whilst maintaining or enhancing D$_2$-receptor and β_2-adrenoceptor agonist activity. This was investigated using **3c** as the lead structure by modification of the substituent on the amino group of the benzthiazolone-ethylamine.

These changes could introduce steric and/or electronic effects that might differentially affect agonist-receptor interactions. A number of studies have shown that the introduction of alkyl substituents adjacent to the amino group in adrenoceptor agonists reduces α-adrenoceptor agonist activity,[22] eg isoprenaline (a selective β_2-adrenoceptor agonist) being derived from adrenaline. In the series related to **3c** the expected effect was seen and both **4a** and **4b** have reduced α_1-adrenoceptor agonist activity, unfortunately D$_2$-receptor agonist activity (**4a**) or activity (**4b**) was also reduced (see Figure 4 and Table 2).

The incorporation of an O, SO$_2$ or NHSO$_2$ moiety into the (CH$_2$)$_6$ portion of the ethylamine nitrogen substituent afforded greater success as seen with compounds **5**

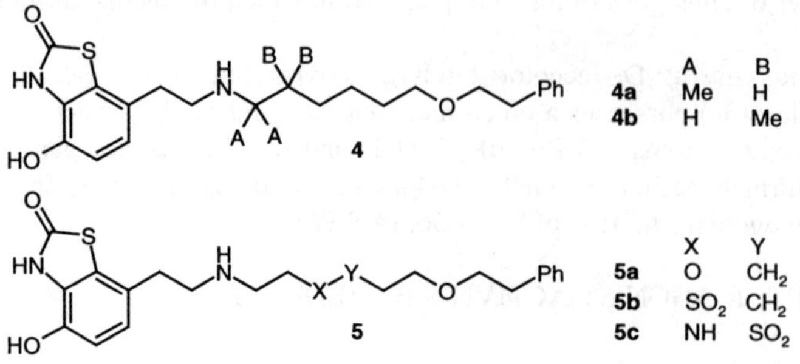

Figure 4 *Structures of Compounds 4 and 5*

Table 2 *In Vitro Biological Activity of 3c, 4a-b, 5a-c and salmeterol (6)*

Entry	D_2LB pK_H	D_2REA $p[A]_{50}$ (IA)	β_2GPT $p[A]_{50}$ (IA)	α_1REA $p[A]_{50}$ (IA)
3c	9.02	$\sim 7.8^a$	7.89 (0.40)	7.56 (0.96)
4a	7.09	-	8.42 (0.46)	7.00 (0.0)
4b	8.27	-	7.01 (0.38)	5.50 (0.0)
5a	8.21	7.41 (0.86)	7.59 (0.47)	6.88 (0.12)
5b	8.61	8.94 (0.90)	7.95 (0.69)	6.08 (0.08)
5c	9.06	8.76 (0.79)	7.39 (0.46)	6.74 (0.06)
6	-	-	7.52 (0.61)	5.49 (0.43)

a α_1-adrenoceptor agonist activity does not allow a reliable
estimate of D_2-receptor agonist activity to be made

(see Figure 4 and Table 2). In all of these compounds α_1-adrenoceptor agonist activity is significantly weaker than that seen in **3d**. The introduction of this heteroatomic unit could have had several effects:

a) Modification of the pKa of the amino group. This effect does seem to provide some evidence for the observed changes in activity with the more weakly basic amines being weaker α_1-adrenoceptor agonists (i.e., **5a, 5b, 5c < 3c**).

b) Introduction of a sterically/electronically demanding substituent. This also has some merit as a rationalisation when the agonist-receptor interactions of this class of compounds are investigated using models of the transmembrane region of the D_2-receptor and β_2-adrenoceptor vs the α_1-adrenoceptor. In our models of these receptors the central portion of the $(CH_2)_6$ moiety is close to helix 7 (see Figure 5). In this region the D_2-receptor and the β_2-adrenoceptor contain relatively hydrophilic amino acids (D_2 - Thr 412, β_2 - Asn 312) whereas the α_1-adrenoceptor contains the hydrophobic Phe 312. Having few steric or electronic requirements it might be expected that **3c** can interact reasonably well with all three receptor types. In the case of **5b** the introduction of the sulphone unit could have reduced the productive interactions in the α_1-adrenoceptor. This could be for steric reasons - the increased size of Phe 312 in the α_1-adrenoceptor vs the relatively smaller Thr 412 (D_2-receptor) or Asn 312 (β_2-adrenoceptor). Alternatively, electronic factors could be important - the binding of the sulphone in the relatively hydrophobic environment of the α_1-adrenoceptor created by Phe 312 being less favourable than that at the more hydrophilic D_2-receptor (Thr 412) and β_2-adrenoceptor (Asn 312). Experiments need to be performed to see if these interactions can be verified, e.g., the binding of compounds such as **3c** and **5b** at receptors containing the appropriate point mutations.

Compounds **5a**, **5b** and **5c** all satisfy the original criteria in that they possess dual D_2-receptor and β_2-adrenoceptor agonist activity (albeit with different D_2 : β_2 ratios, see below)

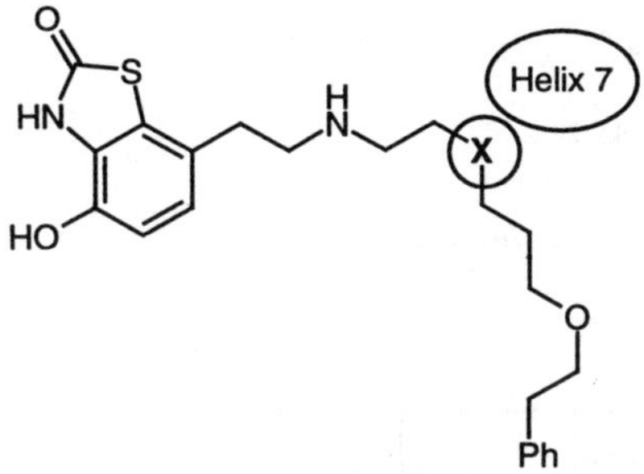

Receptor	D_2	β_2	α_1
Helix 7	Thr 412	Asn 312	Phe 312
X = CH₂ (3c)	±	±	+
X = SO₂ (5b)	+	+	-

Figure 5 *Reduction in α_1-adrenoceptor activity: agonist-receptor interactions*

and they are as potent as salmeterol (**6**). An important structural feature of this series of benzothiazolone derivatives is that they are achiral. Even though they do not contain the benzylic hydroxy group present in the phenylethanolamine component of established β_2-adrenoceptor agonists (e.g. salbutamol and salmeterol), they are still potent β_2-adrenoceptor agonists. This can be rationalised by considering the possible interactions between the these compounds and a model of the β_2-adrenoceptor (see Figure 6). The phenol and the thiazolone NH can act as a catechol isostere and interact with Ser 204 and Ser 207 on helix 5. The amino group would form the expected ionic interaction with Asp 113 on helix 3. Additional interactions are possible: the benzothiazolone sulphur atom with Tyr 308 on helix 7, the benzothiazolone carbonyl group with Asn 293 on helix 6 and the ether with Ser 120 on helix 3 and/or Asn 316 on helix 7. Point mutation studies could help to examine this hypothesis.

5 *IN VITRO* D_2 : β_2 RATIO

The studies described so far also provided information that indicated how the D_2 : β_2 ratio can be controlled. This is exemplified by the compounds shown in Table 3 (see also Figure 7):

a) **7** - a potent selective β_2-adrenoceptor agonist (β_2GPT: p[A]$_{50}$ 8.86, IA 0.35); the gem-dimethylation has reduced α_1-adrenoceptor activity and the addition of a 4-amino group in the distal phenyl group enhanced β_2-adrenoceptor activity,

b) **5a-c** - more subtle changes to the (CH$_2$)$_6$ moiety have provided a range of compounds with *in vitro* D$_2$: β_2 ratios in the 1.5 - 0.04 range,

c) **8** - a potent, selective D$_2$-receptor agonist (D$_2$REA: p[A]$_{50}$ 9.68, IA 0.93); given the D$_2$-receptor selectivity seen with the tertiary amine (**2**), it was expected that alkylation of the ethylamine nitrogen atom would reduce β_2-adrenoceptor activity.

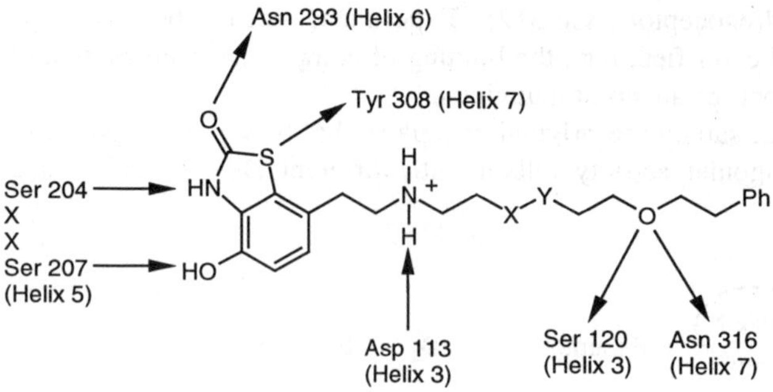

Figure 6 *β_2-adrenoceptor agonist activity: agonist-receptor interactions*

Table 3 *In Vitro D$_2$: β_2 Ratios of **5a-c**, **7** and **8***

Entry	D$_2$REA p[A]$_{50}$ (IA)	β_2GPT p[A]$_{50}$ (IA)	D$_2$: β_2
5a	7.41 (0.86)	7.59 (0.47)	1.5
5b	8.94 (0.90)	7.95 (0.69)	0.1
5c	8.76 (0.79)	7.39 (0.46)	0.04
7	NA	8.86 (0.35)	-
8	9.68 (0.93)	NA	-

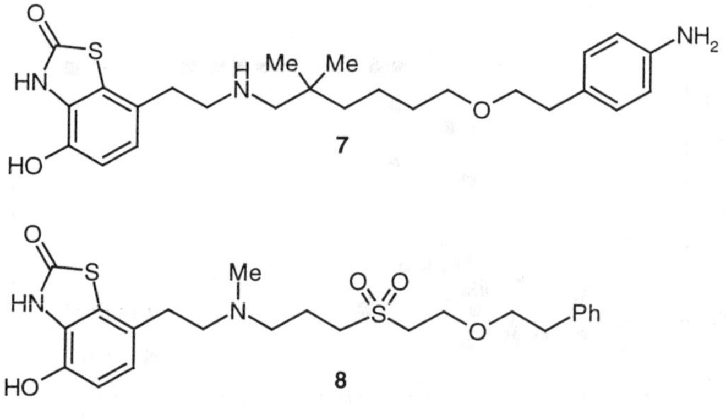

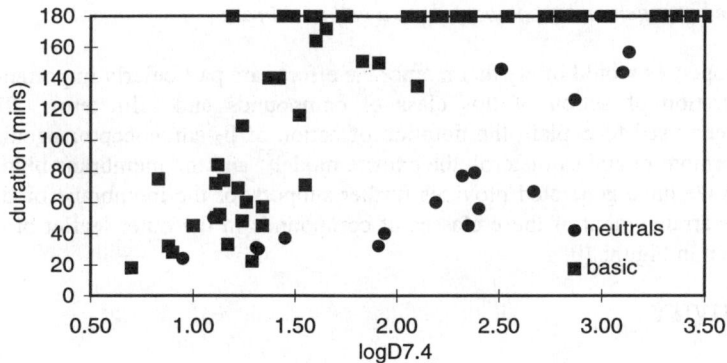

Figure 7 *Structures of Compounds 7 and 8*

6 DURATION OF ACTION

Salmeterol is a long-acting β₂-adrenoceptor agonist both *in vitro* and *in vivo*[23] and studies that have compared salmeterol with salbutamol and formoterol have provided some useful insights into the SAR for duration of action.[24] The guinea-pig superfused isolated trachea has been used as the *in vitro* assay system for duration of action of β₂-adrenoceptor agonist activity.[25] Using the compounds that we have synthesised there appears to be a poor correlation between *in vitro* duration of action and logD₇.₄. However, separating the compounds into those that have a pKa > 7.4 (and would be expected to be significantly charged at physiological pH) and those that have a pKa < 7.4, two distinct groups emerged that do show SARs with LogD₇.₄, see Figure 8. It is significant that these different SARs merged (see Figure 9) when the *in vitro* duration of action was correlated with an index of membrane binding (LogK$_{iam}$, binding to an immobilised artificial membrane - Regis

Figure 8 *SAR of neutral and basic amines between in vitro β₂-adrenoceptor agonist duration of action and logD₇.₄*

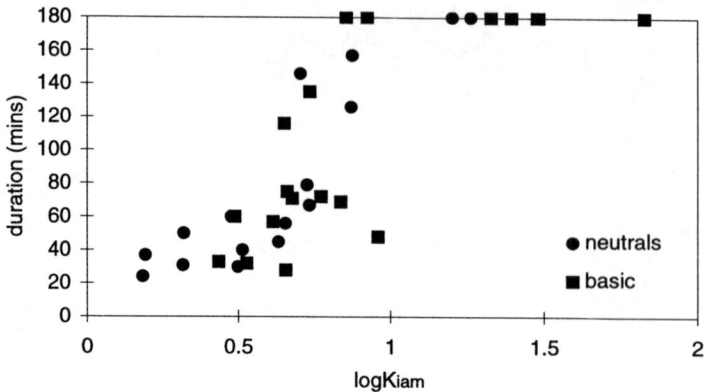

Figure 9 *SAR between in vitro β₂-adrenoceptor agonist duration of action and logKᵢₐₘ*

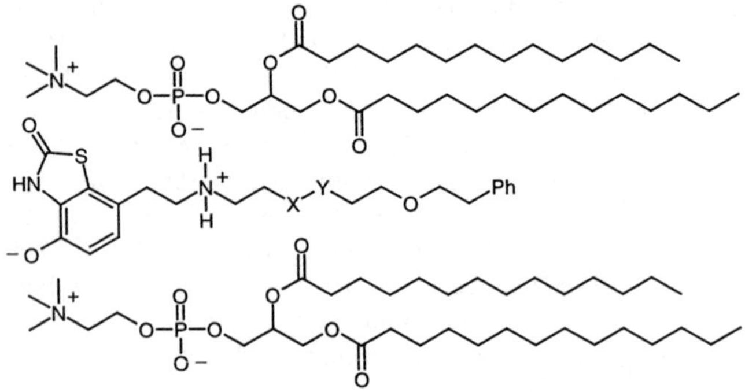

Figure 10 *Possible arrangement of compounds in a cell membrane*

column). These properties would imply that membrane effects are particularly important in controlling the duration of action of this class of compounds and salmeterol. Two hypotheses have been used to explain the duration of action of β₂-adrenoceptor agonists, e.g., salbutamol, formoterol and salmeterol, the exosite model[23] and the membrane binding model.[24] The data we have generated provides further support for the membrane binding model. A possible arrangement of these classes of compounds in the outer leaflet of cell membranes is shown in Figure 10.

7 *IN VIVO* ACTIVITY

The anaesthetised dog has been used to assess the *in vivo* β₂-adrenoceptor and D₂-receptor activity of selected compounds. The reversal of histamine-induced bronchoconstriction following the topical administration of a nebulised aerosol of the test compound being studied was used to estimate β₂-adrenoceptor agonist activity (ED_{50} (µg/kg), propranolol was used as the β-adrenoceptor antagonist to confirm β₂-adrenoceptor agonist activity).[26]

D_2-receptor agonist activity was estimated following the topical administration of a nebulised aerosol of the test compound using the histamine-induced tachypnoea model in the propranolol-treated dog (ED_{50} (µg/kg), domperidone was used as the peripheral D_2-receptor antagonist to confirm D_2-receptor agonist activity).[27] Compounds **5a-c** are potent agonists in these *in vivo* models and provide compounds with D_2 : β_2 ratios in the 10 - 0.1 range (Table 4). **5b** satisfies our objective in that it is a dual D_2-receptor and β_2-adrenoceptor agonist where D_2 = β_2 (D_2 : β_2 ratio of 1.3) where the β_2-adrenoceptor activity is greater than that of salmeterol (**6**). The *in vivo* duration of action of the β_2-adrenoceptor effect has also been evaluated in the histamine-induced bronchoconstriction model in the anaesthetised dog, where an ED_{50} dose of **5a-c** and **6** produced anti-bronchoconstictor activity for >6 hours.

5b (AR-C68397AA, hydrochloride salt) was selected for a more detailed *in vivo* evaluation. When given by nebulised aerosol to the conscious dog **5b** induces emesis at a dose of 29 µg/kg. This compares with an ED_{50} of 0.74 µg/kg in the histamine-induced tachypnoea in the anaesthetised dog and affords an encouraging 'therapeutic ratio' of 39. In addition to its ability to reduce histamine-induced tachypnoea in anaesthetised dog **5b** has been evaluated in additional D_2-receptor models that are relevant to diseases of the airways such as COPD and asthma. **5b** (ED_{50} ~1 µg/kg) affords a reduction in ammonia-induced mucus production in the anaesthetised dog[28], a reduction of capsaicin-induced

Table 4 *In Vivo Biological Activity and D_2 : β_2 Ratios of 5a-c and salmeterol (6)*

Entry	D_2 ED_{50} (µg/kg)	β_2 ED_{50} (µg/kg)	D_2 : β_2
5a	5.0	0.80	6.4
5b	0.74	0.58	1.3
5c	1.3	13	0.1
6	-	15	-

neurogenic inflammation in the anaethetised rat[29] and a reduction in capsaicin-induced cough in the conscious dog.[30] This spectrum of *in vivo* activities makes **5b** an attractive candidate for further development and it is showing promising activity in Phase II clinical studies in COPD.

8 CONCLUSIONS

We have discovered of family of novel, achiral, dual D_2-receptor and β_2-adrenoceptor agonists that are able to modulate the hyperreactivity of pulmonary sensory nerves and also possess anti-bronchoconstrictor activity. An inappropriate level of α_1-adrenoceptor activity present in the early compounds has been reduced whilst maintaining the required dual D_2-receptor and β_2-adrenoceptor agonist activity, changes in activity that can be rationalised by consideration of agonist-receptor binding interactions. It is possible to manipulate the ratio of D_2 : β_2 activity both *in vitro* and *in vivo*. Duration of action can be controlled and would appear to support the membrane binding hypothesis.

The most promising compound (AR-C68397AA, **5b**) is a potent dual D_2-receptor and β_2-adrenoceptor agonist when evaluated in animal models that are relevant to diseases of the airways such as COPD and asthma. AR-C68397AA has been selected for further development as a potential treatment for COPD and asthma and it is showing promising activity in Phase II clinical studies in COPD.

Acknowledgements

It is a pleasure to be able to acknowledge the contributions from the members of the following Astra Charnwood Research Departments. Medicinal Chemistry: Roger Bonnert, Roger Brown, Pete Cage, Dave Chapman, Dave Cheshire, Larry Coe, John Dixon, Liz Kinchin, Amanda Lyons, Garry Pairaudeau, Anil Patel; Physical and Metabolic Sciences: Rupert Austin, Pat Barton, Andy Davis, Darren Flower, Nigel Gensmantel, Carol Manners, Ian Beattie, Anne Cooper, Neil Entwistle, Andy Gray, Craig Lambert, Graeme Moody, Heather Seddon; Biochemistry: Kay Hallam, Steve Harper, Hemant Mistry, Alan Wallace; Pharmacology: Alan Blackham, Iain Dougall, Malbinder Fagura, Diane Harper, Dale Jackson, Simon Lydford, Ken McKechnie, Shahad Mohammed, Caroline Taylor, Keith Vendy, Carol Weyman-Jones, Alan Young; Wilfred Simpson (deceased).

References

1. B. Sibbald, In 'Asthma Basic Mechanisms and Clinical Management', P. J. Barnes, I. W. Rodger, and N. C. Thomson (Eds.); Academic Press: London, 1992, p. 21. J. Du Toit, A. J. Woolcock, and C. M. Salome. *Am. Rev. Resp. Dis.*, 1986, **134**, 498.
2. J. G. Widdicombe and M. Fillenz, In 'Enteroceptors', E. Neil (Ed.); Springer Verlag: Berlin, 1972, p. 81.
3. G. P. Andersen, In 'New Drugs for Asthma Therapy', *Agents and Actions Supplement* **34**. G. P. Anderson, I. D. Chapman and J. Morley, (Eds.); Birkhauser Verlag: Basel, 1991, p. 97.
4. P. J. Barnes, *Pulmonary Pharmacol.*, 1997, **10**, 3.
5. P. J. Barnes, In 'Asthma Basic Mechanisms and Clinical Management', P. J. Barnes, I. W. Rodger and N. C. Thomson (Eds.); Academic Press: London, 1992, p. 359. J. G. Widdicombe. *Agents and Actions - Supplements*, 1989, **28**, 213.
6. M. Ilhan and J. P. Long, *Arch. Int. Pharmacodyn.*, 1975, **216**, 4.
7. S. R. Sampson, M. J. Aminoff, R. A. Jaffe and E. H. Vidruk, *Am. J. Physiol.*, 1976, **230**, 1494.
8. C. Missale, S. R. Nash, S. W. Robinson, M. Jaber and M. G. Caron, *Physiol. Rev.*, 1998, **78**, 189. P. G. Strange, *Adv. Drug Res*, 1996, **28**, 313.
9. A. J. Lawrence, E. Krstew and B. Jarrott, *Br. J. Pharmacol.*, 1995, **114**, 1329.
10. G. X. Xie, K. Jones, S. J. Peroutka and P. P. Palmer, *Brain Research*, 1998, **785**, 129.
11. H. G. Jeffs, *Br. Med. J.*, 1974, **5905**, 454.
12. D. M. Jackson, unpublished observations.
13. D. R. Sibley, A. De Lean and I. Creese, *J. Biol. Chem.* 1982, **257**, 6351.
14. J. P. Heible, S. H. Nelson and O. S. Steinsland, *J. Autonom. Pharmacol.*, 1985, **5**, 115.
15. I. G. Dougall, D. Harper, D. M. Jackson and P. Leff, *Br. J. Pharmacol.*, 1991, **104**, 1057.
16. M. S. Fagura, S. J. Lydford and I. G. Dougall, *Br. J. Pharmacol.*, 1997, **120**, 247.

17. B. G. Main, In 'Comprehensive Medicinal Chemistry', C. Hansch, P. G. Sammes and J. B. Taylor (Eds.), Pergamon Press, Oxford, 1990, Vol. 3, p. 187.
18. R. A. Brown, J. Dixon, J. B. Farmer, C. Hall, R. G. Humphries, F. Ince, S. E. O'Connor, W. T. Simpson and G. W. Smith, *Br. J. Pharmacol.*, 1985, **85**, 599.
19. J. Weinstock, D. E. Gaitanopoulus, O. D. Stringer, R. G. Franz, J. P. Heible, L. B. Kinter, W. A. Mann, K. E. Flaim and G. Gessner, *J. Med. Chem.*, 1987, **30**, 1166.
20. P. D. Snashall, F. A. Boother and G. A. Sterling, *Clin. Sci. Mol. Med.*, 1978, **54**, 283.
21. A. L. Hyman, H. L. Lippton and P. J. Kadowitz, *Fed. Proc.*, 1986, **45**, 2336.
22. P. Pratesi and E. Grana, *Adv. Drug Res.*, 1965, **2**, 127.
22. M. Johnson, *Med. Res. Rev.*, 1995, **15**, 225.
23. G. P. Anderson, A. Linden and K. F. Raber, *Eur. Resp. J.*, 1994, **7**, 569.
24. R. A. Coleman and A. T. Nials, *J. Pharmacol. Methods*, 1989, **21**, 71
25. A. Young, A. Blackham and C. V. Taylor, unpublished observations.
26. A. Young and C. V. Taylor, unpublished observations. See also: R. J. Phipps and P. S. Richardson, *J. Physiol.*, 1976, **261**, 563.
27. A. Young, A. Blackham, C. V. Taylor, K. Vendy, S. T. Harper, H Mistry and C. Hallam, unpublished observations.
28. A. Young and C. Weyman-Jones, unpublished observations.
29. D. M. Jackson, unpublished observations.

Growth Factors

DISCOVERY OF A SMALL, NON-PEPTIDYL MIMIC OF GRANULOCYTE-COLONY STIMULATING FACTOR

J. I. Luengo[†], S.-S. Tian[‡], P. Lamb[‡], A. G. King[†], S. Miller[‡], L. Kessler[‡], L. Averill[†], R. K. Johnson[†], J. G. Gleason[†], S. B. Dillon[†] and J. Rosen[‡]

[†]SmithKline Beecham Pharmaceuticals, 1250 South Collegeville Road, Collegeville, PA 19426
[‡] Ligand Pharmaceuticals, 10255 Science Center Drive, San Diego, CA 92121

Granulocyte colony-stimulating factor (G-CSF) is a 21-kDa hematopoietic cytokine secreted by bone marrow stroma cells, macrophages, fibroblasts and endothelial cells. G-CSF plays a key role in the proliferation and differentiation of granulocyte precursors as well as the activation of mature neutrophils.[1,2] Recombinant human G-CSF, available in both glycosylated and non-glycosylated forms, has become an important therapeutic agent for the treatment of a variety of human neutropenias, including those resulting from chemotherapy, congenital defects and bone marrow transplantation.[3] The biological effects of G-CSF are mediated through a specific transmembrane receptor,[4] which belongs to the type I cytokine receptor superfamily[5] and appears to consist of a single polypeptide chain. The extracellular region of the G-CSF receptor has a modular structure containing an Ig-like domain, a cytokine receptor homologous region, and three fibronectin type III repeats.[6] Binding of G-CSF to its receptor results in receptor homodimerization and leads to the rapid activation and tyrosine phosphorylation of JAK1 and JAK2, two members of the JAK kinase family which are associated with the cytoplasmic domain of the receptor.[7] Activated JAKs phosphorylate tyrosine residues on the cytoplasmic domain of the receptor, which then serve as the binding site for a variety of signaling proteins. The JAKs are then presumed to phosphorylate the receptor-associated proteins, among which are the STATs (signal transducers and activators of transcription). Upon phosphorylation on a C-terminal tyrosine, the STATs dimerize and translocate to the nucleus, where they regulate transcription through binding to specific promoter sequences in target genes.[8,9] Binding of G-CSF to its receptor also promotes the tyrosine phosphorylation of SCH, an adapter protein involved in activation of the Ras/MAP kinase signaling pathway.[10]

Genetically engineered G-CSF, like any other recombinant growth factors, must be administered either subcutaneously or intravenously. Although other agents, such as bivalent receptor antibodies[11,12] and dimeric peptides,[13,14] have been shown to activate cytokine receptors by oligomerization, no small-molecule cytokine mimics, with potential for oral delivery, have been previously reported.

We designed an assay to identify non-peptidyl compounds that activate the G-CSF receptor based on activation of STATs, which are known to play a central role in the G-CSF-mediated responses. Our high-throughput, cell-based screen[15] relied on a G-CSF-responsive reporter gene that was stably transfected into the murine myeloid cell line NFS60. The reporter construct consisted of four copies of a synthetic STAT-binding element linked to a minimal promoter and the gene for luciferase and was transfected into an IL-3 independent variant of the NFS60 cell line, together with a plasmid conferring resistance to neomycin. From the drug resistant clones responsive to G-CSF, a single clone, which exhibited a 20-fold induction of luciferase activity by G-CSF and same pattern of JAK and STAT activation as the parental cells, was selected to screen a library of synthetic organic compounds. In the screen, the cells were incubated for 2.5 hours with individual compounds at a concentration of 10 µM in a 96 well plate format. Compound SB-247464 (Figure 1) was identified as a hit in the assay and showed a dose-response effect with maximum efficacy of 30% that of G-CSF at 1 µM. Activity of SB-247464 was specific for the G-CSF receptor, as the compound had no effect on a second NFS60-based stable cell line, RSVluc, with stably integrated copies of a reporter plasmid that produces luciferase constitutively; as expected, the luciferase levels in this line were not affected by G-CSF either. Likewise, SB-247464 showed no activity on stable cell lines containing STAT responsive reporters that induced luciferase in response to either erythropoietin, interferon α or interferon γ.

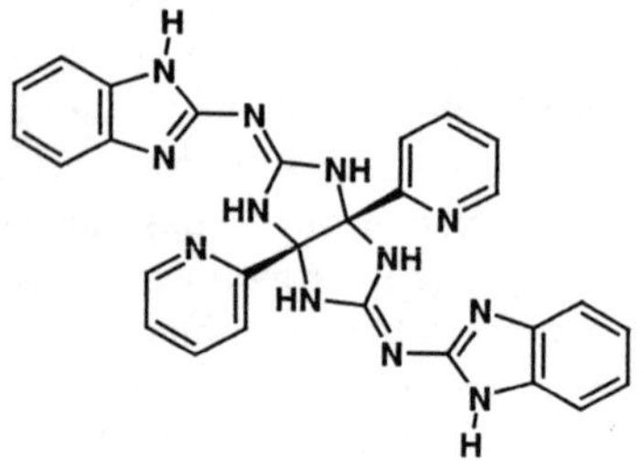

SB-247464

Figure 1 *Structure of SB-247464; the molecule is arbitrarily shown in the C_2-symmetrical configuration, with the exocyclic guanidine C-N bonds in a trans stereochemistry*

As expected from the STAT-based primary screen, SB-247464 induced activation of G-CSF signal transduction pathways. Thus treatment of NFS60 cells with SB-247464 resulted in rapid tyrosine phosphorylation of the G-CSF receptor, JAK1, JAK2, SHC, STAT3 and STAT5, with a similar kinetic profile to that of G-CSF. Efficacy of SB-247464 in these assays was approximately 25-50% that of G-CSF, consistent with the data from the luciferase assay. The action of SB-247464 is not restricted to proteins directly involved in STAT activation, as the adapter protein SHC, involved in activation of the MAP kinase pathway, was also phosphorylated in the NFS60 cells. Activation of signaling events was specific, since no phosphorylation of the β-chain of the IL3 receptor by SB-247464 could be detected. Likewise, activation was dependent on the presence of the murine G-CSF receptor;[16] thus expression of the receptor into non-responsive cells, such as HepG2 and UT7Epo lines, was sufficient to confer sensitivity to SB-247464, as demonstrated by a STAT-responsive luciferase reporter as well as the induction of STAT-DNA complexes.

To assess the activity of SB-247464 in supporting the proliferation and differentiation of cells of the granulocytic lineage, we tested the compound in colony-forming unit-granulocyte (CFU-G) assays from murine bone marrow.[17] SB-247464 stimulated the production of granulocytic colonies, with an efficacy of 20-80% of that of G-CSF at 0.3-3 μM; the colonies appeared uniformly smaller than those promoted by G-CSF, but were consistently larger than 30 cells. Likewise, SB-247464 was able to mimic the activity of G-CSF *in vivo*. Thus subcutaneous administration of SB-247464 twice a day to normal mice caused a dose-dependent increase in peripheral blood neutrophils after 4 days. Efficacy of SB-247464 at 30 mg/kg was comparable to that of 50 μg/kg of G-CSF, elevating neutrophil counts approximately 400% over baseline. The magnitude of the increase was equivalent to that effected by administration of 5-30 μg/kg/day of G-CSF to normal or neutropenic humans.

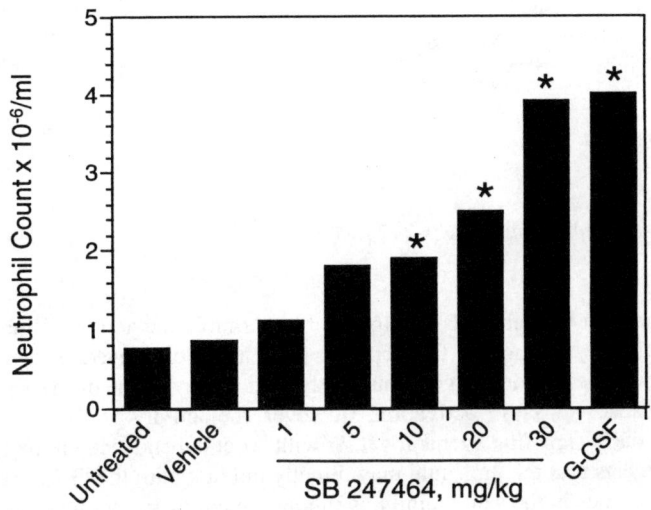

Figure 2 *Granulopoietic activity of SB-247464 in vivo. Female BDF-1 mice were given subcutaneous injections twice daily of either G-CSF (50 □g/kg) or SB-247464 Control animal received only water. After 4 days, neutrophil count was estimated by a Technicon analyzer. Each bar represents the average of five mice. Asterisks indicate neutrophil counts above those of untreated control with a P < 0.001 by analysis of variance*

Synthesis of SB-247464 was based on the condensation of 2,2'-pyridil with 2-guanidinobenzimidazole in the presence of base. The reaction proceeded through the isolable dihydroimidazole **1**, which, depending on the reaction conditions, underwent pinacol rearrangement to afford imidazolinone **2** or condensation with a second equivalent of guanidine to provide the desired SB-247464. Imidazolinones related to **2** were formed exclusively in the reaction with aliphatic guanidines.[18]

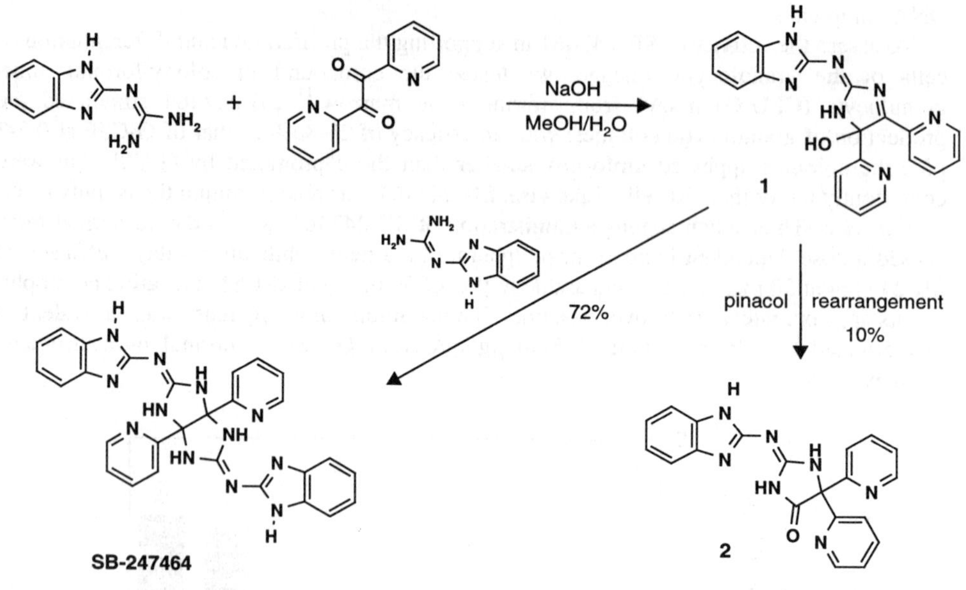

Figure 3 *Synthetic route to SB-247464*

The mechanism by which SB-247464 is able to mimic the action of the protein cytokine G-CSF is currently unknown. The reporter gene assay employed in the high-throughput screen was designed to detect compounds that act at any point in the signal transduction pathway that leads to STAT activation. However, the activity profile of SB-247464, rapid activation of early signaling events together with strict dependence on the expression of the receptor, suggests that the molecule may directly interact with the G-CSF receptor. G-CSF, like other cytokines in the same family, is thought to act by triggering dimerization or high-order oligomerization of its receptor chains.[19,20] The C_2-symmetrical structure of SB-247464 is also suggestive of a mode of action involving dimerization of G-CSF receptor chains, in analogy to the C_2-symmetrical nature of the EPO mimetic peptide recently reported.[21]

The identification of SB-247464 as a G-CSF mimetic provides proof of principle for drug discovery using JAK/STAT based assays, and shows for the first time that a small non-peptidyl molecule can trigger the selective activation of a cytokine receptor. These findings may lead to the development of orally available G-CSF mimics for use in treatment of neutropenia associated with cancer chemotherapies.

References

1. P. Anderlini, D. Przepiorka, R. Champlin and M. Korbling, *Blood*, 1996, **88**, 2819.
2. G. D. Demetri and J. D. Griffin, *Blood*, 1991, **78**, 2791.
3. K. Welte, J. Gabrilove, M. H. Bronchud, E. Platzer and G. Morstyn, *Blood*, 1996, **88**, 1907.
4. A Larsen, T. Davis, B. M. Curtis, S. Gimpel, J. E. Sims, D. Cosman, L. Park, E. Sorensen, C. J. March and C. A. Smith, *J. Exp. Med.*, 1990, **172**, 1559.
5. J. A. Wells and A. M. de Vos, *Annu. Rev. Biochem.*, 1996, **65**, 609.
6. B. R. Avalos, *Blood*, 1996, **88**, 761.
7. J. N. Ihle, B. A. Whitthuhn, F. W. Quelle, K. Yamamoto and O. Silvennoinen, *Ann. Rev. Immunol.*, 1995, **13**, 369.
8. C. Schindler and J. E. Darnell, *Ann. Rev. Biochem.*, 1995, **64**, 621.
9. S.-S. Tian, P. Tapley, C. Sincich, R. B. Roein, J. Rosen and P. Lamb, *Blood*, 1996, **88**, 4435.
10. J. P. de Koning, A. M. Schelen, F. Dong, C. van Buitenen, B. M. T. Burgering, J. L. Bos, B. Lowenberg and I. P. Touw, *Blood*, 1996, **87**, 132.
11. M. Fourcin, S. Chevalier, C. Guillet, O. Robledo, J. Froger, A. Pouplard-Barthelaix and H. Gascan, *J. Biol. Chem.*, 1996, **271**, 11756.
12. T. Takahashi, M. Tanaka, J. Ogasawara, T. Suda, H. Murakami and S. Nagata, *J. Biol. Chem.*, 1996, **271**, 17555.
13. N. C. Wrighton, F. X. Farrell, R. Chang, A. K. Kashyap, F. P. Barbone, L. S. Mulcahy, D. L. Johnson, R. W. Barrett, L. K. Joliffe and W. J. Dower, *Science*, 1996, **273**, 458.
14. S. E. Cwirla, P. Balasubramanian, D. J. Duffin, C. R. Wagstrom, C. M. Gates, S. C. Singer, A. M. Davis, R. L. Tansik, L. C. Mattheakis, C. M. Boytos, P. J. Schatz, D. P. Baccanari, N. C. Wrighton, R. W. Barrett and W. J. Dower, *Science*, 1997, **276**, 1696.
15. J. Rosen, A. Day, T. K. Jones, E. T. Turner-Jones, A. M. Nadzan and R. B. Stein, *J. Med. Chem.* 1995, **38**, 4855.
16. R. Fukunaga, E. Ishizaka-Ikeda, Y. Seto and S. Nagata, *Cell*, 1990, **61**, 341.
17. D. Metcalf and N. A. Nicola, *J. Cell Physiol.*, 1983, **116**, 198.
18. T. Nishimura and K. Kitajima *J. Org. Chem.* 1979, **44**, 818.
19. T. Horan, J. Wen, L. Narhi, V. Parker, A. Garcia, T. Arakawa and J. Philo, *Biochemistry*, 1996, **35**, 4886.
20. O. Hiraoka, H. Anaguchi and Y. Ota, *FEBS Letters*, 1994, **356**, 255.
21. O. Livnah, E. A. Sture, D. L. Johnson, S. A. Middleton, S. L. Mulcahy, N. C. Wrighton, W. J. Dower, L. K. Joliffe and I. A. Wilson, *Science*, 1996, **273**, 464.

EGFR TYROSINE KINASE INHIBITORS IN THE TREATMENT OF CANCER

Andrew J. Barker

Zeneca Pharmaceuticals, Alderley Park, Macclesfield SK10 4TG, UK

1 INTRODUCTION

Conventional cytotoxic agents have efficacy against some tumours, but their toxicity limits their use and they are not particularly effective against the common human solid tumours. Advances in molecular biology have resulted in the identification of important signal transduction processes by which cells communicate with their environment and respond to growth factors.[1] Signalling proteins have proven good targets in the treatment of hormonally driven tumours and the elucidation of non-hormonal signal transduction pathways in mammalian cells presented a range of new potential target proteins for the treatment of uncontrolled cell growth in solid tumours.[2] It was also believed that as interfering with these inappropriate signalling systems should not cause cell death, but stasis, that agents which targeted these proteins would be better tolerated than conventional cytotoxic agents.

2 SIGNALLING ACROSS CELL MEMBRANES

Many of the signal transduction processes were found to act through a family of cell-membrane spanning receptors linked to an internal tyrosine kinase enzyme.[3] On interaction with the ligand the receptors dimerise and tyrosine residues close to the kinase become phosphorylated. This results in activation of the enzyme which now accepts tyrosine bearing substrates which become phosphorylated on this amino acid, the phosphate being derived from ATP (Figure 1). The phosphorylated substrates interact with other proteins and enzymes eventually "passing on" the "information" to the cell nucleus where a variety of effects, including cell division, are initiated. When this process becomes uncontrolled a tumour may result and there is good evidence to suggest that a high proportion of human solid tumours have defects in these signalling systems.[4] In particular, the signalling system acting through the epidermal growth factor receptor tyrosine kinase [EGFRTK] (Figure 2) was of particular interest as it was over-expressed in a high proportion of human solid tumours and its expression was associated with poor patient prognosis (Figure 3).[4,5]

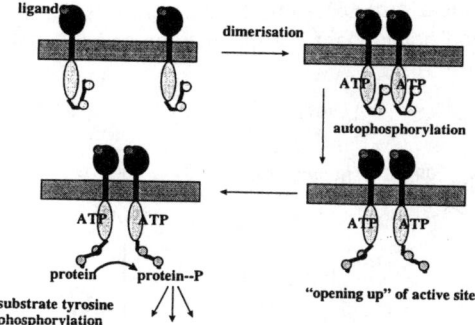

Figure 1 *Activation of Receptor Tyrosine Kinases*

Lesions in this system, such as overexpression of activating ligand, amplification of the receptor, or mutations in the receptor which lead to continuous activation are all known. This information led to the EGFRTK being considered an attractive target for intervention in uncontrolled signal transduction for the treatment of human tumours.

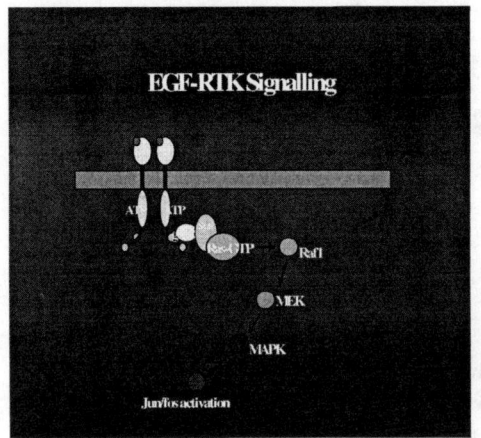

Clinical Targets

Tumours showing high EGFR expression

NSCLC	40-80%
Colorectal	25-77%
Gastric (Advanced)	33%
Pancreatic	30-50%
Ovarian	35-70%

High expression is generally associated with

• Invasion
• Metastasis
• Late-stage disease
• Poor outcome

Figure 2 *EGF-RTK Signalling*   **Figure 3** *EGF-RTK Expression*

3 SAR STUDIES

A high throughput screen (HTS) was established utilising the tyrosine kinase and a synthetic tyrosine containing substrate, the end point being incorporation of radiolabelled phosphate into the peptide substrate.[6] The screen yielded a wide variety of structures capable of inhibiting the kinase enzyme and some of these proved attractive starting points for a medicinal chemistry programme.[7] The focus fell on a group of small molecules containing a 4-anilinoquinazoline pharmacophore. These were simple to synthesise[8] (Figure 4) and exhibited clear structure-activity relationships (SAR) against the kinase. The initial lead was of good potency against the enzyme and also inhibited the growth of EGF stimulated cells (Figure 5).

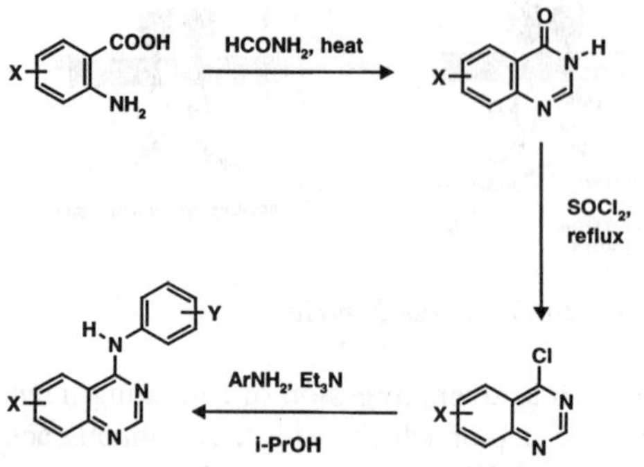

Figure 4 *Synthesis of 4-Anilinoquinazolines*

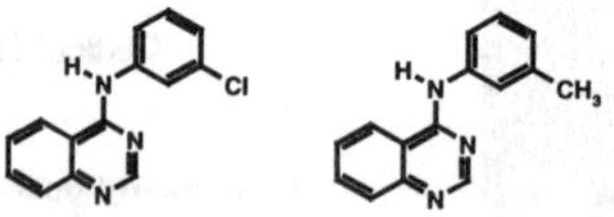

EGF-RTK IC$_{50}$	0.04 µM	0.18 µM
Inhibition of stimulated Cell Growth IC$_{50}$	1.2 µM	2.4 µM

Figure 5 *Anilinoquinazoline Leads*

A rapid investigation of the effects of substitution of the aniline and quinazoline on activity at the kinase revealed some clear messages (Figure 6).[6] On the aniline *meta*-substitution was preferred over *para*, whilst *ortho*-substituents resulted in major losses of activity. Small substituents were preferred on the aniline. Replacement of the quinazoline *N*-atoms with carbon and substitution of the aniline N-H with groups such as methyl resulted in significant loss of kinase potency suggesting that the heterocyclic nitrogen atoms were important in interacting with the kinase. The methyl substitution and *ortho*-aniline substituents caused major conformational changes in the molecule and appeared to make the preferred binding mode less accessible. Similar effects were observed when quinazoline 5-substitutions were made whilst substituents in the 8-position of the heterocycle modified the N-1 basicity and were often detrimental to kinase potency. However, groups in the 6 and 7 positions of the quinazoline were well tolerated and a wide variety of substituents gave highly active kinase inhibitors. Electron donating substituents, especially 6,7-(dimethoxy) groups, gave particularly potent compounds with IC$_{50}$s in the low nanomolar range. These agents, competitive with ATP in the active site of the kinase, also had activity against stimulated cell growth.[6]

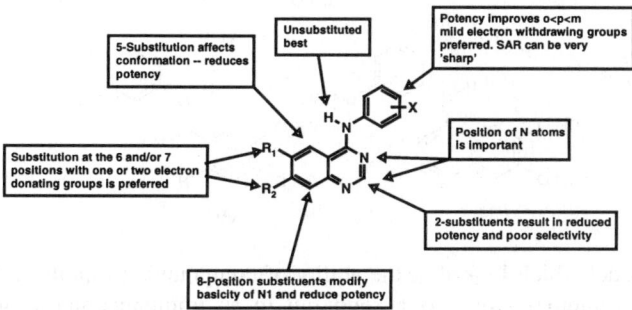

Figure 6 *Anilinoquinazoline in vitro SAR*

It was anticipated that agents inhibiting the kinase would only prevent cell growth that resulted *via* stimulation with EGF. Cell assays were established that produced an increased growth rate of tumour cells on adding exogenous EGF and these compounds consistently inhibited this "increased" growth at concentrations well-below those that began to prevent the "basal" portion (non-EGF stimulated) of cellular proliferation.[6] This was a good sign of selectivity for the kinase in question and alleviated fears that widespread inhibition of kinases might result from ATP competitive agents. In particular, the kinase associated with the insulin receptor was of concern as inhibition of signalling through this receptor would undoubtedly result in serious metabolic side effects. Experiments in similar cell systems looking at the extent of inhibition of insulin driven growth with the compounds of interest showed there to be none at concentrations well above that eliciting a full effect on EGF-driven growth. Detailed experiments looking at levels of phosphorylation of the EGFR kinase in response to EGF in the absence and presence of small molecule inhibitors also established that growth inhibition correlated well with inhibition of activation of the receptor tyrosine kinase itself.

Having potent compounds with activity against EGF-driven cell growth encouraged the testing of these agents in a model of tumour disease. However, pharmacokinetic profiling of an early molecule of interest (1) showed it to have a rapid clearance and short half life in the rat and on examining samples of blood from these animals it quickly became evident that parent compound was being metabolised to produce two new, more polar compounds. Synthesis and detailed chromatography resulted in the identification of two oxidised metabolites, *para*-hydroxy aniline (2) and benzylic alcohol (3). When the methyl group was replaced with the isosteric chlorine atom and the electron rich *para*-position of the aniline was substituted with a fluorine atom, both sites of metabolism were blocked and the pharmacokinetic profile of the new molecule (4) was markedly improved with no major metabolites now observed and extended blood levels seen on oral dosing.

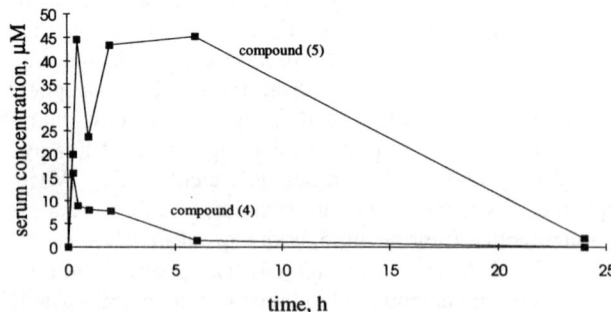

A disease model which looked at the ability of compounds to inhibit a human tumour [A431 vulval carcinoma] grown as a xenograft in an immunocompromised mouse was used to measure in vivo efficacy.[9] Two compounds (4, 5, below) with widely differing activities against the kinase target and against EGF-driven cell growth were tested in this in vivo model. The more potent compound in vitro gave 30% inhibition of the growth of the A431 tumour when dosed at 200 mg/kg orally each day.

Figure 7 *Pharmacokinetics of Compounds (4) and (5) Following 50 mg/kg p.o. Dose to the Mouse*

The second less potent agent when given at similar doses produced an 86% inhibition of tumour growth. This apparent discrepancy was quickly found to be related to the pharmacokinetic profiles of the two agents (Figure 7). The more potent compound, despite having potentially metabolically labile positions blocked, which reduced clearance, did not have blood levels extending over the 24 h period following a single dose. In contrast, the less potent agent did have good blood levels over the 24 h period from a single dose and it became apparent that the inhibitor was required to be present continually for good activity in this model. Examination of more compounds revealed that, in addition to metabolism being a cause of rapid clearance, another major factor was the lipophilicity of the molecule with less lipophilic molecules, in general, having prolonged half lives and reduced clearances (Figure 8).

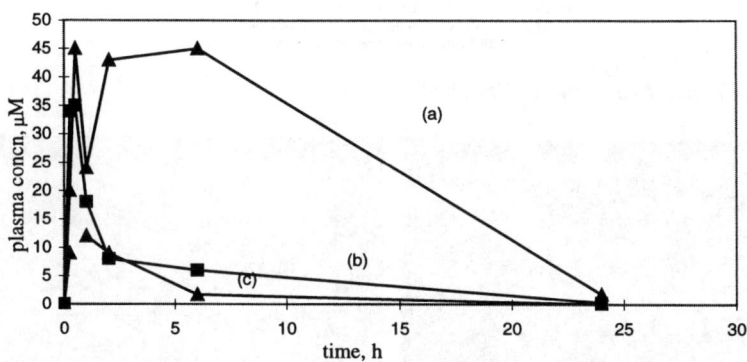

Figure 8 *Effect of Lipophilicity on Pharmacokinetics in Mouse (50 mg/kg p.o. dose)*

Through testing larger numbers of compounds in the in vivo disease model, a SAR for activity was built up. This built on the in vitro relationship with halogen substituents in the aniline reducing metabolism and increasing exposure whilst exchanging the aniline for a heterocycle resulted in increased C_{max} in the pharmacokinetic profile.[6,8] In the 6- and 7-position of the quinazoline, electron donating and polar substituents reduced lipophilicity and clearance resulting in extended blood levels on oral dosing. Following from these and other studies, ZD 1839 was identified as a compound with properties worthy of development. It was highly potent against the EGF RT kinase and whilst it had activity against other class I RTKs, it did not inhibit a variety of other kinases indicating good selectivity. Inhibition of EGF driven tumour cell growth was affected at 0.08 μM whilst effects on "basal" growth were not seen until concentrations of 3.6 μM were reached (Figure 9). The pharmacokinetic profile of ZD 1839 following oral dosing to mice and rats was excellent with good blood levels extending to late time points and these properties are reflected in the activity of the compound against the human tumour xenograft A431 in nude mice where at doses of 200 mg/kg/day p.o. the growth of the tumour was prevented as long as dosing continued (Figure 10). Removal of the compound resulted in rapid growth of the tumours, but growth could be halted and reversed if dosing was resumed. In addition, the compound had activity against advanced tumours in the same animals. In some experiments, dosing was extended for over 100 days indicating how well ZD 1839 was tolerated.

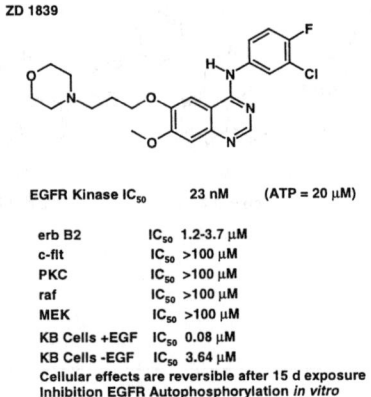

Figure 9 *in vitro Properties of ZD 1839*

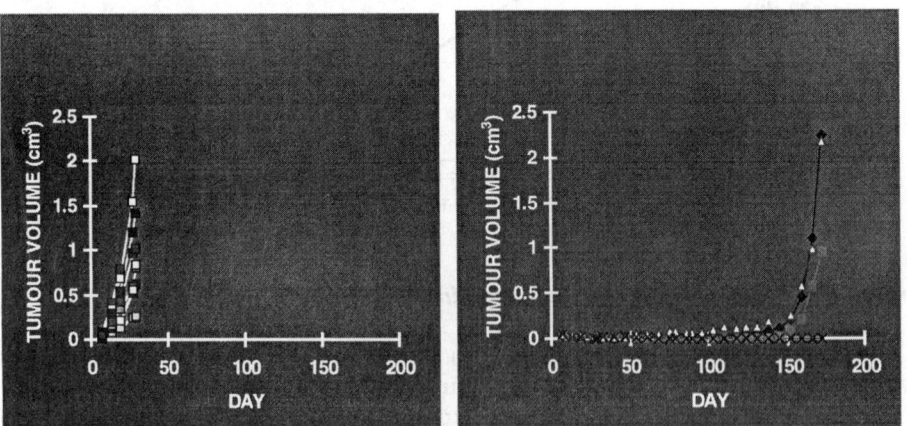

Figure 10 *Activity of ZD 1839 Against A431 Human Tumour Xenografts Grown in Nude Mice (200 mg/kg/day p.o. dose)*

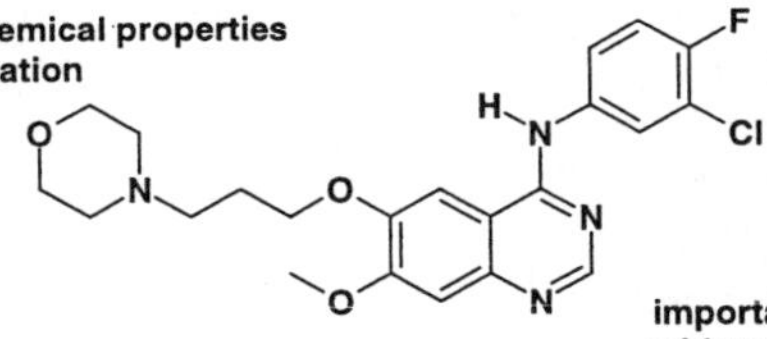

conformational 'lock'
extending aniline into
specificity pocket

physicochemical properties
modification

specificity, protection
from metabolism

important interactions
with target kinase ATP site

potency vs kinase

Figure 11 *Features of the ZD 1839 Molecule*

4 CONCLUSIONS

A HTS was established that yielded a wide variety of structures capable of inhibiting the kinase enzyme, and some of these were attractive starting points for a medicinal chemistry programme. Thus, the 4-anilinoquinazoline pharmacophore was simple to synthesise and exhibited useful SAR against the kinase. The initial lead was of good potency against the enzyme and also inhibited the growth of EGF stimulated cells. In conclusion, various parts of ZD 1839 can be seen to confer specific attributes to the compound (Figure 11), whether this be in vitro potency, protection from metabolism, or good pharmacokinetic properties. The project illustrates that small molecule kinase inhibitors competitive with ATP, but which have selectivity, can be found even though the ATP binding site of kinases is well conserved. Much of the selectivity of the compound is due to the specific conformation of the anilinoquinazolines. ZD 1839 has progressed to the clinic and results from man are eagerly awaited.

References

1. S. E. Egan and R. A. Weinberg, *Nature*, 1993, **365**, 781-783.
2. S. A. Aaronson, *Science*, 1991, **254**, 1146-1153.
3. (a) T. Hunter and J. A. Cooper, *Ann. Rev. Biochem.*, 1985, **54**, 897-930; (b) D. L. Cadena and G. N. Gill, *FASEB J.*, 1992, **6**, 2332-2337.
4. E. M. Dobrusin and D. W. Fry, *Ann. Rep. Med. Chem.*, 1993, **27**, 169-178; (b) A. F. Wilks, *Adv. Cancer Res.*, 1993, **60**, 43-73.
5. A. Pandelia, L. Beguinot, L. M. Vicentini and J. Meldolesi, *TiPS*, 1989, 411-414.
6. A. E. Wakeling, A. J. Barker, D. H. Davies, D. S. Brown, L. R. Green, S. A. Cartlidge and J. R. Woodburn, *Breast Cancer Research and Treatment*, 1996, **38**, 67-73.
7. P. Traxler and N. Lydon, *Drugs Fut.*, 1995, **20**, 1261-1274.
8. K. H. Gibson, W. Grundy, A. A. Godfrey, J. R. Woodburn, S. E. Ashton, B. J. Curry, L. Scarlett, A. J. Barker and D. S. Brown, *Bioorg. Med. Chem. Lett.*, 1997, **7**, 2723-2728.
9. A. E. Wakeling, A. J. Barker, D. H. Davies, D. S. Brown, L. R. Green, S. A. Cartlidge and J. R. Woodburn, *Endocrine Related Cancer*, 1997, **4**, 351-355.

Intracellular Signalling

THE INERADICABLE IMPACT OF IRREVERSIBLE INHIBITORS: CAN ERBBICIDE CONTRIBUTE TO THE TUMOURICIDE OF EGFR-DEPENDENT CANCERS?

A. J. Bridges, W A. Denny,* E. M. Dobrusin, A. M. Doherty, W. L. Elliott, D. W. Fry, K. Hook, W. R. Leopold, D. J. McNamara, J. W. Nelson, B. D. Palmer,* S. Patmore, G. W. Rewcastle,* B. J. Roberts, H. D. H. Showalter, V. Slintak, J. B. Smaill,* A. M. Thompson,* S. Trumpp-Kallmeyer, P W. Vincent, R. T. Winters and H. Zhou.

Parke-Davis Pharmaceutical Research, Division of Warner-Lambert Corporation, 2800 Plymouth Rd., Ann Arbor, MI 48176 and *Cancer Society Research Laboratory, University of Auckland Medical School, Auckland, New Zealand.

1 INTRODUCTION

Cancer has been recognized from ancient times as one of the most relentless of diseases, leading almost inexorably to wasting away and a painful death. The rise of modern medicine has ironically appeared to make cancer more dreaded, as it has risen dramatically to become a leading cause of death as other diseases have been conquered, and the population has grown older, and thus more susceptible to the ravages of cancer. Improved diagnosis seems to have had a major effect of emphasizing how many people actually get the disease and then die of it. The first generation of treatments has made something of a dent in the disease, but not enough to stop the rising death rates, and these treatments have all been associated with low efficacy and severe toxicities that make the low probability cure often seem as bad as the inevitable disease. In the USA the death rate from cancer has increased by 1.4% a year for the last 10 years, and in 1997 560,000 people died in the USA of cancer. [1] As many billions of dollars have been spent in the highly publicized war on cancer, the mood of the public has recently been very pessimistic, and even in 1998 public pronouncements have been made, conceding defeat. However, the mood of those involved in researching new cures has become increasingly upbeat over the last few years, and this excitement and optimism is now beginning to reach the public debate, albeit in a somewhat less guarded form than in the research community.

The source of the optimism has not until recently been based on any clinical results, although these are just beginning to glimmer on the horizon, but from an increase in the fundamental understanding of the molecular basis of the disease. Twenty years ago it was obvious that cancer cells were picking up several mutations and then proliferating excessively, and then frequently going on a rampage through the body, but there was no understanding as to why the mutations led to these phenomena. The treatments involved either excision, which frequently does not get remove all of the tumour, so it, or its too frequent satellite tumours would regrow, or chemical or radiation regimes which were especially toxic to rapidly dividing cells. As the cell was a black box, such agents were found largely by chance, and were usually associated with severe toxicities, as there are bystander cells which must divide quickly in the healthy body. Furthermore, some tumours, for example prostatic carcinomas, tended to be almost immune to the toxic effects of the cytotoxic agents. In retrospect this is not surprising as these tumours tend to grow more slowly than most normal cell types, and their principal defects seem to be in cells not dying when they should, and migrating to sites where they should not be.

A careful study of the cellular biochemistry of tumour cells has ironically led to a considerable, but still very incomplete, understanding of how normal cells make the decision to proliferate, and has found much of the cellular machinery which allows them to replicate themselves. This understanding is, on reflection, not paradoxically gained because the genes identified as mutational targets in cancers often occupy crucial roles in the cellular reproductive machinery. If the genes produce proteins which push cells into proliferation, called oncogenes, they tend to be found in a form or excessive amounts which makes them more active than they should be. If the genes produce proteins which prevent proliferation, they are called tumour suppressor genes, and are often found inactivated by mutations or completely deleted in tumours. As cellular proliferation is an incredibly complex operation, and has many checks and balances to ensure that it occurs whenever needed, and never when not needed, it usually requires mutations in several key proteins to allow a tumour cell to proliferate uncontrollably. Interestingly, many of the mutations, which encourage unchecked proliferation, also allow normally sedentary cells to become mobile and invade other tissues. Our new knowledge of this process, incomplete as it is, allows one to rationally plan new anticancer strategies. If it is overactive, and enhances proliferation inappropriately, block its action, and if it is inactivated, but normally arrests cellular proliferation, restore and enhance its function. This talk describes one such approach, which has already advanced to clinical trials elsewhere.

2   CHOICE OF TARGET FOR CELLULAR PROLIFERATION

Even two decades ago, much of the debate on cellular proliferation centred on questions such as whether cells were natural proliferators, normally externally restrained from so doing, or naturally comatose, requiring goading into proliferation by external sources. Today such a debate would seem to be the equivalent of asking whether a car is stopped by red lights or forced to accelerate by green ones. The role of externalities was correctly recognized, but the rest simply reflected a complete lack of knowledge of the mechanics of the system. The incredible progress in protein identification and gene discovery has allowed us to find many of the components of the control system for cellular proliferation, and now we are trying to construct models of the proliferation engine, and find out how it actually runs. What signals are transduced, and how are they acted upon? Once one is in the position to ask such questions the philosophical points have a habit of receding in importance.

Normal cells do not reproduce unless instructed to by mitogenic signals, which are secreted by *other* cells under almost all circumstances. Among the commonest and most potent of mitogens are the peptidic growth factors, such as Epidermal Growth Factor (EGF), Platelet-derived Growth Factor (PDGF) and Insulin-like Growth Factor I (IGF-1). These and many other growth factors bind to ligand binding domains, with high affinity (10^{-10}-10^{-8} M) on the N-terminal regions of membrane-spanning cognate receptors on the target cell. This causes the receptors to dimerize, and leads to activation of hitherto inactive tyrosine kinase (TK) moieties in the C-terminus of the receptors on the cytoplasmic side of the cell.[2] The tyrosine kinase moiety transfers the γ-phosphate from ATP to the tyrosine hydroxyls of cytoplasmic proteins with appropriate recognition sequences around the tyrosine. A major target of these receptor tyrosine kinases (RTKs) is tyrosines in their own C-terminal sequences, and quite frequently also in their own catalytic domains, (autophosphorylation). There appear to be two major consequences of this tyrosine phosphorylation. Proteins can have hitherto cryptic enzyme activity revealed, turning on a signaling pathway directly, as appears to happen for example with PLC γ. Alternatively, these newly phosphorylated tyrosines can bind to specialized phosphotyrosine-recognising modular domains on other proteins, and change their cellular localization as a result. For example the Grb-2/SOS complex binds via the Grb-2 SH2 domain to phosphotyrosines on the C-termini of RTKs, and once localized at the plasma membrane turns on the Ras pathway. Combinations of both of these processes are also common. Although only 1% of total protein phosphorylation is on tyrosine, it is clear that this phosphorylation, especially in the periplasmic region, leads to the activation of most of the mitogenic signal transduction pathways into the nucleus.[3]

In the hunt for oncogenes, TKs have turned up with greater frequency than any other class of protein. In fact the first and second oncogenes identified, v-Src and v-erbB, were from tumour viruses, and turned out to be TKs, derived from cellular TKs, in such a form that they had become intrinsically active. The RTKs which act as receptors for EGF[4] PDGF and IGF-1 have all been implicated in cancer, the latter recently being suggested to be predictive for prostate cancer.[5] Abnormally heavy protein phosphorylation, especially on tyrosine is found in most, if not all tumours. Because kinase overactivity appears to be an integral part of the oncogenic mechanism of transforming TKs, there has been great interest in finding ways of inhibiting TK activity, in the expectation that the pathways activated by the TKs would be closed down.[6] From our current knowledge of cellular signaling, we would expect such inhibition to be cytostatic, stopping growth, rather than cytotoxic, causing cell death, (unless these TKs are involved in suppression of apoptosis.)

The EGF receptor (EGFr) is a type I RTK, belonging to the erbB family which also includes erbB-2, -3 and -4.[7, 8] There are several other activating ligands for EGFr, notably TGF-α, and three different gene products, and multiple splicing isoforms of the heregulins/neuregulins, which act as the principal ligands for the other erbBs. Interestingly, this family not only activates its TK moieties by forming homodimers, but also becomes activated by forming any of the possible heterodimers between family members. In fact, in cell types which express multiple Type I RTKs, it appears that most of the growth factors preferentially lead to the formation of heterodimers rather than homodimers.[9] Overactivity of one or more members of the erbB (EGFr) family of receptor tyrosine kinases (RTKs) is one of the commonest biochemical lesions found in cancer, being found in one form or another in at least 60% of all solid tumours.[7] This can occur through overexpression of the receptor, so that an appropriate level of signal is overresponded to, by ligand overexpression, so that there is too much growth factor present, mutation of the receptor so that the TK is switched on regardless of ligand status, or autocrine stimulation, whereby the cell manufactures both ligand and receptor. All four of these disregulations are regularly found in solid tumours. Furthermore, overactivity of this pathway, as measured by receptor overexpression has frequently been associated with poor prognosis, non-responsiveness to chemotherapy, short relapse times and dramatically shorter survival times.[10] This has led to more intense development of TK inhibitors for this family than any other.

3   THE SEARCH FOR EGFR FAMILY TYROSINE KINASE INHIBITORS

Two approaches to inhibiting RTK activity have been made.   Inhibitors of the kinase enzyme activity have been sought. Enzyme inhibition is usually a fruitful field for small molecule pharmaceuticals, and one such quest is the major subject of this talk.   The second approach has been to prevent the activating dimerization of the receptors by interfering with ligand-receptor binding.   As the receptor-ligand interaction tends to be made up of many small binding determinants spread over a large area, this small molecule approach has been very unsuccessful, but preventing such interactions by blocking antibodies appears to be intrinsically feasible.   Many antibodies have been raised to both EGFr and ErbB-2, and once one gets away from the agonistic binding antibodies, which mimic the ligand too well, antibodies have been found which can completely suppress both EGFr and ErbB-2 TK activity in vitro. Some of these antibodies show excellent in vivo efficacy against EGFr/ErbB-2-driven tumours, in nude mice xenografts, and several antibodies are currently in clinical trials.[11] Perhaps even more significantly, some of these antibodies show strong synergy in tumour killing with conventional cytotoxic agents.[12, 13]   Several of these antibodies are in clinical trials, and the results of one anti-erbB-2 antibody Herceptin® have recently been published.[14, 15]   These results were very exciting, as they were obtained against erbB-2 overexpressing breast cancers, a logical target for such therapy, but a notoriously difficult type of tumour to treat with very poor prognosis. The antibody produced complete or partial remissions in 15% of patients as a single agent. In combination with classical cytotoxic approaches it produced complete or partial remissions in about 60% of patients, an improvement of 55-130% over the cytotoxic agent alone. This study probably provides the first unequivocal clinical proof of concept, not only of blocking EGFr family TK activity, but also of the whole signal transduction approach to cancer therapy.

First generation EGFr inhibitors were usually identified in cellular assays.   They generally turned out to be difficult to develop, as most turned out to be alkylating or thiating species, and either did not directly inhibit the EGFr or turned out to be very unselective kinase inhibitors.[16]   An excellent example of how difficult it can be to know what your true target is in a cellular assay is Geldanamycin **1**.[17]   (Figure 1). This compound was first identified as a very potent EGFr TK inhibitor, which worked after a prolonged period of inhibition. It was certainly true that after 24 hours incubation with Geldanamycin, no EGFr kinase activity could be detected.   However closer inspection showed that many other kinases were equally inactive, and that the inactivity of all of them was due to the fact that the cells no longer contained these kinases. The true target for **1** is a chaperone protein GRP94, and when Geldanamycin binds to it, the many kinases it helps to transport from the Golgi apparatus to the plasma membrane are redirected to lysosomes, where they are destroyed.   Thus, as old receptor is degraded by normal turnover at the membrane it is no longer replaced, and kinase activity is no longer detectable.

Another early error turned out to be a very understandable belief that one should design kinase inhibitors to exploit the most variable regions of kinases, in order to have some chance of getting selectivity of inhibition between the desired target, and the other 2000 kinases.[18]   Kinases contain an ATP-binding domain, which is generally highly conserved, and a substrate binding domain, which contains the specificity determinants for the enzymes, and is therefore much more variable. Therefore substrate-based inhibition

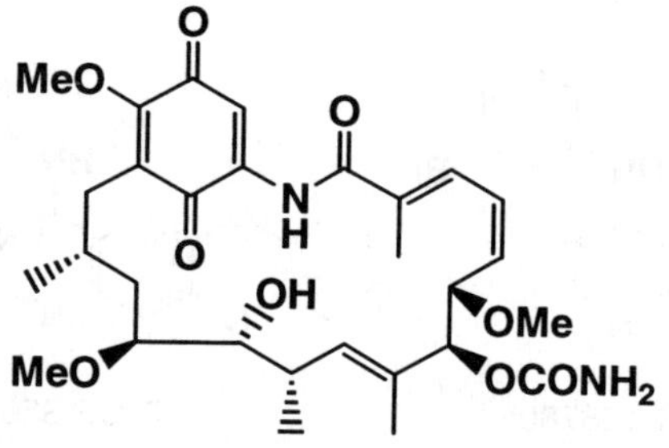

1

Figure 1 *Geldanamycin*

seemed to be the logical way to go, and a series of tyrosine mimics, "tyrphostins" based on the natural product Erbstatin **2**, and further exemplified by **3** and **4** (see Figure 2) were developed.[19] These compounds turned out to be very disappointing, as they are probably weakly targeted alkylators or crosslinkers, and show very little of the anticipated specificity, and usually have very shallow SARs which are very difficult to optimize.

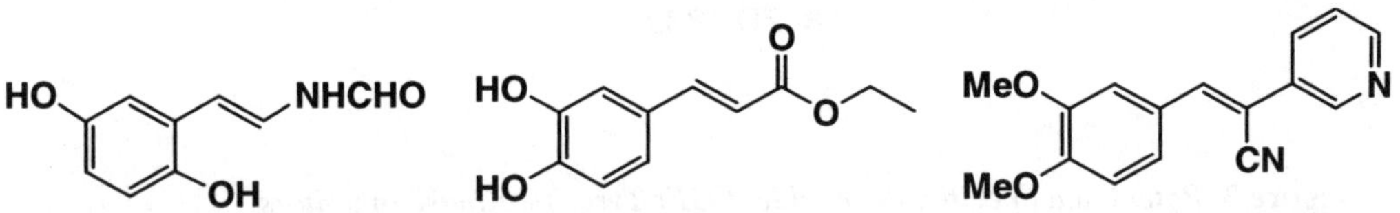

2   **3**   **4**

Figure 2 *Erbstatin and Representative Tyrphostins*

However, in the past five years we and others have published on reversible TK inhibitors, based on the 4-anilinopyrimidine pharmacophore which can show extraordinary potency and selectivity for inhibiting members of the erbB family, exemplified by the inhibitors **5**,[20] **6**,[21] **7**,[22] and **8**.[23] (See Figure 3). In our assay system these compounds have EGFr IC_{50}s of 0.025, 0.006, 0.71 and 1.05 nM respectively, and cellular IC_{50}s against EGF-stimulated autophosphorylation in A431 cells of 14, 13, 13 and 9 nM respectively. The high degree of selectivity for the EGFr family is best illustrated by **6**, which in kinase assays to date has never been shown to have a selectivity below 625,000 for EGFr over any non-Type I TK examined.[21] This selectivity does not extend to the other RTK family members, with **6** having 50-60 nM cellular IC_{50}s against erbB-2 and erbB-4 autophosphorylation. Compounds **7** and **8** are currently in clinical trials, with the initial phase 1 results for **8** recently published. This compound, unlike any conventional anti-cancer drug, is non-toxic enough to have had its Phase 1 carried out in normal volunteers.[24]

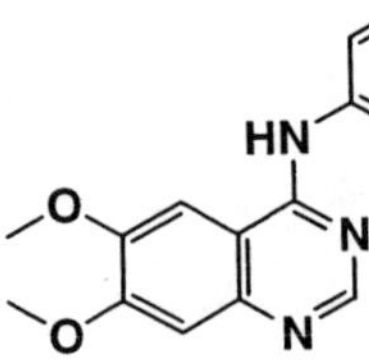

5 PD 153035 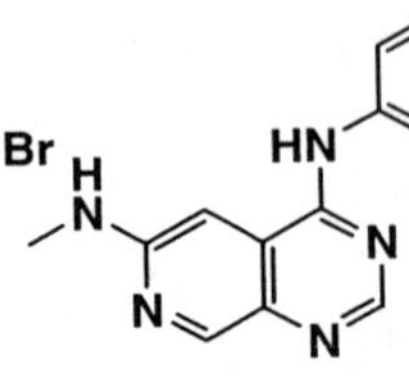   **6 PD 158780** 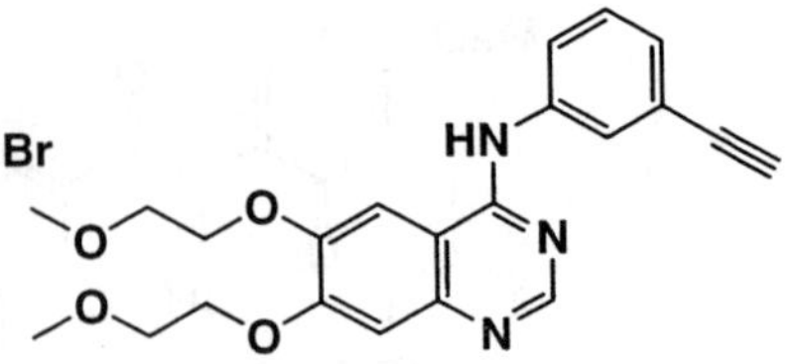   **7 CP 358,774**

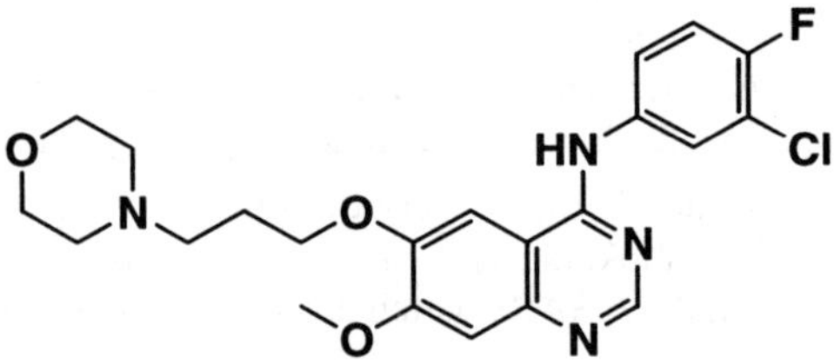

8 ZD 1839

Figure 3 *Potent and Specific, Reversible, EGFr Tyrosine Kinase Inhibitors*

Rather surprisingly, these inhibitors are ATP-competitive, which means that they bind to the most highly conserved domain in the entire kinase superfamily. Despite this, the selectivity first found in the EGFr family is being rapidly extended to other kinases. It appears that ATP-competitive inhibitors are very sensitive to minor structural changes in kinases, and it is often possible to strongly exploit only one or two changed residues in the ATP binding site to produce great selectivity for inhibition of one kinase over another. This is beautifully illustrated by a recent Zeneca patent application which claims **9**, (see Figure 4), an isomer of **8** as a potent, selective, inhibitor of the VEGF RTK.[25] The sensitivity of ATP-competitive inhibitors to minimal structural changes in binding sites has been recently explained in the case of the p38 MAP kinase inhibitors from Smith-Beecham.[26] SB 220025, **10**,[27] is a 19 nM inhibitor of p38α, and an 18 μM inhibitor of erk2. The crystal structures of **10** bound to p38α and erk2 show a hydrophobic pocket for the fluorophenyl ring, which is an excellent fit in p38α, thanks to T^{106}.[28, 29] In erk2, the corresponding residue Q^{105} severely impedes binding into the pocket.[28] Mutating $Q^{105}>T$ makes erk2 a nanomolar substrate of **10**, and mutating $T^{106}>M$ (the corresponding residue in the SB-insensitive p38γ&δ isoforms) destroys the affinity of p38α for **10**, illustrating quite unequivocally that a single residue change can lead to pharmaceutically useful 1000-fold differences in inhibitory potency.

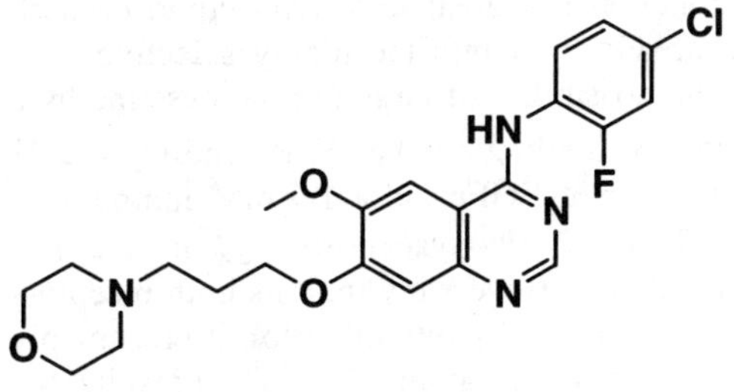

9 Zeneca VEGF Inhibitor 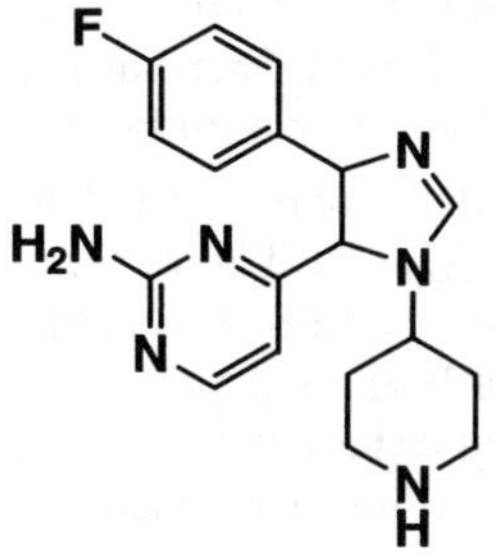   **10** SB 220025

Figure 4 *Zeneca VEGFR Inhibitor and SB220025 p38 MAP Kinase Inhibitor*

The high potencies of inhibition of the original second generation inhibitors did not immediately translate into in vivo activity, as the original compounds tended to have very poor physicochemical properties, which translated into very poor pharmacokinetic profiles. Two approaches have been taken to overcome this problem. The first is the classical medicinal chemistry approach of adding functional groups which add water solubility but do not adversely affect potency, and then look for individual compounds which contain this desirable combination of properties and which also demonstrate good pharmacokinetics. Compounds **7** and **8** are very successful exemplars of this approach. Both have shown good pharmacokinetic parameters and excellent oral efficacy in xenograft studies.[22, 23] Both show IC_{50}s against A431 (a human epidermoid carcinoma, strongly overexpressing the EGFr) xenografts around 12.5 mg/kg/day, and at high doses can lead to complete tumouristasis, and even tumour regression. Tumours tend to regrow after dosing is discontinued, and the compounds are much less toxic than conventional chemotherapy agents. The other approach is to make the binding so tight that the inhibitors would be pseudo-irreversible, and not be limited by Michaelis-Menten kinetics of inhibition, because very high binding affinity could lead to such slow off rates (on the order of hours) that poor pharmacokinetics might be overcome by producing long term enzyme inhibition, even if drug levels were only temporarily high enough to produce adequate inhibition for in vivo efficacy. This approach did work to some extent in our hands. Pyridopyrimidine **6** produces measurable enzyme inhibition in cells for many hours after the external drug has been washed off, and can produce similar good anti-tumour efficacy to **7** and **8**, but only at higher doses (IC_{50} 100 mg/kg/day vs A431).[21]

4   THE DEVELOPMENT OF HIGHLY SELECTIVE, IRREVERSIBLE, TYROSINE KINASE INHIBITORS

To carry this approach to its logical conclusion, we examined the possibility of making these inhibitors truly irreversible. Homology modeling showed that Cys[773] of EGFr*16* should carry out the H-bonding role of Glu[127] of PKA, which has been shown in its X-ray structure to H-bond to the 2'-hydroxyl of adenosine.[30] Furthermore, this use of cysteine as the 2'-hydroxyl H-bond donor is very unusual amongst kinases, suggesting that if it were targeted by a thiol selective alkylating functionality great selectivity could be built into the irreversible step, in addition to that already built into the highly selective non-covalent binding. An initial study confirmed the possibility of targeting the cysteine by a thiol reactive group targeted to the exact site by binding in the ATP binding site.[31] Adenosine **11**, (see Figure 5), is a 195 µM inhibitor of EGFr and a 1.6 mM inhibitor of erbB-2, and is fully reversible. In contrast, 2'-thio-2'-deoxyadenosine **12** is a 2 µM inhibitor of EGFr and a 45 µM inhibitor of erbB-2, and irreversibly inhibits both receptors with second order kinetics. As this inhibition is reversible by dithiothreitol, it presumably involves forming a covalent disulfide bond to the sulfur atom of Cys^{773}, (despite the inhibitor being at an inappropriate oxidation level for disulfide formation). As anticipated, **12** was not a better inhibitor of PDGFr than was **11**, indicating the need of the cysteine in close proximity to the 2'-position in the ribose binding site. However, simple thiols, which have no binding determinant for EGFr, were not (irreversible) inhibitors of it at all, demonstrating that disulfide bond formation occurred (presumably with the help of ambient oxygen) because the non-covalent binding held the adenosine thiol in close proximity to Cys^{773}, effectively boosting the concentration terms in the second order rate equation enormously. We have called this Affinity-Induced Irreversible Inhibition, mainly because this translates to AI^3, a really nifty moniker.

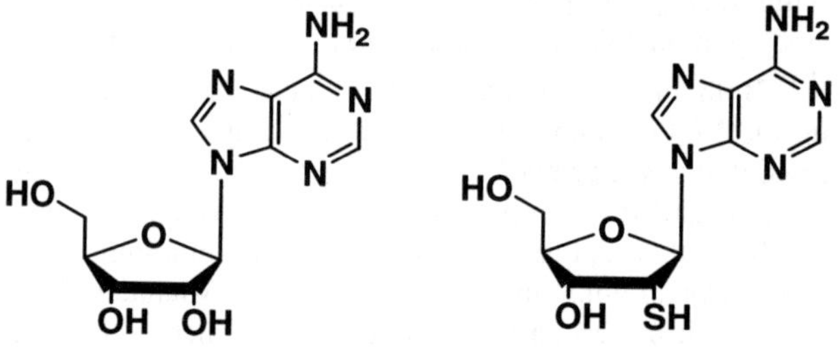

Figure 5 *Adenosine and 2'-Thio-2'-deoxyadenosine*

Initial exploration in our SAR was carried out by attaching various alkylating functionalities onto the anilinoquinazoline pharmacophore.[32] Our first success, compound **13**, (see Figure 6), a 7-acrylamidoquinazoline with an IC_{50} of 490 pM for EGFr and 14 nM for A431 autophosphorylation, also proved to be an irreversible inhibitor of EGFr and erbB-2, as hinted at by cellular wash off studies.[33] Later radiolabeling studies and electrospray mass spectroscopy confirmed the covalent nature of the attachment. Molecular modeling studies, using a binding mode derived from the SAR of our reversible inhibitors,[34] suggests that the terminus of the acrylamide Michael acceptor was placed about 8 Angstroms from the sulfur atom, when the anilinoquinazoline was bound in our proposed mode into the ATP-binding site. Kinetic studies showed that the alkylation was relatively slow, with a half-life of about 20 minutes, and the in vivo activity against an

EGFr-transfected 3T3 fibroblast line turned out to be comparable to compound **6**. This made the point that if one is to compensate for poor pharmacokinetics by irreversible reaction, the alkylation must be fast enough to completely inactivate the receptor in the relatively short time that the local concentration of the inhibitor is high enough to do so. Fortunately, the 6-acrylamido analogue **14**, with enzymatic and cellular IC_{50}s of 630 pM and 2.7 nM respectively, also proved to be fully irreversible, as measured by cellular and enzyme assays, and mass spectroscopy demonstrated both the alkylation of EGFr and the position of alkylation as Cys^{773}. When the enzyme, with the point mutation $C^{773}{>}S$, was incubated with **14**, it was still inhibited by **14** in the nanomolar range, but the interaction was no longer irreversible. In addition, kinetic studies showed that alkylation of EGFr is very rapid, being 100% complete within 1 minute, and modeling suggested that the Michael acceptor terminus is held very close to Cys^{773} with a terminal acryloyl carbon-sulfur distance of only 3.5 Angstroms. Under similar conditions, control cysteine-containing peptides were only about 1% alkylated by **14** in 16 hours, confirming the intrinsically low reactivity of the Michael acceptor. We have found some irreversible inhibitors, with what we assume are better Michael acceptors than acrylamide, which show excellent enzymatic activity, and very poor cellular activity. In these cases we *speculate* that too reactive a Michael acceptor leads to indiscriminate reaction with nucleophiles in both the medium and the cell, so that virtually none of the inhibitor reaches its target intact. If correct, this suggests that there is a "Goldilocks" requirement for useful Michael acceptors, neither too hot nor too cold.

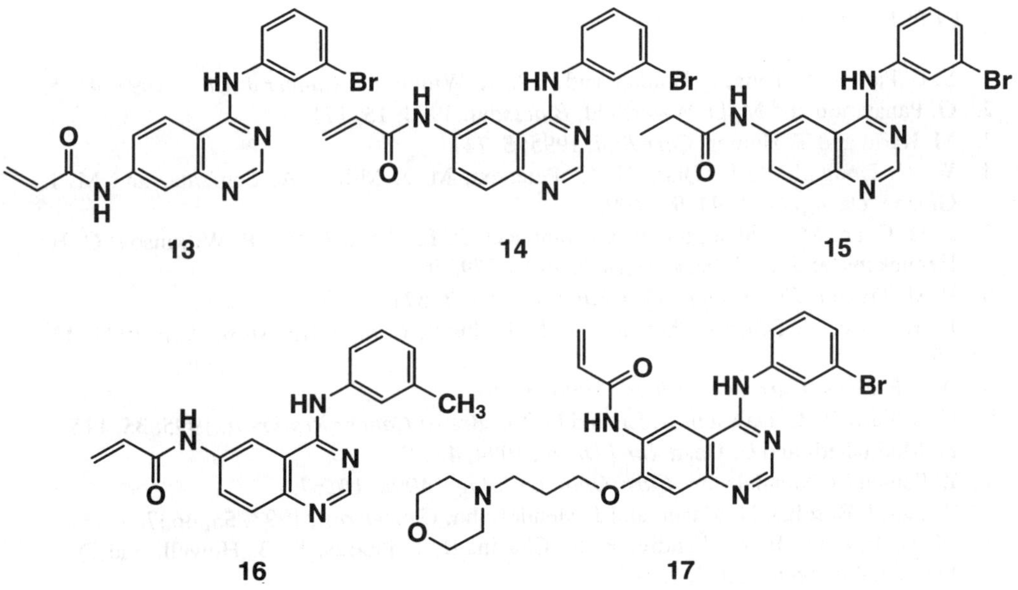

Figure 6 *Some Irreversible Inhibitors of the EGFr Tyrosine Kinase*

As anticipated from this combination, **14** shows markedly better in vivo activity than both **13**, (the slow alkylator analogue) and the 7-propanamide **15**, its saturated, and therefore fully reversible, analogue despite the similar IC_{50}s for EGFr (**15**, enzyme 520 pM, cellular 15 nM).[33] In fact, despite its extremely poor solubility, **14** has about half of the potency of **8** in oral dosing against A431 xenografts. A somewhat less insoluble analogue **16**, with enzymatic and cellular IC_{50}s of 420 pM and 4.8 nM respectively shows excellent activity against A431 xenografts when dosed orally and good activity dosed ip and as an sc infusion.[35] In addition it showed moderate to good activity against EGFr transfected 3T3 fibroblasts, UISO-BCA-1 and MCF-7 human breast cancers, HA 125 human non-small cell lung and SK-OV-3 human ovarian cancer xenograft models, but was without activity against CAKI-1 and Pan-03 xenografts. As some of these models are thought to be driven more by erbB-2 than EGFr, this is indicative that these compounds have good erbB-2 inhibitory activity, and in heregulin-stimulated cellular autophosphorylation assays in MDA-MB-453 human breast cancer cells **14** and **16** have IC_{50}s of 7 and 22 nM respectively. A solubilized analogue of **14**, **17**, (IC_{50}s, EGFr 3.6 nM, A431 5.3 nM, MDA-MB-453 6.4 nM) shows excellent in vivo activity against A431 cells, with IC_{50}s below 10 mg/kg/day for sc infusion and oral dosing and about 10 mg/kg/day by bid ip injection.[36] Orally at 5 mg/kg/day compound **17** not only caused complete tumouristasis during treatment, but caused 3/8 tumours to regress completely, and 1/8 to shrink to less than half of its original size. Such potency in a standard tumour line suggests that irreversible inhibitors of the EGFr family of RTKs will be clinically useful against a wide variety of human tumours.

REFERENCES

1. S. L. Parker, T. Tong, S. Bolden, and P. A. A. Wingo, *CA Cancer J. Clin.*, 1997, **47**, 5.
2. G. Panayotou and M. D. Waterfield, *Bioessays*, 1993, **15**, 171.
3. M. Karin and T. Hunter, *Curr Biol*, 1995, **5**, 747.
4. W. C. Dougall, X. L. Qian, N. C. Peterson, M. J. Miller, A. Samanta, and M. I. Greene, *Oncogene*, 1994, **9**, 2109.
5. J. M. Chan, M. J. Stampfer, E. Giovannucci, P. F. Gann, J. Ma, P. Wilkinson, C. H. Hennekens, and M. Pollack, *Science*, 1998, **279**, 563.
6. P. M. Traxler, *Expert Opin Ther Patents*, 1997, **7**, 571.
7. K. Khazaie, V. Schirrmacher, and R. B. Lichtner, *Cancer Metastasis Rev*, 1993, **12**, 255.
8. A. J. Bridges, *Curr. Med. Chem.*, 1996, **3**, 167.
9. H. S. Earp, T. L. Dawson, X. Li, and H. Yu, *Breast Cancer Res Treat*, 1995, **35**, 115.
10. H. Modjtahedi and C. Dean, *Int J Oncol*, 1994, **4**, 277.
11. Z. Fan and J. Mendelsohn, *Curr. Opin. Oncology*, 1998, **10**, 67.
12. Z. Fan, J. Baselga, H. Masui, and J. Mendelsohn, *Cancer res.*, 1993, **53**, 4637.
13. M. D. Pegram, B. M. Fendly, B. R. Chazin, R. J. Pietras, S. B. Howell, and D. J. Slamon, *Oncogene*, 1994, **9**, 1829.
14. M. A. Cobleigh, C. L. Vogel, D. Tripathy, N. J. Robert, S. Scholl, L. Fehrenbacher, V. Paton, S. Shak, G. Lieberman, and D. Slamon, *Proceedings of ASCO*, 1998, **17**, 97a Abstract 376.

15. D. Slamon, B. Leyland-Jones, S. Shak, V. Paton, A. Bajamonde, T. Fleming, W. Eiermann, J. Wolter, J. Baselga, and L. Norton, *Proceedings of ASCO*, 1998, **17**, 98a. Abstract 377.
16. M. R. Myers, W. He, and C. Hulme, *Curr. Pharm. Design*, 1997, **3**, 473.
17. C. Chavany, E. Mimnaugh, P. Miller, R. Bitton, P. Nguyen, J. Trepel, L. Whitesell, R. Schnur, J. D. Moyer, and L. Neckers, *J Biol Chem*, 1996, **271**, 4974.
18. S. K. Hanks and T. Hunter, *FASEB J*, 1995, **9**, 576.
19. A. Levitzki, *FASEB J*, 1992, **6**, 3275.
20. D. W. Fry, A. J. Kraker, A. Mcmichael, L. A. Ambroso, J. M. Nelson, W. R. Leopold, R. W. Conners, and A. J. Bridges, *Science*, 1994, **265**, 1093.
21. D. W. Fry, J. M. Nelson, V. Slintak, P. R. Keller, G. W. Rewcastle, W. A. Denny, H. R. Zhou, and A. J. Bridges, *Biochem Pharmacol*, 1997, **54**, 877.
22. J. D. Moyer, E. G. Barbacci, K. K. Iwata, L. Arnold, B. Boman, A. Cunningham, C. DiOrio, J. Doty, M. J. Morin, M. P. Moyer, M. Neveu, V. A. Pollack, L. R. Pustilnik, M. M. Reynolds, D. Sloan, A. Theleman, and P. Miller, *Cancer Res*, 1997, **57**, 4838.
23. J. R. Woodburn, A. J. Barker, K. H. Gibson, S. E. Ashton, A. E. Wakeling , B. J. Curry, L. Scarlett, and L. R. Henthorn, *Proceedings of the 88th AACR Annual Meeting*, 1997, **38**, 633 Abstract 4251.
24. H. C. Kelly, A. Laight, C. Q. Morris, J. R. Woodburn, and G. H. P. Richmond, *Ann. Oncology*, 1998, **9, Suppl. 2**, in press.
25. A. P. Thomas, L. P. A. Hennequin, and C. Johnstone, *World Pat. App.*, 1997, WO 97/32856.
26. P. R. Young, M. M. McLaughlin, S. Kumar, S. Kassis, M. L. Doyle, D. McNulty, T. F. Gallagher, S. Fisher, P. C. McDonnell, S. A. Carr, M. J. Huddleston, G. Seibel, T. G. Porter, G. P. Livi, J. L. Adams, and J. C. Lee, *J Biol Chem*, 1997, **272**, 12116.
27. J. L. Adams, J. C. Boehm, and D. Lee, *World Pat. App.*, 1997, WO 97/25045.
28. E. Goldsmith, NMHCC's 2nd Annual International Conference on Cell Signaling, San Diego, CA, 1998.
29. K. P. Wilson, P. G. McCaffrey, K. Hsiao, S. Pazhanisamy, V. Galullo, G. W. Bemis, M. J. Fitzgibbon, P. R. Garon, M. A. Murcko, and M. S. S. Su, *Chem Biol*, 1997, **4**, 423.
30. J. Zheng, D. R. Knighton, L. F. Ten Eyck, R. Karlsson, N.-H. Xuong, S. S. Taylor, and J. M. Sowadski, *Biochemistry*, 1993, **32**, 2154.
31. J. Singh, E. M. Dobrusin, D. W. Fry, T. Haske, A. Whitty, and D. J. McNamara, *J Med Chem*, 1997, **40**, 1130.
32. A. J. Bridges, H. Zhou, D. R. Cody, G. W. Rewcastle, A. Mcmichael, H. D. H. Showalter, D. W. Fry, A. J. Kraker, and W. A. Denny, *J Med Chem*, 1996, **39**, 267.
33. D. W. Fry, A. J. Bridges, W. A. Denny, A. M. Doherty, K. D. Gries, J. L. Hicks, K. E. Hook, P. R. Keller, W. R. Leopold, J. A. Loo, D. J. McNamara, J. M. Nelson, V. Sherwood, J. B. Smaill, S. Trumpp-Kallmeyer, and E. M. Dobrusin, *Proc. Natl. Acad. Sci. USA*, 1998, **95**, in press.
34. B. D. Palmer, S. TrumppKallmeyer, D. W. Fry, J. M. Nelson, H. D. H. Showalter, and W. A. Denny, *J Med Chem*, 1997, **40**, 1519.
35. P. W. Vincent, B. E. Atkinson, Zhou, Z., D. Dykes, W. R. Leopold, S. J. Patmore, B. J. Roberts, A. J. Bridges, and W. L. Elliott, *Proceedings of the 89th AACR Annual Meeting*, 1998, **39**, 560. Abstract 3807.

36. W. A. Denny, *215th National Meeting of the American Chemical Society*, 1998, MEDI 118.

P38 MAP KINASE INHIBITORS: PROGRESS, PITFALLS AND POSSIBILITIES

Jerry L. Adams,[*] Timothy F. Gallagher, Jeffrey C. Boehm, Shouki Kassis, Peter D. Gorycki, Rebecca J. Gum, Edward F. Webb, Margaret E. Sorenson, Juanita M. Smietana, Ravi S. Garigipati, Ralph F. Hall, Andrew Ayrton, Alison Badger, Don E. Griswold, Peter R. Young and John C. Lee

SmithKline Beecham Pharmaceuticals, P.O. Box 1539, King of Prussia, PA 19406

1 INTRODUCTION

The pyridinylimidazoles (e.g. SK&F 86002 and SB 203580) are representative of a novel class of anti-inflammatory agents which inhibit the synthesis of several important proinflammatory proteins.[1] Research initiated to elucidate the molecular target of these compounds resulted in the identification of two novel protein kinases, CSBP1 and CSBP2.[2] We subsequently published additional data demonstrating that selective inhibition of this kinase by the pyridinylimidazoles was correlated with their ability to block the synthesis of proinflammatory cytokines.[3] Independently, p38, the mouse ortholog of CSBP2, was discovered as a protein kinase activated in mouse macrophages in response to lipopolysaccharide (LPS).[4] CSBP2, more commonly referred to as p38/p38α MAP kinase, is a member of the MAP kinase family. In response to a variety of stress stimuli (heat, UV light, LPS, high osmolarity) specific upstream kinases (MKK3 and MMK6) are activated, which subsequently activate p38 by bisphosphorylation on the TGY sequence in the activation loop. Once activated, p38 phosphorylates a number of downstream substrates, including kinases and transcription factors, thereby regulating the synthesis of a number of important proinflammatory proteins.[5] The excessive and dysregulated production of proinflammatory cytokines, such as IL-1, TNFα and IL-6, is believed to initiate events leading to inflammation and tissue destruction. As p38 MAP kinase inhibitors block these early signalling events in the disease process, they have the potential to alter the underlying pathophysiology which drives both chronic and acute inflammation. Hence, selective inhibition of the p38 MAP kinase pathway is an attractive target for the development of therapeutic agents to treat a variety of cytokine driven conditions, such as rheumatoid arthritis, septic shock, inflammatory bowel disease, and stroke. Herein we discuss recent advances in our understanding of the selective inhibition of p38 MAP kinase by the pyridinylimidazoles and describes some of our efforts to develop second generation inhibitors suitable for clinical evaluation.

2 MECHANISM AND STRUCTURAL BASIS FOR THE SELECTIVE INHIBITION
OF P38 BY THE PYRIDINYLIMIDAZOLES

2.1 Mechanism and selectivity

For the development of a sensitive kinase human, $(His)_6$-tagged p38α was expressed in
yeast and partially purified by Ni^{2+}-NTA chromatography. The T669 peptide derived from
a portion of the EGF receptor intracellular domain was used as the phosphate acceptor in
the kinase assay. Using a double reciprocal plot of initial velocity versus ATP
concentration, the K_M for ATP was determined to be 200 μM.[6] Initial velocity
determinations performed in the presence of several concentrations of SB 203580
demonstrated that the inhibitor competes with ATP and has a K_i value of 21 nM. SB
203580 and related pyridinylimidazole inhibitors also bind with high affinity to the
unphosphorylated, low-activity form of p38α kinase as determined by both a radioligand
binding assay and isothermal calorimetry.

In addition to p38α, four distinct homologues, commonly referred to as p38β, p38β2,
p38γ (SAPK3) and p38δ (SAPK4) have been identified.[7] High selectivity for inhibition of
p38α versus ~20 protein kinases has been reported for SB 203580.[8] Selectivity ratios based
on IC_{50}s often exceed 1,000. Whereas SB 203580 is equipotent against p38α and p38β2,
other closely related MAP kinases such as p38γ, p38δ, JNK1 and Erk2 are not inhibited or
only weakly so.

2.2 Pharmacophore

The important elements of the pyridinylimidazole pharmacophore derived from several
series of tri- and tetrasubstituted imidazole p38 inhibitors have been previously described.[3]
Briefly summarized, all imidazole-based p38 inhibitors contain the elements of a 4-aryl-5-
(pyridin-4-yl)imidazole (Figure 1). Further substitution of the imidazole at N-1 adjacent to
the azaheteroaryl (**1** and **3**) or at N-2 is allowed (**1** and **2**), but substitution of the imidazole
nitrogen adjacent to the aryl group is not. The 1,4,5-substitution pattern of the core
imidazole has also been explored.[9] The central imidazole acts as a scaffold to optimally
present the aryl and pyridinyl groups and may be replaced by a variety of 5 and 6-
membered ring heterocycles.[10]

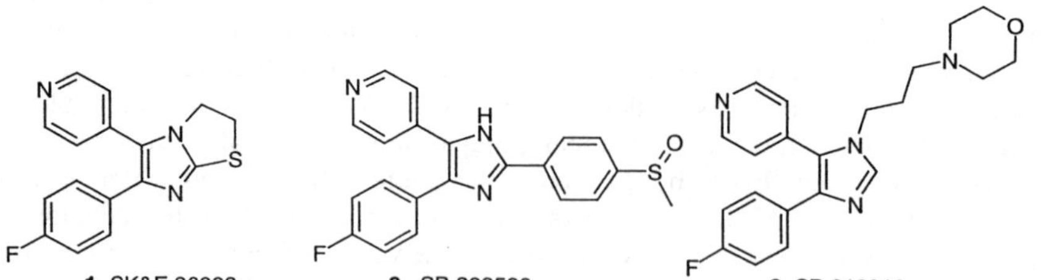

                 1, SK&F 86002   **2**, SB 203580   **3**, SB 210313

Figure 1 *Representative pyridinylimidazole inhibitors of p38 MAP kinase*

2.3 X-Ray Crystallography

Several structures of pyridinylimidazole inhibitor-p38 complexes have been solved.[11] The binding interactions defined by these structures are in good agreement with the published SAR for this class of compounds.[3] Important features observed in these structures are 1) an aryl binding pocket behind and orthogonal to the site normally occupied by the adenine ring of ATP, 2) a hydrogen bond between the pyridinyl nitrogen and the amide N-H of Met109 and 3) H-bond like interaction between lysine 53 and the unalkylated imidazole N-H (Figure 2). The hydrogen bond between the pyridine and the kinase backbone is analogous to the H-bond between the N-1 adenine of ATP and the protein. This hydrogen bond to the N-1 adenine of ATP has been observed in all available kinase crystal structures. Moreover, a hydrogen bond acceptor-donor pair involving the backbone amide NH occurs in all the small molecule inhibitor-kinase crystal structures. The interaction of the imidazole N-H with Lys53 is not a conserved feature of small molecule inhibitor-kinase structures. However, as Lys53 is a totally conserved ATP binding residue, this interaction is unlikely to contribute to selectivity. Whereas the pyridine and imidazole in their role as an adenine mimetic are important for affinity, the fluorophenyl group, which occupies a region of the active site not utilized by ATP, appears to be the key feature contributing to selectivity. The corresponding key feature of the protein which appears to impart selectivity is Thr106 which forms one face of the fluorophenyl binding pocket. A larger side chain at this position would be expected to hinder formation of this binding pocket. For example, ERK2, a close relative of p38 kinase, has the larger glutamine side chain at position 106 and is insensitive to SB 203580.[11a] As a limited number of kinases (10 % of Ser/Thr kinases, 53% of Tyr kinases)[12] have side-chains the size of threonine or smaller (Val or Ser), the relatively rare occurrence of the fluorophenyl binding pocket appears to be an important feature governing selectivity.

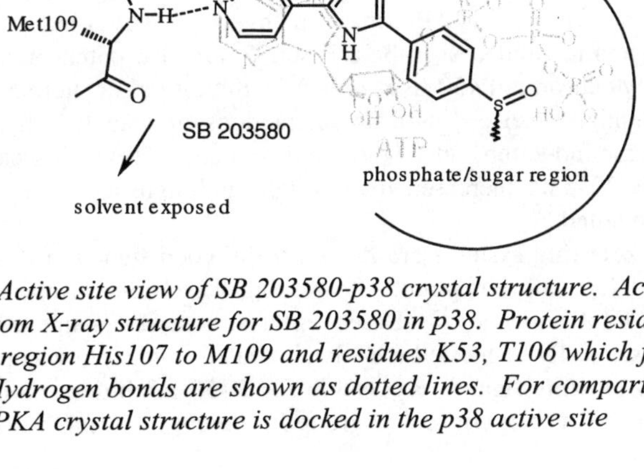

Figure 2 *Active site view of SB 203580-p38 crystal structure. Active site interactions derived from X-ray structure for SB 203580 in p38. Protein residues shown include the hinge region His107 to M109 and residues K53, T106 which form the aryl binding pocket. Hydrogen bonds are shown as dotted lines. For comparison purposes, ATP from the PKA crystal structure is docked in the p38 active site*

2.4 Mutagenesis Experiments

We have used mutagenesis to explore the structural basis for selective inhibition of p38 MAP kinase by the pyridinylimidazoles. All residues whose side-chains projected into the active site within a distance of 3-4 Å of the predicted position of ATP were individually mutated to alanine. The effects of these mutations on kinase activity and the binding of inhibitors were then determined.[13] The data from these experiments are in agreement with the X-ray crystallographic findings demonstrating that the pyridinylimidazoles and ATP occupy different, although overlapping, regions of the ATP pocket. Based upon these initial mutagenesis experiments and the X-ray crystallographic data implicating Thr 106 as a key residue controlling selectivity, additional mutagenesis studies focused on Thr106 and two adjacent residues at the back of the ATP pocket. These same three residues are found in p38β2, the closest homologue to p38α, but are not found in the more distant SB 203580-insensitive homologues, p38γ and p38δ. When these three residues in p38α (Thr106, His107, Leu108)were changed to the Met-Pro-Phe seen in p38γ and p38δ, the mutant proteins were no longer inhibited by SB 203580. A single change of just Thr106 to Met also caused a ten-fold reduction in the sensitivity to SB 203580 in mammalian expressed p38α. Perhaps even more significant was the finding that introduction of the p38α triad (Thr-His-Leu) into p38γ and p38δ or the more distantly related JNK1, was enough to make these kinases as sensitive to SB203580 as p38α. Mutagenesis studies by Eyers et al. have explored the effect of 10 different amino acids at position 106 in p38γ.[14] Native protein (Met 106, IC_{50} >100 μM) was insensitive as were all mutants having residues larger than threonine. Mutants having a threonine (IC_{50} = 0.3 μM) were sensitive to SB 203580, with smaller residues (alanine, IC_{50} = 0.01 μM) showing a further enhancement in sensitivity. These mutagenesis data identify Thr106 as a key residue required for the formation of the aryl specificity pocket and are consistent with the X-ray structural information.

3 THE INHIBITION OF CYTOCHROME P450 BY THE PYRIDINYLIMIDAZOLES

3.1 Pyridinylimidazoles Inhibit P450 Enzymes

SB 203580, a potent and selective inhibitor of p38 MAP kinase, is an orally active inhibitor of cytokine production[15] and has been widely used as a tool compound for studying the importance of the p38 MAP kinase pathway in stress-induced signal transduction.[5] SB 203580 and its predecessor, SK&F 86002, are also potent inhibitors of human and rat hepatic cytochrome P450 isozymes.[16] Inhibitors of human hepatic cytochrome P450 can potentially cause drug-drug interactions or lead to other hepatic changes such as P450 enzyme induction. In 10-day rat dose-ranging toxicological studies with SK&F 86002 and SB 203580, increased liver weight and significant elevations of hepatic P450 enzymes were noted.[17]

The pyridine and imidazole ring systems are known to be good ligands for the ferric heme iron of cytochrome P450, and compounds possessing these heterocyclic rings are often potent inhibitors of P450 enzymes.[18] Based upon structural considerations, the pyridine and not the more sterically congested α,α'-disubstituted imidazole was presumed to be the more likely co-ordinating ligand responsible for P450 inhibition. Subsequent SAR confirmed the importance of the pyridin-4-yl group for potent P450 inhibition, hence a synthetic program was initiated to dissociate p38 MAP kinase inhibition from P450 inhibition.

We chose as our starting scaffold **3** (SB 210313), an orally active p38 MAP kinase inhibitor.[9] Unlike SB 203580 and SK&F 86002, which inhibited a number of human hepatic P450 enzymes, **3** inhibited only one of five human hepatic enzymes (tested at 10 µM; 86 % inhibition for CYP2D6, < 50% inhibition for cytochrome P450s 1A2, 2C9, 2C19 and 3A4). This change in P450 isozyme inhibition profile is attributed to the introduction of a basic side chain and a reduction in overall lipophilicity. These features would be expected to decrease interaction with most P450 enzymes. However, the increased affinity for CYP2D6 with the introduction of the morpholinyl propyl side-chain is consistent with this isozymes preference for a basic amine located some 5-7 Å distal from an aromatic binding pocket.[19] As **3** inhibits only a single P450 isozyme, we were able to focus on CYP2D6 inhibition in subsequent optimization studies.

4 PYRIMIDINYLIMIDAZOLES DEMONSTRATE REDUCED INHIBITION OF P450 ENZYMES

As discussed earlier, SAR studies had established the importance of the pyridin-4-yl group for potent p38 MAP kinase inhibition [3] and subsequent X-ray crystallographic studies have confirmed the key role of the pyridinyl nitrogen as a hydrogen bond acceptor.[11] Assuming that both p38 and P450 inhibition were dependent on the lone pair electrons of the pyridinyl nitrogen, analogs of **3** possessing a 4-azaheteroaromatic group of differing electronic and steric features were prepared in the search for a replacement with the required properties. Our initial approach was to hinder access of the pyridinyl nitrogen to the heme iron of P450 by introduction of sterically demanding alkyl groups. However, p38 kinase inhibition also proved sensitive to steric effects, and potency decreased with the introduction of successive methyl groups (**3** versus **4** and **5**). Substitution of a pyrimidine for the pyridine was considered an attractive alternative as both hydrogen bonding ability and placement of the pyridin-4-yl nitrogen is retained and pyrimidine is known to be a weak P450 inhibitor relative to pyridine.[20]

The majority of the pyrimidine analogs were equivalent to or better than **3** as inhibitors of p38 (Table 1).[16] The most successful pyridine replacements were the 2-protio-, 2-methoxy- and 2-aminopyrimidines **6**, **8** and **10**. Introduction of a heteroatom at the 6 position in **14** and **15** substantially decreased inhibition. The loss of p38 inhibition with the introduction of α,α'-disubstitution in **16** parallels the SAR for the pyridines **5** and suggests a steric constraint in the p38 binding site. Whereas previous SAR established the requirement for a 4-azaheterocycle, the data in Table 1 demonstrate that for compounds meeting this requirement, p38 inhibition can be modulated by both steric and electronic effects.

Compounds demonstrating p38 inhibition equivalent to or better than **3** were examined for inhibition of human cytochrome P450 2D6 (CYP2D6) and oral activity in the mouse (Table 2). Introduction of a methyl group adjacent to the pyridinyl nitrogen reduced inhibition of CYP2D6 (**4**), but proved of limited benefit as oral activity also decreased. However, several of the pyrimidines **6**, **8**, **10** and **11** demonstrated a reduction in CYP2D6 inhibition along with increased oral activity. Particularly noteworthy is the 8-fold increase in oral activity achieved with the 2-aminopyrimidine **10** and the lack of significant

Table 1 *Pyridine replacements*

	R structure	p38 IC$_{50}$, uM[a]		R structure	p38 IC$_{50}$, uM[a]
3		1.3	10		0.48
4		2.1	11		1.9
5		>17	12		3.6
6		0.22	13		5.5
7		1.3	14		>17
8		0.30	15		>17
9		2.0	16		>17

a – IC$_{50}$ values are determined for recombinant human p38α using the procedure described in reference 11a.

CYP2D6 interactions seen with the 2-methoxypyrimidine **8** and the 2-(methylamino)pyrimidine **11**. The reduction in cytochrome P450 interactions achieved with these analogs is consistent with the proposed interaction of the pyridin-4-yl group with the heme iron of P450.

Table 2 *Effect of pyridine replacements on P450 inhibition and oral activity*

	CYP2D6 inhibition[a]	ED$_{50}$ for TNFα[b]		CYP2D6 inhibition[a]	ED$_{50}$ for TNFα[b]
3	86	42 mg/kg	8	7	14 mg/kg
4	51	37 % ***	9	47	43 mg/kg
6	34	12 mg/kg	10	47	5.2 mg/kg
7	19	65 % ***	11	11	19 mg/kg

*a - percent inhibition of human cytochrome 2D6 at 10 uM of test compound; b - the assay was conducted in BALB/c mice using a modification of the published protocol in which TNF levels were determined in the plasma[20]; data are presented as ED$_{50}$ in mg/kg or % inhibition at the screen dose of 50 mg/kg; ***statistically significant from controls at p < 0.001*

5 PYRIMIDINYLIMIDAZOLES: SECOND GENERATION P38 KINASE INHIBITORS

5.1 Optimization of N-1 Imidazole Substitution

In previous studies with the 1-alkyl-4-(aryl)-5-(pyridin-4-yl)imidazoles (**3** and analogs) we noted an improvement in oral activity with those compounds having an alpha-branched N-1 substituent on the imidazole.[9] We have applied this observation, in combination with the 2-aminopyrimidin-4-yl replacement for the pyridinyl group, towards the goal of achieving improved *in vitro* and *in vivo* potency with reduced P450 inhibition. We were particularly interested in preparing compounds with symmetrical, non-chiral, cyclic groups at N-1. Representative compounds in this series are illustrated in Table 3, along with their IC_{50} for inhibition of p38 and ED_{50} for inhibition of LPS-induced TNF synthesis in the mouse. A substantial increase in p38 potency was seen for the piperidinyl (**18, 19**) and cyclohexyl (**20**) compounds. These analogs also demonstrated good *in vivo* activity in the murine assay for inhibition of LPS-induced TNF, however their potency in this assay was no better than the starting morpholinyl propyl compound (**10**).

Table 3 *Effect of N-1 imidazole substitution on p38 oral activity*

		p38, IC$_{50}$ uM^a	ED$_{50}$, mg/kg^b			p38, IC$_{50}$ uM^a	ED$_{50}$, mg/kg^b
10		0.48	5.2	**20**		0.083	7.3
17		0.17	> 50	**21**		0.16	13
18		0.020	8.0	**22**		0.39	>> 50
19		0.070	9.0	**23**		0.23	~ 50

a – same as Table 1; b – murine ED$_{50}$ same as Table 2.

Piperidine **18** was initially chosen for more detailed investigation. However, in rats **18** demonstrated markedly reduced oral bioavailability in comparison to the N-methylaminopyrimidine analog **24** (SB 226882).[22] Hence, in subsequent studies we chose to focus on **24**. Although **24** proved only slightly more potent than SB 203580 as a p38 inhibitor, significant improvement in the overall profile is evident (Tables 4 and 5). Notably, this improved profile includes reduced inhibition of human P450 enzymes.

Table 4 *Comparison of pharmacological profiles*

Assay	2, SB 203580	24, SB 226882
p38 IC_{50}, uM	0.048	0.027
Murine ED_{50} for TNF, mg/kg	15	3.0
IC_{50} in whole blood, uM	6.3 (n=6)	1.0 (n=2)
P450[a]		
1A2	61	0
2C9	75	0
2C19	85	7
3A4	61	25
2D6	67	23

[a] *The potential of compounds to inhibit P450 was determined using isoform selective P450 assays. The assays were performed with heterologously expressed enzymes, substrates present at their Km concentration and test compounds at 10 uM. Measured values are % inhibition relative to control.*

Aminopyrimidine **24** is more potent than **2** in the mouse and the human whole blood assays for inhibition of LPS-induced TNF production. As the human whole blood assay is expected to be a relevant surrogate marker for humans, the increased potency of **24** is particularly encouraging. Turning to animal models of arthritic disease, **24** and **2** were evaluated in the adjuvant arthritic rat and collagen-induced arthritis (CIA) in the mouse. Both compounds were effective in the two models, with **24** appearing to be the more potent compound (Table 5). The improvement in potency is more notable in the CIA mouse model where **24** was as effective as **2** at one fifth of the dose.

Table 5 *Comparison models of arthritis*

	adjuvant arthritic rat[a]			murine collagen-induced arthritis[b]	
	dose mg/kg	paw volume	IL-6	disease severity	SAP
2 (SB 203580)	60	65**	58***	45* @ 50 mg/kg	52***
	30	46**	44***		
	10	11 ns	9 ns		
24 (SB 226882)	30	49**	48***	50* @ 10 mg/kg	56 ns
	10	29*	34***		
	3	15 ns	19*		

a *Dose-dependent suppression of hindpaw inflammation in rats with adjuvant arthritis (AA) by prophylactic administration of SB compounds from days 0-22 (5 days a week, 10 animals per group). Paw inflammation measured on day 16 and IL-6 on day 22. Data are expressed as % inhibition compared to the untreated AA controls.*

b *Mice were dosed for 7 days at indicated dose (p.o., b.i.d.), beginning after the animals had presented with paw or joint edema/swelling. Data are % inhibition of disease severity and serum amyloid P component for drug treated relative to the mean $\pm SE$ from a group of vehicle (0.03N HCl/0.5% Tragacanth).*

Significance relative to controls for both models: ns $p > 0.05$, * $p < 0.05$, ** $p < 0.01$, *** $p < 0.001$. Details of protocols may be found in reference 15.

6 CONCLUSIONS

Our early work with the pyridinylimidazoles established the potent anti-inflammatory properties of these compounds, most important of which is the selective inhibition of proinflammatory cytokine biosynthesis. Subsequently, SB 203580 **2** and related triaryl analogs were prepared and characterized as a key tool compounds used in the identification of p38 as the molecular target regulating cytokine synthesis. SB 203580, a selective inhibitor of p38 kinase, has continued to be an important reagent for the *in vitro* study of signal transduction pathways and for understanding the molecular basis of selective kinase inhibition.[5] However, SB 203580 is a relatively broad spectrum inhibitor of cytochrome P450 enzymes. SB 203580 is also an inhibitor of arachidonic acid metabolism as has been demonstrated for PGE_2 synthesis in HL-60 cells[15] and TXB_2 release from platelets.[23] This latter limitation must be considered when interpreting results obtained from *in vivo* experiments with SB 203580 in which cyclooxygenase-1 inhibition may be contributing to the observed pharmacology.

We have previously described successful efforts to reduce the direct effects of this class of p38 inhibitors on the synthesis of arachidonate metabolites.[9] This report describes our efforts to separate p38 from cytochrome P450 inhibition and hence further improve the profile of this compound class. Based upon the hypothesis that the pyridinyl nitrogen and not the more sterically congested nitrogens of the imidazole was the more likely co-ordinating ligand responsible for P450 inhibition, a synthetic program was initiated to dissociate p38 MAP kinase inhibition from P450 inhibition. This effort led to the identification of several substituted pyrimidine replacements for the pyridin-4-yl group which effectively dissociate p38 kinase from P450 inhibition and furthermore achieved an increase in oral activity. Further optimization of this series to enhance potency and oral activity afforded compound **24** (SB 226882), a 2-aminopyrimidinylimidazole having an improved anti-inflammatory profile and devoid of significant interactions with cytochrome P450 isozymes.

Acknowledgements

We thank Prof. Betsy Goldsmith of the University of Texas Southwest Medical Center for providing the SB 203580-p38 co-ordinates, Dr. Brian Smith of for providing pharmacokinetic data on SB 203580 and SB 226882 and Dr. Michael Bower for preparation of Figure 2. We would also like to thank Dr. John Gleason and Dr. Brian Metcalf for their continued support and encouragement.

References

1. J. C. Lee, A. M. Badger, D. E. Griswold, D. Dunnington, A. Truneh, B. Votta, J. R. White, P. R. Young and P. E. Bender, *Ann. New York Acad. Sci.*, 1993, **696**, 149.
2. J. C. Lee, J. T. Laydon, P. C. McDonnell, T. F. Gallagher, S. Kumar, D. Green, D. McNulty, M. J. Blumenthal, J. R. Heys, S. W. Landvatter, J. E. Strickler, M. M. McLaughlin, I. R. Siemens, S. M. Fisher, G. P. Livi, J. R. White, J. L. Adams and P. R Young, *Nature*, 1994, **372**, 739.
3. T. F. Gallagher, G. L. Seibel, S. Kassis, J. T. Laydon, M. J. Blumenthal, J. C. Lee, D. Lee, J. C. Boehm, S. M. Fier-Thompson, J. W. Abt, M. E. Sorenson, J. M. Smietana, R. F. Hall, R. S. Garigipati, P. E. Bender, K. F. Erhard, A. J. Krog, G. A. Hofmann,

P. L. Sheldrake, P. C. McDonnell, S. Kumar, P. R. Young and J. L. Adams, *Bioorg. Med. Chem.*, 1997, **5**, 49.

4.   J. Han, J.-D. Lee, S. L. Bibb and R. J. Ulevitch, *Science*, 1994, **265**, 808.

5.   P. Cohen, *Trends in Cell Biol.*, 1997, **7**, 353.

6.   P. R. Young, M. M. McLaughlin, S. Kumar, S. Kassis, M. L. Doyle, D. McNulty, T. F. Gallagher, S. Fisher, P.C. McDonnell, S. A. Carr, M. J. Huddleston, G. Seibel, T. G. Porter, G. P. Livi, J. L. Adams and J. C. Lee, *J. Biol. Chem.*, 1997, **272**, 12116.

7.   (a) S. Kumar, P.C. McDonnell, R. J. Gum, A. T. Hand, J. C. Lee, P. R. Young, *Biochem. Biophy. Res. Comm.*, 1997, **235**, 533; (b) M. Goedert, A. Cuenda, M. Craxton, R. Jakes and P. Cohen, *EMBO J.*, 1997, **16**, 3563.

8.   A. Cuenda, J. Rouse, Y. N. Doza, R. Meier, P. Cohen, T. F. Gallagher, P. R. Young and J. C. Lee, *FEBS Lett.*, 1995, **364**, 229.

9.   J. C. Boehm, J. M. Smietana, M. E. Sorenson, R. S. Garigipati, T. F. Gallagher, P. L. Sheldrake, J. Bradbeer, A. M. Badger, J. T. Laydon, J. C. Lee, L. M. Hillegass, D. E. Griswold, J. J. Breton, M. C. Chabot-Fletcher and J. L. Adams, *J. Med. Chem.*, 1996, **39**, 3929.

10.   (a) S. E. de Laszlo, D. Visco, L. Agarwal, L. Chang, J. Chin, G. Croft, A. Forsyth, D. Fletcher, B. Frantz, C. Hacker, W. Hanlon, C. Harper, M. Kostura, B. Li, S. Luell, M. MacCoss, N. Mantlo, E. A. O'Neill, C. Orevillo, M. Pang, J. Parsons, A. Rolando, Y. Sahly, K. Sidler, W. R. Widmer and S. J. O'Keefe, *Bioorg. Med. Chem. Lett.*, 1998, **8**, 2689, (b) J. R. Henry, K. C. Rupert, J. H. Dodd, I. J. Turchi, S. A. Wadsworth, D. E. Cavender, B. Fahmy, G. C. Olini, J. E. Davis, J. L. Pellegrino-Gensey, P. H. Schafer and J. J. Siekierka, *J. Med. Chem.*, 1998, **41**, 4196; (c) G. J. Hanson, *Exp. Opin. Ther. Patents*, 1997, **7**, 729.

11.   (a) Z. L. Wang, B. J. Canagarajah, J. C. Boehm, S. Kassis, M. H. Cobb, P. R. Young, S. Abdel-Meguid, J. L. Adams and E. J. Goldsmith, *Structure*, 1998 **6**, 1117, (b) K. P. Wilson P. G. McCaffrey, K. Hsiao, S. Pazhanisamy, V. Galullo, G. W. Bemis, M. J. Fitzgibbon, P. R. Caron, H. A. Murcko and M. S. S. Su, *Chem. Biol.*, 1997, **4**, 423, (c) L. Tong, S. Pav, D. H. White, S. Rogers, K. M. Crane, C. L. Cywin, M. L. Brown and C. A. Pargellis, *Nature Structural Biology*, 1997, **4**, 311.

12.   S. Hanks and A. M. Quinn, *Methods in Enzymology*, 1991 **200**, 38.

13.   R. J. Gum, M. M. McLaughlin, S. Kumar, Z. L. Wang, M. J. Bower, J. C. Lee, J. L. Adams, G. P. Livi, E. J. Goldsmith and P. R. Young, *J. Biol. Chem.*, 1998, **273**, 15605.

14.   P. A. Eyers, M. Craxton, N. Morrice, P. Cohen and M. Goedert, *Chem. Biol.*, 1998, **5**, 321.

15.   A. M. Badger, J. N. Bradbeer, B. Votta, J. C. Lee, J. L. Adams and D. E. Griswold, *J. Pharm. Exp. Ther.*, 1996, **279**, 1453.

16.   J. L. Adams, J. C. Boehm, S. Kassis, P. D. Gorycki, E. F. Webb, R. Hall, M. Sorenson, J. C. Lee, A. Ayrton, D. E. Griswold and T. F. Gallagher, *Bioorg. Med. Chem. Lett.*, 1998, **8**, 3111.

17.   (a) M. O. Howard, L. W. Schwartz, J. F. Newton, C. W. Qualls Jr., L. A. Yodis and J. R. Ventre, *Toxicol-Pathol.*, 1991, **19**, 115; (b) L. W. Schwartz and B. Short unpublished observations.

18.   B. Testa and P. Jenner, *Drug Metab. Rev.*, 1981, **12**, 1.

19. R. Mackman, R. A. Tschirret-Guth, G. Smith, G. P. Hayhurst, S. W. Ellis, M. S. Lennard, G. T. Tucker, C. R. Wolf and P. R. Ortiz-de-Montellano, *Arch. Biochem. Biophys.*, 1996, **331**, 134.
20. S. G. Kim and R. F. Novak, *Toxicol. Appl. Pharmacol.*, 1993, **120**, 257.
21. D. E. Griswold, L. M. Hillegass, J. J. Breton, K. M. Esser and J. L. Adams, *Drugs Exptl. Clin. Res.*, 1993, **XIX**, 243.
22. Brian Smith, SmithKline Beecham Pharmaceuticals, Department of Drug Metabolism and Pharmacokinetics, unpublished results.
23. A. G. Borsch-Haubold, S. Pasquet and S. P. Watson, *J. Biol. Chem.*, 1998, **273**, 28766.

INHIBITION OF T-CELL TYROSINE KINASES

Peter D. Davis*, Rodger A. Allen, Daniel A. Berg, Jeremy M. Davis, Martin C. Hutchings, Richard Martin, David Moffat and Martin J. Perry

Celltech Chiroscience, 216 Bath Road, Slough SL1 4EN, UK

1 INTRODUCTION

Engagement of the T-cell antigen receptor (TCR) by antigenic peptides presented within MHC I or II is a fundamental event in the cell-mediated immune response. The first steps in the ensuing complex signal-transduction cascade include activation of the tyrosine kinases p56lck and ZAP70 leading to initiation of signals propagated through the calcium/calcineurin, PKC, Raf-1/MEK, and Rho/Rac pathways. Activation of these pathways results in gene expression, activation/proliferation or anergy/apoptosis depending on the stimulus and co-stimuli.[1]

p56lck is recruited into the ligated TCR complex due to its interaction with CD4 which is coligated by the presenting MHC molecule. Also recruited is the tyrosine phosphatase CD45 which is able to dephosphorylate p56lck and relieve the suppression of activity maintained by the phosphorylated Y505. The enzyme is then able to phosphorylate a variety of substrates including itself, leading to further activation. The structural basis for the activating and suppressing phosphorylations of p56lck has been revealed by crystallographic studies of lck and other src-family kinases.[2]

Activated p56lck phosphorylates the TCR ζ-chain which subsequently binds ZAP70 *via* SH2 domain interactions. The recruited ZAP70 is phosphorylated, possibly by p56lck and is then able to propagate the cascade by phosphorylation of substrates probably including SLP76 and LAT. A third kinase, p59fyn, is present in the TCR complex and its role is less well defined, but it may have some function which overlaps with that of p56lck.[1]

A body of evidence points to an essential role for p56lck and ZAP70 in T-cell development and activation. A T-cell line (JCAM.1) lacking p56lck is unable to signal on stimulation of the TCR complex with an anti-CD3 antibody as shown by the lack of a calcium transient.[3] Transfection of JCAM.1 with p56lck reconstitutes signalling, but transfection with a kinase-dead mutant does not, indicating a role for the catalytic activity of the kinase. -Disruption of the lck gene in mice results in a phenotype in which mature peripheral T-cells are reduced in number and those that remain are hyporesponsive to mitogenic and antigenic stimuli.[4] A double lck/fyn knockout has a more profoundly immuno-suppressed phenotype, but single fyn knockouts are only mildly affected. Similarly, ablation of the ZAP70 gene in mice leads to defects in thymic development and in T-cell activation.[5] In humans also the lack of functional ZAP70 leads to a severely immuno-suppressed state.[6]

2 DEVELOPMENT OF INHIBITORS AS POTENTIAL THERAPIES FOR AUTOIMMUNE DISEASE AND TRANSPLANT REJECTION

We have developed inhibitors of both p56lck and ZAP70 as potential therapies for autoimmune disease and transplant rejection. In order to generate leads, we constructed a model of the ATP-binding sites of the enzymes based on that of PKA [7] (these models were refined in the course of this work as other kinase crystal structures became available) and we designed compounds containing known kinase inhibitor motifs that gave a reasonable fit to our model. Examples of leads generated for p56lck are shown in Figure 1. The SAR elucidated around one of these leads, **CT 3559**, is shown in Figure 2. This SAR was fully consistent with our model of binding as disruption of putative hydrogen bonds (by changing the anilino nitrogen to oxygen, or changing one of the pyrimidine nitrogen atoms to a CH) ablated activity, but removal of the other pyrimidine nitrogen, unimportant in our model, had little effect. Also, the model has no room for the 6-methyl group of the inactive **CT 3596**, but accommodates the 5-methyl group of the active **CT 4425**. The relative orientation of the aromatic rings in the model was used to guide a conformational-restriction approach to potency optimisation shown in Figure 3. A six-fold increase in potency was obtained by constraining the phenyl and pyrimidine rings to give **CT 4394**. Addition of a methoxy group, previously observed to increase potency in the acyclic series, further optimised activity. Reference to the model suggested the accessibility of a glutamate residue at the bottom of the pocket in which the trimethoxyphenyl group was bound, so a positively charged group was introduced at the required distance from the aromatic ring resulting in a 20-fold increase in potency. Finally, more hydrophobic space was filled with a gem-dimethyl substituent giving a compound, **CT 5269**, with an IC_{50} of 1.1 nM, over three orders of magnitude more potent than our original lead.

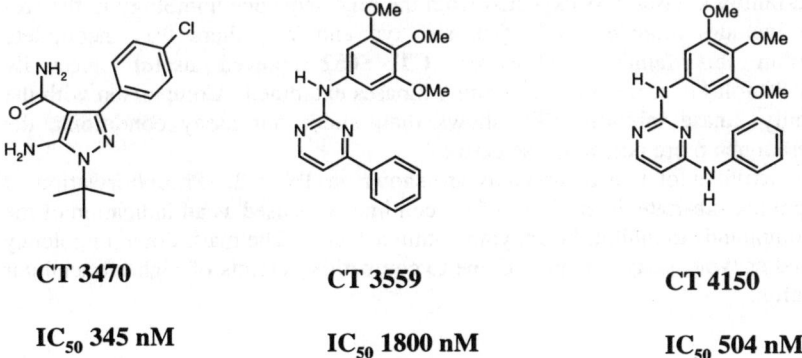

CT 3470	**CT 3559**	**CT 4150**
IC_{50} 345 nM	IC_{50} 1800 nM	IC_{50} 504 nM

Figure 1 *Leads generated for p56lck inhibition*

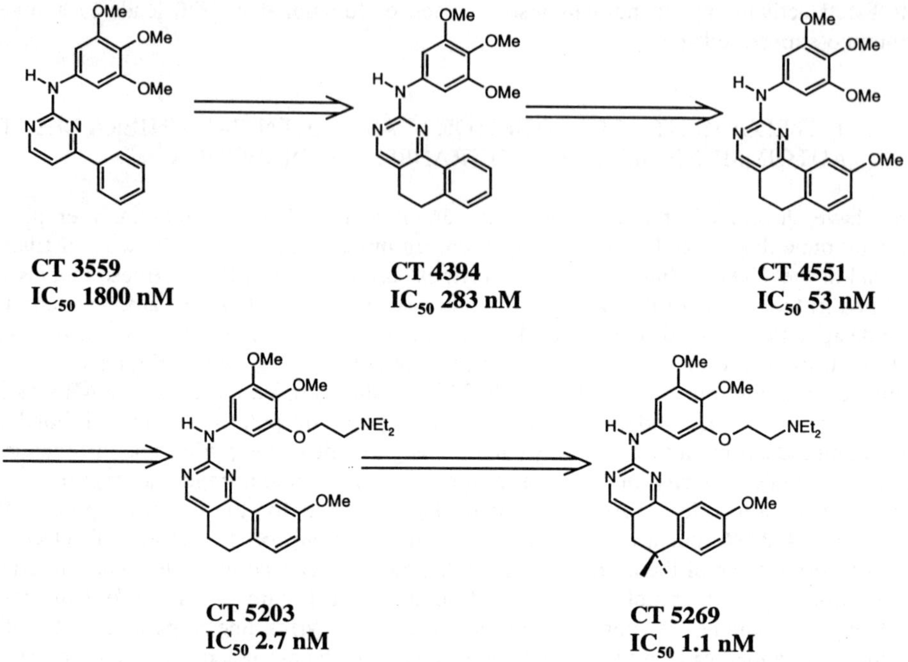

Figure 2 *Optimisation of CT 3559 for p56lck inhibition*

In Table 1, we outline the selectivity for p56lck inhibition of three representative compounds and compare their profiles with that of a published inhibitor, **PP2**.[8] Our compounds were selective over all the non-src-family kinases examined. Particularly important is the selectivity shown over inhibition of csk as this kinase is a negative regulator of src-family kinases. As expected from the high sequence homology in the src-family (which includes *inter alia* lck, fyn, src, lyn and yes) there was incomplete specificity within this family. However, **CT 5652** showed useful selectivity (approximately 80-fold) over the other src-family kinases examined. Comparison with the known src-family kinase inhibitor **PP2** shows that, under our assay conditions, the compounds were much more potent and selective.

The cellular activities of two compounds are shown in Table 2. Phosphorylation of p85, a known p56lck substrate, in E6.1 cells (a T-cell line) was used as an indication of the ability of the compounds to inhibit the enzyme within a T-cell. The mark-down in potency from the isolated enzyme assay is expected due to competitive effects of high intracellular ATP concentration.

Table 1 *Kinase inhibition profiles of p56lck inhibitors*

	CT 5065	CT 5269	CT 5652	PP2
lck	2.5	1.1	52	171
ZAP70	2900	>10000	n.d.	>10000
PKC	1550	n.d.	>10000	8900
EGFR	4616	3155	>10000	700
csk	571	86	>10000	281
cdc2	282	8464	n.d.	n.d.
src	21	6.4	n.d.	n.d.
fyn	24	27	4390	n.d.
lyn	6.5	13	4231	51
yes	5.2	4.1	n.d.	n.d.

Table 2 *Cellular activities of p56lck inhibitors*

Stimulus	Cell	Readout	IC_{50} 5065 (nM)	IC_{50} 5269 (nM)
α-CD3	E6.1	p85 phosphorylation	95	36
α-CD3	E6.1	Calcium flux	196	63
α-CD3	hPBMC	IL2 production	650	35
α-CD3	hPBMC	Proliferation	510	130
PHA	hPBMC	Proliferation (early addn.)	260	n.d.
PHA	hPBMC	Proliferation (late addn.)	>8100	n.d.
MLR	hPBMC	Proliferation	440	960
MLR	mPBMC	Proliferation	57	150
none	JY	Proliferation	2300	2300

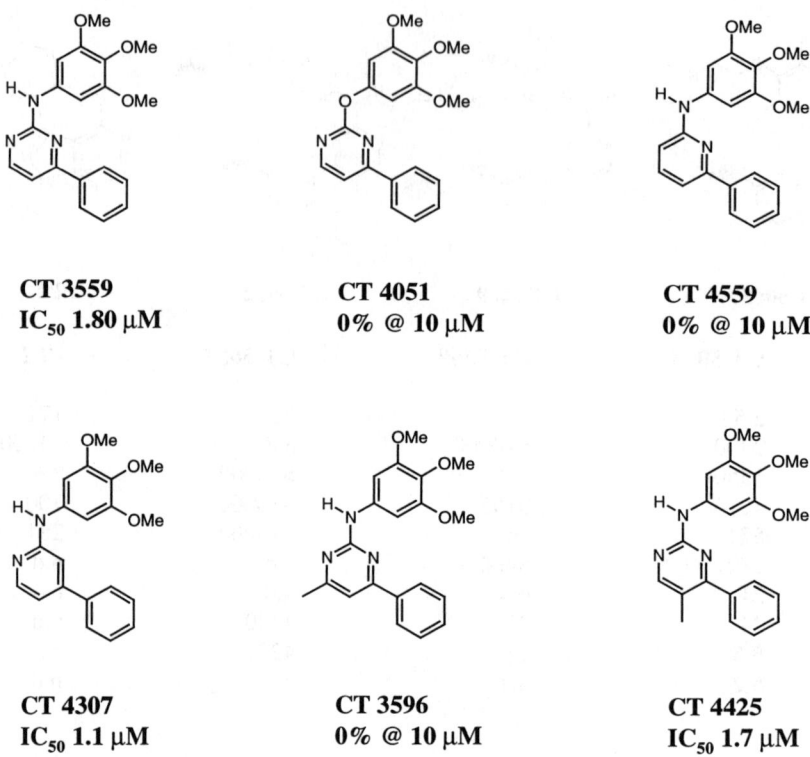

CT 3559 IC$_{50}$ 1.80 μM	**CT 4051** 0% @ 10 μM	**CT 4559** 0% @ 10 μM
CT 4307 IC$_{50}$ 1.1 μM	**CT 3596** 0% @ 10 μM	**CT 4425** IC$_{50}$ 1.7 μM

Figure 3 *SAR of CT 3559 analogues as inhibitors of p56lck*

Anti-CD3 activated calcium flux was inhibited at similar concentrations to *in situ* p56lck inhibition, fully consistent with the predicted effects of inhibition of this enzyme from biochemical and genetic studies. In human peripheral blood mononuclear cells (hPBMC), the production of IL2 and subsequent proliferation was inhibited. Although the precise potencies in these, and other, functional assays can vary, for example between donors, there was a broad correlation with enzyme inhibition. The compounds also inhibited mitogen-driven (PHA) proliferation when added to the cultures at the same time as the mitogen, but not when added towards the end of the assay, ruling out any direct effects on thymidine uptake or DNA synthesis. The murine mixed lymphocyte reaction (mMLR) was strongly inhibited by these compounds, but the human equivalent was less sensitive. Proliferation of a B-cell line, JY, was also inhibited, albeit at higher concentrations, possibly reflecting the incomplete src-family kinase specificity. Overall, the profiles of the compounds were as expected, but the potency in the human MLR was disappointing.

sensitive. Proliferation of a B-cell line, JY, was also inhibited, albeit at higher concentrations, possibly reflecting the incomplete src-family kinase specificity. Overall, the profiles of the compounds were as expected, but the potency in the human MLR was disappointing.

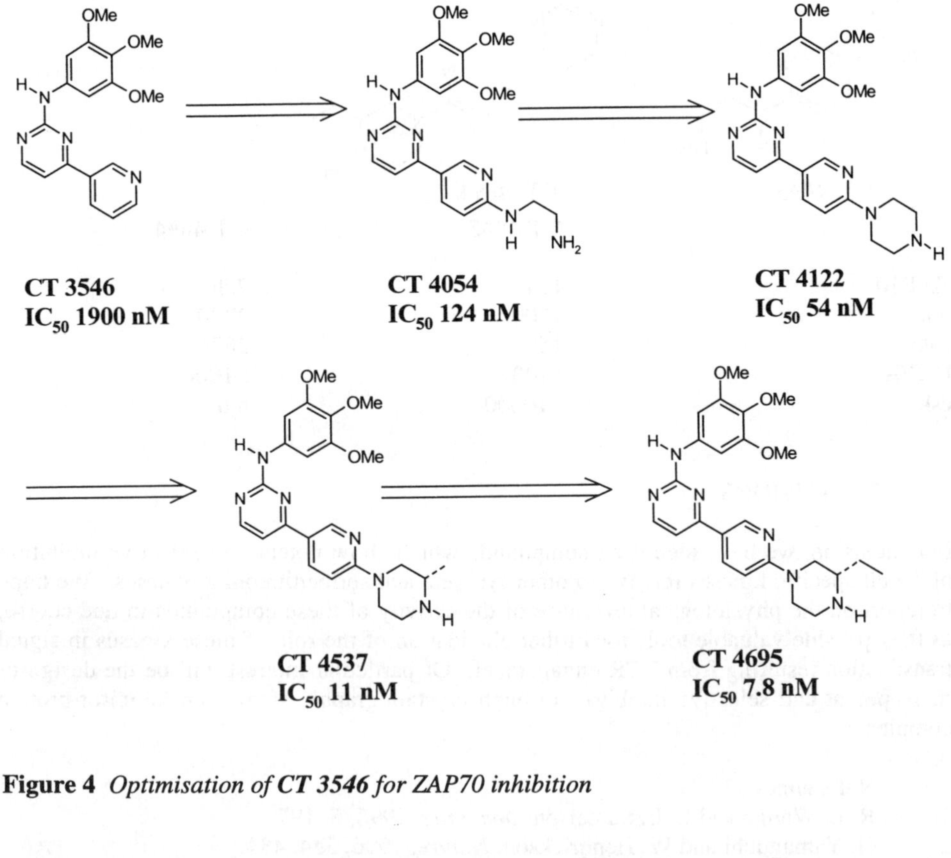

CT 3546
IC_{50} 1900 nM

CT 4054
IC_{50} 124 nM

CT 4122
IC_{50} 54 nM

CT 4537
IC_{50} 11 nM

CT 4695
IC_{50} 7.8 nM

Figure 4 *Optimisation of CT 3546 for ZAP70 inhibition*

One of our leads generated for ZAP70 inhibition, **CT 3546**, had a similar structure to the p56lck lead described above. The optimisation of this lead is shown in Figure 4. Our model suggested that the residues that have been shown in crystal structures to bind metal cations might be accessible from the pyridine ring so this putative site was probed with cationic groups and indeed a potency increase was observed for **CT 4045**. Conformational restriction of the side-chain improved potency further and addition of small lipophilic groups in the (*S*)-orientation on the piperazine ring gave **CT 4695** with an IC_{50} of 7.8 nM against ZAP70 acting on the polyGluTyr substrate. Selectivity was also optimised during this process as **CT 4694** showed only weak inhibition of p56lck, PKC, and EGFR (Table 3).

Table 3 *Kinase inhibition profiles of ZAP70 inhibitors*

	CT 4045	CT 4694
ZAP70	124	7.8
lck	1200	2200
PKC	150	2874
EGFR	5100	>10000
csk	>10000	n.d.

3 CONCLUSIONS

In conclusion, we have identified compounds which show potent and selective inhibition of T-cell specific kinases relative to other tyrosine and serine/threonine kinases. We hope to report on the physiological relevance of the activity of these compounds in due course, as they provide valuable tools for further elucidation of the role of these kinases in signal transduction resulting from TCR engagement. Of particular interest will be the design of more potent and selective inhibitors through crystallographic studies on inhibitor-protein complexes.

References
1. R. L. Wange and L. E. Samelson, *Immunity*, 1996, **5**, 197.
2. H. Yamaguchi and W. Hendrickson, *Nature*, 1996, **384**, 484.
3. D. B. Strauss and A. Weiss, *Cell*, 1992, **70**, 585.
4. T. J. Molina, K. Kishihara, D. P. Siderovski, W. van Ewijk, A. Narendran, E. Timms, A. Waheham, C. J. Paige, K.-U. Hartmann, A. Vaillette, D. Davidson and T. W. Mak, *Nature*, 1992, **357**, 161.
5. I. Negishi, N. Motoyama, K. Nakayama, S. Satoru, H. Shigetsugu, Q. Zhang, A. C. Chan and D. Y. Loh, *Nature*, **376**, 435.
6. A. C. Chan, T. A. Kadlecek, M. E. Elder, A. H. Filpovich, W.-L., Kuo, M. Iwashima, T. G. Parslow and A. Weiss, *Science*, 1994, **264**, 1599.
7. D. R. Knighton, J. Zheng, L. F. Ten Eyck, V. A. Ashford, N. Xuong, S. Taylor and J. M. Sowadski, *Science*, 1991, **253**, 407.
8. J. H. Hanke, J. P. Gardner, R. L. Dow, P. S. Changelian, W. H. Brissette, E. J. Weringer, B. A. Pollok and P. A. Connelly, *J. Biol. Chem.*, 1996, **271**, 695.

GENE REGULATING KINASES AND THEIR ROLE IN INFLAMMATORY DISEASES

Alan J. Lewis and Anthony M. Manning

Signal Pharmaceuticals, Inc
5555 Oberlin Drive
San Diego, CA 92121

1   INTRODUCTION

Phosphorylation of proteins by protein kinases regulates virtually all aspects of cell activity including growth, and differentiation, metabolism and secretion, immune response, fertilization and even memory acquisition. Protein kinases use the γ–phosphate of ATP (or GTP) to phosphorylate serine and threonine or tyrosine residues on proteins. This can result in protein conformation changes leading to altered function, subcellular localization or even to degradation.[1,2] The protein kinases belong to a very large superfamily that are related by their kinase domains (or catalytic domains) which consist of 250-300 amino acids. Protein phosphorylation cascades are used to sense changes in the extracellular environment and allow messages to be sent via an intracellular signalling network to the nucleus for suitable genomic responses. The human genome encodes between 1000-2000 protein kinases and since each kinase phosphorylates between 15-30 substrates the impact on cellular regulation is immense.[2,3]

With the discovery of this important enzyme class, and with the realization that abnormal protein phosphorylation may be the cause or consequence of numerous diseases, the discovery of kinase inhibitors became of great interest. Initial attempts focused on ATP-competitive inhibitors or serine/threonine kinases such as protein kinase C (PKC) and protein kinase A (PKA). However, these early efforts suggested that ATP-competitive inhibitors would not be selective.[4] The catalytic domains of serine/threonine kinases and tyrosine kinases had significant homology and drugs had to compete with mM cellular concentrations of ATP. It was thought that the discovery of potent, efficacious, selective and safe kinase inhibitors would be extremely difficult. Nevertheless, several protein kinase inhibitors are currently under clinical investigation mainly for anti-cancer indications. One compound SU101, also referred to as leflunomide (HWA-486) has antiproliferative and tyrosine kinase inhibitory activity and is in phase II clinical development.[5] It has also recently received FDA approval for treatment of rheumatoid arthritis (RA). However, its mechanism may not be exclusively due to kinase inhibition since it also inhibits dihydrooratate dehydrogenase, an enzyme in the uridine pathway. The increasing number of reports of potent ATP-competitive inhibitors that demonstrate relative selectivity has allayed early fears that kinase selectivity would not be possible.[4,6]

2   GENE REGULATING KINASES

With the surge of information generated by the discovery of new genes there has been an increased interest in mechanisms that control known and new genes in disease. Altered gene transcription is a fundamental process associated with numerous diseases including the inflammatory and autoimmune diseases. Gene transcription is regulated by the binding of activation factors, including transcription factors, to specific gene regulatory regions.[7] Transcription factors bind as monomers, dimers or multimers with other members of their families or with unrelated transcription factors. The ability of transcription factors to bind DNA and modulate gene transcription is tightly regulated in normal cells. Specific intracellular signal transduction pathways regulate transcription factor activity through changes in the level of phosphorylation of key residues and domains within the transcription factor.[8] The signalling pathways are highly diverse, yet display an extraordinary degree of specificity for a given transcription factor or transcription factor family. The identification and molecular cloning of multiple gene regulatory kinases which control key steps of these signal transduction pathways has sparked a new wave of drug discovery, based on using these regulatory enzymes and the transcription factors themselves as targets.

Four important transcription factor families, AP-1/ATF2, NF-κB, NFAT and STAT, appear to be the critical regulators of inflammatory genes including the cytokines, growth factors, inducible enzymes and cell adhesion molecules associated with inflammatory diseases. The activity of these transcription factors is regulated directly or indirectly by mitogen-activated protein kinases (MAPK) pathways.

2.1   MAP Kinase Pathways and the Regulation of Transcription Factor Function

Diverse extracellular stimuli modulate inflammatory gene transcription via phosphorylation cascades that utilize MAPKs.[8,9] MAPK cascades consist of three- or four-tiered signalling modules in which the MAPK is activated by a MAP kinase kinase (MAPKK), which in turn is activated by a MAP kinase kinase kinase (MAPKKK) (Figure 1).

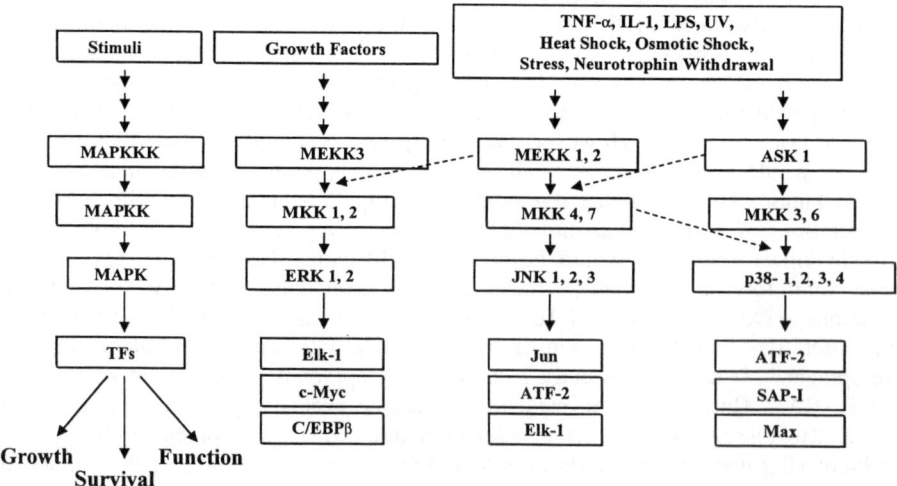

Figure 1 *Mitogen Activated Protein Kinase Pathways and Their Role in Gene Regulation*

The MAPKKK is itself activated by a small G protein such as Ras, either directly or via another upstream kinase.[10]   Three such MAPK signalling cascades, culminating in activation of the ERK, JNK and p38 MAP families of MAPK have been investigated in detail.   The JNKs and p38 kinases are activated in response to the pro-inflammatory cytokines TNF-α and IL-1, and by cellular stress (e.g. heat shock, osmotic shock, reactive oxygen metabolites, protein synthesis inhibitors, UV irradiation).   The MAPKs are proline-directed serine/threonine kinases, which are activated by phosphorylation on closely-spaced threonine and tyrosine residues within the activation loop; the activation sequences characteristic of the ERK, JNK and p38 MAP kinase families are TEY, TPY, and TGY, respectively.   These sequences are targets for phosphorylation by specific MAPKKs, dual specificity threonine/tyrosine kinases which are themselves activated by MAPKKK-mediated phosphorylation at a pair of serine residues in the activation loop.[8,9,10]

While there is some crosstalk between the major MAPK pathways, cells maintain exquisite specificity with extracellular or signals only activating their proper targets.   This is a result of several factors that include preferred interactions between kinases within a module and between MAPKs and their substrates.[11]   Recently, scaffold proteins that bind multiple components of the signaling cascade have been described.   For example, JIP-1 (JNK-interacting protein-1) first characterized as a cytoplasmic inhibitor of the JNK pathway has been shown to selectively bind the MAPK module, MLK → MKK7 → JNK.[12,13]   It has no binding affinity for a variety of other MAPK cascade enzymes.   Different scaffold proteins are likely to exist for other MAPK signaling cascades to preserve substrate specificity.   Many of the studies that have reported cross-talks between the different pathways have employed over-expression of members of the signaling cascades.   This often leads to the erroneous conclusion that no fidelity exists between the cascades.

MAPK pathways regulate the transcriptional activities of a variety of transcription factors, as well as the activities of a number of other cellular proteins involved in gene expression.   For example, MAPK pathways directly regulate AP-1-dependent transcription, both at the level of de novo synthesis of AP-1 family proteins and by controlling their transactivation function.[8,9,13,14]   Activator Protein-1 (AP-1) is a pivotal transcription factor which regulates T-cell activation, cytokine production, and production of matrix metalloproteinases.[15]   AP-1 includes members of the Jun and Fos families of transcription factors, which are characterized by basic region-leucine zipper (bZIP) DNA-binding domains.   AP-1 proteins bind to DNA and activate transcription as Jun homodimers, Jun-Jun heterodimers, or Jun-Fos heterodimers.   There are multiple Jun and Fos family members (c-Jun, JunB, JunD and c-Fos, FosB, Fra-1, Fra-2) which are expressed in different cell types and mediate the transcription of both unique and overlapping genes. AP-1 is also a component of the nuclear factor of activated T cells (NFAT) complex responsible for the transcription of the IL-2 gene and other cytokine genes in activated T cells.[16]

All three MAPK pathways are involved in the transcriptional regulation of Fos- and Jun-family genes.   The ERKs, JNKs, and p38 MAPKs each contribute to upregulation of c-Fos gene transcription, by phosphorylating and activating the Ets-family transcription factors Elk-1 and SAP-1.[15,17]   A major component of AP-1 regulation is a consequence of post translational modification; for example, c-Jun is regulated by phosphorylation at two N-terminal serines in the transactivation domain (amino acids 63 and 73).   This is accomplished by the c-Jun N-terminal kinases JNK1 and JNK2, although JNK2 binds c-Jun with a 10-fold higher affinity than JNK1,[18] and may be the physiologically relevant activator of AP-1.

The JNK protein kinases are encoded by three genes: JNK1, JNK2 and JNK3. JNK1 and 2 are ubiquitously expressed whereas JNK3 is selectively expressed in the brain, heart and testis.[11]   Gene transcripts are alternatively spliced to produce four-JNK1 isoforms, four-JNK2 isoforms and two-JNK3 isoforms.   Inhibitors of JNK-mediated AP-1 activation may prove to be novel anti-inflammatory/immunosuppressive agents that will inhibit inducible expression of inflammatory genes, without affecting AP-1 mediated housekeeping functions.   In T cells JNK activation by co-stimulation through the antigen and CD28 receptors correlates with IL-2 induction.[19]   Recently, the examination of JNK-deficient mice revealed that the JNK pathway is induced in Th1 cells (producers of IFN-gamma and TNFβ) but not in Th2 effector cells (producers of IL-4, IL-5, IL-6, IL-10 and IL-13) upon antigen stimulation.[20]   This suggests that JNK1 and JNK2 do not have redundant functions in T cells and that they play different roles in the control of cell growth, differentiation and death.

Mice with a homozygous disruption for the JNK3 gene are viable.[21]   However, they are resistant to the excitotoxic stress response elicited by kainic acid, a glutamate receptor agonist.   Kainic acid causes neuronal damage especially within the hippocampus.   The neurotoxicity of kainic acid possibly results from the induction of c-Jun and increased AP-1 DNA binding activity.   JNK3 inhibitors may be potentially useful in treating epilepsy and other neurodegenerative diseases such as stroke.   Recently the x-ray crystal structure of the unphosphorylated form of JNK3 was reported and should provide assistance in designing selective JNK3 inhibitors.[22]   JNK3 reveals similarities to the structures of cAMP-dependent protein kinase and ERK2 and p38.

2.2   MAPK Inhibitors

PD098059 (see Figure 2) was discovered in a screen for inhibitors of the ERK cascade.[23] PD098059 binds to the inactive form of MEK1, a primary MAPKK in the ERK cascade, with an IC50 of ~4μM and blocks the phosphorylation required for MEK activation. Specificity of action was indicated by the inability of PD098059 to inhibit phosphorylation mediated by c-Raf, JNK, p38, PKA, PKC, v-Src, active MEK1 and several other serine/threonine and protein tyrosine kinases (including receptor tyrosine kinases).[54] U0126 is another MEK inhibitor that has potent *in vitro* and *in vivo* efficacy in models of inflammation and delayed type hypersensitivity.[24]

While no specific JNK inhibitors have been described, isoform specific inhibitors are being sought by several pharmaceutical companies.   The greatest attention and most progress has been made in the discovery of p38 MAPK inhibitors.[25]   The important role of the p38 pathway in inflammatory processes resulted from studies using a series of pyridinyl imidazoles exemplified by SK&F 86002 and SB203580.[26]   These potent inhibitors of p38 activity block IL-1 and TNF production in lipopolysacharide stimulated human monocytes.   SB 203580 competes with ATP for binding to p38 and is remarkably selective.[27]   It does inhibit JNK2 but with a 10-20 fold lower potency than p38.   SB 203580 binds p38 by inserting into the ATP-binding pocket.   While the 4-fluorophenyl ring of the compound does not make contact with residues in the ATP-binding pocket it is in near proximity to the Thr 106 of the enzyme.   Mutation of this amino acid to Met 106 makes p38 insensitive to SB 203580.[28]   Thr 106 is conserved in p38β, another isoform of p38, that is sensitive to SB 203580, but is replaced by methionine in p38γ, p38δ, JNK1 and JNK2 (all much less sensitive to SB 203580).   Mutation of p38β Thr106 to Met 106 rendered p38β almost resistant to SB 203580 and the reverse mutation of Met 106 to Thr106 in p38γ and p38δ resulted in SB 203580 sensitivity.

Some of the earlier p38 inhibitors, including SB 203580, are inhibitory towards several cytochrome p450 isoforms (1A2, 2C9, 2C19, 3A4, 2D6). This is due to the high-affinity binding of the 4-pyridyl group to heme iron. As a consequence, replacements for the 4-pyridyl ring were sought and the pyrimidine analog as represented by SB 226882 has equivalent p38 inhibitory potency *in vitro* and is effective in *in vivo* mouse models measuring circulating TNF levels. Several other p38 inhibitors have been reported including L-167,307,[29] VK 19911,[30] and SC-102 RWJ 67657 and RWJ 68354.[31]

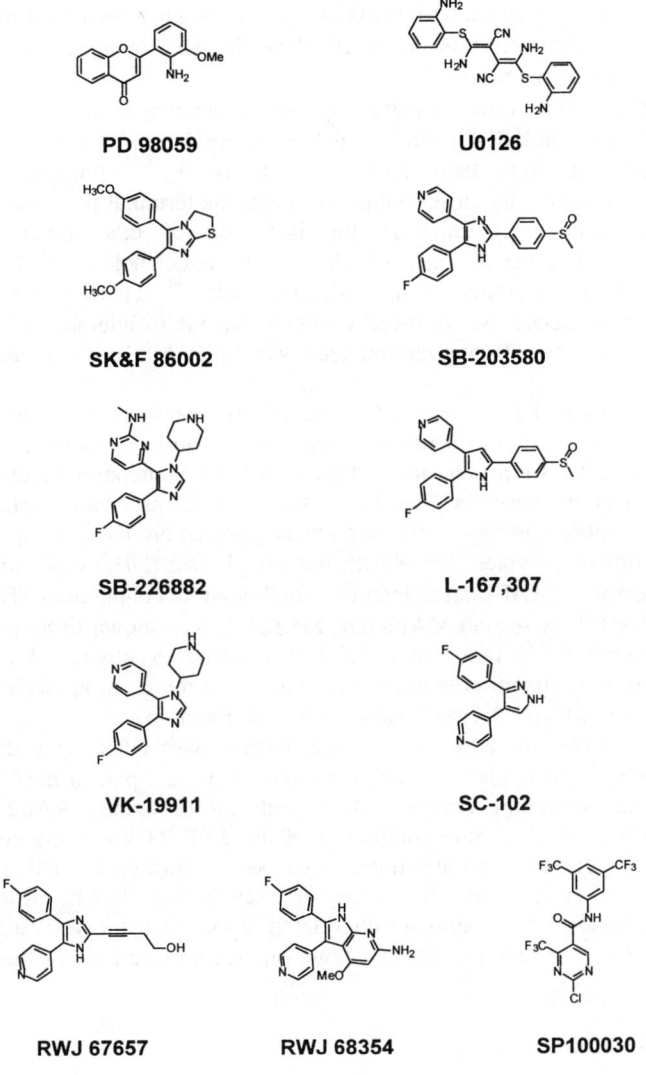

Figure 2 *Selective MAP Kinase Inhibitors and Inhibitors of NF-κB and AP-1 Activation*

2.3   NF-κB Pathways

Nuclear factor-κB (NF-κB) was first described as a B-cell-specific factor which bound to a short DNA sequence motif located in the immunoglobulin κ light chain enhancer, but it is now clear that NF-κB is expressed in all cell types and plays a broader role in gene transcription.[32-34] NF-κB plays a key role in the expression of many genes central to the inflammatory response and has been detected in a variety of inflammatory settings *in vivo* including in atherosclerotic and restenotic lesions, in septicemia in humans, in rheumatoid synovium, and in UV-damaged skin.[35] NF-κB exists in the cytoplasm in an inactive form associated with inhibitory proteins termed IκB, of which the most important may be IκBα, IκBβ, and IκBε.   Activation is achieved through the signal-induced proteolytic degradation of IκB in the cytoplasm.

Extracellular stimuli initiate a signalling cascade leading to activation of two IκB kinases, IKK-1 (IKKα) and IKK-2 (IKKβ) which phosphorylate IκB at specific N-terminal serine residues (S32 and S36 for IκBα, S19 and S23 for IκBβ).[36-39] Phosphorylated IκB is then selectively ubiquinated, by an E3 ubiquitin ligase, the terminal member of a cascade of ubiquitin-conjugating enzymes.   In the last step of this signalling cascade, phosphorylated and ubiquinated IκB, which is still associated with NF-κB in the cytoplasm, is selectively degraded by the 26S proteasome.[40]   This process exposes the nuclear localization sequence (NLS), thereby freeing NF-κB to interact with the nuclear import machinery and translocate the nucleus, where it binds its target to initiate transcription.

The IκB kinases IKK-1 and IKK-2 are related members of a new family of intracellular signal transduction enzymes, containing an amino-terminal kinase domain and a C-terminal region with two protein interaction motifs, a leucine zipper and a helix-loop motif.   There is strong evidence that IKK-1 and IKK-2 are themselves phosphorylated and activated by one or more upstream activating kinases, which are likely to be members of the MAPKKK family of enzymes.[38-39] NF-κB inducing kinase (NIK) was identified by its ability to bind directly to TRAF2, an adapter protein thought to couple both TNF-α and IL-1 receptors to NF-κB.[41]   A second MAPKKK, MEKK-1, was shown to be present in the IKK signalsome complex.[38,42] IKK-1 and IKK-2, their upstream activating kinases MEKK and NIK, and their downstream effector, the E3 ligase, all represent attractive targets for the discovery of drugs which selectively regulate NF-κB function.

In addition there are additional kinase targets associated with the upstream activation of NF-κB.[43] Receptor – interacting protein (RIP) is a part of the TNF-receptor (TNF-R1) associated signalling complex along with TRADD and TRAF2.   RIP is a serine/threonine kinase and the only component of the TNF-R1 signalling complex with enzymatic activity.   However, no substrates have been identified for RIP.   Analogous signaling components exist for the IL-1 receptor, they include IRAK, which is also a serine/threonine kinase that is autophosphorylated upon receptor activation.   IRAK associates with TRAF6 to activate NF-κB; however, recombinant IRAK has no kinase activity.

2.4   NF-κB Inhibitors

There are multiple targets that are amenable to small molecular blockade within the NF-κB activation pathway. No specific inhibitors of the kinases mentioned in Section 3.0 have been described. However, several NF-κB blockers have been reported and are reviewed elsewhere.[33,34] Antioxidants and free radical scavengers such as N-acetylcysteine, curcumin and caffeic acid block NF-κB activation. The 26S proteasome contains a chymotrypsin-like activity that degrades IκB, an activity that can be blocked by peptide aldehydes including MG115, MG341 and Z-LLF-CHO.[44,45]

Several clinically important anti-inflammatory drugs such as salicylates, gold and glucocorticoids also inhibit NF-κB induced gene expression. Glucocorticoids may exert some of their anti-inflammatory effects by inducing the synthesis of IκB.

2.5 NF-κB and AP-1 Inhibitors

A novel class of T cell-specific inhibitors of NF-κB and AP-1 activity was recently identified in a cell-based screening effort to identify modulators of inflammatory gene expression.[46,47] The most potent inhibitor in this series, SP100030, inhibits NF-κB- and AP-1-dependent reporter gene expression in stably-transfected Jurkat T cells with an IC50 of 30 nM. In stimulated Jurkat T cells, SP100030 inhibited the induced transcription of the IL-2, IL-8, TNF-α and GM-CSF genes with a similar IC50. The effects of SP100030 were specific for human T cells, demonstrating activity in 4 separate T cell lines and in primary T cells isolated from whole human blood. SP100030 displayed no activity in non-T cell lines, including monocytes, epithelial cells, fibroblasts, synoviocytes, osteoblasts and endothelial cells. SP100030 demonstrated efficacy in preclinical models of autoimmune disease, including adjuvant-induced arthritis, delayed type hypersensitivity, inflammatory bowel disease and allograft rejection. SP100030 represents a new class of T cell specific dual inhibitors of NF-κB and AP-1 mediated inflammatory gene expression, and suggests that cell-specific inhibitors of inflammatory gene expression can indeed be identified. An interesting implication, possibly related to the crossregulation of NF-κB and AP-1 activity by MAP kinase pathways, is that T cells possess a common target protein that controls the functions of both NF-κB and AP-1.

3 STRATEGIES FOR KINASE INHIBITORS

While protein kinases are challenging drug targets the development of multiple kinase inhibitors for major clinical conditions is anticipated. Despite the conventional dogma that the catalytic site inhibitors are non-specific the identification of very selective kinase inhibitors has created considerable optimism for the future.

In this age of increased chemical diversity it is anticipated new kinase inhibitor templates will emerge from chemical libraries natural product screening as well as the availability of massive combinatorial libraries. For example, selective protein kinase inhibitors have recently been described based upon the unexpected binding of 2,6,9-trisubstituted purines to the ATP-binding site of human cyclin-dependent kinase 2 (CDK2).[48] Such a combinatorial approach to modifying the purine scaffold creates an invaluable tool for identifying potentially selective kinase inhibitors. Natural products such as staurosporine and flavonoids such as quercetin and genistein have provided some of the earlier kinase inhibitor templates and will continue to be a valuable additional source of novel backbones. Finally, the rapid progress made in the production of crystal structures of a number of serine/threonine and tyrosine-specific protein kinases has identified the catalytic core of these important enzymes. We now know that this core consists of a small largely composed of an antiparallel β-sheet, and a large C-terminal domain which is mostly an α-helix. The ATP is bound in the deep cleft that exists between both lobes and exposes its phosphate groups to the opening of the cleft where the substrate also binds. Conserved residues of the C-terminal domain interact with the γ-phosphate of ATP and provide the catalytic machinery for the kinase reaction. Often these protein kinases are activated by phosphorylation of sites in an activation region near the opening of the cleft. As greater knowledge of active site inhibitors for p38 as well as other kinases emerges it is becoming clearer how to modify compounds to create greater potency and specificity.[22,27,30,49,50] Since protein kinases are intracellular enzymes, issues related to cell penetration, selectivity and *in vivo* efficacy and safety remain the challenge for the medicinal chemist. Biologically, enhanced screening capacities resulting from high throughput screening as well as availability of multiple recombinant human kinases has expedited selective kinase inhibitor identification. Since there are so many kinases within the cell rapid and broad profiling remains an important goal the secondary and tertiary events that are modified by selective kinase inhibition will be greatly facilitated by gene chip methodologies that will allow transcript profiles to be obtained. Such profiles will be extremely useful in evaluating the selectivity of drug candidates.

REFERENCES

1. S.K.Hanks and T. Hunter, *FASEB J.*, 1995,**9**, 1255.
2. T.Hunter, *Seminars Cell Biol*, 1994.
3. P.Cohen and M.Goddert, *Chem. Biol*,1998,**5**,R161.
4. M.R.Myers,W.He and C.Hulme, *Curr.Pharmaceut.Des.*,1997,**3**,502.
5. H.T.Silva, and R.E.Morris, *Expert Opin.Invest.Drugs*,1997,**6**,51.
6. J.C.Lee and J.L.Adams, *Curr.Opin.Biol.*,1995,**6**,657.
7. A.M.Manning and A.Rao, "Inflammation: Basic Principles and Clinical Correlates", Lippincott-Williams, New York, 1998, p.1159.
8. M.Karin and T.Hunter, *Curr.Biol.*,1995,**5**,747.
9. B.Su and M.Karin, *Curr.Opin.Immunol.*, 1996,**8**,402.
10. G.R.Fanger,P.Gerwins,C.Widmann et al, *Curr.Opin.Genet.Devel*,1997,**7**,67.
11. Y.T.Ipp and R.J.Davis, *Curr.Biol.*,1998,**10**,205.
12. M. Dickens, J.S. Rogers, J. Cavanagh, et al., *Science*, 1997, **277**,693.
13. A.J.Whitmarsh, J. Cavanaugh, C. Tournier, et al., *Science*,1998,**281**,1671.
14. A.J.Whitmarsh and R.J.Davis, *J.Molec.Med.*,1996,**74**,589.
15. .V.C.Foletta, D.H.Segal and D.R.Cohen, J.Leukoc.Biol,1998,**63**,139.
16. A.Rao, C.Luo and P.G.Hogan, *Ann.Rev. Immunol.*,1997,**15**,707.
17. A.J.Whitmarsh, P.Shore,A.D.Sharrocks and R.J.Davis, *Science*,1995,**269**,403.
18. T.Kallunki, B.Su, Tsigelny et al, *Genes Dev.*,1994,**8**,2996.
19. B.Su,E.Jacinto, M.Hibi et al, *Cell*,1994,**77**,727.
20. D.D.Yang, D.Conze, A.J.Whitmarsh et al, *Immunity*,1998,**9**,575.
21. D.Yang,C.Tournier,M.Wysk et al,Nature,1997,**389**,865.
22. X.Xie,Y.Gu,T.Fox,J.T.Coll, et al., *Structure*,1998,**6**,983.
23. D.T.Dudley, L.Pang, S.J.Decker, et al., *Proc. Nat. Acad. Sci. USA*,1995,**95**,7686.
24. M.F. Favata, K.Y. Horiuchi, E.J. Manos, et al., *J. Biol. Chem.*, 1998,**273**,18623.
25. J.C.Lee,J.T.Laydon,P.C.McDonnell, et al., *Nature*,1994,**372**,739.
26. A.M.Badger, J.N.Bradbeer, B.Votta et al, *J.Pharmacol.Exp.Ther.*1996,**279**,1453
27. L.Tong,S.Pav, D.M.White et al., *Nature Struct. Biol.*,**4**,311.

28. P.A.Eyers,M.Craxton,N.Morrice et al, *Chem.Biol.*,1998,**5**,321.
29. S.E. De Losazto, D. Visco, L. Agarwal, et al., *J. Bioinorg. Med.*, 1998,**8**,2689.
30. K.P.Wilson, P.G.McCaffery, K.Hsiao et al, *Chem. Biol*, 1997,**4**,423.
31. J.R. Henry, K.C. Rupert, J.H. Dodd, et al., *J. Med. Chem.*, 1998,**41**,4196.
32. M.J.May and S.Ghosh, *Immunol. Today*,1998,**19**,80.
33. P.A.Baeuerle and V.R.Baichwal, *Adv.Immunol.*1997,**65**,111.
34. A.S.Baldwin, *Adv.Immunol.*,1996,**14**,649.
35. A.M.Manning and D.C.Anderson, *Ann. Rev. Med. Chem.*1994,**29**,235.
36. H.Regnier,H.Song,H.Gao et al,Cell,1997,**90**,373.
37. J. Didonato,M.Hayakawa,D.M.Rothwarf et al, *Nature*, 1997,**388**,853.
38. F.Mercurio,H.Zhu,B.W.Murray et al, *Science*,1997,**278**,860.
39. J.D.Woronicz, X.Gao,Z.Cao et al, *Science,*1997,**278**,866.
40. Z.Chen, J.Hagler, V.J.Palombella et al, *Genes Dev*,1995,**9**,1586.
41. N.L.Malinin,M.P.Boldin,A.V.Kovalenko and D.Wallach, *Nature*,1997,**385**,540.
42. F.S.Lee, J.Hagler, Z.J.Chen and T.Maniatis, *Cell,* 1997,**88**,1586.
43. V.J.Baichwal and P.A.Baeuerle, *Ann.Rept. Med. Chem.*,1998,**33**,233.
44. V.J.Palombella, O.L.Rando, A.L.Goldberg and T.Maniatis, *Cell*,1994,**78**,773.
45. B-M.E.Traenckner, S.Wilk and P.A.Baeuerle, *EMBO J,*1995,**13**,5433.
46. M.J.Suto and L.J.Ransone,*Curr.Pharmaceut.Des*,1997,**3**,515.
47. R.W.Sullivan, C.G.Bigam,P.Erdman et al, *J.Med.Chem.* 1998,**41**,413.
48. N.S.Gray, L. Wodicka, A-M,W.H. Thunnissen et al, *Science*, 1998,**281**,533.
49. S.S.Taylor and E.Radzio-Andzelm, *Structure,*1994,**2**,345.
50. J.Z. Hang, F. Zhang, D. Ebert, et al., *Structure*, 1995,**3**,299.

THE 6-HYDROXY GROUP OF INOSITOL 1-PHOSPHATE SERVES AS AN H-BOND DONOR IN ITS CATALYTIC HYDROLYSIS BY INOSITOL MONOPHOSPHATASE: DESIGN AND SOLID-PHASE SYNTHESIS OF MECHANISM BASED INHIBITORS.[1]

David Gani,* Mahmoud Akhtar, Martin Beaton, David Miller, Roger Pybus, Duane Stones, Trevor Rutherford and John Wilkie.

School of Chemistry and Centre for Biomolecular Science, The Purdie Building, The University, St. Andrews, Fife, KY16 9ST, UK

1 SUMMARY

Inositol monophosphatase plays a pivotal role in the biosynthesis of the secondary messengers and is believed to be a target for lithium therapy. It is established how lithium works but details of the mechanism for the direct magnesium ion activated hydrolysis of the substrate remain unclear. It is known that substrates require a minimal 1,2-diol phosphate structural motif which in D-*myo*-inositol 1-phosphate relates to the fragment comprising the 1-phosphate ester and 6-hydroxy groups. We have determined that inhibitors which are D-*myo*-inositol 1-phosphate substrate analogues possessing 6-substituents larger than the 6-hydroxy group of the substrate, for example, the O^6-methyl analogue, are able to bind to the enzyme in a congruous manner to the substrate. It is demonstrated, however, that such compounds show no substrate activity whatsoever. It it also shown that a 6-amino group is able to fulfil the role of the 6-hydroxy group of the substrate in conferring substrate activity and that a 6-methylamino group is similarly able to support catalysis. The results indicate that a 6-substituent capable of serving as a hydrogen bond donor is required in the catalytic mechanism for hydrolysis. These results help to distinguish between two mechanisms for hydrolysis which have been proposed: one in which a water molecule associated with a buried magnesium ion ($Mg^{2+}1$) attacks the phosphate ester *via* an inline displacement of the inositol moiety, and one in which adjacent displacement occurs from a water molecule associated with a second magnesium ion ($Mg^{2+}2$). The results obtained here fit very well with the latter mechanism and further indictate that the role of the 6-hydroxyl group of the substrate is to position and stabilise the formation of hydroxide ion on $Mg^{2+}2$ such that it is orientated correctly for the adjacent displacement of inositolate from the phosphorus atom. Methods that exploit the mechanism of catalysis in the design and solid-phase synthesis of inhibitors are presented and new protocols for quantifiying the extent of resin loading using ^{19}F gel-phase NMR spectroscopy are described. It is also shown the ytterbium triflate serves as an excellent catalyst for resin-bound oxirane solvolysis and that stannic chloride can be used as for the facile cleavage of benzyl ethers derived from Merrfield resin, even in the presence of acid labile functional groups.

2 INTRODUCTION

Mammalian brain inositol monophosphatase (IMPase, EC 3.1.3.25) provides inositol for the biosynthesis of the key secondary messenger precursor, phosphatidylinositol 4,5-bisphosphate. Phosphatidylinositol 4,5-bisphosphate is hydrolysed by phosphatidylinositidase C, in response to receptor occupation to give diacylglycerol (DAG) and inositol 1,4,5-trisphosphate (Ins 1,4,5-P$_3$) each of which mediate signal transduction through specific interactions with their own targets.[2-4] DAG activates protein kinase C[5,6] which modulates the activity of many enzymes through phosphorylation,[7] while Ins 1,4,5-P$_3$ causes the release of calcium ions from an intracellular store.[8] Brain cells vary in their ability to take-up inositol[3,9] and a series of phosphatases exist to sequentially hydrolyse Ins 1,4,5-P$_3$ and other inositol polyphosphates *via* the *bis*-phosphates to give inositol 1- and 4-monophosphates, substrates for IMPase. The effect of blocking IMPase with the inhibitor Li$^+$ cation leads to the depletion of free inositol in brain cells[10,11] and thus, several groups have suggested that IMPase might be the target for lithium ion in manic depression therapy. More recently the kinetics of inhibition by Li$^+$ have been probed[12-18] and there is now substantial evidence to show that the activity of IMPase would be very low at therapeutic concentrations of Li$^+$ ion (~1 mM). Importantly, the sensitivity of IMPase to Li$^+$ has been shown to be acutely dependent upon phosphate dianion concentration. This reaction product is present in brain cells at high concentration, indicating that the efficacy of Li$^+$ is greater in cells than was originally thought.[18] Subsequent studies have defined how Li$^+$ ions interact with the enzyme, see below.

IMPase catalyses the hydrolysis of a range of phosphate esters including both enantiomers of *myo*-inositol 1-phosphate (Ins 1-P, **1**) and Ins 4-P,[14] ethane-1,2-diol phosphate[19] and 2'-ribonucleoside[13,19] and 2'-ribofuranoside phosphates.[20] IMPase shows an absolute requirement for divalent metal ions[13] such as Mg^{2+}, and it is now known that two Mg^{2+} ions bind at each active-site of the homodimer.[19-22] The reinterpretation of kinetic data,[18,23,24] taking account of the requirement for two Mg^{2+} ions, suggests that one metal ion (Mg^{2+}1) binds to the enzyme before the substrate and the second metal ion (Mg^{2+}2) binds after the substrate.[25] Lithium ion appears to replace Mg^{2+}2 in a phosphate product complex in its uncompetitive inhibition of the enzyme.[25]

Synthetic transformations of Ins 1-P (**1**) revealed that the 3-OH and 5-OH groups are not necessary for binding or catalysis.[26-28] Indeed, 3,5-dideoxyinositol 1-P (**2**) is a good substrate. Further probing showed that the 4-OH and 2-OH groups were important for binding, whereas the 6-OH group was essential for catalysis. Deletion[26,27] or alkylation[28] of the 6-OH group in Ins 1-P (**1**) lead to tight binding competitive inhibitors of IMPase, for example, compounds **3** and **4**. While phosphate ^{18}O-ligand exchange studies established that the enzyme did not operate *via* a substituted enzyme mechanism, but rather, that water displaced the phosphate ester group directly,[18,29] it was only recently that proposals emerged on how this might be achieved.

Figure 1 *Showing the two proposed mechanisms for hydrolysis:*
A) adjacent attack and B) in-line displacement by the nucleophile

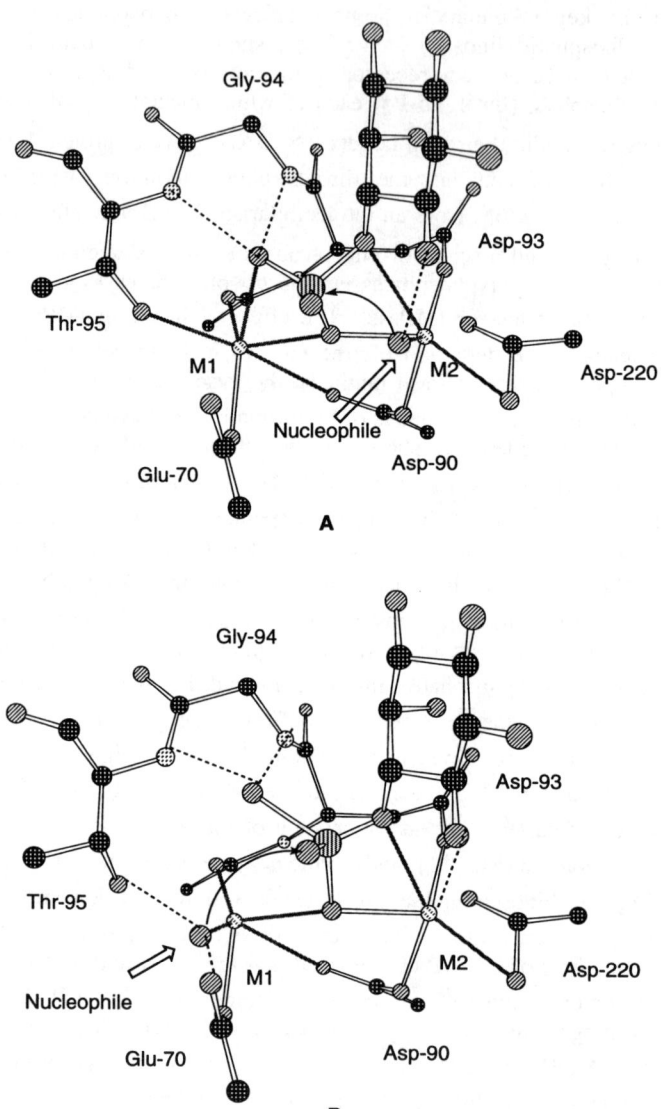

On the basis of kinetic data for hydrolysis of different substrates, data for ^{18}O-phosphate ligand exchange and for inhibition,[18,30,31] together with the X-ray crystal coordinates of a Gd^{3+} sulfate complex of the enzyme,[32] and the results of extensive modelling studies, we proposed a 3-D structure for the active complex in which the second ion to bind, $Mg^{2+}2$, coordinates to and activates the attacking nucleophilic water molecule or hydroxide ion (Figure 1A).[19,25]

According to this model, the role of the catalytic 6-OH group of Ins 1-P (**1**) is to H-bond to the nucleophile so that it is properly positioned to attack the phosphate P-atom *via* the adjacent displacement of the inositol moiety.[19,25] This mechanism differs significantly in detail from a proposal put forward by the Merck, Sharp and Dohme (MSD) group,[22,24,33] which was derived largely from X-ray crystal data for different enzyme metal ion complexes, see Figure 1B. Here, the more deeply buried metal ion ($Mg^{2+}1$) coordinates to, and activates the attacking nucleophile such that it would replace the inositol moiety through an inline displacement, with inversion of configuration at the P-atom.[22] A major problem with the mechanism proposed by the MSD group is that it does not explain why the 6-OH group of the substrate is essential for catalysis. Nevertheless, at a *structural level*, the positions of the substrate binding groups and the metal ions within the active complex are virtually identical in the two models. Recent theoretical studies on the mechanism of phosphoryl transfer indicate that the transition state energy differences for adjacent displacement, with retention of configuration, and inline displacement, with inversion, are small.[34] There is chemical precedent for both types of mechanism,[35] hence, it is not possible, *a priori,* to favour one over the other.

3 RESULTS

3.1 Stereochemical Course of Phosphoryl Transfer

Determination of the stereochemical course of phosphoryl transfer from the substrate to water would distinguish between the two mechanisms. Such a strategy requires the synthesis of samples of the substrate analogue, inositol phosphorothioate, stereospecifically labelled with two isotopes of oxygen and enzymic hydrolysis with water derived from a third isotope of oxygen, to give samples of chiral $[^{18}O,^{17}O]$-thiophosphate. Inositol phosphorothioates are known to be substrates[36] and appropriately chirally labelled samples have been prepared in our laboratory.[37] However, the rate of hydrolysis is very low and so far, it has not been possible to isolate sufficient quantities of the chiral $[^{18}O,^{17}O]$-thiophosphates for determination of the stereochemical course.[38]

3.2 Modelling the Role of the 6-Hydroxy Group

A major difference between the two mechanisms depicted in Figure 1 is the involvement of the 6-hydroxy group. There is no direct role for this group in the model put forward by MSD, but in the alternative mechanism, the 6-OH group must form a hydrogen bond with the attacking nucleophile if it is to align the nucleophile properly for attack on the phosphate ester P-atom.

The inhibitor (1R,2R,4R,6R)-6-methoxy-3,5-dideoxyinositol 1-phosphate **4** was prepared from (1R,2R,4R,6S)-1,6-epoxy-2,4-*bis*-benzyloxy-cyclohexane **5**[39] as described previously for the racemic material.[40] Incubation of the compound **4** with vast excesses of IMPase resulted in the formation of no inorganic phosphate whatsoever, as determined by ^{31}P-NMR spectroscopy. Even after several days, no hydrolysis was detected and the result was verified using a highly sensitive malachite green spectrophotometric assay.[18] This finding indicates that methyl ether **4** is not a slow, tight-binding substrate for the enzyme. The result also suggests that the 6-OH group in the substrate does not serve as an hydrogen bond acceptor. However, it could not be ruled-out that the presence of the methyl ether group was either too large to allow the cyclitol ring to bind in a productive fashion, or that the lone-pairs on the ether O-atom were displaced from their normal position in the substrate through a rotation about the C6-O bond.

Modelling studies on the protein-6-OMe-inhibitor **4** complex using the coordinates generated in fitting the original mechanistic data to the X-ray crystal structure[25] indicated that small 6-O-alkyl ether groups, such as those present in compounds **4** and **6**, would not disturb any interaction between the cyclitol and the protein, Figure 2.

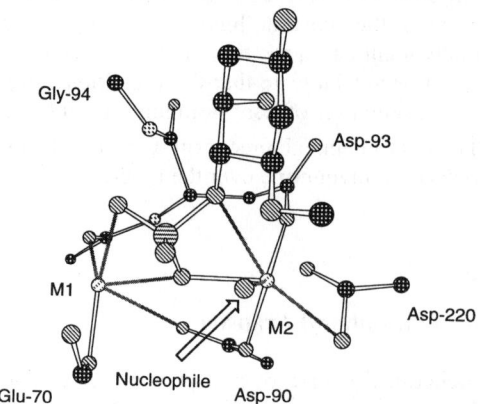

Figure 2 *Showing the modelled binding mode of inhibitor (4)*

It was also evident that for the 6-OMe-inhibitor, a lone-pair of electrons on the O^6-atom could align in the correct orientation to serve as a hydrogen bond acceptor for the putative $Mg^{2+}2$-bound water molecule.

Further analysis performed by re-relaxing structures formed by replacing the O^6-atom by an N-atom indicated that there was enough room for 6-alkylamino groups to bind in the active-site. For the free base forms of the 6-methylamino moiety, two low energy structural arrangements were possible in which there was a hydrogen bond between the nucleophilic water molecule and the N^6-atom. These differed only in whether the secondary NH proton served as an H-bond donor to $Mg^{2+}2$-coordinated hydroxide ion or, whether the $Mg^{2+}2$-bound water molecule (or hydroxide ion) served as an H-bond donor to the N^6-atom, Figure 3.

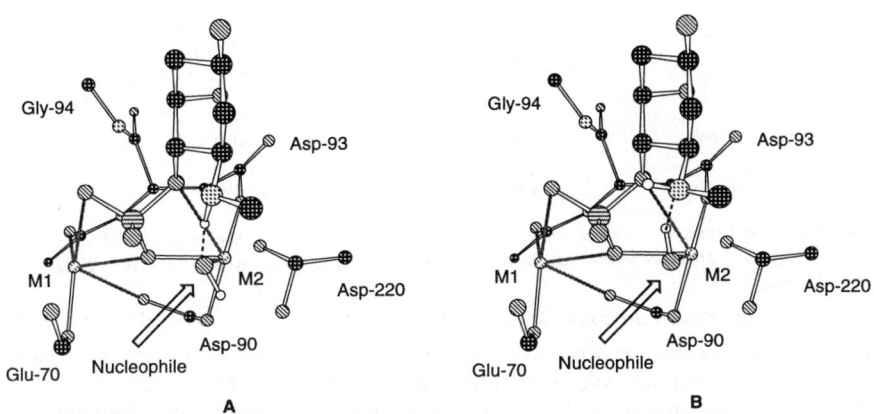

Figure 3 *Showing two possible conformations for the bound 6-methylamino compound (8).*

Such interactions appeared to be able to fulfil the precieved requirements for catalysis in the natural inositol phosphate substrates, for example, Ins 1-P **1**, and thus it appeared that 6-amino-3,5-dideoxyinositol 1-phosphates should serve as substrates. No previous work had investigated the properties of 2-aminocyclitol 1-phosphates and so it was necessary to prepare and test the parent 6-amino-3,5-dideoxyinositol 1-phosphate **7** as a substrate first, before addressing the question of the steric influence of the N^6-alkyl group.

3.3 Synthesis of 6-amino 3,5-dideoxyinositol 1-phosphates

Recently, we reported on the preparation of (1R,2R,4R,6S)-1,6-epoxy-2,4-*bis*-benzyloxy-cyclohexane **5** from (-)-quinic acid.[39] The absolute stereochemistry of this intermediate faciltated its conversion, with inversion of configuration at C-6, to 6-alkyloxy derivatives that could be subsequently converted to 1-phosphate ester derivatives, that were known to be the most active stereoisomers for inositol monophosphatase inhibition,[39,40] for example, compound **6**. It, therefore seemed expedient to employ the epoxide **5** in the synthesis of the amine derivatives **7** and **8**. Accordingly, using a modification of Crotti's procedure[41] for epoxide aminolysis, epoxide **5** was heated with either aqueous ammonia or methylamine in a sealed tube in the presence of 0.2 equivalents of ytterbium (III) triflate to afford each of the amino alcohols **9** and **10** in quantitative conversion,[42] Scheme 1.

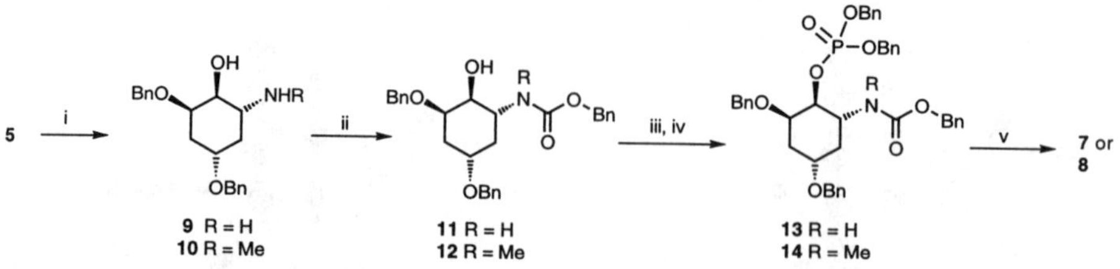

Scheme 1 *Reagents and conditions* : i. aqueous amine, 0.2eq. Yb(III)(OTf)$_3$, 65 °C, Sealed tube, 12-48 h; ii. 2 mol dm^{-3} NaOH, benzylchloroformate, 0 °C, 2 h; iii. diphenylchlorophosphate, TEA, DMAP, DCM, 20 °C, 12 h; iv. BnONa, THF, -70 °C, 2 h; v. Pd/C, H$_{2(g)}$, MeOH, 24 h.

The compounds displayed the expected analytical and spectroscopic properties and analysis of the ^{1}H-NMR spectra showed that the reaction had proceeded with inversion of configuration at C-6 and that no or very little reaction had occurred *via* attack at C-1.

The 6-amino cyclohexanols **9** and **10** were treated with benzyl chloroformate to give the benzyl ureathanes **11** and **12** in yields of 57 and 65%, respectively, and the products were phosphorylated with diphenyl chlorophosphate to give the diphenyl phosphate triesters.   Transesterification with sodium benzyloxide to give the dibenzyl phosphate esters **13** and **14** and catalytic hydrogenolytic removal of all five benzyl protecting groups, to give the required amines **7** and **8**, was acheived using procedures previously optimised for the 6-alkyloxy derivatives (*e.g.* **4**).[40] The amines **7** and **8** were obtained in yields of 88 and 89%, respectively, from the dibenzyl esters **13** and **14**, and were stored as their cyclohexylammonium salts.[42] The salts and all of their synthetic precursors displayed the expected analytical and spectral data (full details will be reported elsewhere) and were ready to be tested for biological activity.

3.4 Determination of biological activity

Buffered deuterium oxide solutions of each of the amino phosphate esters (15 mM) containing magnesium chloride and IMPase were monitored by ^{1}H-NMR spectroscopy and spectra were aquired at 15 minute intervals. The 6-amino phosphate **7** was hydrolysed completely within 6 h to give the expected product, indicating that it is a moderate substrate. Note that the physiological substrate **1** was hydrolysed completely within 15 minutes under these conditions. The result indicates that a 6-amino group is able to fully support catalysis and, indeed, is the first example of a compound possessing any substituent other than a hydroxyl group at C-6 to have been shown to display substrate activity.

The 6-methylamino phosphate **8** took 10 h before an obvious time-dependent change could be observed. The result indicates that the 6-methylamino phosphate **8** is a poor substrate for IMPase. Albeit encouraging, the very low rate of reaction precluded a firm conclusion on the basis of the observation alone because the presence of trace levels of contaminating non-specific phosphatases, which can act upon the 6-methylamino phosphate **8**, could not be excluded. Furthermore, it is known that inorganic phosphate is a good competitive product inhibitor (K_i = 0.3 mM).[14,18] Thus, a very slow observed rate of reaction for the substrate could have been indicative of a high substrate K_m value and, or, a low value of k_{cat}.

In order to resolve the ambiguities the V_{max} and K_m values of both 6-aminocyclitol 1-phosphates **7** and **8** were determined using standard assay conditions.[18] The primary amine **7** displayed a relative V_{max} value of 112 and a K_m value of 515 μM and the methyl homologue **8** gave values of 56 for V_{max} and 140 μM for K_m, respectively. Thus, it is established that both amino groups can provide the necessary interactions with the nucleophile to support catalysis. It is also established that the reason the methyl ether **4** does not serve as a substrate is not because it could not bind in the appropriate manner, in accord with the theoretical predictions.

4 MECHANISM

In the absence of information on the stereochemical course of the IMPase reaction with respect to phosphorus, our attentions have been focussed on the design and evaluation of structural probes for the coordination sphere of $Mg^{2+}2$, see Figure 1. According to the mechanism proposed in Figure 1A,[25] extension of the 6-OH group of the substrate by an ethylene bridge, as in compound **15**, places the 2'-OH group of the 'pendant arm' into the position of the nucleophilic molecule. Recently, the biological activity of both enantiomers of **15** and some closely related inositol monophosphate analogues (compounds **4** and **6**) including the 2'-phosphate **16** and the 1,2'-cyclic phosphate **17** were assessed within the context of the two plausible mechanisms.[40,43] The results showed that the racemic 6-methyl ether **4** was a competitive inhibitor (K_i = 2.5 μM) of almost identical efficacy to the 6-deoxy analogue.[26]

15	**16**	**17**

The racemic 6-propyl ether analogue **6** proved to be slightly more potent and displayed a K_i value of 1.2 μM.[43] This finding was expected and accords with earlier results which showed that the presence of a large lipophilic side-chain appended to C-6 enhanced inhibitor binding.[28,39] In the light of this analysis it might have been expected that the potency of the racemate of the 6-hydroxyethyl ether **15**, with its hydrophilic side-chain, should be very much lower than the isosteric hydrophobic propyl ether **6**. However, previous modelling studies had shown that groups attached to the cyclitol at the 6-position can access either of two quite distinct and contrasting regions of the active site:[25] the lipophilic pocket formed by Val-40 and Leu-42 and to a lesser extent, Trp-219 and Ile-216 (the space occupied by the adenine moiety in the substrate 2'-AMP [and used previously by Merck in the design of inhibitors][28]) and, the hydrophilic site near the $Mg^{2+}2$ ion normally occupied by a nucleophilic water molecule or hydroxide ion, according to the hypothesis, Figure 1A.

When tested, the racemic hydroxyethyl ether **15** behaved as a competitive inhibitor, displayed a K_i value of 1.8 μM[43] and showed no tendency to serve as a substrate. Furthermore, the compound did not undergo transesterification to give compound **16**. This pleasing result accords with the idea that the hydroxyethyl arm can access the coordination sphere of the $Mg^{2+}2$ ion, although clearly more information was required to confirm this notion.

From the earlier modelling work[25] it was possible to predict that, if the 2'-hydroxyethyl ether **15** was able to bind with its side-chain in contact with $Mg^{2+}2$, the (1R,2R,4R,6R)-enantiomer should be a better inhibitor than its (1S,2S,4S,6S)-antipode. Indeed, both displayed competitive inhibition and the K_i value of the (+)-(1S,2S,4S,6S)-hydroxyethyl ether was 60 μM while that for the (-)-(1R,2R,4R,6R)-hydroxyethyl ether **2** was 120-fold lower at 0.5 μM.[43] These results were in accord with theoretical predictions[25] and it appeared that the 2'-hydroxy group could bind in the coordination sphere for $Mg^{2+}2$. It was noted that the (-)-(1S,2R,4S)-antipode of the 6-deoxy analogue **3** which possesses the same spatial configuration as (-)-**15**, was found to be much more potent than its (+)-antipode.[26] It was further reasoned that compound **3** might behave as an inhibitor because there is no 6-hydroxy group available to H-bond to the water molecule in directing it to attack the P-atom. Also in accord with this finding was the observation that the (-)-(1R,2R,4R,6R)-antipode of the propyl ether **6** in which the absolute stereochemistry is defined by the stereochemistry of the starting material, (-)-quinic acid, is more active as an inhibitor than the racemate with a K_i value of 0.87 μM.[39]

In order to further investigate the hypothesis that the side-chain of the hydroxyethyl ether **15** could displace the nucleophilic water molecule, the theoretical intramolecular transesterification product, inhibitor **16**, was prepared. It was argued that it should only be possible to maintain all of the interactions of the peripheral ring hydroxy groups with the enzyme simultaneously with all of the interactions of the phosphate group with the enzyme bound metal ions, if the bridging ester O-atom in the phosphoethyl group of compound **16** could interact with $Mg^{2+}2$. Evidence in support of this arrangement was obtained when compound **16** was shown to be a good competitive inhibitor with a K_i value of 8.5 μM, only five times higher than the racemate of the isomeric 6-hydroxyethyl ether 1-phosphate **15** and the lowest K_i value, by far, for any known primary alkyl phosphate inhibitor for IMPase.[40,43]

Inspite of the weight of evidence which suggested that hydrophilic side-chains appended to the 6-position of the cyclitol could interact with $Mg^{2+}2$, and that this was possible because hydrophilic groups could force the water molecule from the coordination sphere of $Mg^{2+}2$, the catalytic role of the 6-OH group in the substrate remained obscure. If water was indeed, the nucleophile, why should it not act as a hydrogen bond donor in, for example, the 6-methyl ether **4**? Since it is established here that the compound shows no substrate activity whatsoever, arguably, it must be because the electron density on the 6-oxygen atom is not optimally located for H-bond formation due to steric interactions caused by the methyl group. This latter explanation was not in accord with the low observed K_i value for the compound[26] or, the modelling studies presented here, the results of which indicate that there is plenty of space for an extra methyl group to exist in the correct conformation to facilitate H-bond formation between the O^6-atom and the water molecule.

Clearly, a $Mg^{2+}2$-bound water molecule possesses only one available lone-pair and cannot serve, simultaneously, as a hydrogen bond acceptor and a nucleophile. So if water is the nucleophile, it must be an H-bond donor and there appears to be no rational explanation. On the other hand, if a $Mg^{2+}2$-bound hydroxide ion is the nucleophile, its O-atom can simultaneously bind to the metal ion, act as a nucleophile and serve as an H-bond acceptor in interacting with a proton from the 6-OH group of the substrate **1**. Thus, the problem appeared to reduce to two simple questions. Can a C-6 amino group serve as an H-bond donor and confer substrate activity? And, if it can, can a bulky C-6 methyl amino group serve as an H-bond donor in supporting catalysis? If the answer to both questions was yes, there would be no doubt that H-bond formation between the $Mg^{2+}2$-bound hydroxide ion (or water molecule, see below) and the 6-OH proton was essential for catalysis. However, it would not prove that the $Mg^{2+}2$-bound species was the nucleophile, although this point is argued further below.

The results presented here clearly demonstrate that an H-bond donor at C-6 is absolutely essential for catalysis. Both the 6-aminocyclitol phosphate **6** and the 6-methylamino analogue **7** are respectable substrates, whereas the 6-methyl ether **4** shows no activity whatsoever. If hydroxide ion is the nucleophile, as certainly appears to be the case now, other features of the mechanism fall into place.

First, V/K for Ins 1-P increases with increasing pH and does not plateau below pH 9.2, the upper limit of the study, although V_{max} peaks at 8.5.[18] This indicates that a deprotonation step with an apparent pK_a value of greater than 9.2 is required for catalysis in either the *free* enzyme, or other *unbound* species, the cofactor, Mg^{2+}, the substrate, Ins 1-P, or the other substrate, water.[44] The substrate is known to be 'sticky',[17] to bind to the enzyme before the $Mg^{2+}2$ ion, and to fully ionise as the dianion with a pK_a value of 6.2. Thus, it seems likely that the true second substrate which is known not be 'sticky'[18] is Mg^{2+}-OH_2 which possesses a pK_a value of 12.6.[45] Second, the role of the $Mg^{2+}2$ ion becomes apparent, to locate and reduce the pK_a value of the initially bound water molecule, if it is not already bound as hydroxide ion. Donating a lone-pair of electrons to the 6-OH proton in the substrate further reduces the pK_a value of the water molecule and stabilises hydroxide anion. H-Bonding also fixes the position of the ion and presumably stops it from rolling around on the Mg^{2+}-ion which becomes important as the ground state progresses towards the structure of the transition state.

In accord with these ideas, it is known that Zn^{2+} forms of the enzyme (previously referred to as zinc-dependent acid phosphatase)[46] operate best at much lower hydroxide ion concentrations (*ca.* pH 5.0 to 6.5) than the Mg^{2+} forms (*ca.* pH 6.7 to >9.0) for the substrates Ins 1-P and 4-nitrophenyl phosphate (4-NPP). Obviously 4-NPP does not possess a group equivalent to that of the 6-OH group of Ins 1-P **1** and it is interesting to note that with the Mg^{2+} form of the enzyme its activity is 30-times lower than that of Ins 1-P, and its optimal activity is shifted to higher pH. The effect of changing from Mg^{2+} to Zn^{2+} seems to reduce the requirement for hydroxide ions by a factor of *ca.* 10^3-fold for 4-NPP which is consistent with the proposed role of the metal ion in activating the nucleophile[19,25] and consistent with the differences in the pK_a values of 9.0^{47} and 12.6^{45} for water molecules in the complexes $Zn^{2+}(H_2O)_6$ and $Mg^{2+}(H_2O)_6$, respectively.

Whether or not deprotonation of the Mg^{2+}2-bound water molecule occurs before or after the binary complex has entered the active-site is difficult to assess. Double reciprocal plots of initial rate versus Mg^{2+} concentration are curved due to the requirement for two Mg^{2+} ions and therefore, V/K for Mg^{2+} is concentration dependent and dependent on the structure of the substrate.[17,18] The Mg^{2+}2 ion is known to bind last, after all of the other species,[18] and there is a large water-filled cavity linking the enzyme bound position of the Mg^{2+}2 ion to bulk solvent.[25] Therefore, it is quite possible that solvent mediated deprotonation occurs after binding. If a group on the enzyme does serve as a base, it must be the carboxylate group of Asp220 although it seems more likely that it should remain anionic and repel the hydroxide ion towards the P-atom as the structure compresses towards the transition state, Figure 4.

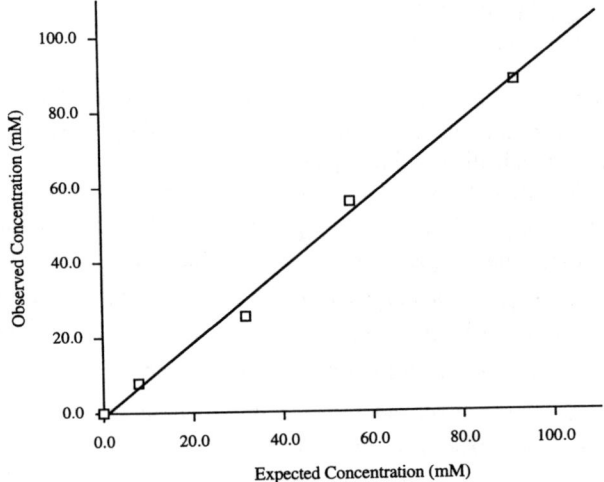

Figure 4 *Correlation of expected signals with those observed for known samples*

5 SIGNIFICANCE

The results presented here are completely in accord with the predictions of our earlier molecular modelling work[25] and provide support for the structural detail of the active complex, Figure 1A. The results indicate that there is a nucleophilic hydroxide ion bound to Mg^{2+}2 and simultaneously hydrogen bonded to the 6-OH proton of Ins 1-P. The results do not rule-out the alternative in-line mechanism depicted in Figure 1B[22] but, if such a mechanism is followed, it is extremely difficult to understand why the replacement of the 6-OH group in a substrate by an alkoxy group or by a hydrogen atom should give a tight-binding non-hydrolysable compound. Clearly such inhibitors would not prevent a water molecule from binding to (or remaining on) Mg^{2+}2 in ternary complexes. Other evidence, including the high rates of ^{18}O-label incorporation from ^{18}O-water into inorganic phosphate in the presence of inositol,[18,29] favours the location of the nucleophile on

$Mg^{2+}2$, rather than on $Mg^{2+}1$, and some of this has been summarised previously.[25] The finding that the 6-amino phosphate esters serve as substrates was expected and implies that hydroxide ion chelated to $Mg^{2+}2$ and H-bonded through to the C-6 heteroatom serves as the nucleophile in an adjacent displacement reaction on the phosphate P-atom. Such a mechanism would proceed with retention of configuration and, to the best our our knowledge, no single step retentive mechanism has been reported for any other enzyme catalysed phosphoryl tranfer reaction.[35,38,48]

6 SYNTHESIS OF INHIBITORS

Given that the optimum cyclitol steochemistry was now well established, it was of interest to probe the ability of structural variants to bind tightly to the enzyme. For example, it was evident from the bound structures of the amines (**7**) and (**8**) that there was sufficient space to introduce larger groups onto the amine N-atom. It was also apparent that one of the three non-bridging phosphate O-atoms could be alkylated to both access the hydrophobic site formed from protein residues Val-40 and Leu-42 and reduce the negative charge on the phosphate group from 2 to one. Therefore, a fast efficient route to phosphate ester, phosphate diesters and phosphonate esters was sought taking advantage of the availability of the oxirane (**5**). Since the immediate precursor to oxirane (**5**) lacks an O^2-benzyl group, it was of interest to alkylate the 2-OH group with Merrifield resin and assess the possibility of preparing homochiral 6-substituted cyclitol 1-phosphates and 1-phosphonates on the solid-phase. However, it was evident that new methodologies would need to be developed to monitor reaction progress and cleave the rather sensitive products from the solid-phase support.

6.1 Solid-Phase Synthesis

The advancement of solid-phase organic synthesis (SPOS) has been driven by a desire to rapidly produce structurally diverse molecules,[49] and the range of reactions that can be performed in the solid phase has grown.[50-52] These advances in SPOS have required both the refinement of existing and the development of new analytical techniques to both monitor the progress of reactions and characterise the products of solid-phase syntheses.[53] Indeed, of the conventional methods, only FT-IR has been of generic use in providing direct, semiquantitative information on the extent of the conversion of the solid-phase material to a resin-bound product in a given reaction. We present here a method capable of providing quantitative information on the extent of the conversion of a solid-phase material which utilises ^{19}F gel phase NMR spectroscopy and which is applicable to a wide range of reaction chemistries and resin-support materials.[54] We also present a new method for cleaving Merrifield ethers.[54]

6.1.1 Gel-Phase NMR Spectroscopic Assay. Given that ^{19}F-NMR signals are of proven utility in monitoring solid-phase reactions,[53,55] ^{19}F gel-phase NMR spectroscopy appeared to offer a good method for quantifying the extent of the derivatisation of a functionalised polymer material with a fluorinated substrate. To test the idea, the rate of the reaction of Merrifield resin with 2-fluorophenol in DMF at 60 °C in the presence of sodium hydride were examined. At various time intervals portions of the reacting resin were removed and were reacted, exhaustively, with excess 4-fluorophenol in DMF at 60 °C in the presence of sodium hydride, Scheme 2.

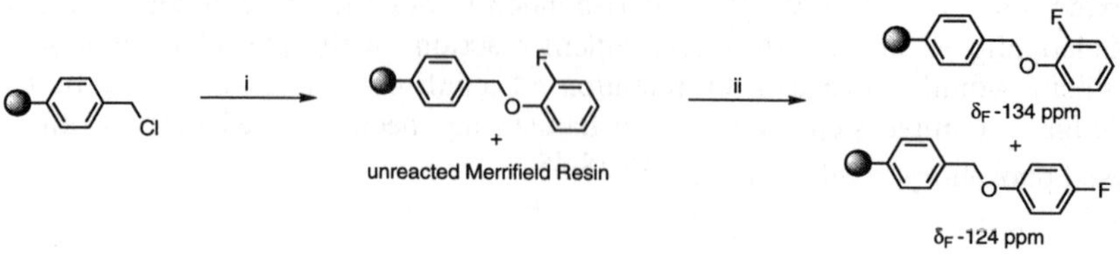

Scheme 2 *Reagents and conditions:* i. 2-fluorophenol, NaH, DMF, 60 °C; ii. 4-fluorophenol, NaH, DMF, 60 °C

Examination of the resin samples by ^{19}F gel-phase NMR spectroscopy showed a major broad signal corresponding to the 2-fluorophenol derivative at -134 ppm (from external CFCl$_3$ at 0 ppm), for extended initial reaction times, and a major broad signal corresponding to the 4-fluorophenol derviative at -124 ppm, for short initial reaction times. Thus, it was evident that ^{19}F gel-phase NMR spectroscopy was useful in providing information directly on the extent of the reaction.

In these experiments it was only possible to obtain qualitative, rather than quantitative, information on the extent of reactions. In order to quantify the amount of fluorine (and hence compound) on the resin we needed to measure the relative intensities of resin bound fluorine and a known concentration of an internal fluorine standard in solution (*e.g.* fluorobenzene) in the ^{19}F-NMR spectrum. To obtain meaningful measurements, there needed to be sufficient delay between RF pulses, such that the solid phase and solution phase nuclei would be similarly relaxed. Experiments were therefore set up to determine the spin-lattice relaxation time constant (T$_1$) for the 2-fluorophenyl Merrifield ether.

The optimal solvent system for simultaneously swelling and dispersing the resin in a homogeneous form within the RF coils was found to be deuterated chloroform:benzene:fluorobenzene (50:49.8:0.2). The resin would either sink or rise up and float in other mixtures of these solvents. For a 100%-loaded 2-fluorophenyl Merrifield ether sample,[56] (30 mg, 1.84 mmol F g^{-1} resin), a T$_1$ value of 4.2±0.6 s was measured for the fluorophenyl ether signal using the inversion recovery technique. In order to accurately quantify the fluorine concentrations from the NMR data, the interpulse relaxation delay was set to 25 s which is at least five times the longest spin-lattice relaxation time constant (T$_1$). Integration of the two ^{19}F NMR signals at -113 ppm and -134 ppm due to the solution-phase fluorobenzene reference standard and gel-phase fluorophenyl ether, respectively, gave an excellent correlation with the expected fluorine content for the Merrifield derivative.[56] In order to determine whether it would be possible to quantify the extent of a solid-phase reaction in 'blind' samples, portions of the 100% -loaded 2-fluorophenyl Merrifield ether,[56] were blended with unfunctionalised Merrifield resin to give samples which contained varying amounts of resin-bound fluorine. NMR spectra were recorded and Figure 4 shows a plot of the theoretical (solid line) *versus* the experimentally determined (data points) resin-bound F-atom content in samples containing 30 mg of the blended resin.

The excellent correlation for the determined F-atom content provides a high level of confidence in monitoring loading reactions and in quantifying the number of sites available in resin samples. This fluorophenol based assay was used to optimise the attachment of an epoxide to Merrifield resin.

6.1.2. Resin-Bound Epoxide. Recently we reported on the preparation of the *bis*-O-benzylated dihydroxycyclohexene oxide, **5**, starting from quinic acid, and its conversion to the inositol monophosphatase inhibitor **6**.[39,40] In order to pursue the analogous solid-phase synthesis of inhibitor **6**, the monobenzyl protected dihydroxycyclohexene oxide **18** was reacted with Merrifield resin to give the Merrifield ether, **19** in 91% yield, Scheme 3, under conditions optimised by removing a small portion of the reacted resin and assessing the extent of reaction through treatment with excess *p*-fluorophenolate and ^{19}F gel phase NMR spectroscopic analysis of the product.

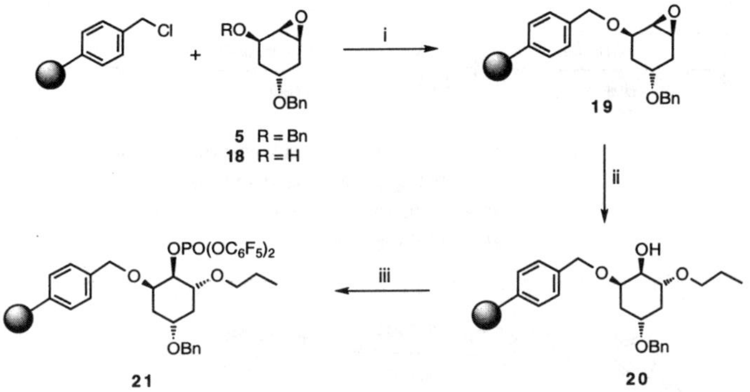

Scheme 3 *Reagents and conditions:* i. NaH, DMF, 60 °C, 3 h, then RT for 16 h; ii. 1-propanol, Yb(OTf)$_3$, dichloroethane, 80 °C, 3 h; iii. ClPO(OC$_6$F$_5$)$_2$, TEA, DMAP, 0 °C→RT, 24 h

The epoxide **19** was treated with propan-1-ol in the presence of ytterbium triflate, under conditions previously optimised for the analogous solution-phase chemistry,[42] to give the required resin bound alcohol **20**, (FTIR; u_{max} OH = 3400 cm^{-1}) in quantitative conversion as judged by the mass increase of the resin. Reaction of the resin-bound alcohol with excess *bis*-pentafluorophenyl chlorophosphate,[57] gave the required resin bound phosphate triester **21**, in quantitative conversion, which displayed the expected IR spectral chacteristics and a ^{31}P-NMR signal at -10 ppm in a gel phase sample.

6.1.3 Cleavage of Merrifield Ethers with Stannic Chloride. Ever since Purdie and his associates first used methyl groups to protect and derivatise hydroxyl groups, the utility of selective protection has been widely applied in natural product synthesis. The O-benzyl group has found particular favour because not only is it easy to introduce, but is also easy to remove. The O-benzylic moieties of Merrifield ethers can be considered as simple protecting groups and should, in principle, be removable by any conventional benzyl group deprotection method. However, in practice, useful solution phase methods for benzyl group removal, such as catalytic hydrogenolysis, are ineffective, and typically Brønsted acidic conditions have been employed for their cleavage. The sensitive phosphate moiety in triester **21** would not survive treatment with HF or triflic acid resin cleavage protocols. However we knew that the Lewis acid SnCl$_4$ had been used to selectively remove benzyl ether groups from carbohydrates,[58] and wished to assess the reagent more generally in cleaving simple primary and secondary alkyl Merrifield ethers to give the required alcohols as a SPOC equivalent to catalytic hydrogenolysis,[54] Figure 5.

Accordingly, primary, secondary and tertiary alcohols and phenols were attached to Merrifield resin, Figure 5. The products gave the expected mass increase and were fully loaded as judged by the fluorophenol based NMR spectroscopy assay and displayed the expected ether IR stretch at ~1100 cm^{-1}. Treatment of each one with 10 equivalents of stannic chloride followed by aqueous work up and extraction into DCM gave the expected alcohol in yields ranging from 70-95%. The one exception to this was the tertiary alcohol derived resin **24** which gave a 50% yield of the olefin **29** upon cleavage, in accord with a mechanism whereby the O-atom chelates to the tin atom to displace chloride ion and generate a benzylic carbocation, or, for resin **24**, the more stable tertiary carbocation which loses a proton to give **29**. Note that $TiCl_4$ was also useful in cleaving Merrifield ethers.

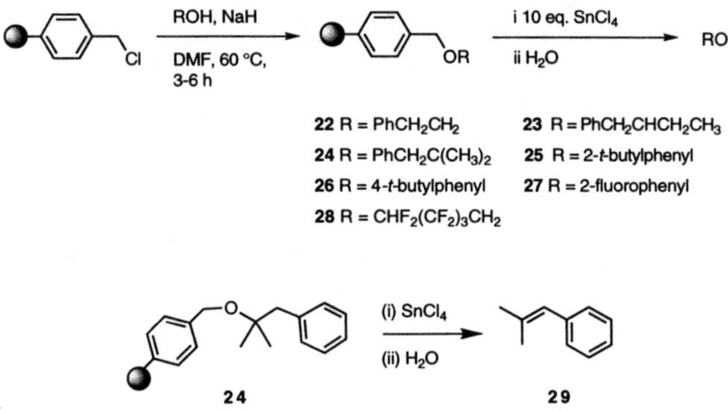
ROH, NaH
DMF, 60 °C, 3-6 h
Cl
OR
i 10 eq. SnCl4
ii H2O
ROH
22 R = PhCH2CH2
23 R = PhCH2CHCH2CH3
24 R = PhCH2C(CH3)2
25 R = 2-t-butylphenyl
26 R = 4-t-butylphenyl
27 R = 2-fluorophenyl
28 R = CHF2(CF2)3CH2
O
(i) SnCl4
(ii) H2O
24
29

Figure 5 *Showing resin loading scheme and details of the structures of "cleavage" substrates*

The rate of the cleavage reaction was investigated for the two fluorinated Merrifield ethers **27** and **28** using gel-phase ^{19}F-NMR spectroscopy. Accordingly, a 1.5-fold excess of stannic chloride was added to each of the resin samples, **27** and **28**, suspended in d-chloroform, and the increase in the intensities of the ^{19}F signals at -142 ppm and -124 ppm, due to the resin cleaved products, respectively were monitored with respect to time. In each case the reaction was rapid and was complete within 1.5 and 4 h, respectively, at 30 °C.

To assess whether the reagent was applicable to the synthesis of inositol monophosphatase inhibitors, the triester **21** was treated with stannic chloride and then the excess tin residues were removed by passage of an aqueous solution of the cleavage product through a cellulose phosphate ion exchange column (Scheme 4).

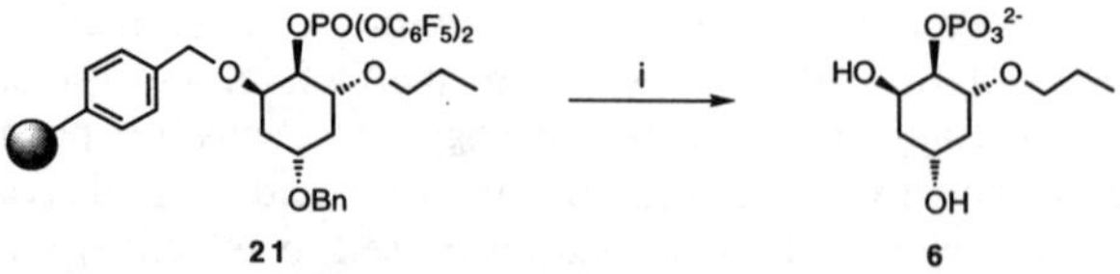
OPO(OC6F5)2
O
O
OBn
21
i
OPO3 2-
HO
O
OH
6

Scheme 4 *Reagents and conditions:* i. 8eq. $SnCl_4$, RT, 16 h; then NH_4OH, then cellulose phosphate cation exchange resin

Greater than 95% pure **6** was obtained in 56% yield as the *bis*-cyclohexylammonium salt from **19** and this (*1R, 2R, 4R, 6R*) phosphatase inhibitor **6** displayed identical physico-chemical and biological properties to the authentic material.[39,40]

We believe that the ^{19}F-NMR spectroscopic assay and the SnCl4 mediated Merrifield resin cleavage reaction will be widely applicable in SPOS and it is evident that the methods described here together with combinatorial chemical protocols[59] will be of considerable utility in the preparation of sensitive inositol monophosphatase inhibitors.

Acknowledgements: We thank the Wellcome Trust for grant 04033, the BBSRC for research grant 49/T07860, the EPSRC for a studentship for MB, and the University of St. Andrews for a studentship to DS.

References

1. This paper is dedicated to Professor Douglas W. Young on the occassion of his 60th Birthday Anniversary.
2. M. J. Berridge and R. F. Irvine, *Nature,* 1989, **341**, 197.
3. D. Gani, C. P. Downes, I. Batty and J. Bramham, *Biochim. Biophys. Acta,* 1993, **1177**, 253.
4. B. V. L. Potter and L. Lampe, *Angew. Chem. Int. Ed. Engl.,* 1995, **34**, 1933.
5. A. Kishimoto, Y. Takai, T. Mori, U. Kikkawa and Y. Nishizuka, *J. Biol. Chem.,* 1980, **255**, 2273.
6. Y. Nishizuka, *Nature,* 1984, **308**, 693.
7. E. H. Fischer, *Angew. Chem. Int. Ed. Engi.,* 1993, **32**, 1130.
8. H. Streb, R. F. Irvine, M. J. Berridge, I. Schulz, *Nature,* 1983, **306**, 67.
9. R. Spector and A. V. Lorenzo, *Amer. J. Physiol.,* 1975, **228**, 1510.
10. J. H. Allison and M. A. Stewart, *Nature,* 1971, **233**, 267.
11. J. H. Allison, M. E. Blisner, W. H. Holland, P. P. Hipps and W. R. Sherman, *Biochem. Biophys. Res. Commun.,* 1976, **71**, 664.
12. L. M. Hallcher and W. R. Sherman, *J. Biol. Chem.,* 1980, **255**, 10896.
13. K. Takimoto, M. Okada, Y. Matsuda and H. Nakagawa, *J. Biochem. (Tokyo),* 1985, **98**, 363.
14. N. S. Gee, C. I. Ragan, K. J. Watling, S. Aspley, R. G. Jackson, G. C. Reid, D. Gani and J. K. Shute, *Biochem. J.,* 1988, **249**, 883.
15. J. K. Shute, R. Baker, D. C. Billington and D. Gani, *J. Chem. Soc., Chem. Commun.* 1988, 626.
16. P. W. Attwood, J. B. Ducep and M. C. Chanal, *Biochem. J.,* 1988, **253**, 387.
17. A. J. Ganzhorn and M. C. Chanal, *Biochemistry,* 1990, **29**, 6065.
18. A. P. Leech, G. R. Baker, J. K. Shute, M. A. Cohen and D. Gani, *Eur. J. Biochem.,* 1993, **212**, 693.
19. A. Cole and D. Gani, *J. Chem. Soc., Chem. Commun.,* 1994, 1139.
20. P. D. Leeson, K. James, I. C. Lennon, N. J. Liverton, S. Aspley and R. G. Jackson, *Bioorg. & Med.. Chem. Lett.,* 1993, **3**, 1925.
21. P. J. Greasley and M. G. Gore, *FEBS Lett.,* 1993, **331**, 114.
22. S. J. Pollack, J. R. Atack, M. R. Knowles, G. McAllister, C. I. Ragan, R. Baker, S. R. Fletcher, L. L. I. Iverson and H. B. Broughton, *Proc. Natl. Acad. Sci. USA,* 1994, **91**, 5766.
23. S. J. Pollack, M. R. Knowles, J. R. Atack, H. B. Broughton, C. I. Ragan, S.-A. Osborne and G. McAllister, *Eur. J. Biochem.,* 1993, **217**, 281.

24. R. Bone, L. Frank, J. P. Springer, S. J. Pollack, S. Osborne, J. R. Atack, M. R. Knowles, G. McAllister, C. I. Ragan, H. B. Broughton, R. Baker and S. R. Fletcher, *Biochemistry,* 1994, **33**, 9460.
25. J. Wilkie, A. G. Cole and D. Gani, *J. Chem. Soc., Perkin Trans. 1,* 1995, 2709.
26. R. Baker, P. D. Leeson, N. J. Liverton and J. J. Kulagowski, *J. Chem. Soc., Chem. Commun.,* 1990, 462.
27. R. Baker, J. J. Kulagowski, D. C. Billington, P. D. Leeson, I. C. Lennon and N. Liverton, *J. Chem. Soc., Chem. Commun.,* 1989, 1383
28. R. Baker, C. Carrick, P. D. Leeson, I. C. Lennon and N. J. Liverton, *J. Chem. Soc., Chem. Commun.,* 1991, 298-300.
29. G. R. Baker and D. Gani, *Biorg. Med. Chem. Lett. 1,* 1991, 193.
30. A. G. Cole and D. Gani, *J. Chem. Soc., Perkin Trans. 1,* 1995, 268.
31. A. G. Cole, J. Wilkie and D. Gani, *J. Chem. Soc., Perkin Trans. 1,* 1995, 2695.
32. R. Bone, J. P. Springer and J. R. Atack, *Proc. Natl. Acad. Sci. USA,* 1992, **89**, 10031.
33. R. Bone, L. Frank, J. P. Springer and J. R. Atack, *Biochemistry,* 1994, **33**, 9468.
34. J. Wilkie and D. Gani, *J. Chem. Soc., Perkin Trans. 2,* 1996, 783.
35. D. Gani and J. Wilkie, *Chem. Soc. Rev.,* 1995, **24**, 55.
36. G. R. Baker, D. C. Billington and D. Gani, *Bioorg. Med. Chem. Lett.,* 1991, **1**, 17.
37. R. Pybus, Ph.D. Thesis, University of St. Andrews, 1998.
38. E. S. Lightcap, C. J. Halkides and P. A. Frey, *Biochemistry,* 1991, **30**, 10307.
39. J. Schulz and D. Gani, *Tetrahedron Lett.,* 1997, **38**, 111.
40. J. Schulz and D. Gani, *J. Chem. Soc. Perkin Trans. I,* 1997, 657.
41. M. Chini, P. Crotti and F. Macchia, *Tetrahedron Lett.,* 1994, **35**, 433.
42. M. W. Beaton and D. Gani, 1998, in the press.
43. J. Schulz, J. Wilkie, P. Lightfoot, T. Rutherford and D. Gani, *J. Chem. Soc., Chem. Commun.,* 1995, 2353.
44. W. W. Cleland, *Methods Enzymol.,* 1982, **87**, 390.
45. D. Herschlag and W. P. Jencks, *Biochemistry,* 1990, **29**, 5172.
46. A. Caselli and G. Ramponi, *Biochim. Biophys. Acta.,* 1996, 241.
47. L. G. Sillen and A. E. Martell, *Spec. Pub. Chem. Soc., London,* 1971, **25**.
48. D. Gani and J. Wilkie. (1997). *'Structure and Bonding',* (Eds. H. A. Hill, P. Sadler and A. Thompson), Springer-Verlag, Heidelberg, 1997, p. 133.
49. K. S. Lam, S. E. Salmon, E. M. Hersh, K. J. Hruby, W. M. Kazmiersky and R. J. Knapp, *Nature* (London), 1991, **354**, 82.
50. P. H. H. Hermkens, H. C. J. Ottenheim and D. Rees, *Tetrahedron,* 1996, **52**, 4527; and H. Maehr, *Bioorg. Med. Chem.,* 1997, **5**, 473.
51. F. Balkenhohl, C. Bussche-Hünnefeld, A. Lansky and C. Zechel, *Angew. Chem., Int. Ed. Engl.,* 1996, **35**, 2288.
52. M. A. Gallop, R. W. Barrett, W. J. Dower, S. P. A. Fodor and E. M. Gordon, *J. Med. Chem.,* 1994, **37**, 1233; E. M. Gordon, R. W. Barrett, W. J. Dower, S. P. A. Fodor and M. A. Gallop, *J. Med. Chem.,* 1994, **37**, 1385.
53. T. Wehler and J. Westman, *Tetrahedron Lett.,* 1996, **37**, 4771; A. Svensson, T. Fex and J. Kihlberg, *Tetrahedron Lett.,* 1996, **37**, 7649; T. Y. Chan, R. Chen, M. J. Sofia, B. C. Smith and D. Glennon, *Tetrahedron Lett.,* 1997, **38**, 2821.
54. D. Stones, D. J. Miller, M. W. Beaton, T. J. Rutherford and D. Gani, *Tetrahedron Lett.,* 1998, **39**, 4875.
55. M. J. Shapiro, G. Kumaravel, R. C. Petter and R. Beveridge, *Tetrahedron Lett.,* 1996, **37**, 4671.

56. This material displayed no available free sites when assayed with excess 4-fluorophenolate, the correct mass increase and gave a satisfactory microanalysis.
57. P. Hormozdiari and D. Gani, *Tetrahedron Lett.*, 1996, **37**, 8227.
58. M. H. Park, R. Takeda and K. Nakanishi, *Tetrahedron Lett.*, 1987, **28**, 3823; J. I. Padron and J. T. Vazquez, *Tetrahedron Assym.*, 1995, **6**, 857.
59. D. Gani, M. Akhtar, F. E. K. Kroll, C. F. M. Smith and D. Stones, *Tetrahedron Lett.*, 1997, **38**, 8577.

Protease Inhibition

DELTA TECHNOLOGY FOR PROTEASE INHIBITION

Bradley A. Katz, Robert M. Stroud,[†] James M. Clark, Thomas E. Jenkins, James W. Janc, William R. Moore and Michael C. Venuti*

Axys Pharmaceuticals, Inc., 180 Kimball Way, South San Francisco, CA 94080, USA, and [†]Department of Biochemistry, University of California, San Francisco, CA 94143, USA

1 INTRODUCTION

Proteases are amongst the most well-understood proteins, both in structure and in function.[1] The four major protease classes - serine, cysteine, aspartyl and metallo-proteases - have all been associated with critical functions in biological pathways. As these enzymes are responsible for regulation of key biological processes, especially including activation and destruction of other proteins, and as they are often misregulated themselves in association with disease pathologies, proteases are ideal biological targets for drug intervention. The successes of drugs such as ACE inhibitors of the metalloprotease class, and, more recently, HIV protease inhibitors of the aspartyl protease class, have only served to strengthen efforts to identify inhibitors of other proteases as drugs.

Contrasting these recent achievements, however, has been the relative lack of success in identifying inhibitors of either serine or cysteine proteases which concomitantly possess drug-like characteristics, especially oral bioavailability. Small molecule inhibitors of these enzymes with required potency and selectivity often require chemically complex peptidomimetic or chiral structures of high molecular weight. These characteristics frequently conspire to compromise oral bioavailability at all steps: absorption, distribution, metabolic destruction, and excretion. Thus, although numerous targets for drug intervention have been identified within these two protease classes, there are no approved drugs, and few clinical development candidates, which exhibit oral bioavailability and other drug-like properties suitable for the treatment of chronic disease. This problem can be approached using the intimate understanding of the three-dimensional structure and associated function of these proteases. The ideal inhibitor would be less peptide-like (fewer amide bonds, achiral), and of low molecular weight (less than 500 Da). The challenge therein is to design a molecular structure which binds with high affinity at the protease active site, preferably in a competitive and reversible manner. Herein we describe one such solution to the problem of identifying reversible non-peptide inhibitors of serine proteases utilizing high affinity scaffolds derived from a new chemical framework compatible with the requirements for potent and selective drugs.

2 ZINC-DEPENDENT PROTEASE INHIBITION DISCOVERY

2.1 Biochemistry

As part of our effort to develop inhibitors of tryptase, a mast cell-derived serine protease implicated in asthma and other inflammatory diseases,[2,3] *bis*(5-amidino-2-benzimidazolyl)-methane (**1**), also widely known as BABIM, was identified from the literature as a useful reference compound for enzyme inhibition assays. BABIM has been described by Tidwell and co-workers as a potent inhibitor of trypsin-like serine proteases, standing out seemingly anomalously as the best inhibitor of trypsin ($K_i = 1.7$ x 10^{-8} M) by a factor of >100-fold in a chemical series of over 60 relatively closely related analogs.[4] BABIM is also reported to inhibit tryptase ($K_i = 1.4$ x 10^{-9} M),[5] tissue kallikrein ($K_i = 1.7$ x 10^{-5} M),[4] thrombin ($K_i = 4.2$ x 10^{-6} M),[4] and urokinase ($K_i = 2.3$ x 10^{-6} M),[6] all at impressive potencies. However, in initial biochemical studies in our hands,[7] the potency of BABIM against trypsin was found to vary substantially under nominally identical assay conditions. Further investigation of this phenomenon[7] led to the surprising finding that the potency of BABIM against trypsin-like serine proteases was modulated by low levels of divalent metal cations, most dramatically by Zn^{2+}.

BABIM (1)   Pentamidine (2)

2.1.1 Zinc-dependency. In preliminary experiments, the apparent inhibition constant (K_i') of BABIM against trypsin was determined to be 50 nM in Tris buffer at pH 8.2 (Table 1). However, the potency of BABIM was enhanced approximately 10-fold, to 4.5 nM, when Zn^{2+} was included in the assay buffer in molar excess over the inhibitor (nominal ratio = 100:1). Other metal ions including magnesium, calcium, manganese, and iron had no effect. Strikingly, the K_i' of BABIM increased to 19 μM when EDTA was included in the assay. Trypsin activity in the absence of inhibitor was not altered by either EDTA, or by the metal ions at the highest concentration (120 μM) used in the inhibition studies. Thus, the potency of BABIM against trypsin *in vitro* was increased >10^3-fold in the presence of Zn^{2+} ions. Zn^{2+}-mediated inhibition was a specific property of BABIM, as pentamidine (**2**), another potent trypsin inhibitor, was not potentiated by Zn^{2+} (Table 1).[8]

In a more extensive study carried out with tryptase, several additional divalent cations were tested for their ability to potentiate inhibition by BABIM. Again, only Zn^{2+} ions significantly enhanced the potency of BABIM (~80-fold), with slight potentiation (< 7-fold) by Cd^{2+}, Co^{2+}, or Mn^{2+} and no effect with other metals tested. Zn^{2+} also significantly enhanced the potency of BABIM against other trypsin-like enzymes, with the magnitude of enhancement ranging from 18-fold for urokinase to >10,000-fold for Factor Xa (Table 2).

Table 1 *Effect of various divalent cations (assay concentration = 120 µM) on BABIM and pentamidine potency against trypsin*

Inhibitor	Ki´ (nM)	Inhibitor	Ki´ (nM)
BABIM	50	pentamidine	39000
BABIM + Zn^{2+}	4.5	pentamidine + Zn^{2+}	49000
BABIM + EDTA	19000	pentamidine + EDTA	44000
BABIM + Mg^{2+}	49		
BABIM + Ca^{2+}	48		
BABIM + Mn^{2+}	40		
BABIM + Fe^{2+}	52		

Table 2 *Zn^{2+}-enhanced activity of BABIM against multiple serine proteases, $K_i´$ (nM)*

Inhibitor	Factor Xa	Thrombin	Urokinase	Tissue Kallikrein
BABIM	40	3800	-	-
BABIM + Zn^{2+}	< 1	20	1500	6600
BABIM + EDTA	10000	119000	270000	320000

BABIM behaves as a competitive inhibitor of tissue kallikrein and thrombin (Figure 1) in the presence of Zn^{2+}, a finding that is consistent with earlier reports describing BABIM as a competitive inhibitor of various trypsin-like enzymes.[4] For competitive inhibitors, K_i is related to $K_i´$ by the formula $K_i = K_i´/(1 + S/K_m)$ where S is substrate concentration and K_m is the Michaelis constant. Under our assay conditions with excess Zn^{2+}, the true K_i of BABIM against thrombin is approximately 7×10^{-9} M. A difference in the amount of free Zn^{2+} present in the two assays readily accounts for the apparent discrepancy in K_i from the previously reported value of 4.2×10^{-6} M.[5]

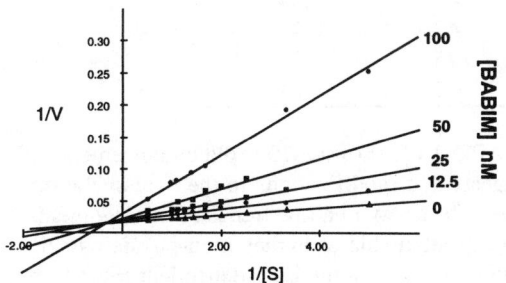

Figure 1 *Lineweaver-Burke plots demonstrating the competitive reversible kinetics of zinc-dependent inhibition of thrombin by BABIM ($K_i´$ = 20 nM) at various inhibitor concentrations (0, 12.5, 25, 50, 100 nM)*

2.1.2 Zinc-dependent Inhibition in Physiological Systems. A number of experiments have demonstrated that Zn^{2+}-mediated inhibition occurs at physiological levels of free

Zn^{2+}. Zn^{2+} ions are ubiquitous in biological tissues and fluids. However, most Zn^{2+} occurs in a tightly bound state as catalytic or structural components of Zn^{2+} metalloproteins or loosely bound to circulating plasma proteins such as albumin.[9,10] Exploiting Zn^{2+}-mediated inhibition as a therapeutic strategy requires validation of the concept in a physiologically-relevant milieu. We chose human plasma as an appropriate and accessible test system since a number of trypsin-like plasma serine proteases in the coagulation, fibrinolytic, and complement cascades represent potential therapeutic targets that should be susceptible to Zn^{2+}-mediated inhibition. In initial experiments, we assessed the exchangeability of Zn^{2+} by filtering heparinized human plasma to remove proteins. Filtration was required to remove interfering protease activities from the plasma, but also removed albumin and other Zn^{2+}-binding proteins. The $K_i{'}$ of BABIM against tryptase in filtrate was 293 nM, increasing to 1770 nM when 1 mM EDTA was added to the filtrate. The decrease in potency seen with EDTA suggests that Zn^{2+}-dependent inhibition occurs in filtered serum. Thus, the level of Zn^{2+} available after filtration appears to be sufficient to support Zn^{2+}-mediated inhibition, although the maximal potency ($K_i{'}$ = 12 nM) attained in the presence of 150 μM Zn^{2+} was not achieved. The $K_i{'}$ of pentamidine against tryptase was 2450 nM in filtered plasma and was unaltered by the addition of either EDTA or Zn^{2+}.

For a Zn^{2+}-dependent inhibitor to work against plasma enzymes *in vivo*, it must compete for Zn^{2+} in the presence of high levels of endogenous Zn^{2+} binding proteins. We developed an assay in which the free Zn^{2+} concentration, buffered by physiological levels of human serum albumin (HSA), was titrated to 2 - 12 nM by the addition of exogenous Zn^{2+}. The $K_i{'}$ of BABIM against trypsin in the HSA assay with no exogenous Zn^{2+} was 47 μM, essentially identical to its $K_i{'}$ in the HSA assay containing EDTA (Table 3). When the free Zn^{2+} levels were increased to 2 or 12 nM by the addition of exogenous Zn^{2+}, BABIM's $K_i{'}$ decreased only slightly to 25 μM and 5 μM, respectively. Thus, BABIM, which shows a 1000-fold enhancement in potency when assayed against trypsin in buffer with excess Zn^{2+}, shows a <10-fold improvement in $K_i{'}$ when assayed at physiological concentrations of free Zn^{2+}.

Table 3   *Potency of BABIM and APD-1 against trypsin at physiological levels of free Zn^{2+}, $K_i{'}$ (μM)*

Assay Conditions	BABIM	APD-1
HSA/EDTA	49	204
HSA/ambient Zn^{2+}	47	32
HSA/2 nM free Zn^{2+}	25	14
HSA/12 nM free Zn^{2+}	5	2.1

In contrast, the mono-*des*-amidino analog APD-1 (**3**) shows a 100-fold enhancement in potency under the same conditions. This suggests that modifications to the core chelation scaffold can yield molecules that function more effectively in a low free Zn^{2+} environment, although **3** does not achieve the level of potency attainable at higher (non-physiological) levels of free Zn^{2+}. We also observed that the potency of some Zn^{2+}-dependent inhibitors in the HSA/Zn^{2+} assay increased significantly with time, as much as 500-fold over 6 h. These observations are consistent with the concept of a moderately available Zn^{2+} pool, buffered largely by albumin, accessible to Zn^{2+}-mediated inhibitors.

2.1.3 Zinc-dependent Inhibitor Selectivity. Serine proteases are ubiquitously distributed and play central roles in many important biological processes. Thus, the selectivity of any protease inhibitor, including Zn^{2+}-dependent compounds, is a critical issue in determining their potential therapeutic utility. BABIM and **3** are not particularly selective compounds, as they both potently inhibit a number of trypsin-like serine proteases. Thrombin, a key enzyme in coagulation, and mast cell tryptase, an enzyme implicated in asthma,[2,3] are representative serine protease targets for pharmaceutical intervention. Compounds **4-6** are examples of potent, Zn^{2+}-dependent inhibitors of tryptase or thrombin that exhibit 10 - 1000-fold selectivity against other trypsin-like enzymes (Table 4). Such compounds act as zinc-dependent scaffolds for further substitution, and can be synthetically elaborated to produce inhibitors of high potency and selectivity for numerous serine proteases. Our complete investigation into the general biochemistry and kinetics of zinc-mediated tryptase inhibition has recently been published.[7]

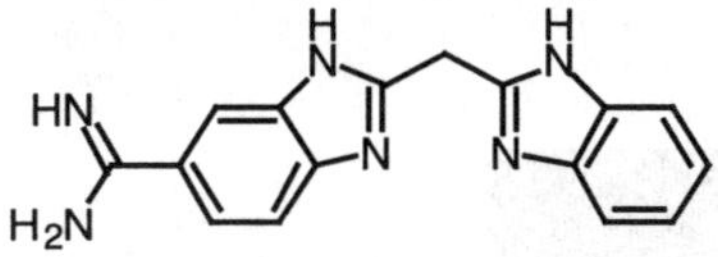

APD-1 (**3**)

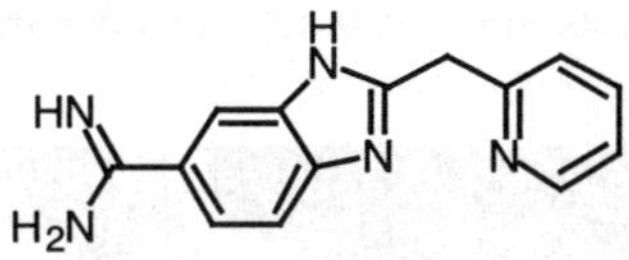

APD-5 (**4**)

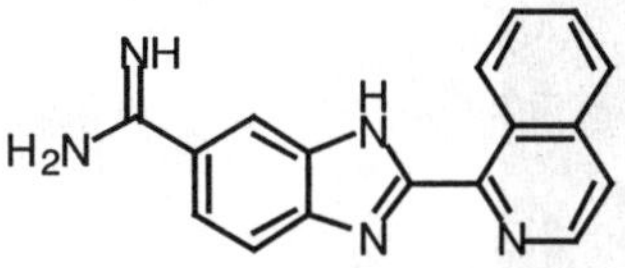

APD-6 (**5**)

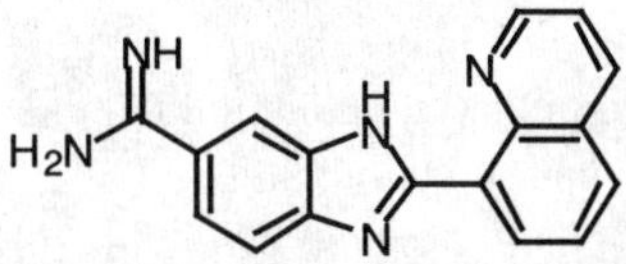

APD-7 (**6**)

Table 4 *Selectivity of inhibitors in the presence of 150 μM Zn^{2+}; K_i' (nM)*

Compound	Trypsin	Tryptase	Thrombin
BABIM (**1**)	4.5	17	20
APD-5 (**4**)	136000	310	10500
APD-6 (**5**)	22500	54500	41
APD-7 (**6**)	8040	11900	321

2.2 X-Ray Crystallography

In-depth structural understanding of many classes of proteins has prompted the design of organic ligands with high affinity and selectivity. Elucidation of the exact structural features of zinc-dependent inhibition for serine proteases might then be expected to suggest the design of new classes of non-peptide organic compounds tailored for potency and specificity, and simultaneously able to take advantage of the zinc-dependent inhibition mode. To determine the three-dimensional structure of the zinc-dependent inhibition of trypsin by BABIM and its keto-analog (**7**), crystals of trypsin-BABIM, pH 8.2, were prepared by soaking trypsin-benzamidine crystals in synthetic mother liquor (1.62 M $MgSO_4 \cdot 7\ H_2O$, 100 mM Tris saturated with BABIM, and containing 1.0 mM or 5.0 mM $ZnSO_4$). Trypsin-keto-BABIM-Zn^{2+} was co-crystallized from solutions saturated in 1.0 mM in Zn^{2+} and keto-BABIM. The structures of the resulting complexes were determined as described previously.[11]

2.2.1 Structure of the enzyme-Zn^{2+}-inhibitor binding motif. The Zn^{2+}-mediated inhibitor binding motif, crystallographically deter-mined at pH 8.2 in 1.0 mM Zn^{2+} for trypsin-BABIM-Zn^{2+} and trypsin-keto-BABIM-Zn^{2+} is shown in Figure 2. A peak of 28 ± 1 electrons and temperature factor (B) of 28 Å^2, corresponds to a Zn^{2+} ion tetrahedrally co-ordinated by one nitrogen from each of the two benzimidazoles of keto-BABIM, by N$\square 2_{His87}$ and by O$\square_{Ser195}$, as depicted in Figure 3. Bond distances for Zn^{2+} are similar to those observed in other Zn^{2+}-containing proteins[12,13] and small molecules.[14-16] The Ser195 hydroxyl may be deprotonated due to co-ordination by Zn^{2+}; lowering of the pK_a of water co-ordinated to Zn^{2+} by as much as 9 units can occur.[12-15]

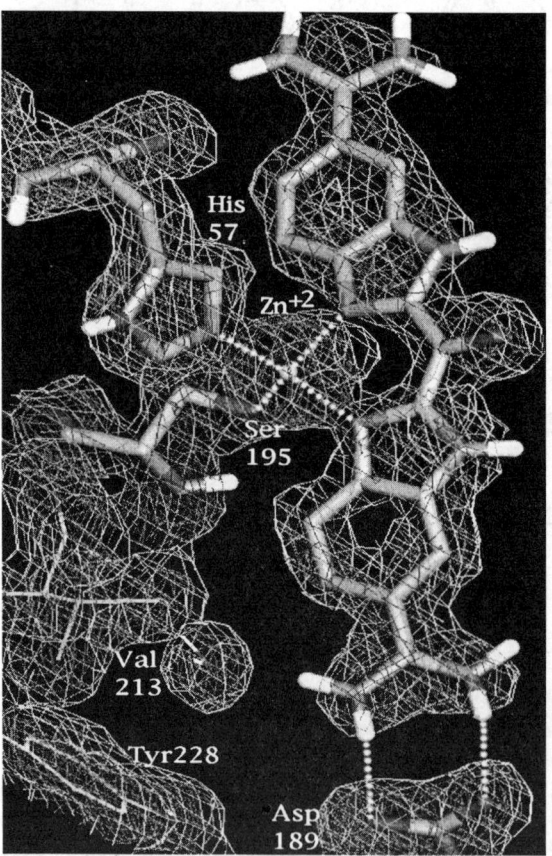

Figure 2 *(2|F_o|-|F_c|),* $\square_c$ *map superimposed on the structure of trypsin-keto-BABIM-Zn^{2+}, pH 8.2*

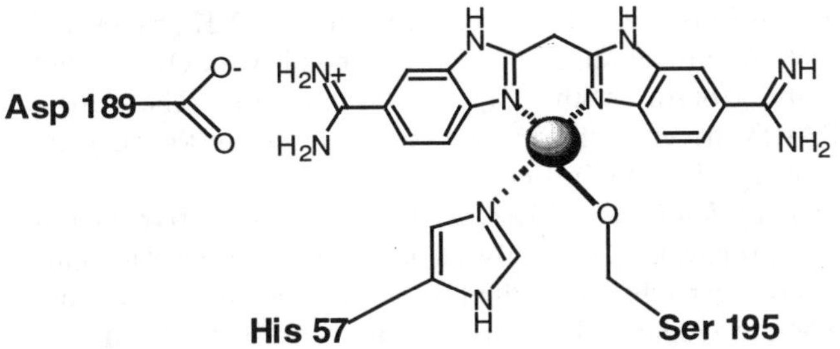

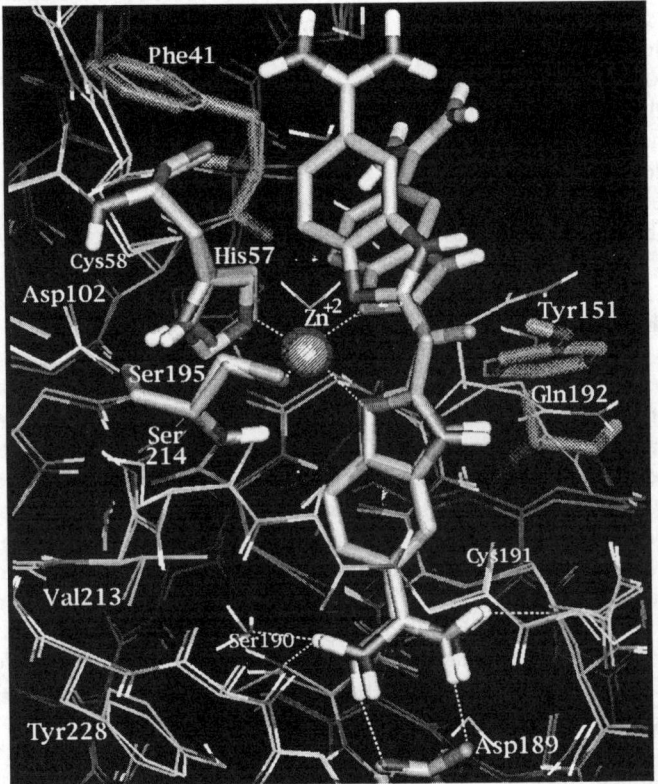

Figure 3 *Canonical re-presentation of the zinc-dependent binding mode, showing chelation of zinc between Ser195, His57 and the inhibitor nitrogen atoms*

Figure 4 *Superposition of trypsin-BABIM-Zn^{2+} and trypsin-keto-BABIM-Zn^{2+}*

The entire bound keto-BABIM molecule is well defined by density. As expected, the amidinobenzene portion of BABIM and keto-BABIM superimpose nearly exactly on benzamidine in the structure of trypsin-benzamidine.[17,18] Two hydrogen bonded salt bridges are formed, each in the optimal *cis* orientation,[19] between amidino NH groups and the carboxylate oxygens of Asp189. A third amidino NH hydrogen bonds with $O\gamma_{Ser190}$ and with a bound water, and the fourth with O_{Gly214} (Figure 4). The distal amidinobenzimidazole of keto-BABIM occupies the S1′ site on the surface of the enzyme, its position and orientation tightly coupled to the Zn^{2+} co-ordination.

2.2.2 *pH and Ion Dependence of Inhibition*. The crystal structure of trypsin co-crystallized with keto-BABIM and 1.0 mM Co^{2+} is nearly identical to that of trypsin-keto-BABIM-Zn^{2+}. Thus, Co^{2+} is also capable of co-inhibition, but assays show that concentrations higher than 25 mM are required. The structure of trypsin-keto-BABIM in the absence of Zn^{2+} or Co^{2+} shows a direct hydrogen bond between the N3 atom of the S1 benzimidazole and $O\gamma_{Ser195}$ showing that high concentrations of Mg^{2+} in the crystallization solution (1.68 M) do not mediate inhibitor binding. The crystal structure of trypsin-BABIM at pH 5.9 in the absence of Zn^{2+}, shows that binding of BABIM is mediated by a sulfate ion. The sulfate oxygens form hydrogen bonds with protonated nitrogens from two of the imidazole groups of BABIM, with $N\epsilon2_{His57}$, $O\gamma_{Ser195}$, N_{Gly193}, and with two water molecules. Addition of 25 mM SO_4^{2-} or PO_4^{3-} (lacking control of ionic strength, however) improves inhibition by 6- and 10-fold, respectively at pH 5.9. Thus, we have visualized crystallographically three fundamentally different binding mechanisms involving a cation (Zn^{2+} or Co^{2+}), an anion (SO_4^{2-}), and no ion, allowing for the design of organic molecules capable of zinc-dependent inhibition. The observed different binding modes reflect the amphoteric nature of the imidazoles of the inhibitors and of His57, and of the hydroxyl of Ser195, whose hydrogen bond donor- acceptor character is dictated by pH and environment.

2.3 Medicinal Chemistry

2.3.1 *Inhibitor Design.* Zn^{2+} ions are important in a variety of biological processes as a consequence of their role as structural or catalytic components of proteins such as metalloenzymes and Zn^{2+} finger proteins.[9,10] Zn^{2+} has also been shown to mediate interactions between macromolecules such as human growth hormone and the prolactin receptor,[20] and to modulate the activity of recombinant trypsin mutagenized to contain a metal binding site.[21] Zinc-dependent protease inhibition is yet another unique role for Zn^{2+} ions in which the inhibitory interaction between a low molecular weight organic molecule and a serine protease is potentiated by Zn^{2+}. This phenomenon, hypothesized to result from the formation of a ternary complex involving Zn^{2+}, BABIM, and the enzyme, was structurally confirmed by determination of the X-ray crystallographic structure of the trypsin-BABIM-Zn^{2+} complex. Since the spatial arrangement of the catalytic triad is highly conserved among trypsin-like serine proteases, Zn^{2+}-potentiated inhibition by BABIM should be a general phenomenon within this enzyme family.[22] This is indeed the case, as has been demonstrated for a number of trypsin-like enzymes of interest as therapeutic targets, where both high potency (nM-pM) and selectivity (10^3 - 10^6-fold) have been achieved with specifically-designed inhibitors (Table 5).

Potent and selective inhibitors have been discovered using a combination of approaches based on the zinc-dependent paradigm, including classical iterative medicinal chemistry, directed structure-based design (utilizing either actual protease crystal structures or homology models), and library screening under zinc-dependent and zinc-depleted assay conditions. The structural simplicity of such zinc-dependent scaffolds has allowed synthesis of widely varied analogs with substitution patterns that exhibit further enhanced affinity and specificity for targets. Specificity is achieved through capture of favourable interactions in the target versus loss or impairment of such interactions in a non-target, *e.g.* by steric clash, or electrostatic incompatibility, or through steric restraints at specificity sites that modulate the Zn^{2+} co-ordination geometry at the active site. The generalization of the design paradigm has also suggested combinatorial approaches to scaffold optimization currently in use. Elaborations in the distal portions of the scaffolds from substituents directed to sites that vary among different proteases have enabled development of potent and selective, Zn^{2+}-mediated inhibitors of specific protease targets such as tryptase, thrombin, factor Xa, and chymase. The technology has also been extended to the protease-like enzyme, β-lactamase (both Tem and ampC variants). Structural details of our efforts to develop inhibitors of numerous serine protease drug targets will be published subsequently.

Table 5 *Potency and selectivity of zinc-dependent inhibitors for various serine proteases*

Primary target	Potency (nM)	Selectivity*
tryptase	0.068	> 20,000
chymase	0.096	> 8,000,000
fXa	0.024	> 18,000
fVIIa	1	> 10
thrombin	2.19	> 100
urokinase	42	> 100
hepatitis C NS3	62	> 1000
CMV protease	4000	> 1000
☐-lactamase	120	> 1000

* Selectivity measured against most highly related protease target

2.3.2 Drug Development. Zn^{2+}-potentiated inhibition represents a unique approach to the design of novel therapeutics directed at serine protease targets. However, at least two major factors must be addressed in the application of this technology to drug design. First, Zn^{2+}-dependent inhibitors must function effectively in environments where most of the Zn^{2+} is sequestered by proteins or other endogenous chelators, a requirement unique to this class of compounds. Second, the inhibitors must be selective for their enzyme target, a requirement shared by all protease inhibitors intended for therapeutic use. Estimates of free Zn^{2+} levels in plasma range from 200 pM in equine plasma,[23] to 1 nM in human plasma.[24] However, the total Zn^{2+} concentration in plasma is approximately 15 μM,[25] of which two-thirds is loosely bound by albumin and is considered exchangeable.[26,27] We modelled the *in vivo* situation by including albumin in the assay to provide buffered Zn^{2+} at physiological levels. In this assay, APD-1 shows a significantly larger Zn^{2+} enhancement than BABIM, indicating that improved inhibitor performance at physiological free Zn^{2+} levels can be engineered via structural modifications around the core Zn^{2+} chelation scaffold. The mechanistic basis for improved Zn^{2+}-dependent inhibition by APD-1 in the presence of HSA is uncertain, but probably does not involve differences in the intrinsic affinity of BABIM and APD-1 for Zn^{2+} as the K_d for Zn^{2+} is similar for both compounds (~ 100 μM). Alternative possibilities include a higher affinity of the Zn^{2+}•APD-1 complex for the enzyme or reduced non-specific protein binding by APD-1. Potent, Zn^{2+}-dependent inhibitors have also been synthesized that are highly selective for thrombin or tryptase, as exemplified by APD-5, APD-6, and APD-7. Thus, two of the major issues in applying this concept to drug development are amenable to resolution through structural modifications of the core Zn^{2+} chelation scaffold.

The availability of Zn^{2+} and the thermodynamic stability of the enzyme•Zn^{2+}•inhibitor ternary complex are the dominant factors affecting the utility of Zn^{2+}-mediated inhibition. A simple model system incorporating these factors in describing the equilibrium involved in forming the ternary complex for an inhibitor in plasma can be defined (Scheme 1).

$$\text{Albumin•}Zn^{2+} \rightleftharpoons \underset{\text{Albumin}}{\overset{\text{Inhibitor}}{Zn^{2+}}} \rightleftharpoons \text{Inhibitor•}Zn^{2+} \overset{\text{Enzyme}}{\rightleftharpoons} \text{Inhibitor•}Zn^{2+}\text{•Enzyme}$$

Scheme 1 *Formation of the enzyme·zinc·inhibitor ternary complex in the presence of physiological zinc buffered by human serum albumin*

Using this model, we have attempted to predict the effectiveness of Zn^{2+}-mediated inhibitors in plasma. This analysis assumes that the system is closed, at equilibrium, and that the starting concentrations of albumin, Zn^{2+}, inhibitor, enzyme, and K_d for the Zn^{2+}•albumin complex are constants. We also assume that Zn^{2+} is buffered exclusively by albumin, ignoring the relatively minor contributions of other plasma Zn^{2+} ligands such as histidine and cysteine.[27,28] Thus, the concentration of the Zn^{2+}•inhibitor species is a function of the inhibitor's affinity for Zn^{2+} and is defined by the concentrations of Zn^{2+}, albumin, and inhibitor. The fraction of enzyme that is inhibited is assumed to be a function of the concentration of the Zn^{2+}•inhibitor complex and the affinity of this species for the enzyme. Although more complex equilibria can be envisioned for the assembly of the ternary complex, this simplified theoretical analysis, together with our experimental data, suggests that effective inhibition can be achieved in plasma *in vivo*, given sufficient intrinsic metal binding affinity and intrinsic potency of the Zn^{2+}•inhibitor complex. It is clear that compounds intended for therapeutic use clearly must incorporate both high selectivity and the ability to compete for Zn^{2+} in the appropriate tissue or plasma environment.

Preliminary indications from model studies indicate that the zinc-dependent potency observed for these compounds can be indicative of their relative *in vivo* efficacy. For example, in a study comparing the efficacy of both zinc-dependent and non-zinc-dependent inhibitors of tryptase in a sheep model of bronchoconstriction, the required dose to achieve efficacy correlated with the zinc-dependent K_i of the inhibitor, as calibrated *in vivo* against the non-zinc-dependent inhibitor K_i (Table 6). Further evidence confirming the *in vivo* viability of zinc-dependent serine protease inhibitors has been obtained in animal models of thrombosis and tumour angiogenesis, amongst others.

It is also important to note that, thus far, we have observed no systematic compound toxicity or barriers to absorption related to any zinc-dependent protease inhibitor motif. Metabolism occurs along well-defined pathways considered normal for organic heterocyclic compounds. Properly designed, compounds also achieve oral bioavailability levels and circulating half-lives commensurate with the very desirable dosing regimens required for the treatment of chronic diseases.

Table 6 *Evidence for zinc-dependent efficacy in a sheep model of bronchoconstriction*

Compound	Ki´ for Tryptase		Sheep Efficacy (mg/dose)*
	+ EDTA	+ Zn^{2+}	
BABIM (1)	4 µM	89 nM	9
Zinc-dependent	7 µM	10 nM	1
Non-zinc-dependent	10 nM	10 nM	1

*Minimum effective whole animal dose, n = 2.

3 CONCLUSIONS

Identification of potent and selective inhibitors of serine proteases has been accomplished many times, but conversion of these inhibitors to orally efficacious drugs remains a major challenge. Although there are now a few examples of peptide-like structures that have been transformed into small organic molecules suitable as drugs (*e.g.* integrin antagonists), the problem of converting a substrate-like peptide, highly dependent on chirality and amide hydrogen-bonding patterns for enzyme recognition, binding potency and selectivity, into a viable drug-like structure lacks a general solution. The requirements of low molecular weight for bioavailability, and eventually, cost-of-goods in manufacturing, also weigh heavily against the use of peptide-like structures, especially in situations where peptide recognition sequences for effective potency and selectivity require over four amino acid equivalents.

The discovery of the zinc-dependent serine protease inhibition mode for low molecular weight organic heterocyclic compounds provides some breakthrough solutions to these problems. The scaffolds are simple to manipulate in synthesis and scale-up, are of reasonable molecular weight (even when elaborated for potency and specificity), and require no amino acid equivalents, peptide bonds, or chiral centres to achieve potencies and selectivities matching or exceeding known protease inhibitors of any class. Classical development issues, including absorption, distribution, metabolism and excretion affecting bioavailability and toxicity, do not exhibit any adverse dependence on the zinc-chelator mode of action for the inhibitors. We continue to pursue this novel approach to the design and synthesis of potent and selective serine protease inhibitors in multiple therapeutic target areas to identify new drug candidates.

Acknowledgements

The authors wish to acknowledge the contributions of the numerous members of the Departments of Structural Chemistry, Biochemistry, Pharmacology and Medicinal Chemistry at Axys Pharmaceuticals, Inc., to the development of the Delta Technology, and to Drs. Michael J. Ross and Heinz W. Gschwend of Axys Pharmaceuticals, Inc., and to Professors Daniel Rich, The University of Wisconsin - Madison, Ralph Hirschmann, The University of Pennsylvania, James Travis, The University of Georgia, and Charles Craik, The University of California, San Francisco, for their early encouragement and support of this work.

References and Notes

1.   D. H. Rich, "Comprehensive Medicinal Chemistry," Pergamon Press, Oxford, England, 1990, Vol. 2, Chapter 8.2, p. 391.

2.   K. Sekizawa, G. H. Caughey, S. C. Lazarus, W. M. Gold and J. A. Nadel, *J. Clin. Invest.*, 1989, **83**, 175.

3.   J. M. Clark, W. M. Abraham, C. E. Fishman, R. Forteza, A. Ahmed, A. Cortes, R. L. Warne, W. R. Moore and R. D. Tanaka, *Am. J. Respir. Crit. Care Med.*, 1995, **152**, 2076.

4.   R. R. Tidwell, J. D. Geratz, O. Dann, G. Volz, D. Zeh and H. Loewe, *J. Med. Chem.*, 1978, **21**, 613.

5.   G. H. Caughey, W. W. Raymond, E. Bacci, R. J. Lombardy and R. R. Tidwell, R., *J. Pharmacol. Exp. Ther.*, 1993, **264**, 676.

6.   J. D. Geratz, S. R. Shaver and R. R. Tidwell, *Thromb. Res.*, 1981, **24**, 73.

7.   J. W. Janc, J. M. Clark, R. L. Warne, K. C. Elrod, B. A. Katz and W. R. Moore, *Biochemistry*, 2000, **39**, 4792.

8.   The zinc-dependent shift in potency has been conveniently referred to as the "delta" for the inhibitor, thus spawning use of the nickname "Delta Technology" to describe the design of zinc-dependent serine protease inhibitors.

9.   B. L. Vallee and D. S. Auld, *Biochemistry*, 1990, **29**, 5647.

10.   J. M. Berg and Y. Shi, *Science*, 1996, **271**, 1081.

11.   B. A. Katz, J. M. Clark, J. S. Finer-Moore, T. E. Jenkins, C. R. Johnson, M. J. Ross, C. Luong, W. R. Moore and R. M. Stroud, *Nature*, 1998, **391**, 608.

12.   W. N. Lipscomb and N. Sträter, *Chem. Rev.*, 1996, **96**, 2375.

13.   D. W. Christianson, *Adv. Protein Chem.*, 1991, **42**, 281.

14.   E. Kimura, T. Shiota, T. Koike, M. Shiro and M. Kodama, *J. Am. Chem. Soc.*, 1990, **112**, 5805.

15.   J. T. Groves and J. R. Olson, *Inorg. Chem.*, 1985, **24**, 2715.

16.   A. Vedani and D. Huhta, *J. Am. Chem. Soc.*, 1990, **112**, 4759.

17.   M. Krieger, L. M. Kay, and R. M. Stroud, *J. Mol. Biol.*, 1974, **83**, 209.

18.   B. A. Katz, J. S. Finer-Moore, R. Mortezaei, D. H. Rich and R. M. Stroud, *Biochemistry* 1995, **34**, 8264.

19.   J. A. Ippolito, R. S. Alexander and D. W. Christianson, *J. Mol. Biol.*, 1990, **215**, 457.

20.   W. Somers, M. Ultsch, A. M. De Vos and A. A. Kossiakoff, *Nature*, 1994, **372**, 478.

21.   J. N. Higaki, B. L. Haymore, S. Chen, R. J. Fletterick and C. S. Craik, *Biochemistry*, 1990, **29**, 8582.

22.   J. M. Clark, T. E. Jenkins, B. A. Katz and R. M. Stroud, U.S. Patent 5,693,515 (2 Dec 1997).

23.   G. R. Magneson, J. M. Puvathingal and W. J. Ray, Jr., *J. Biol. Chem.*, 1987, **262**, 11140.

24.   G. Berthon, C. Matuchansky and P. May, *J. Inorg. Biochem.*, 1980, **13**, 63.

25.   H. A. Schroeder and A. P. Nason, *Clin. Chem.*, 1971, **17**, 461.

26.   A. F. Parisi and B. L. Vallee, *Biochemistry*, 1970, **9**, 2421.

27.   E. L. Giroux, *Biochem. Med.*, 1975, **12**, 258.

28.   W. R. Harris and C. Keen, *J. Nutr.*, 1989, **119**, 1677.

STRUCTURE-BASED DESIGN OF IRREVERSIBLE, PEPTIDOMIMETIC HUMAN RHINOVIRUS 3C PROTEASE INHIBITORS

P. S. Dragovich,* S. E. Webber, S. A. Fuhrman, A. K. Patick, D. A. Matthews, T. J. Prins, R. Zhou, J. T. Marakovits, C. E. Ford, J. W. Meador, III, R. A. Ferre, and S. T. Worland

Agouron Pharmaceuticals, Inc., 3565 General Atomics Court, San Diego, CA 92121

1 INTRODUCTION

The human rhinoviruses (HRVs) belong to the picornavirus family and are the single most significant cause of the common cold.[1,2] These viruses translate a positive-strand RNA genome into a large polyprotein which undergoes co- and post-translational processing by virally encoded proteases to produce the structural and enzymatic proteins required for viral replication.[3] Specifically, cleavage between several glutamine-glycine residues is effected by the human rhinovirus 3C protease (3CP),[3] a cysteine protease with structural similarity to the trypsin protein family but possessing minimal homology to prevalent mammalian enzymes.[4,5] Importantly, the activity of 3CP is essential for viral replication, and its active site residues are believed to be highly conserved throughout the more than 100 known rhinovirus serotypes.[4-7]

Currently, no effective therapy exists to directly treat rhinoviral infections, and the numerous viral serotypes make the development of a vaccine unlikely.[1a,1e] Previous approaches toward the identification of antirhinoviral therapeutics include the use of interferon,[1a,1c,1e] disruption of virus-receptor (ICAM-1) interactions,[1b] the examination of capsid-binding antipicornaviral compounds,[8] and the use of other miscellaneous agents.[7,9] In addition, several examples of 3CP inhibitors have recently been described in the literature, and some of these compounds exhibit antiviral properties.[10]

Recently, we detailed the discovery and development of a new class of HRV 3C protease inhibitors which display in vitro antiviral activity against several rhinovirus serotypes.[11,12] These inhibitors are comprised of a substrate-derived tripeptide binding determinant which provides affinity for the target protease and a Michael acceptor moiety which irreversibly forms a covalent adduct with the active site cysteine residue of the 3C enzyme (e.g., compound 1, Figure 1 and Table 1).[13-15] Although optimization of these tripeptides afforded relatively active antirhinoviral agents such as compound 2 (Table 1), we sought to further improve these "optimized" molecules by (1) reducing their peptidic character, and (2) replacing the N-terminal thiocarbamate moiety present in the most active inhibitors with a simple amide. The results of our efforts to accomplish these two goals are described below.

Table 1 [a]*HRV Serotype-14.* [b]*See ref 11 for assay method and error*

Compd No.	$k_{obs}/[I]$ $(M^{-1}s^{-1})^{a,b}$	EC_{50} $(\mu M)^{a,b}$	CC_{50} $(\mu M)^b$
1	25 000	0.54	>100
2	800 000	0.056	>100
3	17 400	0.36	>100

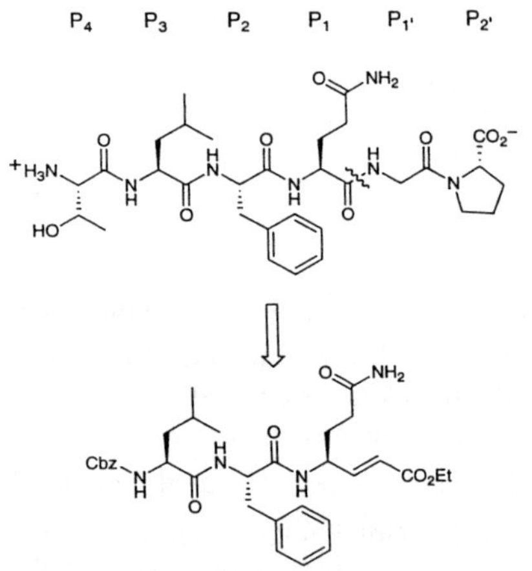

Figure 1 *Design of Irreversible HRV 3CP Inhibitors*

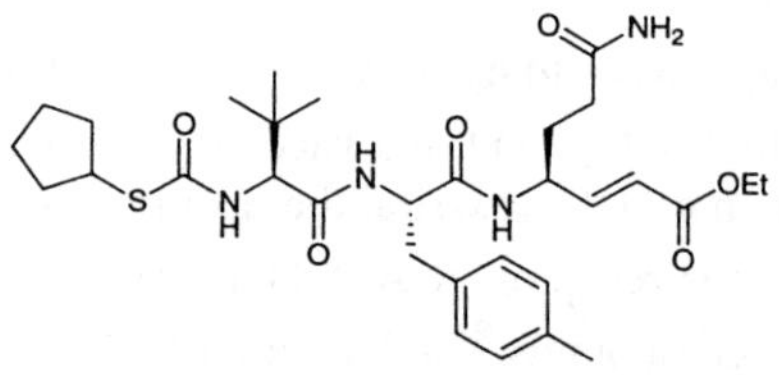

2 PEPTIDOMIMETIC SAR

The goal of reducing the peptidic character of the above inhibitors was addressed by incorporation of a peptidomimetic binding element into the inhibitor design. Replacement of a backbone amide moiety present in peptide-derived enzyme inhibitors with a ketomethylene isostere often affords peptidomimetic compounds with inhibition activities similar to the parent molecules (Figure 2).[16]

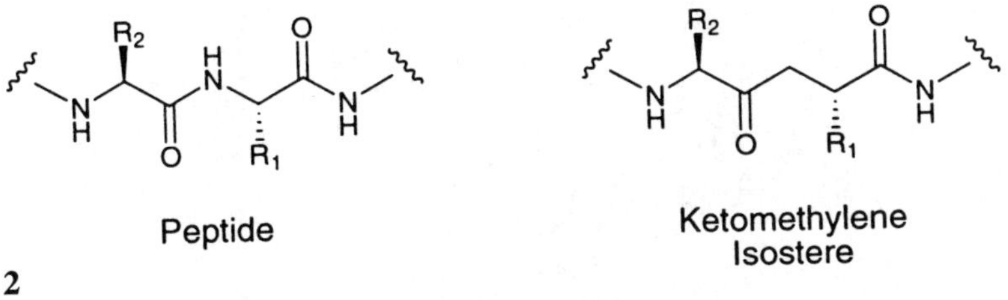

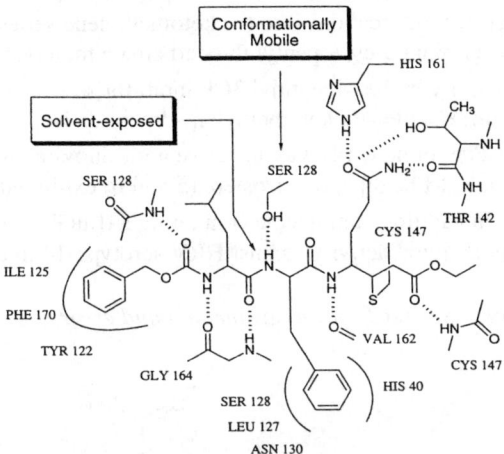

Figure 2

We therefore wished to introduce such an isostere into a representative tripeptidyl 3CP inhibitor (e.g., compound **1**). Analysis of the HRV-2 3CP-**1** crystal structure[11] indicated that the backbone amide NH linking the P2 Phe and P3 Leu amino acid residues of **1** was highly solvent exposed (Figure 3).[14]

Figure 3 *Schematic diagram of **1** bound in the HRV-2 3CP active site. Hydrogen bonds are represented as dashed lines and the residues which make up the enzyme binding subsites are depicted*

In addition, comparison with the uncomplexed 3CP crystal structure[4a] suggested that the 3CP serine residue which formed a hydrogen bond with the P2-P3 amide NH of **1** was somewhat conformationally mobile (Ser-128, Figure 3). In contrast, the inhibitor P1-P2 and P3-P4 amide NH's were only slightly solvent-exposed and formed hydrogen bonds with relatively stationary 3CP residues. Replacement of the P2-P3 amide linkage with a ketomethylene dipeptide isostere was therefore anticipated to have less of an impact on inhibitor 3CP affinity than substitution of either of the two other amide bonds present in the tripeptidyl backbone of **1**.

In the event, the ketomethylene-containing peptidomimetic compound **3** displayed slightly reduced anti-3CP inhibitory activity but *improved* antiviral properties when compared with the tripeptide **1** (Table 1).

3

The improved antiviral activity of **3** relative to **1** was consistent with increased cell membrane permeability due to the reduced potential of **3** to form hydrogen bonds with water.[17] As was observed for related tripeptidyl 3CP inhibitors, the ketomethylene-containing molecule **3** did not exhibit cytotoxicity in cell culture up to its solubility limit (CC_{50} >100 μM).[18] These encouraging results prompted the incorporation of several previously discovered peptide modifications into the ketomethylene series.

Accordingly, an N-terminal S-cyclopentyl thiocarbamate moiety, which afforded the most potent antiviral activity in the tripeptidyl 3CP inhibitor series,[12] was evaluated in combination with several different P2-P3 ketomethylene dipeptide isosteres (Table 2). An inhibitor containing a Leu-Phe mimetic (**4**) was an active antirhinoviral agent, while compounds incorporating a Val-Phe dipeptide isostere (**5** and **6**) exhibited even greater antirhinoviral potency.[19] In addition, a molecule containing a tBuGly-Phe dipeptide mimic (**7**) displayed potent antiviral activity against HRV serotype-14 in cell culture.

Table 2 [a]*HRV Serotype-14.* [b]*See ref 11 for assay method and error*

Compd No.	R_1	R_2	$k_{obs}/[I]$ $(M^{-1}s^{-1})^{a,b}$	EC_{50} $(μM)^{a,b}$	CC_{50} $(μM)^{b}$
4	CH_2Ph	$CH_2CH(CH_3)_2$	55 700	0.19	>100
5	CH_2Ph	$CH(CH_3)_2$	255 000	0.020	>100
6	$CH_2Ph(4\text{-}F)$	$CH(CH_3)_2$	293 000	0.020	>100
7	CH_2Ph	$C(CH_3)_3$	124 000	0.050	>100

A comparison was also made between one of the more potent ketomethylene-containing compounds (**6**) and an analogous tripeptidyl inhibitor (**8**). Examination of broad-spectrum EC_{90} antiviral data clearly indicated that the peptidomimetic **6** was the superior antiviral agent and was at least 10-fold more potent than **8** for each HRV serotype studied (Table 3).[19] The above data indicate that the antirhinoviral activity displayed by tripeptidyl Michael acceptors may be substantially increased by reducing their peptidic characteristics. The full details of this study, including the syntheses of compounds **3-8**, are presented elsewhere.[20]

Table 3 *[a]See ref 11 for assay method and error*

Compd No.	R	X	EC$_{90}$ vs Rhinovirus Serotype (μM)[a]			
			HRV-14	HRV-1A	HRV-2	HRV-10
6		CH$_2$	0.10	0.20	0.25	0.25
8		NH	1.8	2.5	5.4	2.5
13		CH$_2$	0.32	0.63	0.28	0.79

3 THIOCARBAMATE REPLACEMENT

In our earlier work with tripeptidyl 3CP inhibitors, we found that inclusion of an *N*-terminal P4 thiocarbamate moiety in the inhibitor design significantly improved anti-3CP and antirhinoviral activity relative to similar carbamate-containing molecules (compare compounds **1** and **9**, Table 4).[12]

Table 4 *[a]HRV Serotype-14.* *[b]See ref 11 for assay method and error.* *[c]1:1 mixture of diastereomers*

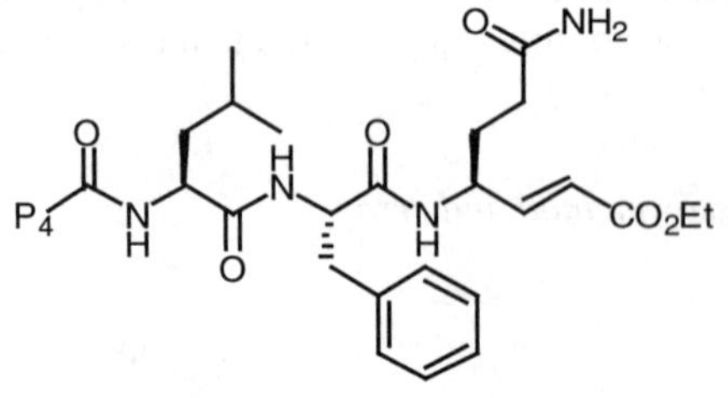

Compd No.	P$_4$	$k_{obs}/[I]$ (M^{-1}s^{-1})^{a,b}	EC$_{50}$ (μM)^{a,b}	CC$_{50}$ (μM)^b
1	PhCH$_2$O	25 000	0.54	>100
9	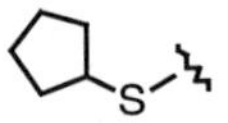	114 000	0.18	>100
12a^c	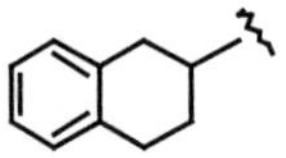	95 000	1.4	>100
12b	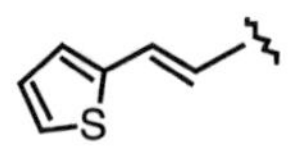	66 000	0.55	>100
12c	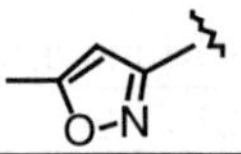	260 000	0.25	>100

However, we were concerned that the thiocarbamate moiety might undergo facile *in vivo* metabolism, and we therefore sought to replace this functionality with amides which also imparted good activity levels to the resulting inhibitors (Figure 4).

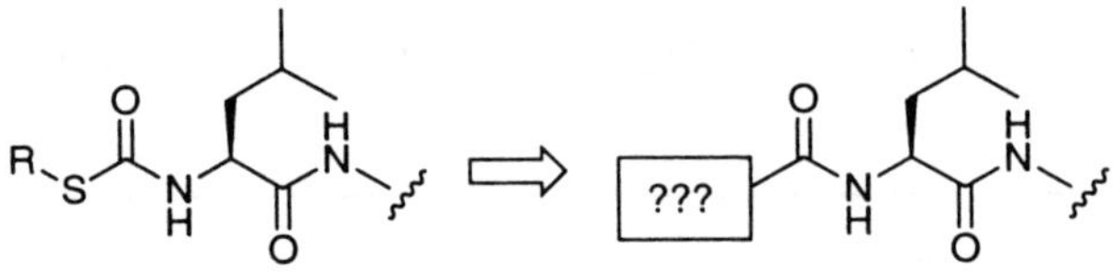

Figure 4 *Design of thiocarbamate replacements*

Solid-phase chemistry was therefore utilized to identify suitable thiocarbamate replacements as indicated in Scheme 1. Again, full experimental details of these studies, including the syntheses of representative molecules, are presented elsewhere.[21] In brief, the tripeptidyl intermediate **10** was assembled on Chiron Multipins™[22] derivatized with Rink Amide[23] linker. In the key synthetic operation, the Fmoc group was removed from **10**, and the newly-formed *N*-terminal amine was independently coupled with a variety of carboxylic acids and acid chlorides to afford approximately 500 unique intermediates **11**. Subsequent exposure of these intermediates to aqueous TFA effected cleavage from the solid support and provided the desired tripeptidyl compounds **12**. The acids and acid chlorides utilized in this study were selected from the ACD[24] based on their commercial availability, low molecular weight (< 300), and lack of highly reactive functional groups other than the desired acyl moiety (e.g., □-bromo-carboxylic acids were not included).

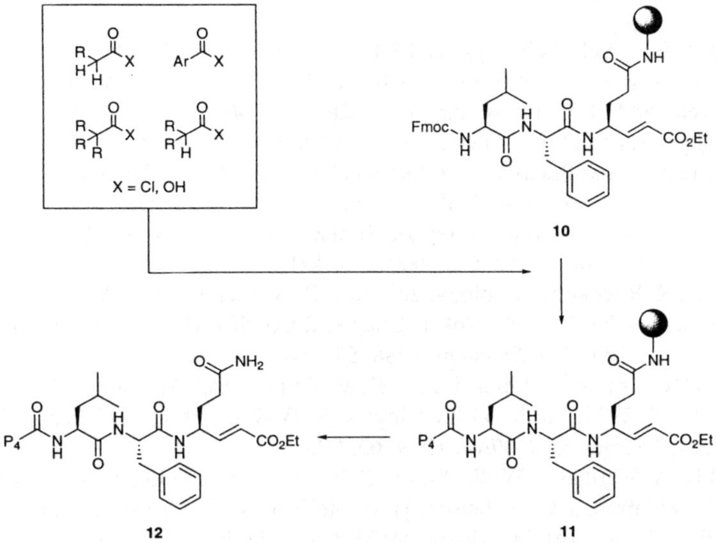

Scheme 1

The completed series of approximately 500 derivatized tripeptide Michael acceptors (**12**) was then screened for HRV-14 3CP inhibition activity, and the most active compounds (**12a-c**) were resynthesized utilizing solution-phase methods. The biological activities of these molecules are depicted in Table 4. These compounds potently and irreversibly inhibited HRV-14 3C protease with activities that nearly equaled or exceeded that displayed by the thiocarbamate-containing compound **9**. Compounds **12a-c** also exhibited various levels of antiviral activity without observed cytotoxicity when tested against HRV-14 in cell culture. Combination of the *N*-terminal isoxazole moiety present in the most active inhibitor **12c** with the ketomethylene peptide isostere described above provided a potent antirhinoviral agent (**13**) which exhibited sub-micromolar EC$_{90}$ antiviral activity against several rhinovirus serotypes in cell culture (Table 3).[19]

4 CONCLUSIONS

The structure-activity studies described above identified Michael acceptor-containing molecules which incorporate ketomethylene dipeptide isosteres as both highly active irreversible inhibitors of the human rhinovirus 3C protease (HRV-3CP) and potent antirhinoviral agents in cell culture. Such compounds typically displayed slightly reduced 3C protease inhibition activity relative to the corresponding peptide-derived molecules, but also exhibited significantly improved antiviral properties. In a related study, the *N*-terminal optimization of tripeptide-derived HRV-3CP inhibitors was accomplished utilizing solid-phase synthesis and high-throughput assay techniques. Such optimization identified a molecules containing an *N*-terminal 5-methylisoxazole-3-carboxamide as a potent 3CP inhibitor and broad-spectrum antirhinoviral agent. Incorporation of such a carboxamide into the ketomethylene series of peptidomimetic HRV-3CP inhibitors also afforded highly active broad-spectrum antirhinoviral agents.

References and Notes

1.   (a) R. B. Couch, 'Virology, 2nd Ed.', B. N. Fields and D. M. Knipe, Eds., Raven Press, New York, 1990; Vol. 1, Chapter 22, p. 607. (b) M. A. McKinlay, D. C. Pevear, and M. G. Rossmann, *Annu. Rev. Microbiol.* 1992, **46**, 635. (c) R. J. Phillpotts, and D. A. J. Tyrrell, *Br. Med. Bull.* 1985, **41**, 386. (d) J. M. Gwaltney, 'Principles and Practices of Infectious Diseases,' G. L. Mandell, R. G. Douglas, and J. E. Bennett, Eds., John Wiley & Sons, New York, 1985, Chapter 38, p. 351. (e) J. M. Gwaltney, 'Viral Infections of Humans,' A. S. Evans, Ed., Plenum Publishing Corp., New York, 1982; Chapter 20, p. 491.
2.   (a) R. R. Rueckert, 'Virology, 2nd Ed.,' B. N. Fields and D. M. Knipe, Eds., Raven Press, New York, 1990; Vol. 1, Chapter 20, p. 507. (b) H.-G. Kräusslich, and E. Wimmer, *Ann. Rev. Biochem.* 1988, **57**, 701.
3.   (a) D. C. Orr, A. C. Long, J. Kay, B. M. Dunn, and J. M. Cameron, *J. Gen. Virol.* 1989, **70**, 2931. (b) M. G. Cordingley, R. B. Register, P. L. Callahan, V. M. Garsky, and R. J. Colonno, *J. Virol.* 1989, **63**, 5037.
4.   (a) D. A. Matthews, W. W. Smith, R. A. Ferre, B. Condon, G. Budahazi, W. Sisson, J. E. Villafranca, C. A. Janson, H. E. McElroy, C. L. Gribskov, and S. Worland, *Cell* 1994, **77**, 761. (b) M. Allaire, M. M. Chernaia, B. A. Malcolm, and M. N. G. James, *Nature* 1994, **369**, 72. (c) J. F. Bazan, and R. J. Fletterick, *Proc. Natl. Acad. Sci. U.S.A.* 1988, **85**, 7872. (d) A. E. Gorbalenya, V. M. Blinov, and A. P. Donchenko, *FEBS Lett.* 1986, **194**, 253.
5.   (a) W.-M. Lee, W. Wang, and R. R. Rueckert, *Virus Genes* 1994, **9**, 177. (b) B. Aschauer, G. Werner, J. McCray, B. Rosenwirth, and H. Bachmayer, *Virology* 1991, **184**, 587. (c) P. J. Hughes, C. North, C. H. Jellis, P. D. Minor, and G. Stanway, *J. Gen. Virol.* 1988, **69**, 49. (d) M. Duechler, T. Skern, W. Sommergruber, C. Neubauer, P. Gruendler, I. Fogy, D. Blaas, and E. Kuechler, *Proc. Natl. Acad. Sci. U.S.A.* 1987, **84**, 2605. (e) G. Werner, B. Rosenwirth, E. Bauer, J.-M. Siefert, F.-J. Werner, and J. Besemer, *J. Virol.* 1986, **57**, 1084. (f) T. Skern, W. Sommergruber, D. Blaas, P. Gruendler, F. Fraundorfer, C. Pieler, I. Fogy, and E. Kuechler, *Nucleic Acids Res.* 1985, **13**, 2111. (g) P. L. Callahan, S. Mizutani, and R. J. Colonno, *Proc. Natl. Acad. Sci. U.S.A.* 1985, **82**, 732. (h) G. Stanway, P. J. Hughes, R. C. Mountford, P. D. Minor, and J. W. Almond, *Nucleic Acids Res.* 1984, **12**, 7859.
6.   V. V. Hamparian, R. J. Colonno, M. K. Cooney, E. C. Dick, J. M. Gwaltney, Jr.; J. H. Hughes, W. S. Jordan, Jr.; A. Z. Kapikian, W. J. Mogabgab, A. Monto, C. A. Phillips, R. R. Rueckert, J. H. Schieble, E. J. Stott, and D. A. J. Tyrrell, *Virology* 1987, **159**, 191.
7.   L. Carrasco, *Pharmac. Ther.* 1994, **64**, 215.
8.   (a) D. A. Oren, A. Zhang, H. Nesvadba, B. Rosenwirth, and E. Arnold, *J. Mol. Biol.* 1996, **259**, 120 and references therein. (b) G. D. Diana, P. Rudewicz, D. C. Pevear, T. J. Nitz, S. C. Aldous, D. J. Aldous, D. T. Robinson, T. Draper, F. J. Dutko, C. Aldi, G. Gendron, R. C. Oglesby, D. L. Volkots, M. Reuman, T. R. Bailey, R. Czerniak, T. Block, R. Roland, and J. Oppermann, *J. Med. Chem.* 1995, **38**, 1355 and references therein.

9.   M. J. Tebbe, W. A. Spitzer, F. Victor, S. C. Miller, C. C. Lee, T. R. Sattelberg, Sr.;
      E. McKinney, and J. C. Tang, *J. Med. Chem.* 1997, **40**, 3937 and references therein.
10.   (a) S. E. Webber, K. Okano, T. L. Little, S. H. Reich, Y. Xin, S. A. Fuhrman, D. A.
      Matthews, T. F. Hendrickson, R. A. Love, A. K. Patick, J. W. Meador, III; R. A.
      Ferre, E. L. Brown, C. E. Ford, S. L. Binford, and S. T. Worland, *J. Med. Chem.*
      1998, **41**, 2786. (b) L. N. Jungheim, J. D. Cohen, R. B. Johnson, E. C. Villarreal, M.
      Wakulchik; R. J. Loncharich, and Q. M. Wang, *Bioorg. Med. Chem. Lett.* 1997, **7**,
      1589. (c) S. E. Webber, J. Tikhe, S. T. Worland, S. A. Fuhrman, T. F. Hendrickson,
      D. A. Matthews, R. A. Love, A. K. Patick, J. W. Meador, R. A. Ferre, E. L. Brown,
      D. M. DeLisle, C. E. Ford, and S. L. Binford, *J. Med. Chem.* 1996, **39**, 5072. (d) T.
      A. Shepherd, G. A. Cox, E. McKinney, J. Tang, M. Wakulchik, R. E. Zimmerman,
      and E. C. Villarreal, *Bioorg. Med. Chem. Lett.* 1996, **6**, 2893. (e) G. M. Brill, W. M.
      Kati, D. Montgomery, J. P. Karwowski, P. E. Humphrey, M. Jackson, J. J. Clement,
      S. Kadam, R. H. Chen, and J. B. McAlpine, *J. Antibiotics* 1996, **49**, 541. (f) H. L.
      Sham, W. Rosenbrook, W. Kati, D. A. Betebenner, N. E. Wideburg, A. Saldivar, J. J.
      Plattner, and D. W. Norbeck, *J. Chem. Soc., Perkin Trans. 1* 1995, 1081. (g) B. A.
      Malcolm, C. Lowe, S. Shechosky, R. T. McKay, C. C. Yang, V. J. Shah, R. J. Simon,
      J. C. Vederas, and D. V. Santi, *Biochemistry* 1995, **34**, 8172. (h) S. W. Kaldor, M.
      Hammond, B. A. Dressman, J. M. Labus, F. W. Chadwell, A. D. Kline, and B. A.
      Heinz, *Bioorg. Med. Chem. Lett.* 1995, **5**, 2021. (i) S. Kadam, J. Poddig, P.
      Humphrey, J. Karwowski, M. Jackson, S. Tennent, L. Fung, J. Hochlowski, R.
      Rasmussen, and J. McAlpine, *J. Antibiotics* 1994, **47**, 836. (j) S. B. Singh, M. G.
      Cordingley, R. G. Ball, J. L. Smith, A. W. Dombrowski, and M. A. Goetz,
      Tetrahedron Lett. 1991, **32**, 5279. (k) J. W. Skiles and D. McNeil, *Tetrahedron
      Lett.* 1990, **31**, 7277.
11.   P. S. Dragovich, S. E. Webber, R. E. Babine, S. A. Fuhrman, A. K. Patick, D. A.
      Matthews, C. A. Lee, S. R. Reich, T. J. Prins, J. T. Marakovits, E. S. Littlefield, R.
      Zhou, J. Tikhe, C. E. Ford, M. B. Wallace, J. W. Meador, III; R. A. Ferre, E. L.
      Brown, S. L. Binford, J. E. V. Harr, D. M. DeLisle, and S. T. Worland, *J. Med.
      Chem.* 1998, **41**, 2806.
12.   P. S. Dragovich, S. E. Webber, R. E. Babine, S. A. Fuhrman, A. K. Patick, D. A.
      Matthews, S. H. Reich, J. T. Marakovits, T. J. Prins, R. Zhou, J. Tikhe, E. S.
      Littlefield, T. M. Bleckman, M. B. Wallace, T. L. Little, C. E. Ford, J. W. Meador,
      III, R. A. Ferre, E. L. Brown, S. L. Binford, D. M. DeLisle, and S. T. Worland, *J.
      Med. Chem.* 1998, **41**, 2819.
13.   M. G. Cordingley, P. L. Callahan, V. V. Sardana, V. M. Garsky, and R. J. Colonno,
      J. Biol. Chem. 1990, **265**, 9062.
14.   The nomenclature used for describing the individual amino acid residues of a peptide
      substrate (P_2, P_1, P_1', P_2', etc.) and the corresponding enzyme subsites (S_2, S_1, S_1',
      S_2', etc.) is described in: I. Schechter and A. Berger, *Biochem. Biophys. Res.
      Commun.* 1967, **27**, 157.
15.   A similar series of HRV 3CP inhibitors has also recently been described. See: J.-s.
      Kong, S. Venkatraman, K. Furness, S. Nimkar, T. A. Shepherd, Q. M. Wang, J.
      Aubé, and R. P. Hanzlik, *J. Med. Chem.* 1998, **41**, 2579.
16.   For selected examples, see: (a) R. G. Almquist, J. Crase, C. Jennings-White, R. F.
      Meyer, M. L. Hoefle, R. D. Smith, A. D. Essenburg, and H. R. Kaplan, *J. Med.
      Chem.* 1982, **25**, 1292. (b) M. W. Holladay, F. G. Salituro, and D. H. Rich, *J. Med.*

Chem. 1987, **30**, 374. (c) R. G. Almquist, W.-R. Chao, A. K. Judd, C. Mitoma, D. J. Rossi, R. E. Panasevich, and R. J. Matthews, *J. Med. Chem.* 1988, **31**, 561. (d) S. L. Harbeson and D. H. Rich, *J. Med. Chem.* 1989, **32**, 1378.

17.   (a) R. A. Conradi, A. R. Hilgers, N. F. H. Ho, and P. S. Burton, *Pharm. Res.* 1991, **8**, 1453. (b) R. A. Conradi, A. R. Hilgers, N. F. H. Ho, and P. S. Burton, *Pharm. Res.* 1992, **9**, 435.

18.   In several cases, the solubility of a given inhibitor was determined to be less than the reported CC_{50} value. However, the most active antiviral agents were clearly soluble and non-toxic to more than 10x the reported EC_{50} values.

19.   The antirhinoviral properties of several inhibitors were determined under similar conditions against many additional HRV serotypes in cell culture (~30). In general, the activity observed for a given compound vs serotypes 1A, 2, and 10 was representative of its broad-spectrum antiviral properties. Serotype 14 nearly always proved to be the most susceptible to the antiviral agents under study. Full details of these experiments will be presented elsewhere (A. K. Patick, manuscript in preparation).

20.   P. S. Dragovich, T. J. Prins, R. Zhou, S. A. Fuhrman, A. K. Patick, D. A. Matthews, C. E. Ford, J. W. Meador, III; R. A. Ferre, and S. T. Worland, *J. Med. Chem.* submitted.

21.   P. S. Dragovich, R. Zhou, D. J. Skalitzky, S. A. Fuhrman, A. K. Patick, C. E. Ford, J. W. Meador, III; and S. T. Worland *Bioorg. Med. Chem.* in press.

22.   (a) H. M. Geysen, R. H. Meloen, and S. J. Barteling, *Proc. Natl. Acad. Sci. U.S.A.* 1984, **81**, 3998. (b) R. M. Valerio, A. M. Bray, and N. J. Maeji, *Int. J. Peptide Protein Res.* 1994, **44**, 158.

23.   H. Rink, *Tetrahedron Lett.* 1987, **28**, 3787.

24.   Available Chemicals Directory, v97.1, Molecular Design Limited, 14600 Catalina Street, San Leandro, CA 94577.

INHIBITION OF THE OSTEOCLAST SPECIFIC THIOL PROTEASE, CATHEPSIN K

Daniel F. Veber[†], Robert Marquis[†], Dennis S. Yamashita[†], Scott K. Thompson[†], Hye-Ja Oh[†], Yu Ru[†], Stacie M. Halbert[†], Mary J. Bossard[‡], Thaddeus A. Tomaszek[‡], Mark A. Levy[‡], David Tew[‡], Thomas Meek[‡], Art Shu[††], J. Richard Heys[††], Karla J. D'Alessio[¶], Michael S. McQueney[¶], Bernard Y. Amegadzie[#], Charles R. Hanning[#], Renee L. DesJarlais[**], Jacques Briand[**], Susanta K. Sarkar[**], Michael J. Huddleston[**], Carl F. Ijames[**], Steven A. Carr[**], Keith T. Garnes[††], Cheryl A. Janson[¶], Baoguang Zhao[§], Ward W. Smith[§], Sherin S. Abdel-Meguid[§],

Departments of Medicinal Chemistry[†], Molecular Recognition[‡], Macromolecular Sciences[§], Protein Biochemistry[¶], Gene Expression Sciences[#], Physical and Structural Chemistry[**], Radiochemistry[††], and Cellular Biochemistry[‡‡], SmithKline Beecham Pharmaceuticals, King of Prussia, PA 19406

1 INTRODUCTION

Cathepsin K is a cysteine protease and a member of the papain superfamily. It is selectively expressed in osteoclasts and has been implicated in the process of bone resorption. Therefore, inhibition of Cathepsin K offers a promising mechanism for the treatment of diseases characterized by excessive bone loss such as osteoporosis.[1] The discovery of inhibitors of this and other cysteine proteases has been the goal of research in our laboratory. In order to minimize potential immunological complications, a key consideration of our protease inhibitor design has been to avoid the incorporation of intrinsically reactive groups that might derivitize the side chain or backbone elements of proteins other than our target enzyme. Notable therapeutic successes toward this end have been seen with inhibitors of the metalloprotease, angiotensin converting enzyme,[2] and inhibitors of the aspartyl protease, HIV protease.[3] No comparable examples exist for the serine and cysteine proteases for which inhibitors generally utilize reactive groups which covalently modify the enzyme for their efficacy. Krantz has discussed the "quiescent affinity" of acyloxy ketones for cysteine proteases,[4] and Rando has discussed mechanism-based enzyme inactivators of the reactive β-lactams antibiotics for inhibition of β-lactamases.[5] Palmer has reported time-dependent, non-selective, irreversible inhibitors of Cathepsin K (and other cysteine proteases) based on a vinyl sulfone[6] first utilized by Hanzlik.[7] Nonetheless, such compounds ultimately become covalently and irreversibly linked such that the protein remains derivatized after denaturation and degradation, thereby offering the opportunity for undesired antigenic and immunogenic responses. We have reported the successful design of selective, "quiescent" yet reversible, inhibitors of Cathepsin K based on a poorly electrophilic diacylaminomethyl ketone scaffold, an approach inspired by the overlay of aldehyde-type structures bound in different modes elucidated by X-ray co-crystallography on papain.[8] The design concept has been expanded to include compounds which undergo a single turnover to give a highly selective, essentially irreversible mechanism of inhibition.[9] Initial examples of these new classes have been found to suffer deficiencies which limit therapeutic application, including rapid clearance from the blood stream and low oral bioavailability. These deficiencies have led us to develop analogs designed to remove each of the H-bond acceptor and H-bond donor groups individually. H-bonding capacity has been postulated to contribute to poor oral bioavailability. [10] Indeed, removal of H-bonding groups has led to improved oral bioavailability compared to the initially prepared compounds.

Our initial design hypothesis has been confirmed by X-ray crystallography of inhibitors bound to Cathepsin K for each mechanistic class. Notably, the crystal state appears to trap the tetrahedral intermediate for each of the turnover inhibitors, while continuation to enzyme acylation is seen in solution studies.

2 DIAMINOKETONES AS CATHEPSIN K INHIBITORS

There has been considerable structural information reported in the literature for members of the papain superfamily both with inhibitors bound and for the free enzymes.[11] Papain, the prototypical member, has yielded X-ray co-crystal data with the irreversible inhibitor, E-64,[12] as well as a reversible, peptide aldehyde inhibitor, leupeptin (**1**).[13] Both provide detailed information relating to inhibitor binding sites which orient on the "unprimed"[14] side of the active site (amino-terminal to the cleaved amide bond). Prior to the availability of cloned and expressed cathepsin K, we undertook an X-ray evaluation of a complex of papain with Cbz-leu-leu-leu aldehyde (**2**) as a close analog of the already known leupeptin structure. An unexpected binding orientation of this compound toward the "primed" side of the active site was observed, an interesting result since the S2' pocket appears more poorly defined relative to the S2 hydrophobic pocket of papain. We capitalized on this observation, and overlayed structures of bound aldehyde **2** and leupeptin (**1**) to achieve structures of inhibitors that span the entire active site(see fig.1). A first design involved removal of the aldehyde hydrogens (circled) and combining the two carbonyl groups into a single structure, the hypothetical ketone which resulted was further simplified by deletion of the side chains of the arginine in **1** and the C-terminal leucine in **2**. The deletion of the leucine was based on modelling, where it appeared that Trp 184 would form a better aromatic-aromatic interaction (on the primed side) if **2** was shortened by one amino acid. This deletion was also consistent with our earlier enzymatic studies which showed the dipeptide aldehyde, Cbz-Leu-Leu aldehyde, to be a potent, time-dependent inhibitor of Cathepsin K (kinact / [I] = 3 x 10^5 M^{-1} s^{-1})[15, 16] . Therefore, we chose the readily accessible symmetrical ketone **3a** as our intial synthetic target (Figure 1). The reduced electrophilicity of a ketone compared to an aldehyde was an especially attractive aspect of this target.

We found that ketone **3a**, although a relatively poor inhibitor of papain with a $K_{i,app}$ >10 μM, had a $K_{i,app}$ of 22 nM against our actual target, Cathepsin K. In addition to being selective relative to papain, it is also a poor inhibitor of the other members of the papain family (K $_{i,app}$ Cathepsin L, 0.34 μM; Cathepsin B ,1.3 μM; Cathepsin S, 0.89 μM).[17]

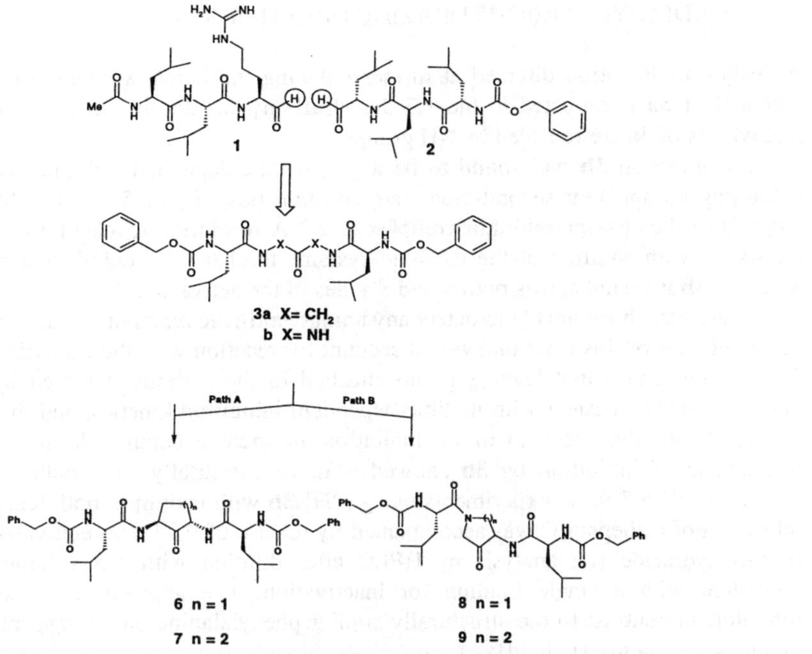

Figure 1 *The Evolution of Diacylaminomethyl Ketone Design and hypotheses for possible cyclization of the ketone region*

We hypothesize that the poor binding to papain may be a consequence of steric interactions of the more bulky Cbz group in P-3 compared to the smaller acetyl group which occupies this site when leupeptin is the inhibitor. The importance of the ketone functionality for inhibition was confirmed by the low potency of alcohol **4** (Table 1) which is seen to lose nearly 3 orders of magnitude in potency in spite of retaining all of the peripheral binding elements of **3a** other than the ketone.

Ketone **3a** was shown to be a competitive inhibitor versus a peptide substrate with Cathepsin K by a Lineweaver-Burk plot[18] and showed no evidence of time-dependent inhibition over the 30 minute progress curve analysis. Rapid reversibility of inhibition was observed upon dilution using the protocol as previously described[15] for determining dissociation rates (k_{off}) of reversible inhibitors bound to Cathepsin K. Furthermore, consistent with reversible inhibition, ketone **3a** was subjected to incubation with Cathepsin K for 5h, denatured, and the resulting protein was then analyzed by electrospray mass spectrometry[19] showing no evidence for formation of covalently modified enzyme. This is in contrast to Cbz-Leu-Leu-CH$_2$Br, a member of the classical α-halomethyl ketone cysteine protease inhibitors, in which a covalent adduct was observed by mass spectroscopy (Cat K + 374.8 Da). This also contrasts with the 1,5-diacyl carbohydrazide inhibitors discussed below which also form a covalent adduct with the enzyme.

The X-ray co-crystal structure of **3a** bound to Cathepsin K was determined and the structure is consistent with formation of a thiohemiketal between the ketone carbonyl and the active site Cys 25.[8]

3   1,5-DIACYLCARBOHYDRAZIDE INHIBITORS

A design modification directed at further reducing the potential reactivity of the central carbonyl of **3a** is reflected in the 1,5 diacylcarbohydrazide **3b** . In this compound the 2 methylenes of **3a** are replaced by NH groups.

Compound **3b** was found to be a potent time-dependent inhibitor of cathepsin K exhibiting an apparent second-order rate constant (k_{obs}/[I]) of 5.2 x 10^6 $M^{-1}s^{-1}$. X-ray analysis of the enzyme-inhibitor complex at 2.2 Å resolution revealed a mode of binding consistent with addition of the catalytic cysteine thiol to the central urea carbonyl. The inhibitor **3b** is bound across both S and S' sides of the active site.[9]

We have been unable to detect any unique, intrinsic reactivity of the central carbonyl of compounds of this class that would account for reaction with the catalytic cysteine. That **3b** contains a potential leaving group attached to the carbonyl targeted by the cysteine nucleophile is consistent with its time-dependent inhibition kinetics, and this prompted us to investigate the mechanism of inhibition in greater detail. Initial studies of the mechanism of inhibition by **3b** showed it to be essentially irreversible on dilution or dialysis at pH 5.5-9. In experiments using [^{3}H]-**3b** with tritium in both leucyl side chains, inhibition of cathepsin K was accompanied by release of 0.92-0.96 equivalents of [^{3}H]-Z-leucinylhydrazide (on analysis by HPLC after dilution with 1.5 volumes of DMSO), consistent with a single binding for inactivation. The apparent irreversibility of this inhibitor, in contrast to the structurally similar phenylalanine-based azapeptide esters and amides reported by Abeles[20] to be slow-turnover papain inhibitors, may also be due to the formation of a more stable cathepsin K/inhibitor complex relative to Abeles' phenylalanine-based papain intermediate. To obtain direct evidence of a stable acyl enzyme species, cathepsin K was inactivated independently with **3b** and the unsymmetric inhibitor **5**, and the mixtures were analyzed by on-line LC-MS. In the case of inactivation with **3b**, the products observed were the Z-leucylhydrazinecarbamoyl-cathepsin K adduct (Mr = M + 305, where M is the molecular weight of free cathepsin K) and Z-leucinylhydrazide (Mr = 279). The Z-leucinylhydrazinecarbamoyl-cathepsin K adduct was also observed by ^{13}C NMR spectroscopic analysis of the complex of cathepsin K with **3b** which was ^{13}C-labeled at the central carbonyl and ^{15}N-labeled at all four hydrazine nitrogens. We are unable to determine in this experiment if the Z-leucinylhydrazine is still bound in the active site, as was observed crystallographically, but without a covalent attachment to the labeled carbonyl. Such complexes, where the released S' side fragment remains bound, have been observed with cyclic, proteinaceous, Kunitz inhibitors of trypsin,[21] but have not been reported for small acyclic molecules. Upon inactivation with **5**, the same Z-leucinylhydrazinyl acyl-cathepsin K adduct was observed by MS analysis as was seen with **3b**. The absence of a benzyloxybenzoylhydrazinyl acyl-cathepsin K adduct is indicative of a single mode of binding and is also consistent with the complex observed crystallographically,[9] where **5** is observed as an apparent tetrahedral adduct spanning the S and S' sites in a single orientation, with the non-peptide portion bound on the S' side of the active site. This compound class, which we had initially presumed might prove less reactive toward the enzyme than a ketone, has in actuality proven to be a single turn-over substrate with an extremely slow or undetectable deacylation rate.

Table 1 *Activities of Diacylaminomethyl Ketones v. Cathepsin K.* a. $k_{obs}/[I]=5.2x10^6 m^{-1} s^{-1}$, b. $k_{obs}/[I]=2.24x10^5 m^{-1} s^{-1}$

#	Structure	$K_{i,app}$ (nM)
3a		23
4		>10000
3b		2.7[a]
5		27[b]
8: X=O;n=1 9: X=O;n=2 10: X= H,H;n=1		2.3; 2.6; 3.5
11: X=H, Y=H, Z=Me; 12: X=H, Y=Me, Z=H; 13: X=Me, Y=H, Z=H		90; 36; 1200
14		67
15		1.8
16		13

4 CYCLIC KETONE INHIBITORS

Initial studies of the *in vivo* disposition of ketones related to **3a** suggested poor oral bioavailability and rapid clearance from the blood stream. Historically, the introduction of a conformational constraint has been used to capture a bioactive orientation of a molecule.[22] Such modifications are also introduced in the hope of altering metabolic or other *in vivo* clearance pathways while retaining the desired inhibitory potency.

Modeling experiments based on the crystal structure of inhibitor **3a** bound to the active site cysteine of cathepsin K revealed that the enzyme could potentially accommodate cyclization of the 1,3-diaminoketone template in two alternative fashions with no clearly deleterious steric interactions with the peptide backbone residues.[23] As seen in Figure 1, cyclization of **3a** along path A would produce the 2,5-diaminocyclopentanone and 2,6-diaminocyclohexanone inhibitors **6** and **7**. No overall reorganization of the original 1,3-diaminoketone template is observed in these analogs. Alternatively, cyclization along path B would provide the 3-aminopyrrolidinone and 4-aminopiperidinone inhibitors **8** and **9** respectively. Molecular modeling revealed that this cyclization produces a change in the conformation of the original 1,3-diaminoketone substructure. All four of the proposed conformationally constrained scaffolds **6, 7, 8** and **9** have been modeled as tetrahedral adducts with the active site cysteine 25 of cathepsin K. The interactions between the inhibitor and enzyme have been optimized by overlapping the tetrahedral adducts with that from the inhibitor **1**/cathepsin K crystal structure and adjusting torsion angles to mimic the conformation of **1** bound to the enzyme. No apparent deleterious enzyme/inhibitor interactions were seen and the bound conformations appear to represent relatively low energy states, especially in **6**. Prediction of the binding of inhibitors using path B is much less obvious than for path A. Indeed, the exceptional differences in potency of analogs derived *via* path B compared to those of path A was not predicted and remains a matter of conjecture in spite of the extensive structural data available.

Compound **6**, which was tested as a mixture of diastereomers, possessed weak time-dependent inhibitory activity with a $k_{obs}/[I]$ of 56 M^{-1} s^{-1}. Each diastereomer of **7** was found to have weak inhibitory activity against cathepsin K with $K_{i,app}$ of 16 µM (R,R) and 15 µM (S,S) respectively. On the other hand, the compounds **8** and **9**, representing the less easily predicted designs were in fact potent inhibitors of cathepsin K with $K_{i,app}$ of 2.3 and 2.6 nM respectively(Cathepsin K inhibition data are also presented in Table 1.[16]). Diastereomers **8** were found to selectively inhibit cathepsin K over cathepsins B and L. As a diastereomeric mixture these inhibitors were greater than 1000 fold selective for cathepsin K over cathepsin B, approximately 17 fold selective over cathepsin L. The six membered ring cyclic diaminoketones **9** were somewhat less selective than **8**. The X-ray crystal structure of inhibitors **8** with cathepsin K confirmed this hypothesis as it is consistent with a covalent interaction of the active site cysteine 25 with the carbonyl group of **8** to form a tetrahedral adduct. The hemithioketal was flanked by the N1-Cbz leucine which was observed to bind on the unprimed side of the active site.[14] The *iso*-butyl group of the leucine was bound in the hydrophobic S2 pocket which was formed from the sidechain atoms of Leu 160, Ala 134 and Met 68. A hydrogen bond between the amide carbonyl of the N1 Cbz-leucine and Gly 66 of the protein backbone is also seen in the X-ray structure of **8** complexed with the enzyme. The N3 Cbz-leucine was bound on the primed side of the active site. The stereochemistry of the diastereomeric C3 center was consistent with the R configuration in the X-ray crystal structure and may therefore represent the more potent of the two diastereomeric forms. Alternatively, this stereochemical preference may be a function of crystal packing forces.

A possible rationalization for the dramatic difference in activities for the compounds resulting from the 2 different modes of cyclization is that, despite favorable modeling predictions and the presence of the correct specificity elements required for initial Michaelis binding in both sets of compounds, inhibitors **8** and **9** have the appended N1 amine group oriented to allow facile attack by the active site cysteine nucleophile on the electrophilic carbonyl of the designed inhibitors. This may not be true for inhibitors **6** and **7** in which the diaminoketone template of **3a** has been substituted in both the α and α' positions. This substitution may cause a steric interaction which was unforeseen in the original modeling experiments on the thiol adducts.

Compound **10,** in which a methylene group has replaced the tertiary amide carbonyl, retains the overall potency of the parent cyclic diaminoketone **3a** (Table 1). The X-ray co-crystal structure of inhibitor **10** with cathepsin K does not show a hydrogen bond in this region of the inhibitor/protein complex. This change to a reduced amide has served to remove the amide linkage and simultaneously increase overall aqueous solubility.

5 REMOVAL OF AMIDIC BONDS TO REDUCE H-BONDING POTENTIAL

Because **3a** possesses high potency and selectivity for Cathepsin K inhibition, it is an excellent lead for further evaluation in a medicinal chemistry program. Our SAR investigation began by retaining one of the Cbz-leucines and varying the other with a variety of amino acid analogs (Table 1). The N-methyl analog of **3a**, ketone **11**, is 4-fold less active than **3a**. The activities of compounds with a methyl group at the alpha carbon of **3a** are dependent on the absolute configuration: Ketone **12**, derived from L-alanine, has potency comparable to **3a**. However, ketone **13**, derived from D-alanine, is 50-fold less potent than **3a**.

Next, the peptidomimetic, 4-phenoxyphenyl group was designed to approximately span the distance of a Cbz amino acid (picking up the Trp 184 aromatic interaction) while eliminating the carbamate functionality completely: amide **14** has good potency; whereas, the 4-phenoxyphenyl sulfonamide **15** is 10-fold more active than our original peptide-based lead. Excellent potency could be retained by replacing the remaining Cbz group with a water solubilizing group such as the 4-pyridyl carbonyl in ketone **16**. This compound was induced to form crystals when bound to the enzyme and an X-ray co-crystal structure of these confirmed our design hypothesis in that the 4-pyridine-carbonyl-leucine portion of **16** overlays with the Cbz-Leu portion of **3a** in the unprimed region of the active site, and the terminal phenyl of the 4-phenoxy phenyl sulfonamide forms an aromatic-aromatic interaction with Trp 184 on the primed side of the active site.[8]

Full peptidomimetic replacement of prime side Cbz-leucine in ketone **3a** has also proven possible with (4-biphenyl)-4-methyl-valeramide (inhibitor **17**).[24] Both the (4-biphenyl)-4-methyl-valeramide and the 4-phenoxyphenyl sulfonamide were incorporated into a single structure, ketone **18**, a modestly potent inhibitor of cathepsin K. The use of computer-aided molecular modeling suggested that the 3-biphenyl regioisomer, ketone **7**, would better engage Tyr 67 on the unprimed side of the active site in an aromatic-aromatic interaction than could be achieved with the 4-isomer. Indeed, ketone **19** is a very potent inhibitor of cathepsin K (Figure 2). Furthermore, an X-ray co-crystal structure of analog **20** was solved bound to cathepsin K and confirmed the binding hypothesis with the (3-biphenyl)-4-methyl-valeramide group binding on the unprimed side of the active site.

A variant of **17** lacking the isobutyl group but having a water solubilizing pyridine is also a potent cathepsin K inhibitor. This compound was found to have an oral bioavailability (%F) of 8.7% in rats whereas **15** which has an oral availability of only 0.8% in rats. Combining the peptidomimetic features of **18** and **21** results in **22** which retains high inhibitory potency but demonstrates oral bioavailability (6.7%) quite similar to **21**.

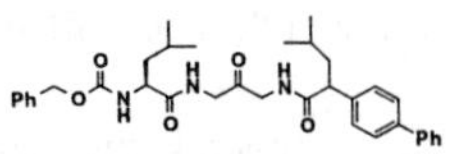

17: 12 nM 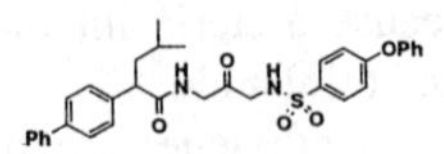   18: 490 nM 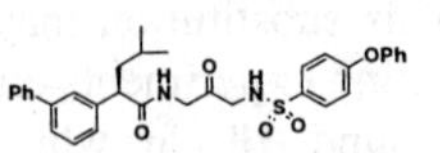   19: 1.4 nM

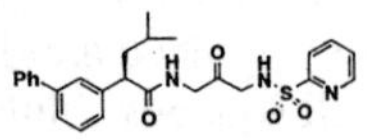

20: 3.5 nM 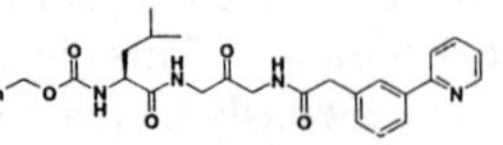   21: 13 nM 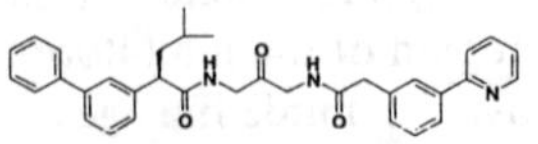   22: 16 nM

Figure 2 *Cathepsin K Inhibitor Structures and Ki,apps*

6 CONCLUSIONS

We have designed 1,3-Bis(acylamino)-2-propanones as new potent, reversible inhibitors of the cysteine protease cathepsin K. Cyclized versions have been discovered which retain high potency. Closely related 1,5-diacylcarbohydrazides are single turnover, effectively irreversible inhibitors. We have designed two isomeric Cbz-leu mimetics which are only effective replacements when placed in the correct context for proper binding to cathepsin K. Thus, the 2-(4-biphenyl)-4-methylvaleryl is an effective Cbz-leu replacement only on the prime side while 2-(3-biphenyl)-4-methylvaleryl is effective only on the unprime side. The presence of these peptidomimetic groups appears to contribute to improved oral bioavailability which will be a requisite of any therapeutically useful inhibitors to find use for treatment of osteoporosis.

References

1.   a) Drake, F.H.; Dodds, R.A.; James, I.E.; Conner, J.R.; Debouck, C.; Richardson, S.; Lee-Rykaczewdki, E.; Coleman, L.; Rieman, D.; Barthlow, R.; Hastings, G.; Gowen, M. *J. Biol. Chem.* **1996**, *271*, 12511-12516. b) Bossard, M.J.; Tomaszek, T.A.; Thompson, S.K.; Amegadzie, B.Y.; Hannings, C.R.; Jones, C.; Kurdyla, J.T.; McNulty, D.E.; Drake, F.H.; Gowev, M.; Levy, M.A. *J. Biol. Chem.* **1996**, *271*, 12517-12524. d) McGrath, M.E.; Klaus, J.L.; Barnes, M.G.; Bromme, D. *Nat. Struct. Biol.* **1997**, *4*, 105-109. e) Zhao, B., Janson, C.A.; Amegadzie, B.Y.; D'Alessio, K.; Griffin, C.; Hanning, C.R.; Jones, C.; Kurdyla, J.; McQueney, M.; Qiu, X.; Smith, W.W.; Abdel-Meguid, S.S. *Nat. Struct. Biol.* **1997**, *4*, 109-111 (references cited therein).

2.   a) Ondetti, Miguel A.; Rubin, Bernard; Cushman, David W. *Science* **1977**, *196*, 441-4. b) Patchett, A. A.; Harris, E.; Tristram, E. W.; Wyvratt, M. J.; Wu, M. T.; Taub, D.; Peterson, E. R.; Ikeler, T. J.; Ten Broeke, J.; Payne, L.G.; Ondeyka, D.L.; Thorsett, E.D.; Greenlee, W.J.; Lohr, N.S.; Hoffsommer, R.D.; Joshua, H.; Ruyle, W.V.; Rothrock, J.W.; Aster, S.D.; Maycock, A.L.; Robinson, F.M.; Hirshmann, R.; Sweet, C.S.; Ulm, E.H.; Gross, D.M.; Vassil, T.C.; Stone, C.A. *Nature* **1980**, *288*, 280-3.

3.   a) Roberts, N. A.; Martin, J.A.; Kinchington, D.; Broadhurst, A.V.; Craig, J.C.; Duncan, I.B.; Galpin, S.A.; Handa, B.K.; Kay, J.; Krohn, A.; Lambert, R.W.; Merrett, J.H.; Mills, J.S.; Parkes, K.E.B.; Redshaw, S.; Ritchie, A.J.; Taylor, D.L.; Thomas, G.J.; Machin, P.J. *Science* **1990**, *248*, 358-61. b) Erickson, J.; Neidhart,

D.J.; VanDrie, J.; Kempf, D.J.; Wang, X.C.; Norbeck, D.W.; Plattner, J.J.; Rittenhouse, J.W.; Turon, M.; Wideburg, N.; Kohlbrenner, W.E.; Simmer, R.; Helfrich, R.; Paul, D.A.; Knigge, M. *Science* 1990, *249*, 527-33. c) Vacca, J. P.; Dorsey, B. D.; Schleif, W. A.; Levin, R. B.; McDaniel, S. L.; Darke, P. L.; Zugay, J.; Quintero, J. C.; Blahy, O. M.; Roth, E.; Sardana, V.V.; Schlabach, A.J.; Graham, P.I.; Condra, J.H.; Gotlib, L.; Holloway, M.K.; Lin, J.; Chen, I.-W.; Vastag, K.; Ostovic, D.; Anderson, P.S.; Emini, E.A.; Huff, J.R. *Proc. Natl. Acad. Sci. USA* 1994, *91*, 4096-100.

4. Smith, R.A.; Copp, L.J.; Coles, P.; Pauls, H.W.; Robinson, V.J.; Spencer, R.W.; Heard, S.B.; Krantz, A. *J. Am. Chem. Soc.* 1988, *110*, 4429-4431.

5. Rando, R.R. *Pharmacological Reviews* 1984, *36*, 111-42.

6. Palmer, J.T.; Rasnick, D.; Klaus, J.L.; Bromme *J. Med. Chem.* 1995, *38*, 3193-96.

7. Thompson, S.A.; Andrews, P.R.; Hanzlik, R.P. *J. Med. Chem.* 1986, *29*, 104-111.

8. Yamashita, D.S.; Smith, W.W.; Zhao, B., Janson, C.A.; Tomaszek, T.A.; Bossard, M.J.; Levy, M.A.; Oh, H.-J.; Y. Carr, T.J.; Thompson, S.K.; Ijames, C.F.; Carr, S.A.; McQueney, M.; D'Alessio, K.J.; Amegadzie, B.Y.; Hanning, C.R.; Abdel-Meguid, S.; DesJarlais, R.L.; Gleason, J.G.; Veber, D.F. *J. Am. Chem. Soc.*, 1997, *119*, 11351-11352.

9. Thompson, S.K.; Halbert, S.M.; Bossard, M.J.; Tomaszek, T.A.; Levy, M.A.; Zhao, B.; Smith, W.W.; Abdel-Meguid, S.S.; Janson, C.A.; D'Alessio, K.J.; McQueney, M.S.; Amegadzie, B.Y.; Hanning, C.R.; DesJarlais, R.L.; Briand, J.; Sarkar, S.K.; Huddleston, M.J.; Ijames, C.F.; Carr, S.A.; Garnes, K.T.; Shu, A.; Heys, J.R.; Bradbeer, J.; Zembryki, D.; Lee-Rykaczewski, L.; James, I.E.; Lark, M.W.; Drake. F.H.; Gowen, M.; Gleason, J.G.; Veber, D.F. *Proc. Natl. Acad. Sci. USA*, 1997, *94*, 14249-14254.

10. Spatola, A.F. *Chem. Biochem. Amino Acids, Pept., Proteins,* 1983, *7*, 267-357.

11. Turk, D.; Guncar, G.; Podobnik, M.; Turk, B. *Biol. Chem.*, 1998, *379*, 137-47.

12. Yamamoto, D.; Matsumoto, K.; Ohishi, H.; Ishida, T.; Inoue, M.; Kitamura, K.; Mizuno, H. *J. Biol. Chem,* 1991, *266*, 14771-7.

13. a) Aoyagi, T.; Takeuchi, T.; Matsuzaki, A.; Kawamura, K.; Kondo, S.; Hamada, M.; Maeda, K.; Umezawa, H. *J. Antibiot.* 1969, *22*, 558-68. b) The $K_{i,apps}$ of leupeptin against Cathepsins B and S are 122 nM and 17 nM, respectively. Leupeptin was also determined to be a time-dependent inhibitor of Cathepsin L with a k_{inact} / [I] = 1.5×10^5 $M^{-1}s^{-1}$.

14. Schecter, I.; Berger, A. *Biochem. Biophys. Res. Comm.* 1967, *27*, 157-62.

15. Votta, Bartholomew J.; Levy, Mark A.; Badger, Alison; Bradbeer, Jeremy; Dodds, Robert A.; James, Ian E.; Thompson, Scott; Bossard, Mary J.; Carr, Thomas; Connor, Janice R.; Tomaszek, Thaddeus A.; Szewczuk, Lawrence; Drake, Fred H.; Veber, Daniel F.; Gowen, Maxine *J. Bone and Min. Res.*, 1997, *12*, 1396-1406.

16. Inhibitors were evaluated for inhibition against purified recombinant Cathepsin K as described in reference 1b.

17. Inhibitors were assayed against human liver Cathepsin B (Calbiochem), human liver Cathepsin L (Calbiochem), and human recombinant Cathepsin S (Shi, G.P.; Munger, J.S.; Meara, J.P.; Rich, D.H.; Chapman, H.A. *J. Biol. Chem.* 1992, *267*, 7258-62) with the following substrates: Cbz-Phe-Arg-AMC at 50 •M (Km=140 •M), Cbz-Phe-Arg-AMC at 5 •M (Km=3 •M), and Cbz-Val-Val-Arg-AMC at 50

•M (Km=70 •M), respectively, in 100 mM acetate, 20 mM cysteine, 5 mM EDTA, pH 5.5 buffer with a final DMSO concentration of 10 %.

18.   Segel, I. H. *Enzyme Kinetics* (John Wiley and Sons, NY) **1975**, pp. 107-109.

19.   Covalent enzyme-ligand complexes were analyzed by nanoflow electrospray (nanospray) mass spectrometry on a PE-Sciex API-III triple quadrupole mass spectrometer as previously described (Wilm, M. and Mann, M *Anal. Chem.* **1996**, *68*, 1-8; Carr, S. A. , Huddleston, M. J., and Annan, R. S. *Anal. Biochem.* **1996**, *239*, 180-192). Incubations were performed by mixing 5 •L each of apo Cathepsin K (27 •M in 20 mM MES, 10 mM NaCl, 2 mM cys, at pH 6.0) and inhibitor (158 •M in 20:80 DMSO:water, 5.8 fold molar excess), vortexing briefly, and incubating at room temperature for either 2.1 h (bromomethylketone) or 5 h (**3a**). Then 90 •L of 1:1:6% water:methanol:formic acid was added, the solution briefly vortexed, and ca. 2 uL analyzed by nanospray MS. The baculovirus expressed apo Cathepsin K consisted of two N-terminal processing variants RAPD...(ca. 55%) and GRAPD...(ca. 45%) with average measured Mr of 23645.6 and 23702.7 (calculated Mr = 23645.6 and 23702.6, respectively). The determined Mr for the bromomethylketone-CatK complexes were 24,018.5 and 24076.1 (calculated Mr = 24020.1 and 24077.1, respectively). No shifts in mass were detected for **3a** reacted with cathepsin K.

20.   Baggio, R.; Shi, Y.-Q.; Wu, Y.-Q.; Abeles, R.H. *Biochemistry* **1996**, *35*, 3351-3353.

21.   Laskowski, M. *Adv. Exp. Med. Biol.* **1986**, *199*, 1-17.

22.   a) Liskamp, R.M.J. *Recl. Trav.Chim.Pays-Bas* **1994**, 113-1-19. b) Marshall, G.R.; Fedric, G.A.; Moore, M.L. *Ann. Rep. Med. Chem.* **1978**, *13*, 227-38.

23.   Marquis, R.W.; Yamashita, D.S.; Ru,Y.; LoCastro, S.M.; Oh, H.-J.; Erhard, K.E.; DesJarlais, R.L.; Smith, W.W.; Zhao, B.; Janson, C.A.; Abdel-Meguid, S.S.; Tomaszek, T.A.; Levy, M.A.; Veber, D.F., *J. Med. Chem.* **1998**, *41*, in press.

24.   DesJarlais, R.L.; Yamashita, D.S.; Oh, H.-J.; Uzinskas, I.N.; Erhard, K.F.; Allen, A.C.; Haltiwanger, R.C.; Zhao, B.; Smith, W.W.; Abdel-Meguid, S.S.; D'Alessio, K.; Janson, C.A.; McQueney, M.S.; Tomaszek, T.A.; Levy, M.A.; Veber, D.F. *J. Am. Chem. Soc.* **1998**, *120*, in press.

Glycochemistry and Glycobiology

GLYCOCONJUGATES AS TUMOR-ASSOCIATED ANTIGENS AND LIGANDS IN REGULATORY PROCESSES

Horst Kunz[1], Beate Liebe[1], Wolfgang Dippold[2], Christine Claus[2], Jörg Habermann[1], Ulrich Sprengard[1,3], Karen Peilstöcker[1] and Markus Rösch[1]

[1]Institut für Organische Chemie der Universität Mainz, Mainz, Germany
[2]I. Medizinische Klinik der Universität Mainz, Mainz, Germany
[3]Zentralforschung der HOECHST AG, Frankfurt/M, Germany.

1 INTRODUCTION

The organized cooperation of cells in multicellular organisms requires multiple and safe transfer of molecular information either by messenger molecules or by direct intercellular communication. Complex glycoconjugates, glycoproteins and glycolipids, exposed on the extracellular surface of the outer cell membrane play decisive roles in these biological communication processes.[1,2] The saccharide side chains influence the physico-chemical properties and the preferred conformation of glycoproteins. They can protect the protein backbone from degradation through proteolysis. The saccharide portions can also act directly as ligands or receptors, for example in infectious processes[3] or in cell adhesions.[4] The investigation of these biological recognition processes requires model glycoprotein ligands of exactly specified structure in both the saccharide and the peptide portion. Such pure glycoproteins can hardly be obtained in sufficient amounts from biological sources because of the biological microheterogeneity of the glycan portions. Therefore, chemical and enzymatic constructions of glycopeptide ligands of defined structure is of particular interest for the elucidation of these biological recognition phenomena on cell membranes and for a subsequent development of drugs that are able to influence these mechanisms. The knowledge about the essential features of the natural ligands is a prerequisite for the design of synthetic pharmaceuticals useful for the therapy of the corresponding diseases.

2 TUMOR-ASSOCIATED GLYCOPEPTIDE ANTIGENS

The expression of cell surface glycoproteins obviously is significant for the status of cells,[5] and the glycan side chains of glycoproteins on normal cells and tumor cells have been found to be quite different. The altered glycoproteins expressed on tumor cells, for example O-glycoproteins containing the Tn-antigen structure **1** or the Thomsen-Friedenreich (T) antigen structure **2**, have early been described to be tumor associated cell surface antigens (Scheme 1).[6]

Aiming at the construction of synthetic antitumor vaccines, the synthesis of such tumor-associated T antigen glycopeptides like **3** has been performed.[7] The glycopeptide **3** contains the N-terminal tripeptide of glycophorin A of sub-blood group M. Glycophorin A is the major transmembrane glycoprotein of the erythrocytes.[8] It exists in two sub-blood

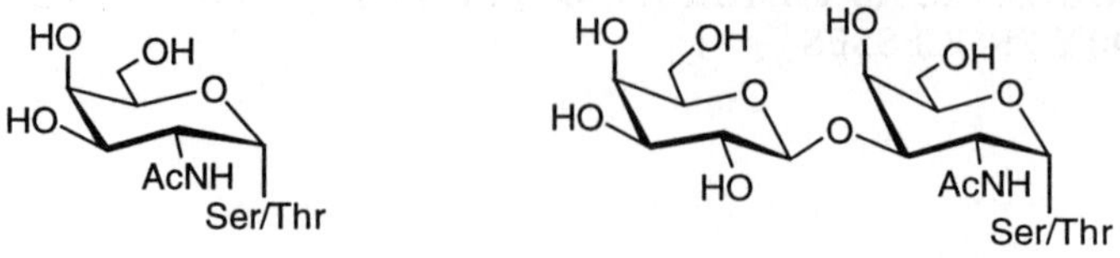

Scheme 1

group specificities which differ in the 131 amino acids sequence in only two amino acid positions. One of these differences concerns the N-terminal amino acid which is serine in glycophorine of blood group M and leucine in blood group N. The synthetic glycopeptide **3** was coupled to bovine serum albumin in order to obtain a synthetic antigen **4** (Scheme 2).

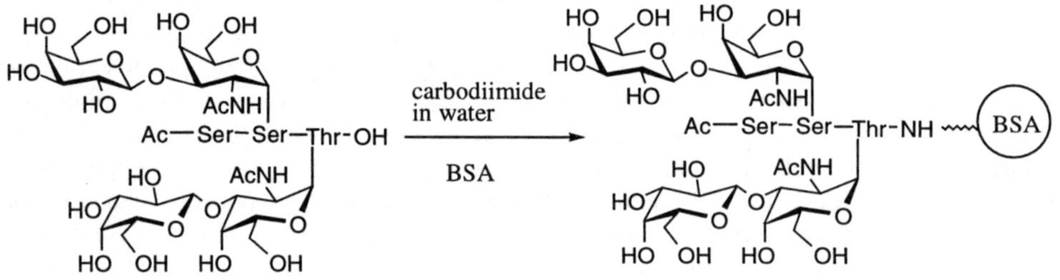

Scheme 2

The formed glycopeptide antigen **4** contained on average about 30 glycopeptides **3** per molecule of protein. Monoclonal antibodies induced with this synthetic vaccine **4** reacted with all investigated epithelial tumor cells. However, control experiments revealed that these antibodies also showed affinity to normal cell of the same tissues. It was concluded from these experiments that the antigen structure in **4** is tumor-associated but not sufficiently tumor-specific[8] and that further structural information are necessary to enable the formation of tumorselective vaccines. Such information could be located within the peptide portion of the cell surface glycoproteins.

Interesting investigations of the tumor-associated polymorphic epithelial mucin MUC-1 delivered this desired information about tumor-relevant peptide portions of glycoproteins[9,10] and stimulated us to synthesize glycopeptide partial structures of the tandem-repeat of MUC-1 with Tn-antigen[11–13] **5**, T-antigen[12,14] **6** and sialyl-Tn-antigen[15,16] **7** saccharide side chains. The syntheses of the Tn and T antigen glycopeptides have been carried out either by fragment condensation reactions using chemical[12] or enzymatic protecting group techniques[11,14] or by solid-phase synthesis (Scheme 3).[13]

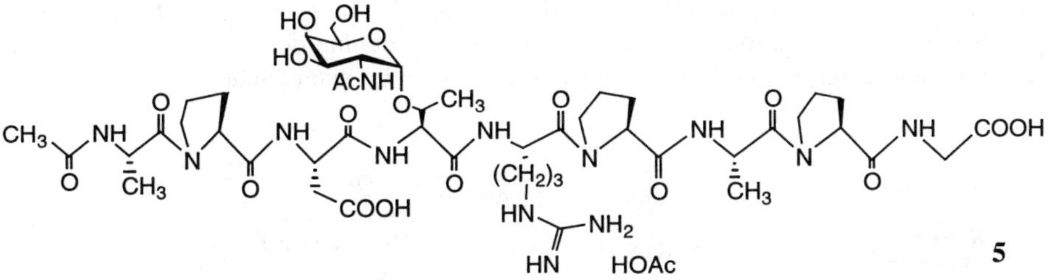

Scheme 3

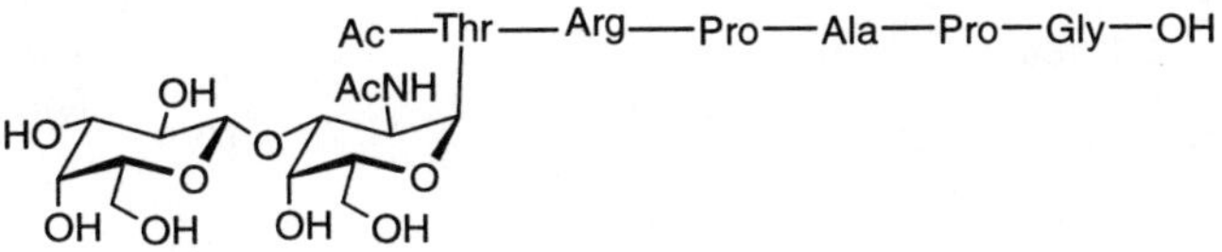

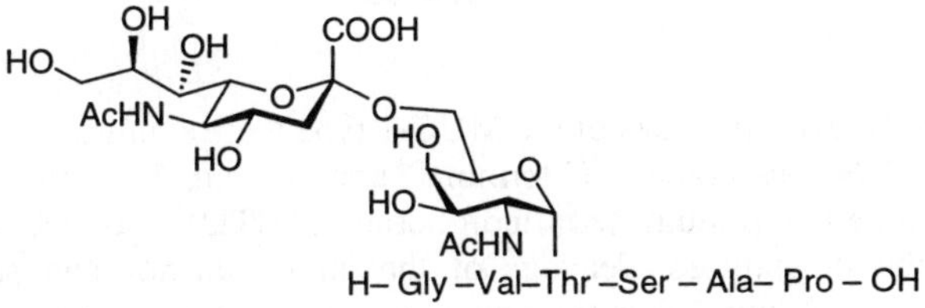

The full tandem repeat of MUC-1 **8** containing a Tn antigen side chain was obtained on solid phase using an allylic anchoring technique (Scheme 4).[13]

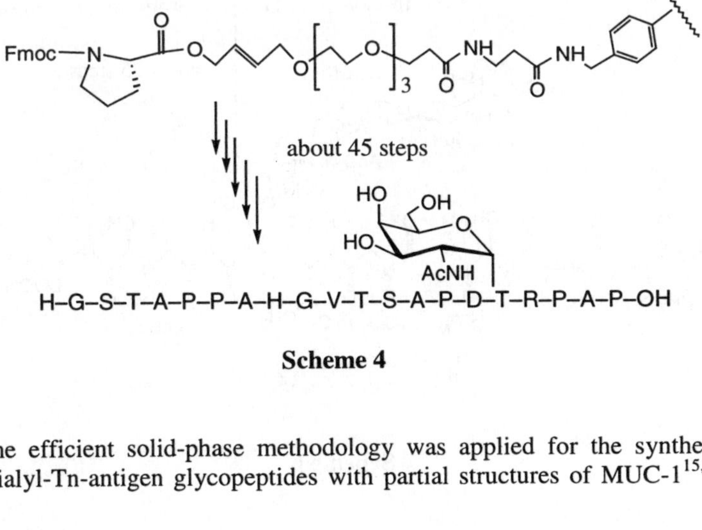

H–G–S–T–A–P–P–A–H–G–V–T–S–A–P–D–T–R–P–A–P–OH **8**

Scheme 4

The same efficient solid-phase methodology was applied for the synthesis of the demanding sialyl-Tn-antigen glycopeptides with partial structures of MUC-1[15,16] using a preformed sialyl- Tn antigen threonin building block **10**. The synthesis of **10** was achieved by regioselective sialylation of the azido-galactosyl threonine derivative **9**, subsequent O-acetylation and careful acidolytic cleavage of the *tert*-butyl ester (Scheme 5).[15]

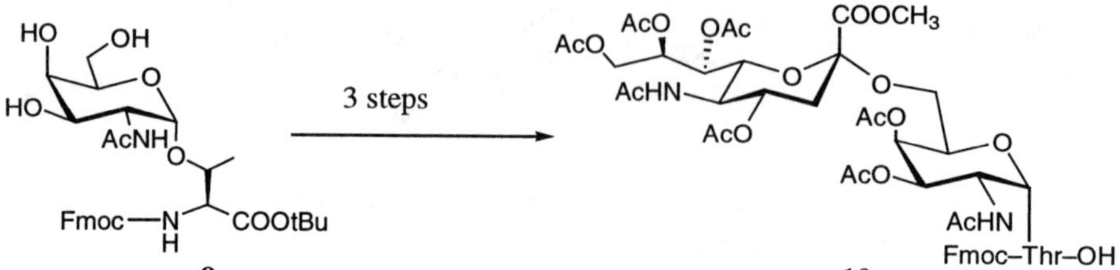

Scheme 5

The sialyl Tn glycopeptides of the MUC-1 type were synthesized on solid-phase using allylic (HYCRON) anchored[13] C-terminal amino acid **11** and O-1*H*-benzotriazolyl-N,N,N',N'-tetramethyluronium tetrafluoroborate (TBTU)[17] as the condensing reagent. After palladium(0)-catalyzed cleavage of the allylic anchor and subsequent acidolytic removal of the trityl and *tert*-butyl protecting groups from amino acid side chains, the sialyl Tn glycoundecapeptide **12** was isolated in an overall yield (25 steps) of 42% (based on **11**) (Scheme 6).[16]

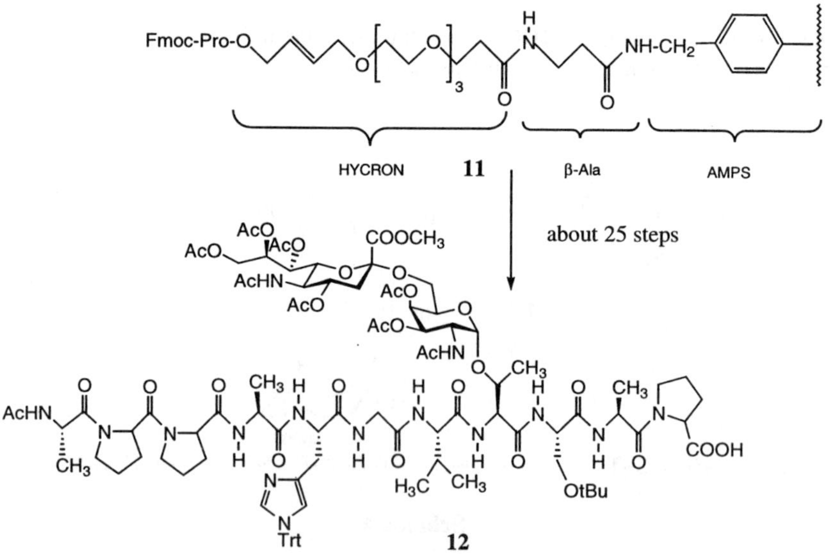

Scheme 6

While the removal of the O-acetyl group from **12** proceeded smoothly at pH 10 in aqueous methanol, the saponification of the methyl ester of the neuraminic acid portion demanded carefully optimized reaction conditions. At pH 11.5 in water, the hydrolysis of the ester group was optimally achieved without β-elimination of the carbohydrate portion or epimerization within the peptide backbone to give the sialyl Tn MUC-1 glycopeptide **13** in analytically pure form. For immunological evaluations, the synthetic peptide and glycopeptide partial structures of MUC-1 were coupled to human serum albumin (HSA) using a carbodiimide-based method which we had elaborated for the synthetic T antigen7 and Lewis[x] antigen glycopeptides (Scheme 7).[18,19]

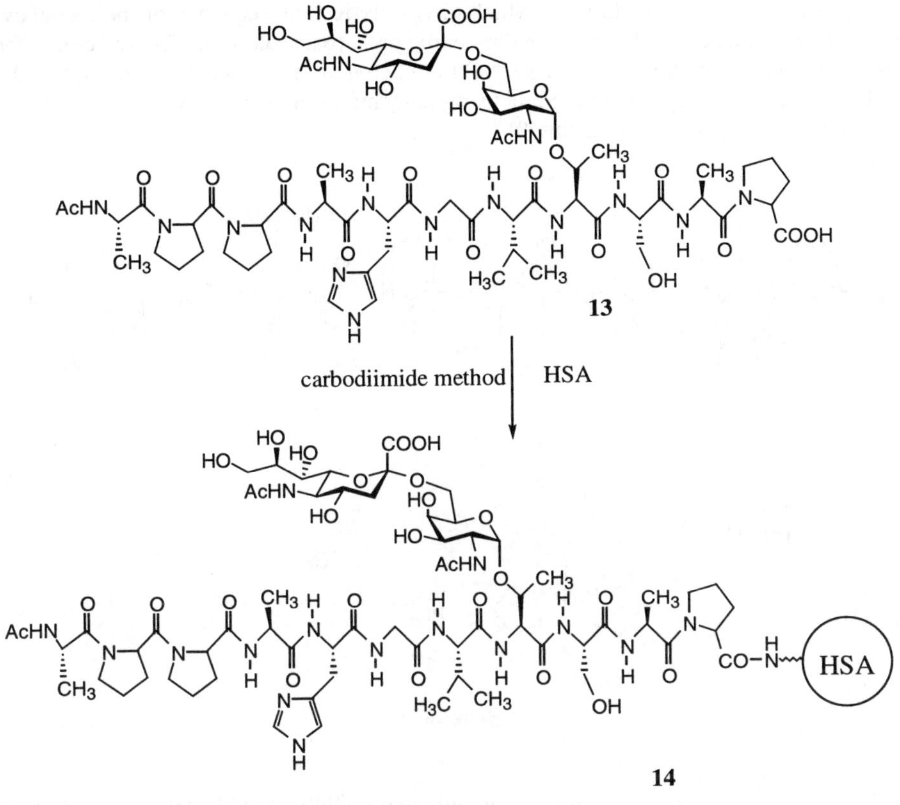

Scheme 7

From **13** a synthetic neoglycoprotein antigen **14** was obtained which contained on average 5 molecules of the synthetic antigen per molecule protein. The neoglycoproteins were subsequently labeled with fluoresceine isothiocyanate. The immunogenicity of 25 synthetic peptide and glycopeptide antigens was investigated in a proliferation test on blood lymphozytes. The HSA conjugate of the partially protected sialyl Tn MUC-1 glycopeptide **12** showed the strongest stimulation of lymphocyte proliferation.

3 CADHERIN GLYCOPEPTIDES

Cadherins are found on tissue-forming cell-types and are involved in cell adhesion, cell morphogenesis, cell targeting and contact inhibition of cell growth.[20,21] Down-regulation of cadherins in tumor cells results in an acquisition of invasiveness.[20] This observation suggests a role of cadherins as tumor suppressing factors. The cadherins interact on the surface of epithelial cells (E-cadherin) with homophilic specificity. The binding occurs between two identical molecules. The homophilic binding domain is mainly formed by two β-strands separated by a tight β-turn. This structure exposes the tripeptide His-Ala-Val which is a suggested binding motif.[21]

The solid-phase synthesis of dodecapeptide **15** using TBTU as the condensation reagent resulted in a low yield (5%) which obviously is a consequence of the tendency of this peptide sequence to form secondary structures (backfolding). To overcome these difficulties, we have introduced O-pentafluorophenyl uronium salts, e. g. PfPyU **16**,[22] which proved to be very efficient condensing reagents in the synthesis of difficult peptide and glycopeptide sequences (Scheme 8).[23]

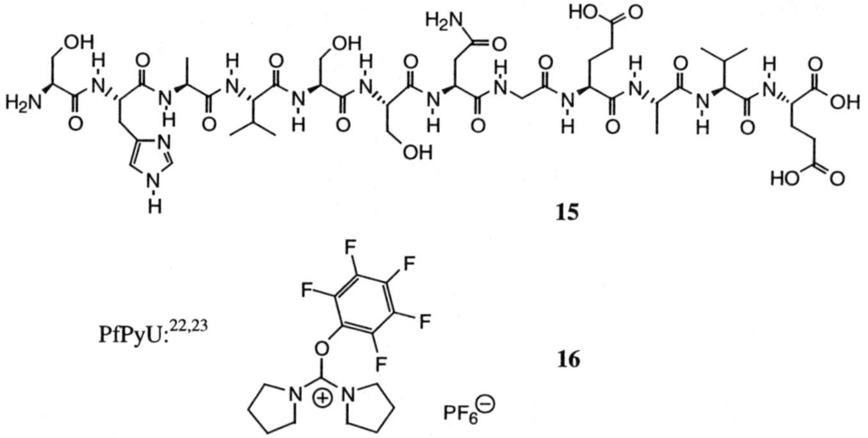

15

PfPyU:[22,23]

16

Scheme 8

Using a preformed Tn antigen serine conjugate, allylic HYCRON anchored starting compound **17** together with this O-pentafluorophenyl uronium reagent, we have synthesized for example, the resin-linked glycododecapeptide **18**. After palladium(0)-catalyzed detachment, the selective carboxy-deblocked glycopeptide partial structure **19** of E-cadherin 1 was obtained in an overall yield of 55% based on **17**.[23]

Complete deprotection gave the free glycodecapeptide **20** which represents the major part (β strands F and G) of the homophilic recognition domain of E-CAD-1 (Scheme 9).

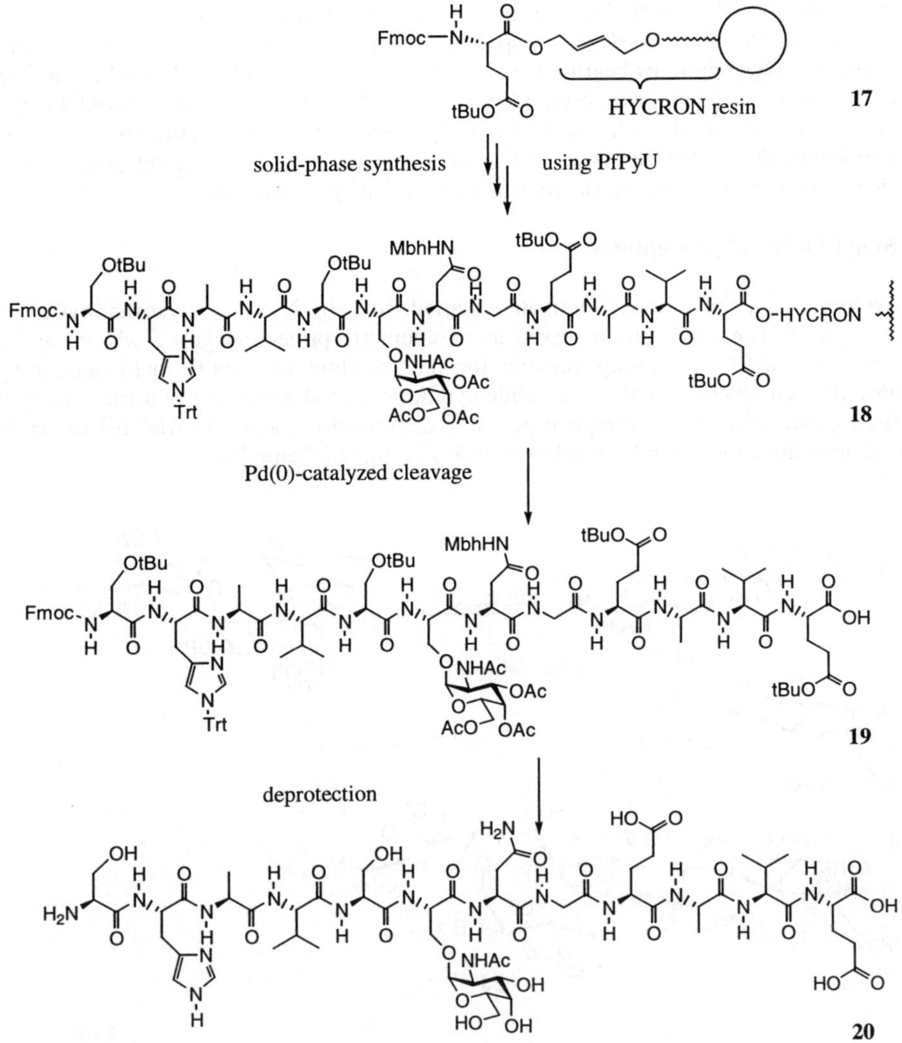

Scheme 9

Glycopeptides of this type showed interesting effects on cultures of transformed keratinocytes. They induce the receptor for contactinhibin,[24] a glycoprotein which plays a major role in the regulation of contactinhibition and cell growth.

4 LIGANDS OF SELECTINS

The recruitment of leukocytes is a central process in inflammatory diseases. An early step in the adhesion of leukocytes to endothelial cells of veins within inflamed tissues consists of the recognition of carbohydrate ligands of glycoproteins exposed on the surface of leukocytes by P- and E-selectins, specific receptors expressed on the vascular endothelium.[25] Glycoproteins bearing the sialyl Lewis[x] epitope, sulfated Lewis[x] ligands or the sialyl Lewis[a] antigen have been identified as effective ligands recognized by the selectins. In particular, the tetrasaccharide sialyl Lewis[x] became an attractive synthetic target molecule during the past years.[26] Our interest was focussed on the influence of the peptide part on the recognition properties of sialyl Lewis[x] glycopeptides.[27]

4.1 Sialyl Lewis[x] Glycopeptides

The synthesis of a sialyl Lewis[x] derivative started from the N-acetyl glycosaminyl azide derivative **21**.[28] The azido group served as an anomeric protecting group which can be converted into an amino group suitable for the coupling to aspartic acid containing peptides after completion of the saccharide synthesis. Introduction of the α-fucoside unit and the β-galactoside and subsequent selective deprotection gave a Lewis[x] trisaccharide azide **22** with three unblocked secondary hydroxy groups (Scheme 10).

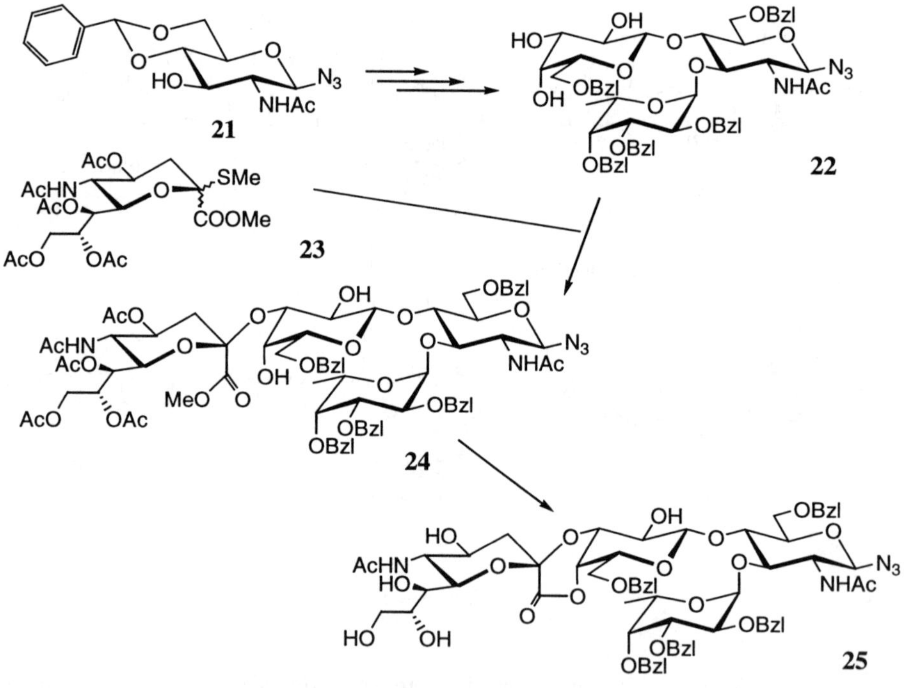

Scheme 10

Sialylation using the xanthogenate **23** as the glycosyl donor resulted in a stereoselective and regioselective formation of the sialyl Lewis[x] tetrasaccharide azide which after acylation gave the fully protected saccharide **24**.[27]

In order to avoid side reactions on the target glycopeptide structures, we decided to remove the ester protecting groups on the stage of the base-stable saccharide. Saponification of **24** with diluted sodium hydroxide (pH 11.5) in aqueous methanol and subsequent treatment with acidic ion-exchange resin did not give the free carboxylic compound, but the 4-lactone **25** in a pure form.

The lactone in **25** was considered a promising internal protecting group, which allows the differentiation of the carboxylic group of the sialyl acid portion from the peptide functionalities. Hydrogenolysis of the anomeric azido group delivered the sialyl Lewis^x lactone amine **26**, a pivotal building block for the construction of sialyl Lewis^x glycopeptides. Condensation of **26** with preformed peptides, e. g. the RGD-tetrapeptide **27**, gave the sialyl Lewis^x glycopeptides, as for example **28**.[27] The lactone remained a stable carboxy protection during these condensation reactions (Scheme 11).

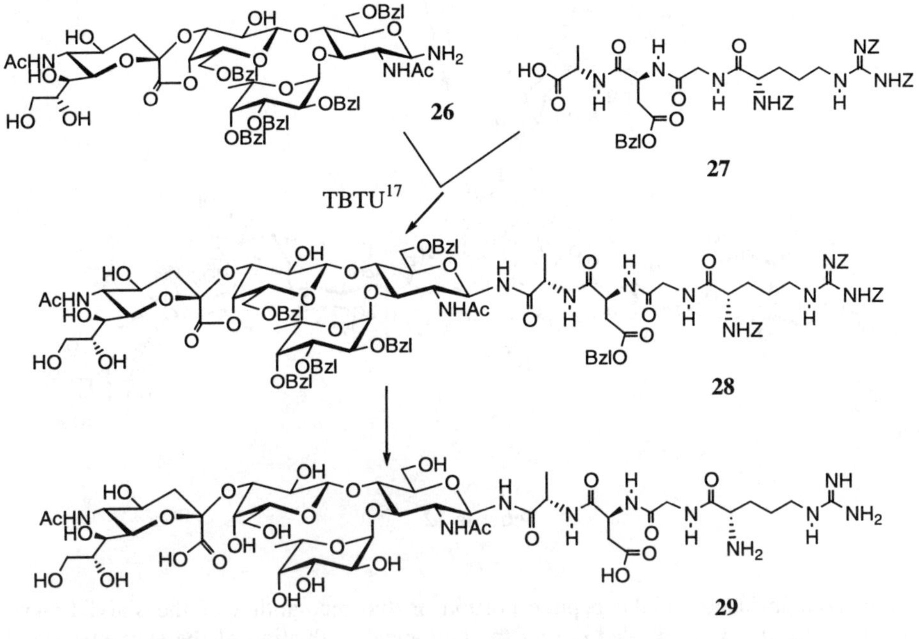

Scheme 11

Complete deprotection of **29** was achieved by hydrogenolytic cleavage of the benzylic groups and subsequent careful hydrolysis of the lactone to furnish the free sialyl Lewis^x-RGD glycopeptide **29**.[27]

Simultaneous multiple condensation of the sialyl Lewis^x lactone amine **26** with preformed cyclopeptides, as for example the cycloheptapeptide **30** containing three free side-chain carboxylic groups of aspartic acid units, gave multivalent sialyl Lewis^x cyclopeptides, e. g. **31** after hydrolytic opening of the lactone rings (Scheme 12).[29]

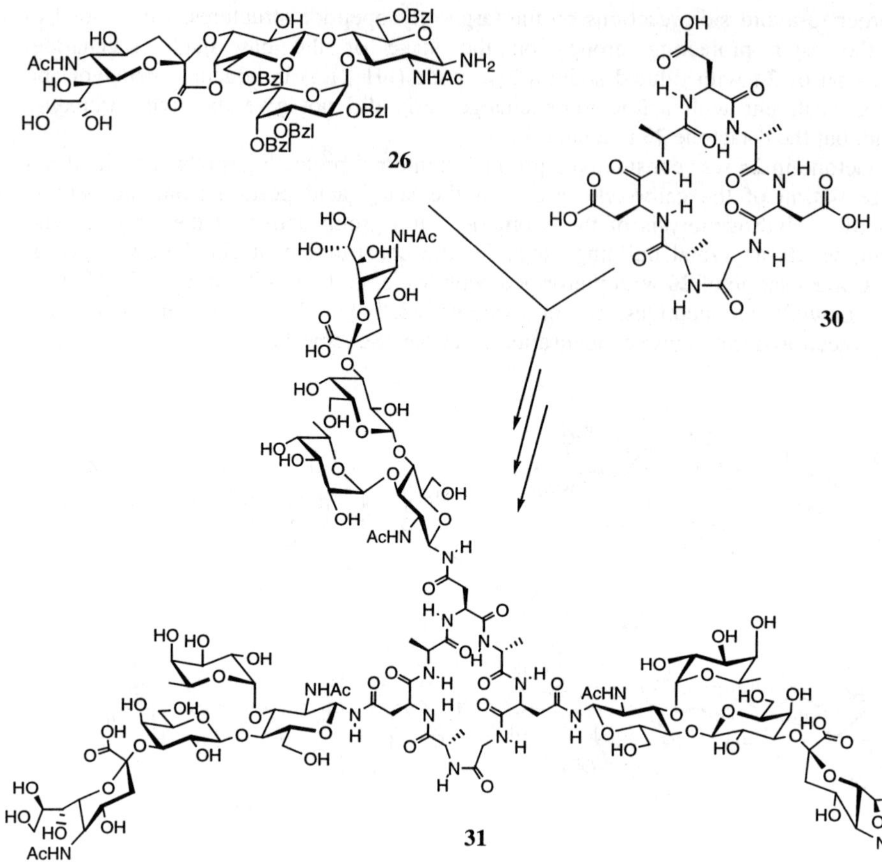

Scheme 12

A marked influence of the peptide portion on the recognition of the sialyl Lewis^x epitope by selectins was revealed during the biological evaluation of the synthetic sialyl Lewis^x glycopeptides using immobilized IgG-selectin fusion proteins. The sialyl Lewis^x-RGD glycopeptide showed high affinity to P-selectin (IC50 11-26 μM) and no affinity to E-selectin (IC50 >2 mM).[27] In contrast, the sialyl Lewis^x cyclopeptide **31** proved to be an efficient ligand to E-selectin (IC50 ~ 0.3 mM) but no ligand to P-selectin (IC50 >10mM).[29] This results suggest that the peptide chain modulates this biological recognition of the saccharide epitope in terms of both, affinity and selectivity.

4.2 Sialyl Lewis^a Glycopeptides

The sialyl Lewis^a epitope, a regioisomer to sialyl Lewis^x, also constitutes an important ligand involved in cell adhesion processes.[30] In addition, this saccharide has been described as a tumor associated antigen (CA 19-9).[31] The synthesis of a protected sialyl Lewis^a azide **32** was carried[32] out using a similar methodology as has been applied for the synthesis of the sialyl Lewis^x azide **24**. However, a decisively modified strategy had to be elaborated.

Moreover, saponification of the ester protecting groups of **32** did not lead to a pure lactone, as was formed from **24**.

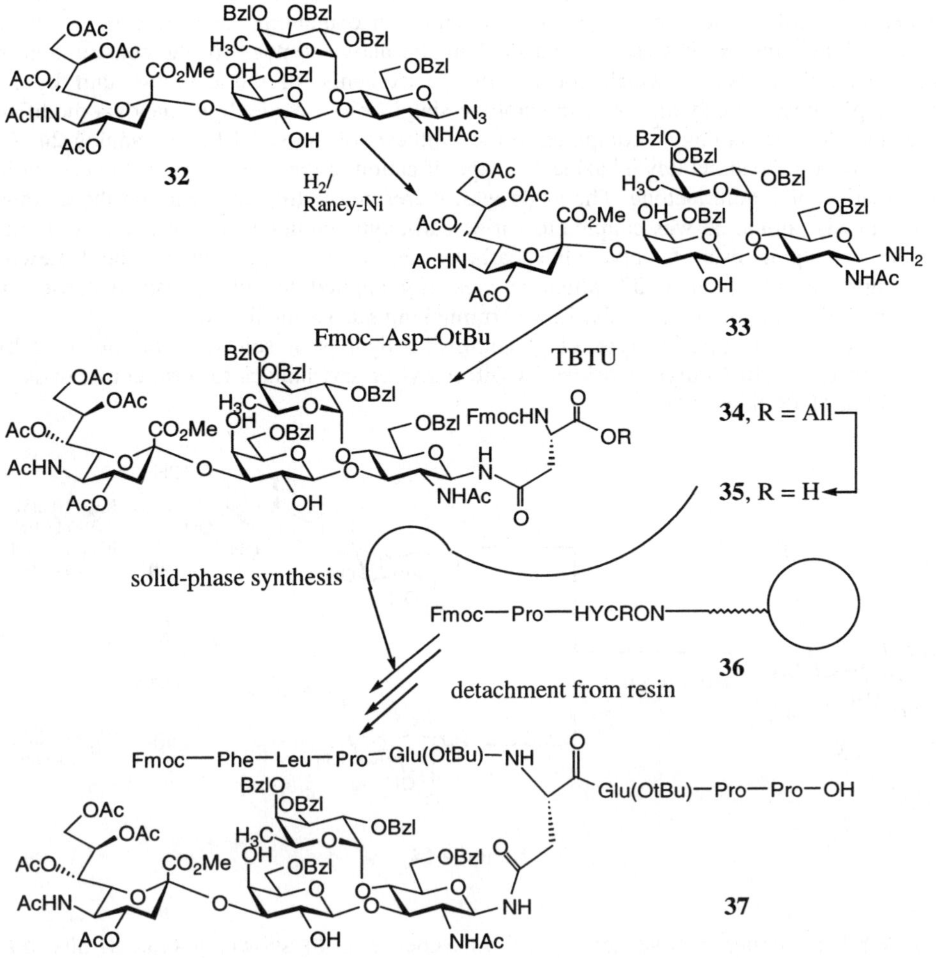

Scheme 13

Therefore, **32** was converted into sialyl Lewis[a] amine **33** which was then condensed with Fmoc aspartic acid α-allyl ester to yield the N-glycosidic conjugate **34**.[32] Palladium(0)-catalyzed removal[33] of the allylic ester gave the sialyl Lewis[a] asparagine conjugate **35** suitable for the application in solid-phase glycopeptide synthesis on a HYCRON-anchored proline **36**. After alternating deprotections, coupling reactions using TBTU,[17] capping reaction and final palladium(0)-catalyzed detachment from the resin, the partly deprotected sialyl Lewis[a] glycooctapeptide **37**, a partial structure of PSGL-1,[30] was obtained. The deprotection of **37** is under investigation and should lead to an interesting ligand to receptors involved in cell adhesion (Scheme 13).

4.3 Sialyl Lewis[x] Mimetics, Arabino Sialyl Lewis[x] Glycopeptides

Biological degradation of saccharide ligands often opposes the applicability of these specific natural ligands as drugs. To overcome these problems, glycomimetics with S-linked or C-linked saccharide structures[33] or nitrogen containing saccharide rings[34] have been designed and synthesized. We pursued an alternative strategy for the construction of sialyl Lewis[x], mimetics, which consists of the exchange of the L-fucose unit by D-arabinopyranose. The synthesis of the arabino sialyl Lewis[x] azide **38** demands a modified strategy and methodology[35] compared to the synthesis of the sialyl Lewis[x] analog **24**. As was shown for the sialyl Lewis[a] azide, the saponification of the ester group did not result in a formation of a pure lactone. The azide was, therefore, hydrogenolyzed and the arabino sialyl Lewis[x] amine **36** was coupled to varied carboxylic components, for example, to the Fmoc asparagine ester to give after carboxy deprotection the arabino sialyl Lewis[x] asparagine building block **37** which is presently applied to solid-phase syntheses of glycopeptides containing the sialyl Lewis[x] mimicking saccharide ligand.

A completely different type of glycomimetic ligands of selectins was obtained by Heck reactions[36] of C-allyl fucosides **38** with vinyl or aryl halides to form compounds of type **39** or **40** (Scheme 14).[37]

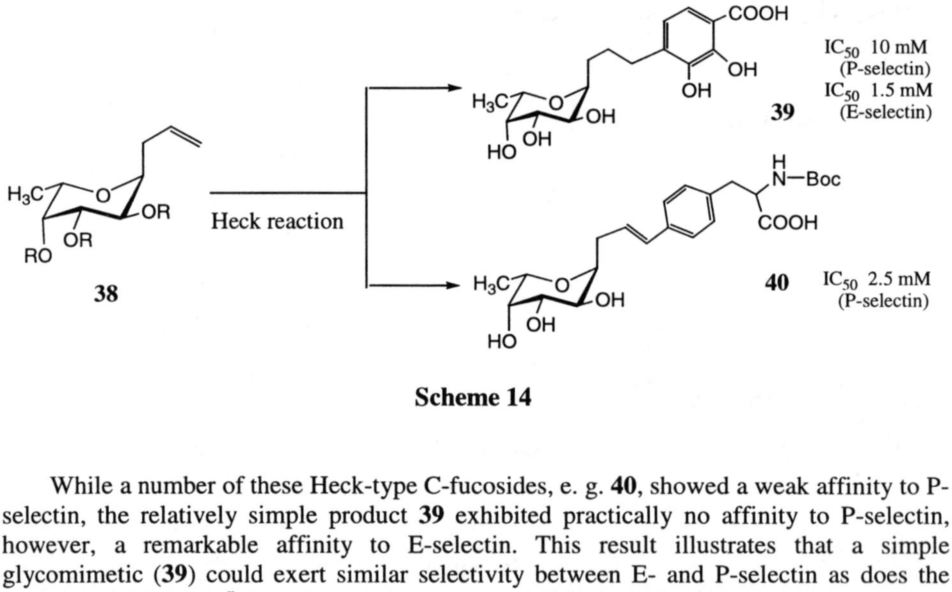

Scheme 14

While a number of these Heck-type C-fucosides, e. g. **40**, showed a weak affinity to P-selectin, the relatively simple product **39** exhibited practically no affinity to P-selectin, however, a remarkable affinity to E-selectin. This result illustrates that a simple glycomimetic (**39**) could exert similar selectivity between E- and P-selectin as does the complex sialyl Lewis[x] **31**. At the same time, the affinity of **39** to E-Selectin is almost as high as that of the sialyl Lewis[x] tetrasaccharide itself (IC50 1.5 mM). As compounds like **39** are stable towards glycosidases and other metabolizing enzymes, the observed effects for **39** constitute a promising basis for the development of anti-inflammatory drugs.

References

1.   A. Kobata, *Acc. Chem. Res.*, 1993, **26**, 319.
2.   H. Lis and N. Sharon, *Eur. J. Biochem.*, 1993, **218**,1.
3.   For an example, see: I. Ofek and N. Sharon, *Infect. Immun.*, 1988, **56**, 539.

4. For review, see: M. Fukuda, *Bioorg. Med. Chem.*, 1995, **3**, 207.
5. B. A. Fenderson, E. M. Eddy, and S. Hakamori, *Bio Essays*, 1990, **12**, 173.
6 G. F. Springer, *Science*, 1984, **224**, 1198.
7. H. Kunz and S. Birnbach, *Angew. Chem. Int. Ed. Engl.*, 1986, **25**, 360; H. Kunz, S. Birnbach, and P. Wernig, *Carbohydr. Res.*, 1990, **202**, 207.
8. W. Dippold, A. Steinborn, S. Birnbach, and H. Kunz, unpublished; Dissertation A. Steinborn, Universität Mainz, 1990.
9. S. Gendler, J. Taylor-Papadimitriou, T. Duhig, J. Rothbard, and J. Burchell, *J. Biol. Chem.*, 1988, **263**, 12820.
10. S. Briggs, M. R. Price, and S. J. B. Tendler, *Eur. J. Cancer*, 1993, **29 A**, 230.
11. P. Braun, H. Waldmann, and H. Kunz, *Bioorg. Med. Chem.*, 1993, **1**, 197.
12. M. Leuck and H. Kunz, *J. Prakt. Chem.*, 1997, **339**, 322.
13. O. Seitz and H. Kunz, *Angew. Chem. Int. Ed. Engl.*, 1995, **34**, 803.
14. P. Braun, G. M. Davies, M. R. Price, P. M. Williams, S. B. J. Tendler, and H. Kunz, *Bioorg. Med. Chem.*, 1998, **6**, 1.
15. B. Liebe, H. Kunz, *Angew. Chem. Int. Ed. Engl.*, 1997,
16. B. Liebe, H. Kunz, *Helv. Chim. Acta*, 1997, **80**, 1473.
17. R. Knorr, A. Trzeciak, W. Baumwarth, *Tetrahedron Lett.*, 1980, **30**, 1927.
18. K. von dem Bruch, H. Kunz, *Angew. Chem. Int. Ed. Engl.*, 1994, **33**, 101.
19. H. Kunz, K. von dem Bruch, *Meth. Enzymol.*, 1994, **267 B**, 3.
20. M. Takeichi, *Curr. Opin. Cell. Biol.*, 1995, **7**, 619.
21. B. Nagar, M. Overduin, M. Ikura, J. M. Rini, *Nature*, 1996, **380**, 360.
22. J. Habermann, H. Kunz, *J. Prakt. Chem.*, 1998, **340**, 233.
23. J. Habermann, H. Kunz, *Tetrahedron Lett.*, 1998, **39**, 265.
24. G. Grandl, D. Faust, F. Oesch, R. J. Wieser, *Curr. Biol.*, 1995, **5**, 526.
25. T. A. Springer, *Cell*, 1994, **76**, 301.
26. a) A. Kameyama, H. Ishida, M. Kiso, A. Hasegawa, *Carbohydr. Res.*, 1991, **209**, C1; b) K. C. Nicolaou, C. W. Hummel, N. J. Bochkovich, C.-H. Wong, *J. Chem. Soc. Chem. Commun.*, **1991**, 870; c) S. J. Danishefsky, J. Gervay, J. M. Peterson, F. E. McDonald, K. Koseki, T. Oriyama, D. A. Griffith, *J. Am. Chem. Soc.*, 1992, **114**, 8329; d) A. Hasegawa, T. Ando, A. Kameyama, M. Kiso, *J. Carbohydr. Chem.*, 1992, **11**, 645; e) Hasegawa, K. Fushimi, H. Ishida, M. Kiso, *ibid.*, 1993, **12**, 1203.
27. U. Sprengard, G. Kretzschmar, E. Bartnik, C. Hüls, H. Kunz, *Angew. Chem. Int. Ed. Engl.*, 1995, **34**, 99.
28. C. Unverzagt, H. Kunz, *J. Prakt. Chem.*, 1992, **334**, 570.
29. U. Sprengard, M. Schudok, G. Kretzschmar, H. Kunz, *Angew. Chem. Int. Ed, Engl.*, 1996, **35**, 321.
30. R. P. McEver, R. D. Cumming, *J. Clin. Invest.*, 1997, **100**, S. 97.
31. A. Takada, Oh. Ohmori, T. Yaneda, K. Tsuynoka, A. Hasegawa, M. Kiso, R. Kannagi, *Cancer Res.*, 1993, **53**, 354.
32. K. Peilstöcker, Diplomarbeit, 1995, and ongoing investigations for Ph. D. thesis, Universität Mainz, to be submitted.
33. T. Eisele, A. Töpfer, G. Kretzschmar, R. R. Schmidt, *Tetrahedron Lett.*, 996, **37**, 1389.
34. M. Kiso, H. Furui, H. Ishida, A. Hasegawa, *J. Carbohydr. Chem.*, 1996, **15**, 1, and references cited therein.
35. M. Rösch, Diplomarbeit and ongoing research for Ph. D. thesis, Universität Mainz, to be submitted.
36. C. B. Ziegler, R. F. Heck, *J. Org. Chem.*, 1978, **43**, 2941.
37. U. Sprengard, W. Schmidt, H. Kunz, unpublished results; Dissertation U. Sprengard, Universität Mainz, 1996.

PROTEIN GLYCOSYLATION IN IMMUNO- AND NEURO-PATHOLOGY

Elizabeth F. Hounsell and David V. Renouf

Department of Biochemistry and Molecular Biology, University College London, Gower Street London WC1E 6BT.

1. INTRODUCTION

Protein and lipid glycosylation present multiple potential targets for therapeutic exploitation. We are now aware of the large diversity of oligosaccharide conjugates, their specificity of structure and recognition. Endogenous carbohydrate binding proteins (selectins of mammals, adhesins of bacteria, viral agglutinins, etc) and antibodies largely increase their avidity by interaction with multivalent coreceptors present on highly branched N-linked chains of glycoproteins or on mucin-type O-linked glycopeptides. Glycolipids (GSPL) and GPI-protein membrane anchors present other clustered targets, particularly in infection (e.g. bacterial toxin binding to the former and novel sequences of the latter in trypanosomes and prions). Already showing medicinal potential are oligosaccharide/glycopeptide tumour associated antigens, influenza binding to sialic acid residues, drugs for GSPL storage diseases, serum amyloid P acting as a lectin and polysulphated oligosaccharide interactions. The last two cases also highlight aspects of glycoproteins and proteoglycans in pathologies associated with protein folding i.e. the amyloidoses, plaque or fibril diseases. Our results in these areas illustrate a) mucin oligosaccharide and polysaccharide structures as tumour associated[1] and bacterial antigens, as allergens,[2] and their exploitation via dendrimer technology,[3] b) proteoglycan oligosaccharide diversity and function,[4] and c) the role of the glycosylation of proteins and their GPI anchors in conformation and stability.[5,6] Table 1 indicates the new areas for drug targeting based on clustered determinants on which this chapter concentrates.

Table 1 *New potential drug targets based in glycochemistry*

Tumour antigens
Allergens
T cell antigens
Innate immune response
Selectins
Sialoadhesins
Proteoglycans
GPI interactions

2 PROTEIN AND LIPID GLYCOSYLATION

Table 2 shows the main types of protein and lipid glycosylation. However, there are a growing number of these. For example O-linked glycosylation can also involve GlcNAcβ1 linked to the hydroxyl group of Ser or Thr (Table 3)[7]. Fuc, Xyl and Glc have been found linked to Ser/Thr of several eukaryote molecules involved in fibrinolysis and coagulation. Novel O- and N- (to Asn) linked oligosaccharides found in bacteria[9] are suggested as virulence factors.[9] Nature's most prolific use of O-linked is in mucins where GalNAc is substituted further by an array of Gal, GlcNAc or GalNAc residues forming different cores[1]. To these cores can be attached Gal-GlcNAc backbones and on these and the cores are arrayed Fuc or sialic acid residues which form conformational epitopes recognised by mammalian

---

Table 2 *Major classes of protein and lipid glycosylation*

	$(Xaa)_n\text{-}NH_2$ |
N-linked chain core	GlcNAcβ1-N-Asn.Xaa.Ser/Thr-$(Xaa)_n$-CO_2H
O-linked chain core	GalNAcα1-O-Ser/Thr
Proteoglycans	Heparan sulphate, Heparin, Chondroitan sulphate
GPI anchors	Protein-CO-Glycan-Phosphatidylinositol
Glycolipids	Oligosaccharide-Glcβ1- ceramide (GSPL)
Lipopolysaccharides (LPS)	PS-Core-Lipid A

Table 3 *Protein O-linked glycosylation*

Multiple glycosylation site neighbours

GalNAcα-O based (extracellular)
    Recognition of sialyl Le[x] by selectins (P, E, L)
    Recognition by antibodies
        Tumour associated antigens
        MUC gene products
GlcNAcβ-O based (intracellular)
    Reciprocal with Ser/Thr phosphorylation
    Signals in nuclear regulation

---

carbohydrate binding proteins (e.g. the P, E and L selectins) and antibodies. This recognition can be exploited in inflammation (eg. sialyl Le[x] mimics which inhibit lymphocyte trafficking to endothelial tissues) and cancer (discussed in the next section). All of these post-translational modifications of proteins are catalysed by specific enzymes, the glycosyltransferases. These are specific for the type of monosaccharide, the particular hydroxyl groups around the glycosidic ring which are linked together and the α or β configuration. Each glycosidic bond imparts a conformational constraint on the oligosaccharide, leading, particularly with branching (i.e. more than two hydroxyl groups occupied), to a potential recognition motif. There is one additional modification which is non-enzymic, called glycation, where circulating Glc reacts with free amino groups of proteins (N-terminus or Lys side chains). The Schiff base formed is stabilised by an Amadori rearrangement and then undergoes further reactions to give advanced glycation end products (AGEs). These are important in pathology and glyco-based drugs are a potential target for therapy.

3 CARBOHYDRATE TUMOUR ANTIGENS AND ALLERGENS

Some of the most promising targets for tumour diagnosis and therapy are the oligosaccharide chains linked in multivalent form (for high affinity) to the MUC gene products (reviewed in[1]). MUC genes code for very large proteins with multiple tandem repeats containing Ser/Thr residues which are O-glycosylated. The first monosaccharide to be added, GalNAc, gives the Tn epitope. This can be sialylated at the C-6 position giving sialyl Tn which predominates in tumour cells. Other glycosyltransferases function better in normal cells adding Gal (giving the T antigen) or GlcNAc (often leading to chain elongation). The T antigen is also found as a tumour associated antigen sialylated at C-3 or C-6 of Gal. Chain elongation allows for more varied glycosylation such as the addition of sialyl Le[a], (a distinct isomer of sialyl Le[x]) which was one of the first carbohydrate tumour associated antigens used in diagnosis and therapy.[10] In addition to the recognition of specific oligosaccharide motifs by antibodies, ways are being explored to raise cytotoxic T cell responses to mono-, and hopefully oligo-, saccharides attached to peptide. It has been shown that T cell clones can distinguish different monosaccharides presented on glycopeptides by MHC[11] and antibody reactivity to mucins can be dependent on the display of the T disaccharide on different amino acid sequences, i.e. is peptide and sugar specific.[12]

Allergens may also be oligosaccharide or glycoprotein in nature. Again the oligosaccharides are presented in multivalent form as mucins (e.g. pigeon mucin in chronic alveolitis of the lung in pigeon fanciers[13]) or polysaccharides (e.g. mannans of yeast in brewers and bakers[2]). Antibodies of patients with Crohn's disease (CD) have also been shown to react to yeast mannan and we have speculated[2] that this may arise due to cross reactivity to mannans of mycobacteria which are associated with CD.[14] On the other hand, non-immunogenic mannans can be used as part of a strategy to raise T cell responses to antigens by targeting antigen presenting cells via the macrophage mannan binding receptor.[15] This and soluble mannose binding proteins mediate the innate immune response (IIR) which functions particularly in the young before they have attained an extensive antibody repertoire. Defects in the IIR lead to multiple infections offering the potential to boost this response as an anti-bacterial policy. Also important here are the new oligosaccharide-to-protein conjugate vaccines based on the oligosaccharides of bacterial LPS (Table 2).

4 CELL TO CELL INTERACTIONS

Of the MUC gene products originally only one, MUC1, was known as a transmembrane protein, the others being characterised in the mucus layer outside the cells. However there is now evidence that MUC3 and MUC4 splice variants occur leading to transmembrane products. These have the potential to be multifunctional molecules at the epithelial cell surface with various roles of their oligosaccharides and protein domains. Other large, highly

O-glycosylated molecules have been characterised on endothelial cells and lymphocytes where they function in interactions.[15] These are called mucin-like molecules because they do not have the tandem repeats of the MUC gene products, but they share the property with mucins that the many closeby O-linked chains interact with the protein backbone to constrain an extended conformation with interesting spatial consequences often depicted like a "bottle (or test tube) brush" . These molecules also carry the longer oligosaccharide chains which bear the sialyl Le^x motif in multivalent form which are the ligands for selectins mediating the initial interaction of lymphocytes with endothelium. This step is necessary to slow down the lymphocytes in the circulation so that they will arrive at sites of infection, but is also involved in an inappropriate response to self tissues in inflammation and hence a target for drug therapy. This may have to take into account that nature uses a multivalent array to increase affinity, together with elements of peptide specificity as discussed above.

There is another class of mammalian carbohydrate binding proteins, called sialoadhesins some of which also have the potential to interact with mucins via the sialic acid linked to the C-6 of GalNAc, and the C-3 or C-6 of Gal. Each sialoadhesin has a different specificity. Others interact with the sialic acid linked to the C-3 or C-6 of Gal on N-linked chains (see Section 7). In humans these are all a type of sialic acid called Neu5Ac (with an alternative Neu5Gc occasionally occurring as part of a tumour associated antigen). There is one additional type of mammalian (and a bacterial cross reactant) sialic acid motif which is a repeating stretch of Neu5Ac. This is found on the neural cell adhesion molecule NCAM, is developmentally regulated and implicated in neural cell interactions. One form of NCAM is anchored to the cell membrane via a GPI anchor (Table 2), as with several other molecules functional in the brain (see Section 6).

5 PROTEOGLYCANS

Before discussing GPI anchored glycoproteins, there is one more class of polyanionic oligosaccharide to be considered, the proteoglycans (Tables 1 and 4), because one family of them, heparan sulphate, is often found associated with GPI anchored proteins and always with

---

Table 4 *Polyanionic glycoses besides DNA/RNA*

Mucins and mucin-type cell membrane molecules
Glycoproteins - sialylated oligosaccharides
e.g. sialyl Tn
    sialyl T
    sialyl Le^a
Neural cell adhesion molecule (NCAM)
    polysialic acids (NeuAc)$_n$
    developmentally regulated
    sometimes GPI-anchored
Proteoglycans (PGs)

---

amyloid plaques (see Section 6). Although the oligosaccharide chains (glycosaminoglycans) of these are made up of only disaccharide repeats, GlcNAc-GlcA in the case of heparan sulphate, there is tremendous diversity in the types of modifications that can occur, leading to different functional domains in these again very large molecules (smaller glycosaminoglycan sequences are also found linked to transmembrane glycoproteins). Thus, in heparan sulphate: GlcA (the C-6 acid of Glc) can be epimerised to IdoA (iduronic acid); the N-acetyl (NAc) of GlcNAc can be replaced by sulphate (GlcNS); and, various of the free hydroxyl groups can be sulphated.[4] This leads to a large potential number of anionic recognition motifs which can interact specifically with different arrays of basic amino acids (Lys and Arg) on the surface of proteins. It is envisaged that each type of protein, be it a cell membrane glycoprotein, growth factor, cytokine, enzyme or virus coat glycoprotein, which has a basic protein patch will interact with different glycosaminoglycan structures. If this is the case, then an anionic mimic could be designed to inhibit each specific interaction. As has been shown for the interaction of heparin with anti-thrombin III and thrombin, after considerable drug developement hexasaccharides can be synthesised with the same affinity as the very high molecular weight mixtures used up to now in anti-coagulant therapy. This effect is functional in the circulation, other PG events occur in the extracellular matrix (e.g. amyloid plaque stabilisation), at the cell surface (e.g. modulating the concentrations of fibroblast growth factor and its interactions with its receptor) and intracellularly (e.g. controlling release of chymases from mast cell granules).

6 GLYCOSYLATION IN PROTEIN FOLDING AND GPI FUNCTION

Many cell signalling events are mediated by glycoproteins which are anchored into the cell membrane by attachment at the carboxy terminal to GPI. As these anchors do not traverse the membrane (the attached lipid is part of the external layer of the lipid bilayer only) they are thought to mediate cell signalling events by associating with integrated transmembrane glycoproteins. In several cases (Table 5) it has been suggested that this association is mediated by N-linked oligosaccharides of the GPI-anchored glycoprotein. So far no role has

Table 5 *Examples of interacting components at the cell surface involving GPI*

GPI anchored glycoprotein	Interacting proteins	Medical area
Prion	Not known	Spongiform encephalopathy
Thy-1	?, but also proteoglycan	Brain function
[a]GDNFRα	GDNFR	Neural cell interactions
CD14	Integrins e.g. [c]CR3: toll	Microbial pathogenesis and apoptosis
CD16	" " "	AIDS
[b]uPAR	" " "	Cancer

[a]Glial cell derived neurotrophic factor receptor α interacts with its glycoprotein receptor, GDNFR, which interacts with GDNF. [b]Urinary plasminogen activator receptor. [c]Complement receptor 3 also binds iC3b and ICAM

been ascribed to the glycan of the GPI or its lipid, but our contention is that these are essential components in intramembrane interactions and intracellular trafficking. This can be explored by conformational studies of the various domains of the molecules, mimicking of the spatial restraints on interacting components, and use of these mimics as inhibitors of cell interactions. We have ·obtained evidence[5] that the GPI anchor of Thy-1 affects the conformation of the protein, such that when released from the cell membrane it would not interact with the same ligand compared to its lipidated form. GPI may also be used to direct intra- and inter-cellular trafficking, the latter by analogy to glycolipids which for example are not synthesised by red blood cells, but migrate to their cell surfaces. This may be particularly important in the prion diseases (Table 6) to facilitate infectivity. In addition, internalisation from the cell membrane is required for the transformation of the protease susceptible form of a prion protein to the protease resistant form when particular amino acid mutations are present. Is this route lipid dependant (Table 7)? Also, amino acid mutations are associated with changes in N-glycosylation. On the other hand small (non-glycosylated) peptides from the prion glycoprotein appear to be all that are necessary to act as a template in the formation of pathogenic plaques.

Table 6 *Transmissable spongiform encephalopathies*

BSE, Creutzfeldt-Jakob disease, scrapie
Prions: N-terminus and repeat domain
      Di-S-bonded loop with two N-glycans
      C-terminal-GPI protein anchor
Amino acid mutations give changes in N-glycosylation
Intracellular trafficking required for pathogenesis
? Intercellular trafficking, c.f. glycolipids
? Interactions with proteoglycans, c.f. HIV, HSV

Table 7 *Lipid variability of eukarytic cell membranes*

Glycerol	CH_2- CH_2- CH_2 -
	OH OH
acyl	-O-C (=O)- $(CH_2)_n$ CH_3
alkyl	-O-C -$(CH_2)_n$ CH_3
unsaturated	*versus* saturated $(CH_2)_n$
cholesterol	

One of the other amyloidogenic diseases also often involves glycosylated molecules. This is primary systemic amyloidosis which occurs in patients with myeloma who synthesise large amounts of immunoglobulins (Ig). In some patients the light chains of IgG form fibrils found deposited systemically in several organs. Because light chains have variable amino acid sequences from their function in binding antigen, the consensus glycosylation sequence (Table 2) for the addition of N-linked chains can randomly occur. We and others[6] have analysed several amyloidogenic light chains to show that when the consensus sequence is present, the site is occupied. For three out of three which we have analysed the oligosaccharide shown below was found. We have characterised these glycosylated light chains (and all the other glycoproteins and oligosaccharides discussed in this article) by the

---

Biantennary, bisected, core fucosylated N-linked chain

$$Gal\beta1\text{-}4GlcNAc\beta1\text{-}2Man\alpha1 \setminus \qquad\qquad Fuc\ \alpha$$
$$6 \qquad\qquad ½\ (1\text{-}6)$$
$$GlcNAc\beta1\text{-}4Man\beta1\text{-}4GlcNAc\beta1\text{-}4GlcNAc\beta1\text{-}Asn$$
$$3$$
$$Gal\beta1\text{-}4GlcNAc\beta1\text{-}2Man\alpha1 /$$

Plus NeuAcα2-6 linked to the Gal residues

---

use of carbohydrate biochemistry techniques[16], NMR spectroscopy[4] and computer graphics molecular dynamics studies. In this way we can explore the structure and conformation of the oligosaccharide chains, the role of the glycosylation in protein function and the accessible motifs for further interactions, e.g. with proteoglycans or homo- and hetero-typic associations. In light chain amyloidosis the particular oligosaccharide may be involved in circulation of the glycoprotein and possible interactions with carbohydrate binding proteins in the systemic tissues. Besides having specific recognition functions glycosylation will also be involved in stabilising the protein and helping in its correct folding and expression. The statistics (Table 8) certainly point to some functional involvement.[17]

Table 8 *Primary systemic amyloidosis*

- 15% of myeloma patients deposit their Ig light chains in fibrils (amyloidosis) in systemic tissues.
- The Ig in fibrils is 4 times more likely to be glycosylated than that of non-amylodogenic myeloma patients.
- There is a reversal from normal Ig glycosylation which occurs mostly in hypervariable regions (60%) to amyloid glycosylation which is found predominantly in the framework region (60%)
- i.e. There is a predisposition to Ig framework glycosylation in amyloid light chain disease of myeloma.

ACKNOWLEDGEMENTS

The authors are grateful to Mrs Gail Evans for preparing this manuscript, to our collaborators Lone Omtvedt, Knut Sletten, Svein Haavik, Roger Morris, Sylvain Lehmann, Mia Young, Roger Barnes, Fraser Stoddart and Narayanaswamy Jayaraman, and for MRC and EU financial support.

References

1. E. F. Hounsell, M. Young and M. J. Davies, Clin. Sci., 1997, **93**, 287.

2. M. Young, M. J. Davies, D. Bailey, B. Smestad-Paulsen, J. Wold, R. M. R. Barnes and E. F. Hounsell, Glycoconjugate. J., In press.

3. P. R Ashton, E. F. Hounsell, N. Jayaraman, T. Nilsen, M. Young, N. Spencer and J. F. Stoddart, J. Org. Chem., 1998, **63**, 3429.

4. E. F. Hounsell and D. Bailey, In 'Glycopeptides and related compounds: Synthesis, Analysis and Applications'. Eds. D.G. Large and C.D. Warren. Marcel Dekker Inc 1997, p. 631

5. E. Barboni, B .P. Rivero, A. J. T. George, S. R. Martin, D. V. Renouf, E. F. Hounsell, P. C. Barber and R. J. Morris, J. Cell Science, 1995, **108,** 487.

6. L. A. Omtvedt, S. Haavik, E. F. Hounsell, H. Barsett and K. Sletten, Amyloid: Int. J. Exp. Clin. Invest., 1995, **2**, 150.

7. K. D. Greis and G. W. Hart, In 'Glycoanalysis Protocols'. Ed. E. F. Hounsell, Humana Press, Totowa, New Jersey, 1998, Chapter 2, p. 219.

8. P. Messner, Glycoconjugate J. 1997, **14**, 3.

9. M. A. Curtis, A. Thickett, J. M. Slaney, M. Rangarajan, J. Adduse-Opose, P. Shepherd, N. Paramonov and E. F. Hounsell, Molec. Microbiol., In press.

10. B. Betchel, A.J. Ward, K. Wroblewski, H. Kaprowski, J. Thurin, J. Biol. Chem. 1990, **265**, 2028.

11. E. Meinjohanns, M. Meldal, T. Jensen, O. Werdelin, L. Galli-Stampino, S. Mouritsen, and K. Bock, J. Chem. Soc. Perkin Trans. 1, 1997, 871.

12. S. Haavik, M. Nilsen, T. Thingstad, H. Barsett, D. V. Renouf, E. F. Hounsell, and J. F. Codington, In preparation.

13. C. I. Baldwin, A. Todd, S. J. Bourke, A. Allen, and J. E. Calvert, Clin. Exp. Aller., 1998, **28**, 349.

14. K. A. Ohene-Gyan, J. Haagsma, M. J. Davies, and E. F. Hounsell, Comp. Immunol. Microbiol. Inf. Dis.,. 1995, **18**, 161.

15. A. N. Barclay, M. L. Birkeland, M. H. Brown, A. D Beyers,. S. J. Davis, C. Somoza, A. F. William, 'The Leucocyte Antigen Facts Book', Academic Press, 1993.

16. E.F. Hounsell, Glycoanalysis Protocols, Methods in Molecular Biology, Humana Press, Totowa, New York, 1998, Vol. 76.

17. L. A. Omtvedt, D. Bailey, D. V. Renouf, M. J. Davies, N. Paramonov, S. Haavik, G. Husby, K. Sletten, and E. F. Hounsell, In preparation.

BACTERIAL POLYSACCHARIDES - BIOLOGICALLY ACTIVE SURFACE POLYMERS

Per-Erik Jansson*[a], M. John Albert[b], Yuriy A. Knirel[a,c], Anthony P. Moran[d], Nina Kocharova[a,c], Sof'ya N. Senchenkova[a,c], Andrej Weintraub[e], and Göran Widmalm[f]

[a] Karolinska Institute, Clinical Research Centre, Huddinge University Hospital, Huddinge, Sweden. [b]International Centre for Diarrhoeal Diseases Research, Bangladesh, Dhaka, Bangladesh, [c]N. D. Zelinsky Institute of Organic Chemistry, Russian Academy of Sciences, Moscow, Russia, [d]Department of Microbiology, National University Galway, Galway, Ireland, [e]Karolinska Institute, Department of Immunology, Microbiology, Pathology and Infectious Diseases, Division of Oral Clinical Bacteriology, Huddinge University Hospital, Huddinge, Sweden. [f]Department of Organic Chemistry, Arrhenius Laboratory, Stockholm University, Stockholm, Sweden

SUMMARY

The chemistry of lipopolysaccharides from *Helicobacter pylori* and *Vibrio cholerae* has been studied with the aim to study correlations structure - biological activity. One interesting feature of these carbohydrate structures, namely mimicry i.e. that identical or nearly identical structures are present on the bacterium and the host, has been a separate interest in the investigations. The similarities are in some cases large and discussed below.

1 INTRODUCTION

Helicobacter pylori is an important gastroduodenal pathogen in humans. Immunological and structural studies have been performed on surface components like phospholipids, lipopolysaccharides (LPS) and some proteins. LPS has in general low immunological activity which may aid the survival of the microorganism in a chronic infection. Nevertheless, *H. pylori* has been found to contribute to gastric disease, and LPS is therefore considered to be a virulence factor.

Vibrio cholerae is a pathogen that has caused several epidemics in different parts of the world. *V. cholerae* O1 has in all previous cases been the cause of an acute, diarrhoeal infection, cholera, but in the latest epidemic a new agent has been the cause and this fact increased the interest for non-O1 Vibrios. Of interest are also bacteria with similar surface carbohydrate components as they may have similar genetics and give information on genesis of new strains.

1.1 Lipopolysaccharides

The lipopolysaccharide (LPS) is a major feature of the cell wall of gram-negative bacteria, and the most prevalent glycolipid in these.[1] Many endotoxic activities have been attributed to the LPS, like pyrogenicity and adjuvant activity. A most significant feature of LPS is that it carries the serological O-specificity, or the O-factors of the bacterial strain. It has also become clear recently that the O-chain (the polysaccharide) modulates the activation of the alternative complement pathway. Gram-negative bacteria can be either capsulated or non-capsulated, the capsule being a polysaccharide as well. The immune response is directed towards surface components and even though the capsular polysaccharide covers a substantial part of the bacterial cell antibodies are normally obtained to both the LPS and the capsular polysaccharide. Among the lipopolysaccharides, most are composed of lipid A, a core oligosaccharide and an O-polysaccharide. The lipid A which is mostly conserved, consists of a phosphorylated glucosamine disaccharide substituted with 5-7 fatty acids. The core oligosaccharide is in general composed of hexose, L-*glycero*-D-*manno*-heptose (LDHep) and 3-deoxy-octulosonic acid (Kdo) residues, the latter constituting the bridge to lipid A. The O-chain polysaccharide which is linked to the core, is a polymer comprised of oligosaccharide repeats, with a typical repeat number of 10-25. The variability between O-polysaccharides is very large and in principle type-specific. Certain species of gram-negative pathogens like *Moraxella catarrhalis* and *Haemophilus influenzae* express a rough-type LPS sometimes referred to as lipooligosaccharide (LOS). This LPS, lacks the O-chains and consists of the lipid A and an oligosaccharide, thereby exhibiting structural similarities with enterobacterial rough type bacteria, i.e. with short oligosaccharides in the LPS and with typical rough colony morphology.

2 HELICOBACTER PYLORI

In 1982, Australian researchers discovered spiral-shaped bacteria in the stomach, later named *Helicobacter pylori*, and proposed that the bacteria were the underlying cause of gastritis and peptic ulcers. Evidence linking *H. pylori* to ulcers mounted over the next 10 years. The bacterium infects 50-60% of the world's population by the age of 50 years, making it the second commonest infection worldwide. At present *H. pylori* is considered the causative agent of chronic atrophic gastritis and is associated with development of duodenal and gastric ulcers and is a cofactor in the development of gastric cancer. Recently, *H. pylori* has been classified as a type I human carcinogen by the World Health Organization (WHO).[2]

It remains unclear precisely how *H. pylori* damages the gastric epithelium. The bacterium can survive in the stomach because it produces the enzyme urease which generates ammonia that neutralizes the stomach's acid. *H. pylori* can penetrate through the stomach's protective lining and produce substances that weaken the protective mucus and make the stomach epithelial cells more susceptible to the damaging effects of acid and pepsin. Within weeks after infection with *H. pylori*, some people develop gastritis, whereas it may take decades for pathology to develop in other individuals and they may remain asymptomatic. However, most people will never have symptoms or problems related to the infection.

It is considered that attachment of *H. pylori* to epithelial cell surface is a prerequisite for its colonization and subsequent pathogenesis.[3] Such attachment is mediated by adhesins (proteins) on the bacterial surface, which recognize proteins or glycoconjugates on the surface of epithelial cells. Differences have been reported in the binding of different *H. pylori* strains to human gastric mucosal cells containing the Lewis b antigen. On the other hand, several host and environmental factors appear to play roles in determining the likelihood of acquiring the infection and manifesting the various presentations of *H. pylori*-associated disease.[4] One host factor is the type of Lewis blood group antigens present, which are known to be abundant in salivary and gastric secretions where they may agglutinate the infecting organism and thereby remove *H. pylori*.

2.1 Lipopolysaccharides of *H. pylori* - Lewis antigens and molecular mimicry

Some strains of *H. pylori* have been investigated with respect to their LPS (Table 1). The variable parts of the *H. pylori* LPS are found in the O-chain which, in all cases investigated, consists of alternating 3-linked galactose (Gal) and 4-linked N-acetyl-glucosamine (GlcNAc) units, (N-acetyl-lactosamine disaccharide, LacNAc) where a large part of the GlcNAc residues are substituted with fucose (6-deoxygalactose, Fuc) groups making up so-called Lewis x elements. In some cases the terminal trisaccharide may be substituted with another Fuc group, then in the 2-position of the Gal residue, thereby making up a Lewis y element. The repeating units are mixtures of blocks of LacNAc and Lewis x elements. In some cases the LacNAc element has been substituted with other side chains, with single glucose (Glc) or Gal units. Initial serotyping assigned six provisional groups[5] of which some have been investigated. (see table). The large proportion of non-typables, however, has lead to that typing with either monoclonal antibodies or commercial Lewis recognizing lectins are prevalent.

Molecular mimicry of host structures has been observed in *Neisseria, Campylobacter, Haemophilus,* and *Helicobacter*.[6] In *Neisseria* the LPSs mimic glycosphingolipids found on human cells. It has been speculated that this may serve to camouflage the bacterial surface from the host. Similarly, *Campylobacter* mimic gangliosides, *Haemophilus* mimic the lacto-neoseries of glycosphingolipids, and *Helicobacter* Lewis antigens. For *H. pylori* it has been known for some time that blood group A, B, and H structures are present on certain gram-negative bacteria but the expression of Lewis blood group antigens is a new phenomenon. Autoantibodies are antibodies directed towards host structures, e.g. Lewis determinants on mucosal surface. The observation that during infection such antibodies occur may be a consequence of the Lewis structures on the surface on *H. pylori* bacteria.

Table 1 *Structures of O-antigen chains in Helicobacter pylori lipopolysaccharides.*[7]

Serogroup/ Strain	Terminus			O-chain/heptan
O:1[8] (NCTC 11637)	Gal→ 4GlcNAc 3 ↑ Fuc		Le^x	[Gal→ 4GlcNAc]$_n$→ [3Gal→ 4GlcNAc]$_n$→ 3 ↑ Fuc
P466[9]	Gal→ 4GlcNAc 2 3 ↑ ↑ Fuc Fuc		Le^y	[Gal→ 4GlcNAc]$_n$→ [3Gal→ 4GlcNAc]$_n$→ 3 ↑ Fuc
O:3[10]	Gal→ 4GlcNAc 2 3 ↑ ↑ ±Fuc Fuc		Le^x/Le^y	[Gal→ 4GlcNAc]$_n$→ [3Gal→ 4GlcNAc]$_n$→ [3DDHep]$_n$→ 3 ↑ Fuc
O:6[10] (MO19)	Gal→ 4GlcNAc 2 3 ↑ ↑ Fuc Fuc		Le^y	Gal→ [3DDHep]$_n$→
UA861[11]	Gal→ 4GlcNAc 2 3 ↑ ↑ Fuc Fuc		Le^y	[Gal→ 4GlcNAc]$_n$→ [3Gal→ 4GlcNAc]$_n$→ [Gal→ 4GlcNAc]$_n$→ 3 6 ↑ ↑ Fuc Glc
Hp471[12]	Gal→ 4GlcNAc 2 3 ↑ ↑ Fuc Fuc		Le^y	[Gal→ 4GlcNAc]$_n$→ [3Gal→ 4GlcNAc]$_n$→ [Gal→ 4GlcNAc]$_n$→ 3 6 ↑ ↑ Fuc Gal

2.2 *H. pylori* core region

The core region in *H. pylori* appears to be a conserved part of the LPS with an inner hexasaccharide Glc-Gal-DD-Hep-LD-Hep-LD-Hep-Kdo as a prevalent structure.[10] Differences found were in the degree of phosphorylation only. A different number of Glc and DD-Hep units substitute this oligosaccharide. A highly unique feature suggested, is that the DD-Hep residues, referred to as a heptan, should constitute an intervening region between the core and the O-chain.[7]

Present study[13,14]. The present study was undertaken to relate the structure of four *H. pylori* LPSs to their biological activities. For LPS extraction, biomass of *H. pylori* 26695, whose complete genome sequence has been published,[15] was grown under two different pH values to induce phase variation in the LPS. The resulting preparations were called S1 and S2. Biomass of *H. pylori* strains AF1 and 007, that differ in their ability to recruit and stimulate neutrophils, and are non-typable as Lewis antigens were also obtained.

LPS of the four *H. pylori* preparations were isolated by extraction with hot aqueous phenol, and degraded with buffer, pH 4.2, for 4 h at 100 °C. The water-soluble carbohydrate portion was fractionated by gel-permeation chromatography in a buffered eluent to give an O-specific polysaccharide and a core oligosaccharide respectively. Similar results were obtained from the LPS from *H. pylori* S2. The 007 and AF1 preparations also gave O-chains and oligomeric fractions. The gel-chromatographic curves are shown in Figure 1.

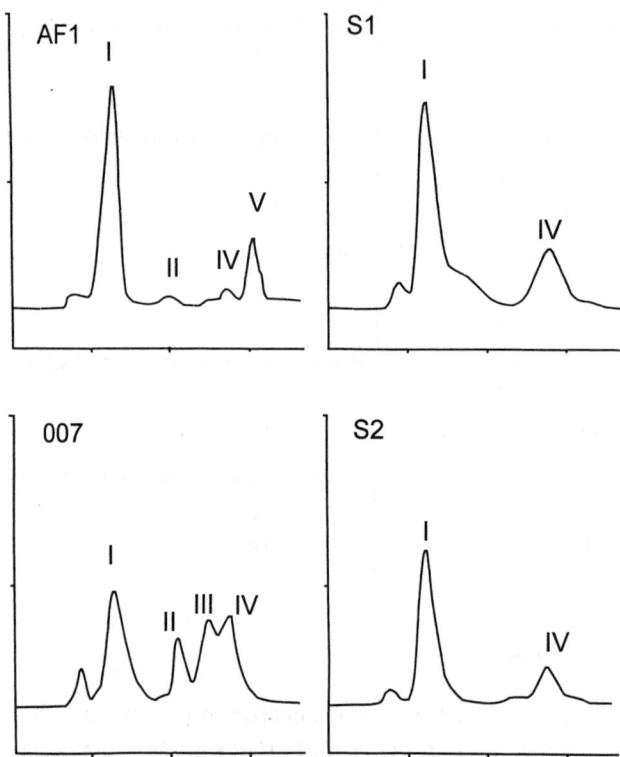

Figure 1   *Gel filtration curves on Sephadex G50 from soluble products after delipidation with dilute acid*

The different fractions were analyzed by methylation analysis i.e. methylation, hydrolysis, and GLC analysis of the methylated monosaccharide derivatives obtained. An absence of a methyl group indicates a linkage, *e.g.* a 2,4,6-tri-O-methyl hexose derivative indicates a 3-linkage (1 and 5 being engaged in the ring) and so on. Methylation analysis of the S1 polymer, followed by GLC-MS analysis, resulted in identification of the major components: terminal Fuc, terminal and 3-substituted Gal, 4-substituted, and 3,4- and 4,6-disubstituted GlcNAc. These data suggested that the polymer is branched with two different lateral sugar residues (Gal and Fuc) and two different GlcNAc residues as branch point residues. Comparison with the methylation analysis data for LPS showed that no terminal Fuc was cleaved during mild acid degradation. A number of minor derivatives in the GLC analysis were detected as well. Most of them were derived from the LPS core region attached to the O-chain, as followed from the comparison with the methylation analysis data for the core. However, two minor products originated from the O-chain. One of them was from 3,4,6-trisubstituted GlcNAc and showed that a minor portion of GlcNAc residues carries two side chains. The analysis of the methylated GlcNAc derivatives are thus a good indicator of how the LacNAc element is substituted. This is shown in Figure 2. The analysis of the S2 preparation showed some significant changes compared to S1, namely that the derivative corresponding to a Gal-substituted GlcNAc residue was absent as well as that in the terminal element, the latter strongly indicating the absence of the Lewis y determinant which requires such a sugar, instead terminating with a Lewis x unit.

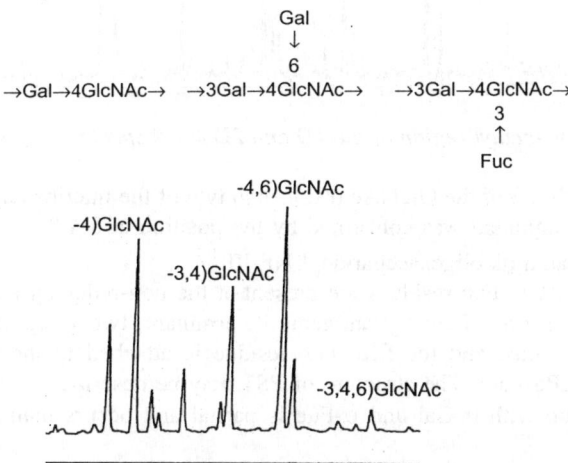

Figure 2   *The amino sugar region in the GLC chromatogram in the methylation analysis of H. pylori S1*

The 600-MHz ^{1}H-NMR spectrum of the O-chain pointed to a high degree of structural heterogeneity due to nonstoichiometric substitution with two lateral monosaccharides, Gal and Fuc. The spectrum also contained signals for *N*-acetyl groups of GlcNAc at δ 2.02 and CH_3-C groups (H6) of Fuc at δ1.14-1.26 with an intensity ratio of 1:0.35. In total there were H6 signals for five Fuc residues, two of which appeared as separate doublets at δ 1.23 and 1.26, and three others as a superposition of three doublets at δ 1.14 ($J_{5,6}$ ~6 Hz for all H6 signals). The two-dimensional DQF-COSY spectrum (Figure 3) showed Fuc H6/H5 cross-peaks at δ 1.14/4.35, 1.14/4.81, 1.23/4.87 and 1.26/4.25. The first cross-peak was from the Fuc residue which is adjacent to the LPS core region, as followed from the presence of a cross-peak at the same co-ordinates, as the only Fuc H6/H5 cross-peak, in the DQF-COSY spectrum of the oligosaccharide. The cross-peak at δ 1.26/4.25 was assigned to the Fuc group substituting position 2 of β-Gal in the terminal non-reducing LacNAc unit (e.g., compare δ$_{H5}$ 4.22 for Fuc in 2'-(α-fucopyranosyl)lactose).[16]

Hence, the cross-peak at δ 1.23/4.87 having a single intensity, should originate from Fuc attached to GlcNAc in the same terminal non-reducing LacNAc unit, whereas the last cross-peak at δ 1.14/4.81 of double intensity was from the Fuc groups

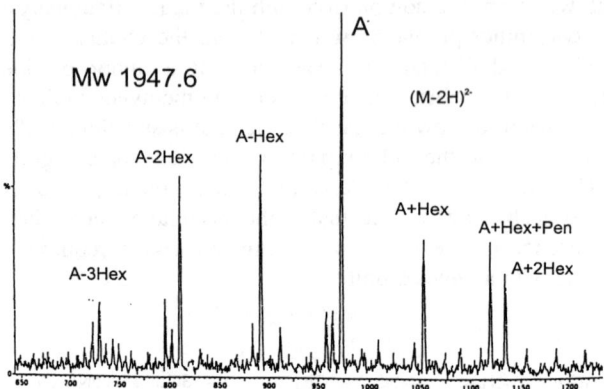

Figure 3   *The methyl region of the 1D and 2D NMR spectrum of H. pylori S1*

substituting position 3 of the GlcNAc residues in two of the interior LacNAc units of PS. The latter assignment was confirmed by the position at δ 4.83 of the signal for Fuc H5 in a human milk oligosaccharide, LNF-III.[17]

Therefore, two Fuc residues are present at the non-reducing LacNAc unit of PS1, where they form a Lewis y antigenic determinant, two more are attached to interior LacNAc units, and the fifth Fuc residue is attached to the unit which is adjacent to the LPS core. The structure of PS1 may be described by the formula **2**, where substitution with α-Gal and α-Fuc is partial and occurs mainly in different LacNAc units.

$$
\begin{array}{ccc}
 & [\alpha\text{-Gal}] & \\
 & 1 & \\
 & \downarrow & \\
 & 6 & \\
\beta\text{-Gal-}(1\rightarrow 4)\text{- }\beta\text{-GlcNAc-}(1\text{-}[\rightarrow 3)\text{-}\beta\text{-Gal-}(1\rightarrow 4)\text{- }\beta\text{-GlcNAc-}(1\rightarrow]_n \\
3 \qquad\qquad\qquad 3 \qquad\qquad\qquad\qquad\qquad 3 \\
\uparrow \qquad\qquad\qquad \uparrow \qquad\qquad\qquad\qquad\qquad \uparrow \\
1 \qquad\qquad\qquad 1 \qquad\qquad\qquad\qquad\qquad 1 \\
\alpha\text{-L-Fuc} \qquad \alpha\text{-L-Fuc} \qquad\qquad\qquad [\alpha\text{-L-Fuc}]
\end{array}
$$

Based on the ratios of the GlcNAc derivatives revealed in methylation analysis and on the Fuc and GlcNAc methyl group signal intensities in the ^{1}H-NMR spectrum, the average degree of polymerisation in PS1 is 14 LacNAc units.

The methylation analysis of the O-chain of AF1 indicated that it contained a Lewis y determinant despite the fact that it could not be serotyped as such. Analogously 007 was indicated to be Lewis x. The ^{1}H NMR spectrum of AF1, was similar to that of S1 and clearly demonstrated that the Lewis y determinant was there. The methylation analysis further showed that the chain-length was approximately as in S1 and S2 and that no side chains other than Fuc were present and also that both contained a heptan. In the same way in 007, the presence of a terminal Lewis x element was demonstrated with ^{1}H NMR spectroscopy. It is also worth noting that all LacNAc units were substituted with Fuc groups thereby making them different from all the previously investigated strains.

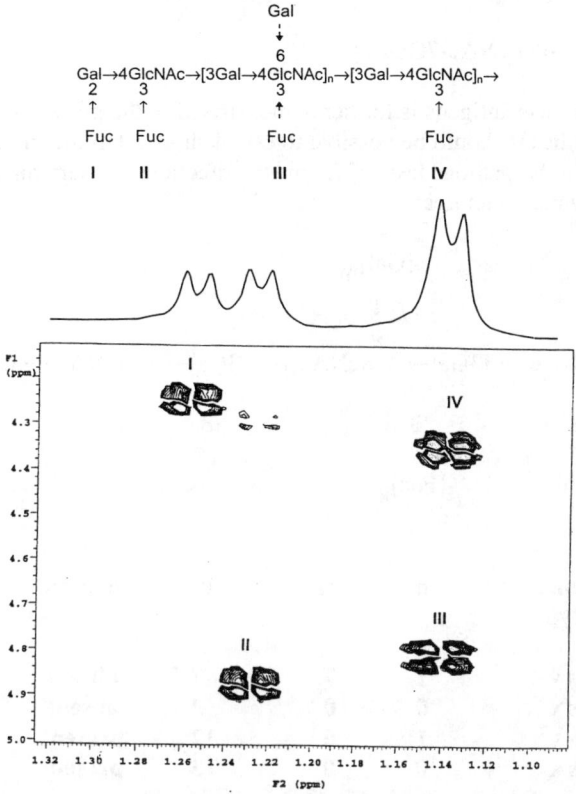

Figure 4   *Electrospray MS from the core fraction of H. pylori S1*

The data on the core elements were in agreement with previous data on the conserved inner part of the core with hexose and heptoses. Electrospray mass spectrometry was very helpful in establishing molecular weights of the core oligosaccharides, as demonstrated in Figure 4. A predominant oligosaccharide A was found, the structure of which we suggest to be the following, based on previous findings. No information has yet been obtained on the location of the heptan.

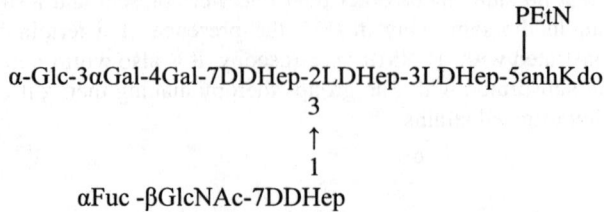

PEtN
|
α-Glc-3αGal-4Gal-7DDHep-2LDHep-3LDHep-5anhKdo
3
↑
1
αFuc -βGlcNAc-7DDHep

The diversity of the Lewis antigens is further demonstrated in the present study. With further biological studies it should be possible to establish in detail the importance of the Lewis elements in the pathogenesis of *H. pylori* infections. In summary we have established the following structures:

$[Gal]_m$
↓
6
Gal→ 4GlcNAc → [3Gal→ 4GlcNAc]$_{12}$→ 3Gal→ 4GlcNAc → Core
[Heptan]
2   3   3   3
↑   ↑   ↑   ↑
$[Fuc]_n$ Fuc $[Fuc]_k$ Fuc

Strain	Antigenic phenotype	n	m	k	heptan
S1	Lewis y	1	7	2	absent
S2	Lewis x	0	0	2	absent
AF1	Lewis y	1	0	12	present
007	Lewis x	0	0	12	present

3 VIBRIO CHOLERAE

3.1 Introduction

Vibrio cholerae contains some 170 different serotypes with the classical O:1 as the most common one. The others (O:2 etc.) are collectively called non-O:1-vibrions. *V. cholerae* is a major cause of diarrhoea and was earlier a common cause for casualties originating from dysfunction of the membrane fluid pumps. Of the non-O:1-vibrions only few have yet been characterized and for the O-antigen polysaccharides that have been investigated a high frequency of unusual sugars is observed. The O-antigens of O76[18] and O144[19] both consist of homopolymers of L-perosamine (see below).

Until a new epidemic of diarrhoea started in India and South Bangladesh in late 1992, *V. cholerae* non-O1 were not considered as epidemic agents of cholera. The new causative agent belonged to none of the then known 138 O-serogroups of *V. cholerae* and hence, was named *V. cholerae* O139 with the synonym Bengal.[20] It shares several properties with and in many aspects is indistinguishable from *V. cholerae* biotype El Tor. *V. cholerae* O1 is divided in two serotypes, Inaba and Ogawa, and in two biotypes El Tor and classic. The main difference between the O1 and O139 serogroups is the architecture of the cell wall. The O-chain in the LPS of O1 is a homopolymer of 4-(3-deoxy-L-glycero-tetronamido)-4,6-dideoxy-D-mannose[21] terminated in Ogawa with the 2-O-methyl derivative of the same monosaccharide.[22] None of the *V. cholerae* O1 strains is encapsulated. In contrast, *V. cholerae* O139 produces a polysaccharide capsule and has an LPS with a short O-chain. Serological and genetic studies suggested that in *V. cholerae* O139 the capsular polysaccharide (CPS) has the same repeating unit as the O-side chain of the LPS. It was suggested that *V. cholerae* O139 arose from a *V. cholerae* O1 strain by genetic rearrangements including deletion of the O1 rfb region and acquisition of a 35 kb region of DNA which encodes the O139 surface polysaccharide.[23]

After a brief period of displacement, *V. cholerae* belonging to the serogroup O1 reappeared rapidly and became the dominating serogroup causing cholera from 1994 onwards. The O1 strains which reappeared were phenotypically similar to those before the genesis of O139 but showed most interesting fluctuations in patterns of resistance to various antibiotics.[24] These rapid shifts in antimicrobial resistance were in contrast to earlier decades when antibiotic resistance was no usual phenomenon. The quick shifts indicate enhanced mobility of genetic elements which confers resistance to antibiotics. In particular there was an increase in a cytotoxin and it has been concluded that there is a high level of genetic exchange and a relatively high level of clonality. There is a possibility that this is a prelude to genesis of new non-O1, non-O139 strains with unrecognized toxic properties. Hence the interest in such, non-O1, non-O139, strains is great.

3.2 Structure of the capsular polysaccharide from *V. cholerae* O139

The CPS from *V. cholerae* O139 contains phosphorus, Gal, GlcNAc, 6-deoxy-GlcNAc (QuiNAc), galacturonic acid (GalA), and Colitose (Col, "3,6-dideoxy-galactose"), the latter a sugar with very acid sensitive glycosidic linkages.[25, 26] Thus, on traditional delipidation with dilute acetic acid a material devoid of this sugar is obtained. But also, a depolymerisation took place by cleavage of the QuiNAc residue due partly to its 6-deoxy group but also to the presence of a β-2-N-acetamido group which is known to be prone to hydrolysis. An FAB mass spectrum of the product clearly established that it contained GlcNac, QuiNAc, Gal, GalA and phosphate. However the molecular weight obtained differed by 18 amu to that calculated suggesting that the phosphate should be cyclic as proposed in a early study of O139.[26] Further evidence of the cyclic structure came from a series of downfield shifts of signals in the ¹H NMR spectrum, especially for the signals from H-4, H-6a and H-6b. The attachment could thus be established as being to positions 4-and 6. It is worth noting the unusually high value for the coupling constant $J_{H6b,P} = 22$ Hz which is evidently due to the anti-periplanar orientation of the coupled atoms (Figure 5).

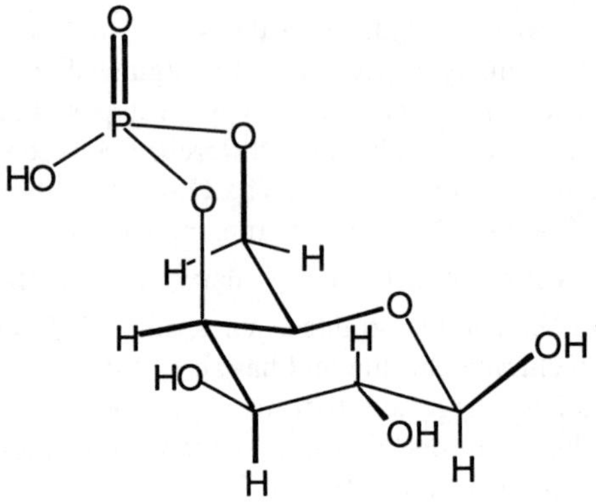

Figure 5   *The structure of β-Galactose-4,6-cyclophosphate*

The cyclic phosphate does not distort the ring as evident form the coupling constants of the ring protons which are typical of β-galactopyranose. The ^{31}P NMR spectrum contained one signal at -1.38 ppm, relative to external 85 % phosphoric acid, appearing as a doublet with a J value of 22 Hz, in agreement with earlier observations. By using 1D and 2D ^{1}H and ^{13}C NMR spectroscopy and methylation analysis, with and without dephosphorylation, it could be established that the tetrasaccharide had the structure

βGal46P → 3βGlcNAc → 4αGalA → 3QuiNAc

Dephosphorylation of the native CPS gave full removal of the colitose residues, dephosphorylation and again depolymerisation. A methylation analysis of the larger polymeric fragments demonstrated that it consisted of the same tetrasaccharide repeats described above polymerized through the 6-position of the GlcNAc residue thus having the structure

βGal
↓
3
→ 6βGlcNAc → 4αGalA → 3βQuiNAc →

And, as the native CPS could be demonstrated to have the Gal residue substituted through position 2 and the GlcNAc through positions 3,4 and 6, where positions 3 and 6 are already assigned, yields the structure

αCol→ 2βGal4,6P
↓
3
→ 6βGlcNAc → 4αGalA→3 βQuiNAc →
4
↑
αCol

3.3 Structure of lipopolysaccharide from *V. cholerae* O139

SDS-polyacrylamide gel electrophoresis showed that the LPS of the capsule-free mutant strain MO10-T4 had a short chain LPS i.e. with no O-chain polysaccharide attached. The LPS of MO10-T4 as well as the CPS of *V. cholerae* O139 is phosphorylated and contains Gal, colitose, GlcNAc, 6-deoxy-GlcNAc, and GalA, while Glc, Fru, LD-Hep, and Kdo are present in the LPS only and typical for the core region of LPS.[27-29]

The LPS was degraded by three different protocols as shown in Figure 6, two involving strong alkaline treatment and one involving degradation with acid. The mixtures arising from the alkaline treatments were purified by High Performance Anion Exchange Chromatography (HPAEC).

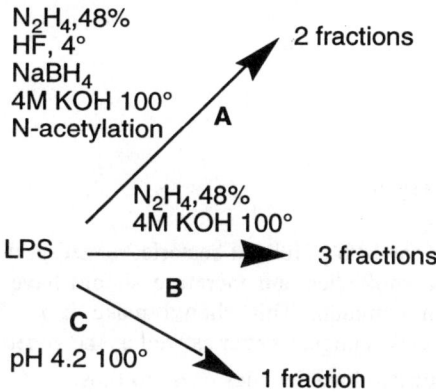

Figure 6 *Degradation scheme for the Vibrio cholerae O139 lipopolysaccharide*

One of the fractions in degradation scheme A had several signals resembling those of the core of *V. cholerae* O1[30] and in addition signals at δ 4.73, 5.60, and 5.75, the latter two indicative of H-1 and H-4 of a 4,5-unsaturated GalA residue arising after the strong alkaline treatment. Together with 2D NMR NOE data and sugar analysis the following structure could be established where Hep is αLD-Hep:

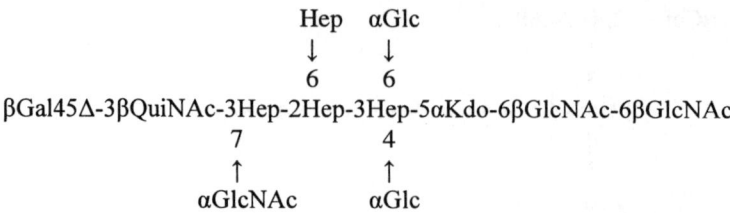

Hep   αGlc
 ↓   ↓
 6   6
βGal45Δ-3βQuiNAc-3Hep-2Hep-3Hep-5αKdo-6βGlcNAc-6βGlcNAc
         7   4
         ↑   ↑
      αGlcNAc   αGlc

    Alkaline degradation of the native LPS showed that there was bis-, tris-, and tetrakis-phosphates present in the LPS. The acidic degradation of the LPS was giving a mixture that was not resolvable by conventional gel chromatography and therefore had to be analyzed as such. The heterogeneity was a result partly from different phosphate content but also from a partial removal of the acid labile sugar colitose. Analysis of the mixture with NMR spectroscopy and methylation analysis showed that another four sugars were present, two colitoses, phosphorylated Gal and GlcNAc, with the same constitution as in the CPS. It was also shown that the GlcN residue in the core was not N-acetylated. The LPS of *V. cholerae* O139 thus has the following structure. It is striking to see that with one repeat of the CPS terminating the core it should be immunogenic just like the CPS.

αCol→ 2βGal4,6P
         ↓
         3
βGlcNAc→4αGalA→3βQuiNAc→CORE
         4
         ↑
       αCol

3.4 Antigens related to *V. cholerae* O139

We have also studied the structure of the LPS of related bacteria, related in the sense that they give rise to cross reactive antibodies and therefore should have similar structures or structural elements in common. This should make it possible to understand the immunochemistry of these antigens better as well as the genetics. The LPS of *V. cholerae* O22[31, 32] has many characteristics close to those of O139 and has a closely related structure of its polysaccharide part of the LPS , the underlined sugars differ from those in O139. The difference is not affecting the ring systems and the conformation should therefore be kept.

αCol→2βGalA3,4Ac
         ↓
         3
αGlcNAc→4αGalA→3βQuiNAc→CORE
         4
         ↑
       αCol

Another species of the Vibrionaceae, *Aeromonas trota*,[33] cross-reacts with the O139. Unlike O139 *A. trota* is a polysaccharide proper and has the following repeats:

$$\rightarrow 3\beta Gal \rightarrow 3\beta GlcNAc \rightarrow 4\alpha Rha \rightarrow 3\alpha GalNAc \rightarrow$$
$$24$$
$$\uparrow\uparrow$$
$$\alpha Col\alpha Col$$

It is easily recognized that four sugars, the Col, Gal, and GlcNAc residues, are the same as in O139, also with anomers and with linkages. The tetrasaccharide is, however, except for the non-reducing terminal, embedded in a polymer chain and it is not obvious that more than the terminal is recognized by antibodies. The response is however positive in the slide agglutination test.

In second group of cross-reactive antigens is found *V. cholerae* O155[34] and *V. mimicus* N-1990.[35] They cross-react though the apparent similarities only go as far as to a Gal46P residue (O155) and the disaccharide element βGal46P-3HexNAc (N-1990). Still they react, possibly because of a favorable exposition of these elements and a large importance for the charged phosphate.

3.5 Mimicry with blood group antigens

The Lewis b blood group antigen has resemblance with the Lewis y antigen in that it has a Fuc-Gal and a Fuc substituent, but with different sites of attachment, and is therefore different as to what regards the three-dimensional appearance. A simplified view of the Lewis b tetrasaccharide is shown in Figure 7:

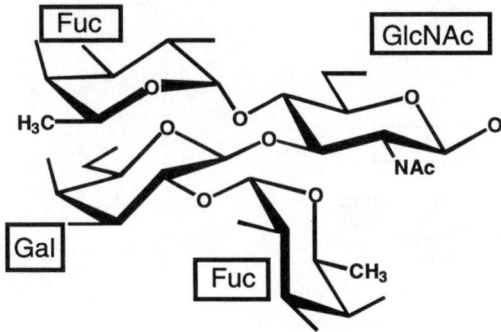

Figure 7   *A simplified three dimensional view of the Lewis b tetrasaccharide. Each stick represents a hydroxy group*

This unit has large similarities to tetrasaccharides found in the polysaccharides from *V. cholerae* O139, and O22, and from *Aeromonas trota*. The pertinent tetrasaccharides are shown in Figure 8 and other sugars are depicted with filled rings. In the O139 LPS (A) in the terminal hexasaccharide, which is one repeat of the capsular polysaccharide, a tetrasaccharide consisting of two colitose, one Gal4,6-phosphate, and one GlcNac residue, have the same three-dimensional appearance as the B tetrasaccharide. The difference lies in the colitose residues which have one oxygen less at the 3-position, and that the Gal residue has a 4,6-phosphate. The *V. cholerae* O22 (B) has the colitoses as well, but instead of the Gal, as in Lewis B, it has a GalA residue which is acetylated at either of positions 3 or 4. For simplicity it is shown in one position only. The *A. trota* polysaccharide (C) has the same tetrasaccharide as O139 but has no 4,6-phosphate. The non-reducing terminal is clearly exposed in the chain it may or may not be the case. A factor in common for all these three ring systems is that they have the same absolute and anomeric configuration, meaning that the overall three-dimensional features will be very similar. The biological consequences for *V. cholerae* have not been as clearly indicated as in *Helicobacter*. Considering previous observations on other bacteria such could easily be foreseen, however.

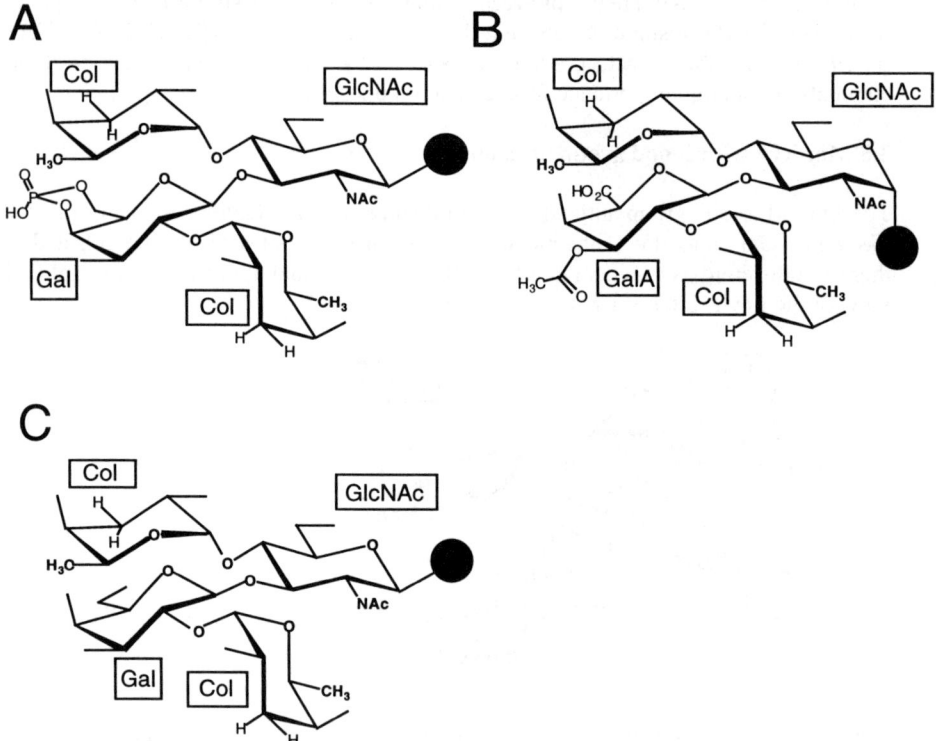

Figure 8   *Simplified threedimensional views of Lewis b-like tetrasaccharides from V. cholerae O139 (A), and O22 (B), and Aeromonas trota (C)*

4 IMPLICATIONS FOR THE FUTURE

The mimicry of Lewis antigens by *H. pylori* is unlikely to be purely coincidental. However, its real role is not yet established, and needs further experiment. If the expression of host structures is a camouflage for the bacterium, how does it differ between LPS with different chain length and different expression of Lewis determinants? Questions that may have an answer in a not too distant future. A long term objective for cure of the infection could be the development of new inhibitory treatments. More detailed examination of *H. pylori* LPS should give further insight to the pathology of this fascinating bacterium.

Structural analysis of surface polysaccharides that are related to the *V. cholerae* O139 antigen may result in the understanding of how this new cholera causing micro-organism developed. Transfer of genetic material between bacteria is a common phenomenon both within the same species as well as between different species. It is not completely clear from where the *V. cholerae* O139 obtained the 35 kb fragment conferring its surface polysaccharides. The antigenic switch happened in an endemic are where *V. cholerae* O1 has been present for many years and the acquisition of the new surface antigens may have been a way to avoid the immunity that is present in the population due to frequent exposure of *V. cholerae* O1. One reason why the *V. cholerae* O139 established itself so successfully and rapidly could be the mimicry of the Lewis b antigen. It is possible that we will face similar attempts from the microbial kingdom in the future since this may be their way of survival.

REFERENCES

1. E. T. Rietschel, J. M. Cavaillon, and N. H. Cavaillon, in '*Bacterial endotoxic lipopolysaccharides*'CRC Press, Boca Raton, 1992., pp 3 & 205.

2. B. E. Dunn, H. Cohen, and M. J. Blaser, *Clin. Microbiol. Rev.*, 1997, **10**, 720-741.

3. A. P. Moran, *FEMS Immun. Med. Microbiol.*, 1995, **10**, 271-280.

4. A. P. Moran, *Aliment. Pharmacol. Ther.*, 1996, **10**, 39-50.

5. S. D. Mills, L. A. Kurjanczyk, and J. L. Penner, *J. Clin. Microbiol.*, 1992, **30**, 3175-3180.

6. A. P. Moran, M. M. Prendergast, and B. J. Appelmelk, *FEMS Immun. Med. Microbiol.*, 1996, **16**, 105-115.

7. G. O. Aspinall, *Carbohydrates in Europe*, 1998 May, 24-29.

8. G. O. Aspinall, M. A. Monteiro, H. Pang, E. J. Walsh, and A. P. Moran, *Biochemistry*, 1996, **35**, 2489-2497.

9. G. O. Aspinall and M. A. Monteiro, *Biochemistry*, 1996, **35**, 2498-2504.

10. G. O. Aspinall, M. A. Monteiro, R. Shaver, L. Kurjanczyk, and J. Penner, *Eur. J. Biochem.*, 1997, **248**, 592-601.

11. M. A. Monteiro, D. Rasko, D. E. Taylor, and M. B. Perry, *Glycobiology*, 1998, **8**, 107-112.

12.   G. O. Aspinall, A. S. Mainkar, and A. P. Moran, *Ir. J. Med. Sci.*, 1997, **166**, 26-27.

13.   Y. A. Knirel, S. N. Senchenkova, P.-E. Jansson, G. Widmalm, and A. P. Moran, Manuscript, 1998.

14.   Y. A. Knirel, S. N. Senchenkova, P.-E. Jansson, G. Widmalm, and A. P. Moran, Manuscript, 1998.

15.   J. F. Tomb, O. White, A. R. Kerlavage, R. A. Clayton, G. G. Sutton, R. D. Fleischmann, K. A. Ketchum, H. P. Klenk, S. Gill, B. A. Dougherty, K. Nelson, J. Quackenbush, L. Zhou, E. F. Kirkness, S. Peterson, B. Loftus, D. Richardson, R. Dodson, H. G. Khalak, A. Glodek, K. McKenney, L. M. Fitzegerald, N. Lee, M. D. Adams, J. C. Venter, et. al, *Nature*, 1997, **388**, 539-547.

16.   K. Hermansson, P. E. Jansson, L. Kenne, G. Widmalm, and F. Lindh, *Carbohydr. Res.*, 1992, **235**, 69-81.

17.   J. Breg, D. Romijn, J. F. G. Vliegenthart, G. Strecker, and J. Montreuil, *Carbohydr. Res.*, 1988, **183**, 19-34.

18.   S. Kondo, Y. Sano, Y. Isshiki, and K. Hisatsune, *Microbiology*, 1996, **142**, 2879-2885.

19.   Y. Sano, S. Kondo, Y. Isshiki, T. Shimada, and K. Hisatsune, *Microbiology & Immunology*, 1996, **40**, 735-741.

20.   G. Nair, M. Albert, T. Shimada, and Y. Takeda, *Rev. Medic. Microbiol.*, 1996, **7**, 43-51.

21.   L. Kenne, B. Lindberg, P. Unger, B. Gustafsson, and T. Holme, *Carbohydr. Res.*, 1982, **100**, 341-349.

22.   T. Ito, T. Higuchi, M. Hirobe, K. Hiramatsu, and T. Yokota, *Carbohydr. Res.*, 1994, **256**, 113-128.

23.   L. E. Comstock, D. Maneval, P. Panigrahi, A. Joseph, M. M. Levine, J. B. Kaper, J. G. Morris, and J. A. Johnson, *Infect. Immun.*, 1995, **63**, 317-323.

24.   A. K. Mukhopadhyay, S. Garg, R. Mitra, A. Basu, K. Rajendra, D. Dutta, S. Battacharya, T. Shimada, T. Takeda, Y. Takeda, and G. Nair, *J. Clin. Microbiol.*, 1996, **34**, 2537-2543.

25.   Y. A. Knirel, L. Paredes, P. E. Jansson, A. Weintraub, G. Widmalm, and M. J. Albert, *Eur. J. Biochem.*, 1995, **232**, 391-396.

26.   L. M. Preston, Q. Xu, J. A. Johnson, A. Joseph, D. R. Manneval Jr, K. Husain, G. P. Reddy, C. A. Bush, and J. G. Morris Jr, *J. Bact.*, 1995, **177**, 835-838.

27.   Y. A. Knirel, G. Widmalm, S. N. Senchenkova, P.-E. Jansson, and A. Weintraub, *Eur. J. Biochem.*, 1997, **247**, 402-410.

28.   A. D. Cox, J. R. Brisson, V. Varma, and M. B. Perry, *Carbohydr. Res.*, 1996, **290**, 43-58.

29.   A. D. Cox and M. B. Perry, *Carbohydr. Res.*, 1996, **290**, 59-65.

30.   E. V. Vinogradov, K. Bock, O. Holst, and H. Brade, *Eur. J. Biochem.*, 1995, **233**, 152-158.

31.   A. D. Cox, J.-R. Brisson, P. Thibault, and M. B. Perry, *Carbohydr. Res.*, 1997, **304**, 191-208.

32.   Y. A. Knirel, S. N. Senchenkova, P.-E. Jansson, and A. Weintraub, *Carbohydr. Res.* In press, 1998.

33.   Y. A. Knirel, S. N. Senchenkova, P. E. Jansson, A. Weintraub, and M. J. Albert, *Eur. J. Biochem.*, 1996, **238**, 160-165.

34.   S. N. Senchenkova, G. V. Zatonsky, A. S. Shashkov , Y. A. Knirel, P.-E. Jansson, A. Weintraub, and M. J. Albert, *Eur J Biochem.*, 1998, **254**, 58-62.

35.   C. Landersjö, A. Weintraub, M. Ansaruzzan, M. J. Albert, and G. Widmalm, *Eur. J. Biochem*, 1998, **251**, 986-990.

STRUCTURE ACTIVITY RELATIONSHIPS OF SYNTHETIC HEPARIN-LIKE OLIGOSACCHARIDES

C.M. Dreef - Tromp, J.E.M. Basten and C.A.A. van Boeckel

N.V. Organon Scientific Development Group,
P.O. Box 20, 5340 BH Oss, The Netherlands

INTRODUCTION

Since 1936, heparin has been used in clinics for the prevention and treatment of thrombosis. Its main antithrombotic activity is explained by its ability to potentiate the activity of the serine protease inhibitor antithrombin III (AT-III), which inactivates a number of serine protease - such as thrombin and factor Xa- in the coagulation cascade[1].

By the end of the 1970's heparin fragments (obtained by chemical or enzymatic degradation) had been isolated by affinity chromatography on immobilised AT-III and the high affinity fractions had been analysed. From these studies it was deduced[2] in 1981 that a unique pentasaccharide fragment, that occurs in about one-third of the heparin polysaccharide chains, constitutes the minimal binding domain for AT-III. The pentasaccharide fragment (also known as the DEFGH part of heparin) was synthesised[3,4] a couple of years later to confirm the earlier proposal.

The synthetic pentasaccharide fragment; (compound **1**: ORG 31540/SR90107; see Figure 1) was found to elicit a very selective antithrombotic mode of action, in that it only accelerates the AT-III mediated inhibition of coagulation factor Xa but not that of thrombin[5].

Structure - Activity
Org 31540/SR90107

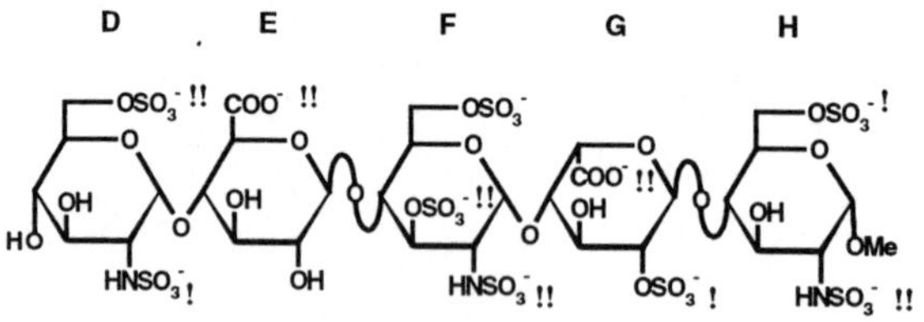

1

Figure 1

1 STRUCTURE-ACTIVITY RELATIONSHIPS (SAR) OF THE UNIQUE PENTA-SACCHARIDE DOMAIN OF HEPARIN

In an early stage it was recognised that the interaction of the unique pentasaccharide (PS) domain of heparin with AT-III should be highly specific in nature since other polyanions at similar concentrations cannot substitute for the interaction of pentasaccharide with AT-III. The specificity of the interaction of the sulphated pentasaccharide with the protein was confirmed when heparin pentasaccharide analogues were synthesised and tested[6] (in a joint venture between N.V. Organon and Sanofi) for inhibition of blood coagulation factor Xa. First it was established which of the charged groups play an important role in the activation of AT-III. It was found that some groups are strictly required (indicated with !! in compound 1) for the activation of AT-III while other groups (denoted ! in compound 1) contribute significantly during the AT-III activation. Taking into account these structure-activity relationships and by contemplating molecular modelling data we postulated a simplified heparin-AT-III interaction model. On the basis of this model we introduced[7] an extra sulphate group at position 3 of unit H of the naturally occurring fragment to give analogue 2 (see Figure 2). This extra-sulphated analogue displays higher affinity towards AT-III and an enhanced AT-III mediated ↑Xa activity (1250 U/mg for 2 vs. 700 U/mg for 1).

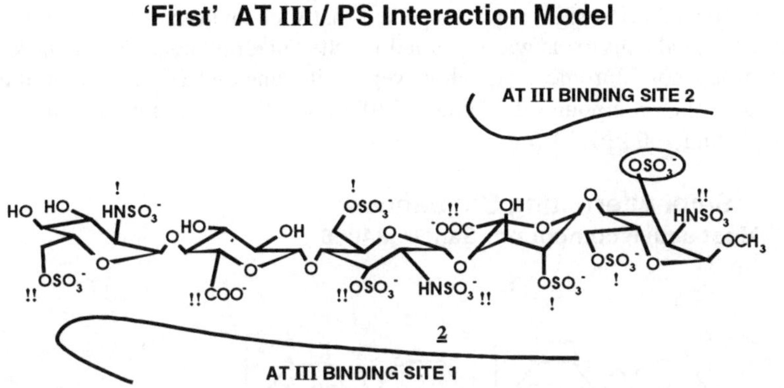

'First' AT III / PS Interaction Model

Figure 2

When the essential, negatively charged groups of the PS fragment were designated and analogues with higher activity were obtained, attention was turned to new simplified series[8,9] in which all hydroxyl groups are methylated and in which all the N-sulphate groups are replaced by O-sulphate groups (*e.g.* pentasaccharide 3 in Figure 3). To our surprise these modifications did not affect the biological activity of the PS. It should be stressed that the synthesis of these methylated analogues is much easier than that of heparin-like fragments, which is a distinct advantage in economically feasible drug development. In order to realise even shorter synthetic routes PS-analogues were designed comprising a "semi-alternating" sequence (*i.e.* disaccharide moiety EF is similar to GH).

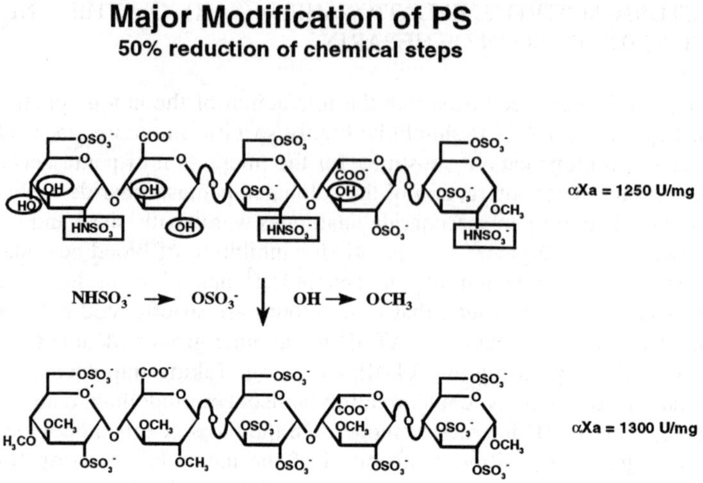

Major Modification of PS
50% reduction of chemical steps

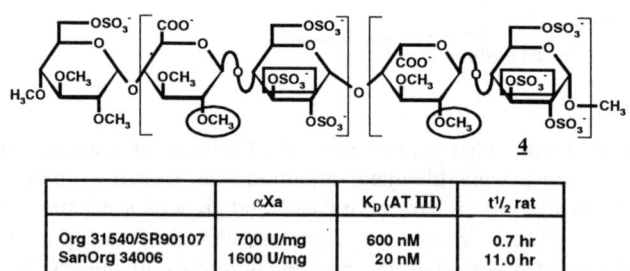

Figure 3

In this series we prepared several analogues methylated at the 2-O and 3-O positions of both uronic acid moieties. At first sight it was expected that such analogues would loose at least half of their biological activity as was observed for the "natural" counterparts lacking the 2-O-sulphate groups of iduronic acid. However, quite unexpectedly, one of these methylated analogues, (*i.e.* compound **4**: SanOrg 34006; see Figure 4) turned out to be highly potent[6], displaying 1600↑Xa U/mg.

Semi-alternating Sequences
Most active compound: SanOrg34006

	αXa	K_D (AT III)	$t^{1/2}$ rat
Org 31540/SR90107	700 U/mg	600 nM	0.7 hr
SanOrg 34006	1600 U/mg	20 nM	11.0 hr

Figure 4

The potent compound **4** not only binds much stronger to AT-III (K_D=20nM), relative to the PS (compound **1**, K_D= 600 nM), but also its elimination half-life (in rat and pig) is about fifteen fold longer. For many PS analogues it was found[9] that the elimination half life is proportional to the affinity of AT-III.

In order to find other possibilities to enhance the anti-Xa activity of PS we introduced different types of charged groups. However, only less active compounds were found. For instance, it turned out that the two essential carboxylate groups of the PS could not be replaced by sulphates[6]. In addition the essential 6-O sulphate group at the D-unit was replaced by various other charged groups (see Figure 5). Thus, replacement of the 6-O sulphate by phosphate[10] or carboxylate led to a complete loss of biological activity[11], while replacement with a methyl- phosphonate or methyl-carboxylate is accompanied[11] by at least 90% reduction of activity.

Substitution of Sulphate Group

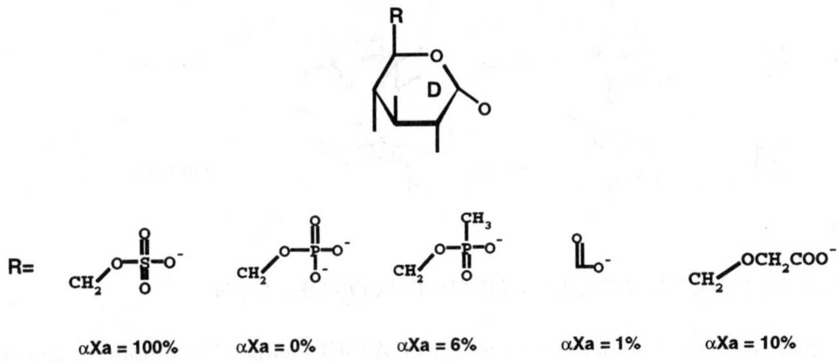

Figure 5

Apparently, the essential sulphate and carboxylate groups of the PS interact in a highly specific mode (probably *via* directed hydrogen-bridges) with AT-III, rather than through an ordinary Coulomb-interaction.

In Figure 6 various modifications are depicted to summarise some studies directed to the role of the carbohydrate moieties, with the purpose to simplify the synthesis. The units D, E, F and H of the PS are rigid glucopyranose moieties occurring exclusively in the $^4C_1$ chair conformation, whereas unit G is a relatively flexible idopyranose moiety which is present in an equilibrium between a skew-boat($^2S_0$) and a chair ($^4C_1$) conformation. As shown in entry *i*) the presence of the rigid D-unit is less critical as the modified analogue[12] with a flexible glycol-O-methyl moiety still shows 50% of the biological activity. On the other hand the rigidity of glucuronic acid moiety E is of crucial importance[6] (see entry *ii*). A similar modification[6] of iduronic acid moiety G is less harmful (see entry *iii*), a phenomenon which is in line with the flexible properties of this carbohydrate. An analogue containing the iduronic acid moiety G in a fixed $^1C_4$ conformation was also prepared[13] and found to be virtually inactive (see entry *iv*), meaning that either the flexibility of this unit or the occurrence of the $^2S_0$ conformation is essential.

Rigidity vs. Flexibility (PS)

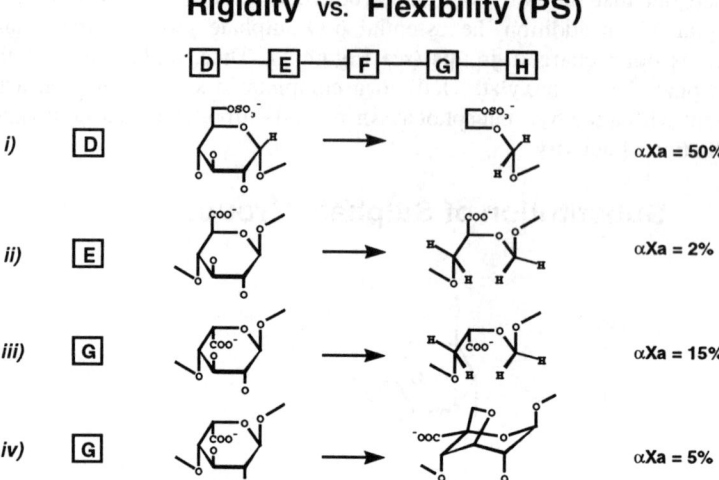

i) D   αXa = 50%

ii) E   αXa = 2%

iii) G   αXa = 15%

iv) G   αXa = 5%

Figure 6

2   SAR OF SYNTHETIC CONJUGATE MIMICS OF HEPARIN

The next challenge was to extend the concept of AT-III mediated inhibition of factor Xa by pentasaccharides towards synthetically feasible derivatives displaying both anti-factor Xa and anti-thrombin activity.

It is known that for AT-III mediated inhibition of thrombin a heparin fragment comprising at least 18 saccharide units is required to facilitate the binding of AT-III and thrombin to the same polysaccharide chain (the so called "bridge" or "template" mechanism).

Design of Synthetic Conjugates

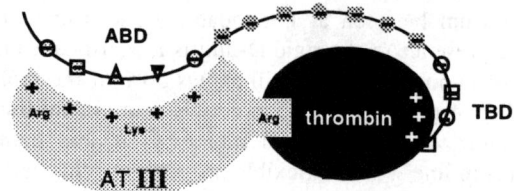

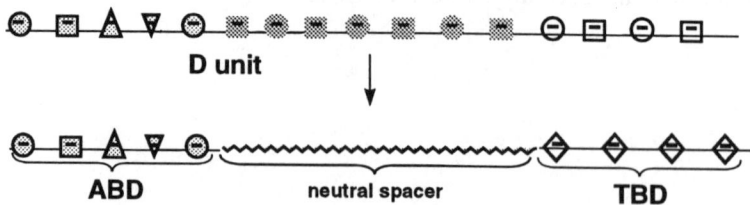

Figure 7

In the formation of the heparin/AT-III/thrombin ternary complex   the unique pentasaccharide   sequence interacts specifically with AT-III, while some sulphated oligosaccharide fragment along the heparin chain interacts in a less specific mode with thrombin. Our model[14]   of the ternary complex (see Figure 7) revealed that heparin analogues may be obtained when a thrombin binding oligosaccharide is tethered to the non-reducing terminus of the AT-III binding pentasaccharide   (*i.e.* the D-unit) with a neutral spacer of about 50 atoms in length. To this end glycoconjugates (*e.g.* compound **5** in Figure 8) were synthesised which comprise a PS as AT-III binding domain (ABD), a linear spacer and a persulphated oligosaccharide as thrombin binding domain (TBD).

Biological Activity

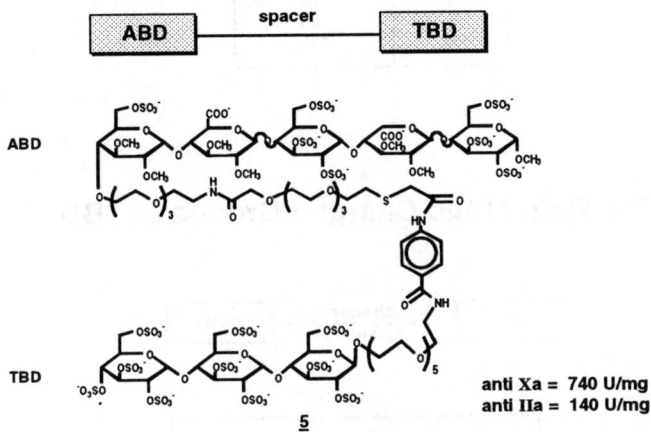

anti Xa = 740 U/mg
anti IIa = 140 U/mg

5

Figure 8

Compound **5** was one of the first conjugates that has been synthesised and which indeed displayed good to strong AT-III mediated anti-thrombin activity (**5** = 140 U/mg; heparin = 160 U/mg), besides the expected anti-factor Xa activity (740 U/mg).

Both the potency and the anti-factor Xa/anti-thrombin ratio of this new type of heparin like molecules can be adjusted in a rational way by varying the AT-III affinity of the PS (ABD), the TBD and the spacer length.

For example in Figure 9 the influence of the chain length of a conjugate on the anti-thrombin activity is shown; obviously the activity increases on elongation   of the molecular spacer thus facilitating the formation of the ternary complex[14].

The Role of the Spacer Length

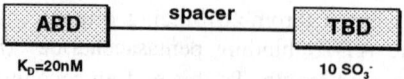

K_D=20nM

10 SO_3^-

spacer length	αIIa(U/mg)
18	1
32	15
46	20
59	120

Figure 9

The Role of the Charged Groups at TBD

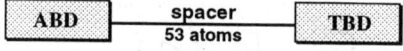

TBD		αIIa(U/mg)
dermatan	* 4 SO_3^-	2
dermatan	* 5 SO_3^-	10
heparin	* 6 SO_3^-	10
cellobiose	* 7 SO_3^-	10
cellobiose	* 7 PO_3^{2-}	26
maltotriose	*10 SO_3^-	65
maltotriose	*10 PO_3^{2-}	167
maltopentaose	*16 SO_3^-	330

Figure 10

In Figure 10 the effect of the charge density at the TBD on the anti thrombin activity is summarised. Whereas the interaction of the ABD moiety with AT-III is highly specific in nature the interaction of the TBD moiety with thrombin is largely determined by the charge density of the TBD[15]. It is noteworthy that phosphate groups at the TBD interact stronger with thrombin than the naturally occurring sulphate groups, whereas for the binding for the ABD (PS) to AT-III a phosphate instead of a sulphate group had a detrimental effect (*vide supra*).

Concerning the SAR of conjugates with heparin-like activity we were also eager to compare the anti-thrombin activity of conjugates containing a flexible polyethylene glycol spacer (*e.g.* compound **5**) with analogues having a rigid spacer between ABD and TBD[16]. To this end compound **6** (see Figure 11) was synthesised, comprising a fully methylated maltose-octasaccharide spacer.

The Role of the Spacer Flexibility

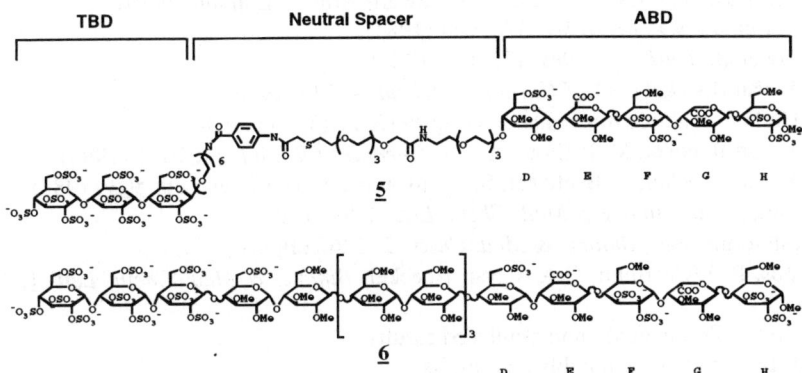

Figure 11

The replacement of the flexible polyethylene glycol spacer (*i.e.* compound **5**) by a rigid oligosaccharide spacer (*i.e.* compound **6**) increases the anti -thrombin activity at least tenfold (anti -thrombin activity: **5** = 140 U/mg, **6** = 1900 U/mg), while only a small difference in anti factor Xa activity was observed.

In addition we compared the interaction of the heparin mimics **5** and **6** in their ability to bind to platelet factor 4 (PF4). This side-reaction may be of relevance for the induction of life threatening heparin associated thrombocytopenia (HAT) since the antigen in HAT consists of complexes of heparin or sulphated oligosaccharides with PF4. As can be seen in Table 1 the rigid analogue **6** interacts less favourably with PF4 than the flexible conjugate **5**. This observation is in line with studies described by Greinacher[17] who demonstrated that the interaction between PF4 and sulphated polysaccharides enhances upon increasing the charge density and flexibility of the saccharide.

Compound	PF4 neutralisation
heparin	1
5	0.4
6	0.05

Table 1

CONCLUSION

Starting from an unique pentasaccharide domain in heparin various well defined synthetic antithrombotics could be obtained displaying a tailor made profile with respect to anti-factor Xa and anti-thrombin activities as well as their half-life in circulation. Several of these compounds are now under clinical investigation.

ACKNOWLEDGEMENT
The authors thank all the people who have been collaborating on this project most of whom are mentioned as (co)authors in the list of references.

REFERENCES

1. Heparin (Eds.:D.A. Lane, U. Lindahl), Edward Arnold, London (1989).
2. J. Choay et al., *N.Y. Acad, Sci.* 370, 644 (1981).
3. P. Sinay et al., *Carbohydr.Res.* 132, C5 (1984).
4. C.A.A. van Boeckel et al., *J.Carbohydr.Chem.* 4, 293 (1985).
5. J.M. Herbert et al., *Cardiovascular Drug Reviews* 15, 1 (1997).
6. C.A.A. van Boeckel, M. Petitou, *Angew.Chem.Int.Ed.Engl.* 32, 1671 (1993).
7. C.A.A. van Boeckel, T, Beetz and S.F. van Aelst, *Tetrahedron Lett.* 803 (1988)
8. G.Jaurand, et al., *Bioorg & Med. Chem. Lett.* 2, 897 (1992).
9. P. Westerduin et al., *Bioorg. & Med. Chem.* 2, 1267 (1994).
10. J.N. Vos, P. Westerduin, C.A.A. van Boeckel, *Bioorg. & Med. Chem. Lett.* 1, 143 (1991).
11. C.M. Dreef-Tromp et al., non published results.
12. J.E.M. Basten et al., non published results.
13. N. Sakairi et al., *Chem, Eur. J.* 2, 1007 (1996).
14. P.D.J. Grootenhuis et al., *Nature Struct. Biol.*, 2 736 (1995).
15. J.E.M. Basten, C.M. Dreef-Tromp, B. de Wijs and C.A.A. van Boeckel, *Bioorg. & Med. Chem. Lett.* 8, 1201 (1998).
16. C.M. Dreef-Tromp et al., *Bioorg. & Med. Chem. Lett.*, in press.
17. A. Greinacher et al., *Thromb. Haemost*, 74, 886 (1995).

Part of this work was done in collaboration with Sanofi Recherche.

GLYCOPEPTIDES: A LINK BETWEEN COMPLEX CARBOHYDRATES AND GLYCOPROTEINS

P. M. St. Hilaire, M. Meldal and K. Bock

Carlsberg Laboratory, Department of Chemistry, Gamle Carlsberg Vej 10, DK-2500 Valby, Denmark

1 GLYCOCONJUGATES

The glycosylation of proteins and lipids is ubiquitous in Nature and is of vital importance to the proper functioning of biological systems. It is well-established that glycoconjugates on cell surfaces play a vital role in a wide variety of biological phenomena including immune response, intercellular recognition, cellular adhesion, intracellular targeting, cell growth regulation, cancer cell metastasis, and inflammation.[1,2] A plethora of viral, bacterial, mycoplasmal and parasitic infections are also mediated by the tight associative interaction of glycoconjugates with a protein receptor. As a consequence, there has been a tremendous impetus to develop carbohydrate-based therapeutics.[3] However, the isolation, purification, structural characterization and subsequent synthesis of complex oligosaccharides, which are responsible for many important biological interactions, represents each a major research project which requires special skills and expertise, particularly when labile residues like sialic acid are present in the parent structure. In contrast to other proteins, specifically glycosylated proteins are not easily available by gene-technology since they are postranslational products resulting from the activity of glycohydrolases and transferases. Chemical synthesis of such complex oligosaccharides presents a serious challenge and the organic synthesis of about 100 mg. of an oligosaccharide larger than a pentasaccharide is a major achievement. In many cases, only a few residues at the non-reducing end of a complex glycan are necessary for the interaction to the receptor. Therefore, the use of simplified synthetic molecules, glycopeptides, that can be rapidly generated and can mimic the natural ligand can give important information about the nature and topology of the ligand-receptor interaction and such compounds may serve as leads in the development of drugs.[4]

1.1 Glycopeptide Mimics of Complex Oligosaccharides

It has previously been demonstrated that synthetic glycopeptides can mimic the interaction of a complex oligosaccharide and its receptor in biological assays.[5-15] These results are in some cases, most likely due to a topological similarity of the two types of compounds. In one example, linear tripeptides bearing two 6'-O-phosphorylated mannose disaccharides linked $\alpha(1\rightarrow2)$ were found to be 10 times tighter binders of the mannose-6-phosphate receptor (MPR) than one of its natural ligands, a branched mannose pentasaccharide.[5] However, when cyclic peptides were used to present 6-O-phosphorylated α-D-Man-1$\rightarrow$2-α-

D-Man ligands to the MPR, the binding affinity was much reduced compared to their linear analogs.[6]

Glycopeptide mimetics of SLex for the inhibition of selectin binding have also been developed. Surprisingly, these glycopeptide ligands show increased binding compared to the original SLex ligand, probably due to favourable interaction of the peptide scaffold with the receptor. A trivalent SLex tetrasaccharide derivative attached to a rigid cyclic heptapeptide scaffold showed a 2-3 fold (on a per mole of SLex basis) enhanced inhibition capability,[7] while linear displays of SLex on different scaffolds showed no increase in selectin binding.[8,9] The best results come from a series of glycopeptide mimetics containing Fuc, Man or Gal residues which demonstrated up to 40-fold increase in potency over the parent oligosaccharide and were much more facile to prepare.[10-14]

2 GLYCOPEPTIDE SYNTHETIC STRATEGIES

The synthesis of glycopeptides has evolved to a highly skilled field with only few synthetic challenges remaining in assembly of natural glycopeptides. In the chemical synthesis of glycopeptides it has been demonstrated that the most versatile methodology for chemical assembly is the use of appropriately glycosylated amino acid building blocks in solid phase peptide synthesis.[16] Strategies developed for the synthesis of various types of glycopeptides are summarized in Figure 1 and discussed below.

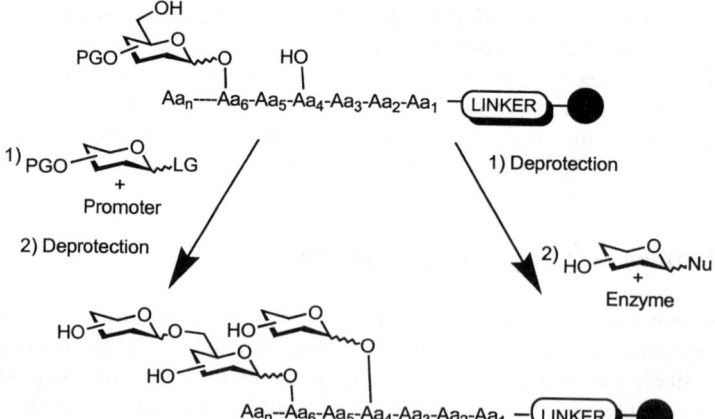

a) **Strategy 1**: The building block approach (X = O,C, S, N); Y = suitable leaving group.

b) **Strategy 2**: Chemical or enzymatic glycosylation of peptide or glycopeptide; Nu = nucleotide.

Figure 1 *Strategies developed for glycopeptide synthesis.*

2.1 Synthesis of *O*-Linked Mucin Core Oligosaccharides

A variety of different schemes have been presented for the synthesis *O*-linked mucin core oligosaccharides all characterized by the presence of a GalNAc residue α-linked to Ser or Thr. These are predominantly based on the introduction of GalN₃ onto a Fmoc-protected amino acid derivative, followed by elongation of the glycan structure[17-19] or alternatively the use an oligosaccharide building block with a GalN₃ reducing end residue as a glycosyl donor.[20,21] The azido protecting groups often employed as a non-participation group in α-glycosylation reactions are readily reduced with DDT[19] or thioacetic acid[22] on the solid phase and can even be used for solid phase peptide and glycopeptide synthesis employing α-azido acids.[23,24] Alternatively, the use of the β-directing 2-*N*-dithiasuccinyl protecting group for amino sugar glycosylation reactions which allowed mild thiolytic reduction with DDT or even selectively in the presence of azides with propanedithiol on the solid phase[19,25-27] to liberate the amino functions afforded a range of the GlcNAc containing mucin core glycosyl amino acids as Fmoc-Ser/Thr-OPfp esters activated and ready to use directly in multiple glycopeptide assembly. In solution on the building block stage both the azido- and Dts-functionalities can most conveniently be converted to *N*-acetates by reduction with Zn in the presence of acetic anhydride.[17] The tetrachlorophthaloyl[28] or the trichloroethoxycarbonyl group may be also used for amino sugar protection and β-glycosylation, especially for *O*-GlcNAc glycopeptides using the Pfp ester approach.[29,30] Another difficulty has been to generate acyl protected α-fucosyl amino acid derivatives which in contrast to the ether protected fucosides are sufficiently stable during the acidic conditions of cleavage from the resin. This problem was solved by a non-stereoselective glycosylation using fucose peracetate and Lewis acid.[31]

2.2 Synthesis of *N*-Linked Oligosaccharides

The synthesis of *N*-linked glycosyl building blocks has been predominantly achieved by coupling of the glycosylamine to aspartic acid. However, the use of the glycosyl azides as precursors for glycosylamines have been found to be a very versatile approach.[32,33] The release of complex glycans by chemical means followed by transformation into building blocks have provided a novel method of preparing large amounts of pure *N*-linked glycopeptides with a single glycoform.[34] Recently, complex *N*-linked glycopeptides have been prepared chemo-enzymatically (see section 2.4) or on solid phase *via* glycal assembly.[35]

2.3 Solid Phase Glycosylation of Peptides

Solid phase glycosylation reactions have been highlighted in the recent years particularly by use of the highly reactive glycosylsulfoxides or trichloroacteimidates which can both be used under conditions of homogeneous catalysis. Glycosylation of saccharides linked to solid phase either directly or as a glycopeptide has been achieved and can be quantitative especially on non-polar resins.[36] However, polar resins are required for solid phase bioassays and methods were developed for glycosylation on a range of polar resins.[36] Direct peptide glycosylation on solid phase has the potential to generate libraries of templated glycopeptides; however, for unresolved reasons, direct glycosylation of peptides have proven much more difficult to achieve. Quantitative peptide glycosylation was achieved on a novel PEG-based resin containing only ether bonds[37] and a small model library comprising four compounds was generated by two consecutive glycosylation reactions.[38,39]

2.4 Chemo-Enzymatic Assembly of Glycopeptides

Chemo-enzymatic glycopeptide assembly has become increasingly more important to facilitate the access to more complex glycan structures, while chemical synthesis is still preferred when larger amounts of the compound is required for biochemical studies. Employing a building block approach for chemical assembly of simple glycopeptides enzymes have been used for the subsequent modification to yield more complex glycans.[32,40] Endoglycosidases,[41-43] particularly Endo M[42] and Endo A,[43] have been used to transfer large pools of oligosaccharides from a natural source of glycoprotein or a purified fragment to a synthetic glycopeptide target. The method requires a fairly uniform glycan structure in order to yield well defined glycopeptides and both high mannose and complex glycans can be transferred. Alternatively the glycan linked to an asparagine residue can be manipulated with glycosyltransferases.[40] The great potential of using glycosyltransferases to build up the saccharides on a simple glycopeptide core was elegantly demonstrated by the construction of ribonuclease B carrying an *N*-linked SLex tetrasaccharide[32] and by the assembly of trivalent SLex glycopeptide templates.[9] Furthermore it was demonstrated that the protein backbone could be enzymatically cleaved and then re-ligated to yield the parent protein. Recently, the ligation of glycopeptide fragments obtained from solid and solution phase synthesis, have been ligated using subtilisin.[44] This technique of enzymatic manipulation, though requiring the availability of large amounts of enzymes and of glycosyl nucleotide donors, has a great potential for the construction of very complex compounds of biochemical interest.

3 GLYCOPEPTIDE LIBRARIES: SYNTHESIS AND ANALYSIS

The glycopeptide building block approach used in portion mixing library synthesis allows the expedient formation of numerous glycopeptides as putative ligands in protein binding assays. However, rapid and unambiguous analysis of modified peptides on solid phase remains challenging. A search of the literature reveals a paucity of papers concerning the synthesis of glycopeptide libraries due in part to the difficulties in identifying the components of a library. One successful glycopeptide library synthesis has been developed and includes an analysis based on *ladder synthesis*, an analytical technique which involves capping a small portion of the growing oligomer during synthesis.[45] The "ladder" of peptide fragments is generated by an *encoded in-situ capping* methodology that allows immediate distinction between glycopeptide residues of identical masses and features rapid *on-bead* MALDI-TOF mass analysis facilitated by use of a photolabile linker. Glycopeptide ligands were identified for the C-type lectin from *Lathyrus odoratus* by screening the fluorescent labelled protein in a solid phase binding assay of the PEG-sarcosine resin-bound glycopeptide library.[46] Of the several glycopeptide ligands detected, most contained mannose and *N*-Acetylglucosamine, glycans that display a specificity for the lectin in solid phase assays. The most active glycopeptides detected were **T(α-D-Man)FFFVNKV** and **T(α-D-Man)LFKGFHV**.

4 GLYCOPEPTIDES INTERACTING WITH MHC CLASS II

The above outlined methodology has been used to synthesize a long series of glycopeptides related to the non immunogenic peptide fragment from haemoglobin 86-96, which has been demonstrated to bind strongly to MHC II, but not been able to trigger T-cell.[47] A systematic scan of this structure with glycosylated amino acid building blocks (GalNAc-α-Ser/Thr) revealed that particularly position 92 with a asparagine residue could be substituted by glycosylated amino acids and at the same time be able to stimulate T-cell proliferation.[48] Several such peptides were therefore synthesized in order to map out the size and specificity of these interactions.[49] It shown that particularly glycopeptides substituted with T- and Tn- antigens were active, whereas larger structures or different basic sugars structures rendered this stimulation absent.[50,51]

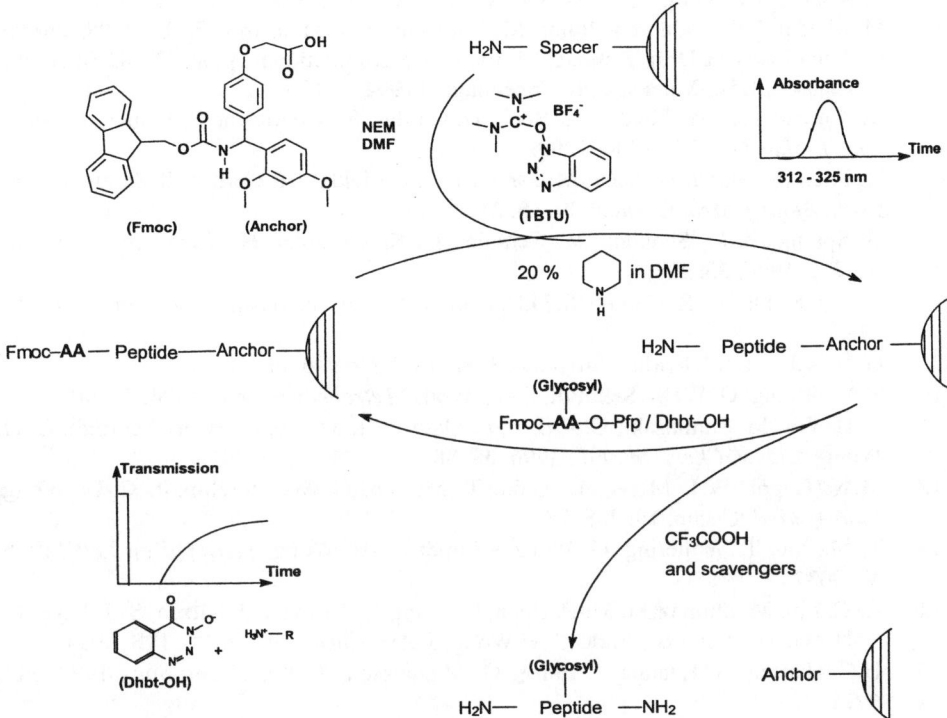

Figure 2 *Automated solid phase synthesis of glycopeptides*

5 CONCLUSIONS

Glycopeptide synthesis has evolved to a high level of sophistication amenable to automated procedures (Figure 2). It has clearly been demonstrated above that access to synthetic glycopeptide structures in relatively large amounts made it possible to study protein-carbohydrate interactions in greater detail and more expediently than using a traditional approach. It remains to be shown whether the application of glycopeptide libraries will speed up this process even more; and substantial resources are currently being allocated for such studies in our laboratory

6 ACKNOWLEDGEMENTS

The many former post doctoral researchers and students over the last 10 years, who have contributed to the progress in glycopeptide synthesis are thanked for their dedication and enthusiasm.

REFERENCES

1.   A.Varki, *Glycobiology,* 1993, **3**, 97.
2.   R. Dwek, *Chem. Rev.*, 1996, **96**, 683.
3.   P. Sears, C.-H. Wong, *J. Chem. Soc., Chem Commun*, 1998, 1161.
4.   M. Meldal, I. Christiansen-Brams, M. Christensen, S. Mouritsen, K. Bock. "Complex Carbohydrates in Drug Research. Structural and Functional Aspects", K. Bock and H. Clausen, Eds.; Munksgaard, Copenhagen, 1994, p. 153.
5.   M. Christensen, M. Meldal, K. Bock, H. Cordes, S. Mouritsen, H. Elsner, *J. Chem. Soc., Perkin Trans. 1,* 1994, 1299.
6.   H. Franzyk, M. Christensen, M. Jørgensen, M. Meldal, H. Cordes, S. Mouritsen, K. Bock, *Bioorg. Med. Chem.*, 1997, **5**, 21.
7.   U. Sprengard, M. Schudok, W. Schmidt, G. Kretzschmar, H. Kunz, *Angew. Chem. Int. Ed.,* 1996, **35**, 321.
8.   A. Liu, K. Dillon, R. Campbell, D.C. Cox, D. M. Huryn, *Tetrahedron Lett.,* 1996, **37**, 3785.
9.   G. Baisch and R. Oehrlein, *Angew. Chem. Int. Ed.,* 1996, **35**, 1812.
10.   T. Woltering, G. Weitz-Schmidt, C.-H. Wong, *Tetrahedron Lett.,* 1996, **37**, 9033.
11.   S.-H. Wu, M. Shimazaki, C.-C. Lin, L. Qaio, W.J. Moree, G. Weitz-Schmidt, C.-H. Wong, *Angew. Chem. Int. Ed.,* 1996, **35**, 88.
12.   M.W. Cappi, W.J. Moree, L. Qaio, T. Marron, G. Weitz-Schmidt, C.-H. Wong, *Bioorg. Med. Chem.*, 1997, **5**, 283.
13.   T. Marron, T. Woltering, G. Weitz-Schmidt, C.-H. Wong, *Tetrahedron Lett.,* 1996, **37**, 9037.
14.   C.-C. Lin, M. Shimazaki, M.-P. Heck, R. Wang, T. Kimura, H. Ritzen, S. Takayama, S.-H. Wu, G. Weitz-Schmidt, C.-H Wong, *J. Am. Chem. Soc.,* 1996, **118**, 6826.
15.   H.C. Hansen, S. Haataja, J. Finne, G. Magnusson, *J. Am. Chem. Soc.*, 1997, **119**, 6974.
16.   For a recent review see: M. Meldal and P.M. St.Hilaire, *Curr. Opin. Chem. Biol.*, 1997, **1**, 552.
17.   E. Meinjohanns, M. Meldal, H. Paulsen, A. Schleyer, K. Bock, *J. Chem. Soc., Perkin Trans. 1,* 1996, 985.
18.   N. Mathieux, H. Paulsen, M. Meldal, K. Bock, *J. Chem. Soc., Perkin Trans. 1,* 1997, 2359.
19.   E. Meinjohanns, M. Meldal, T. Jensen, O. Werdelin, L. Galli-Stampino, S. Mouritsen, K. Bock, *J. Chem. Soc., Perkin Trans. 1,* 1997, 871.
20.   K. Frische, M. Meldal, O. Werdelin, S. Mouritsen, T. Jensen, L. Galli-Stampino, K. Bock, *J. Pept. Sci.*, 1996, **2**, 212.

21. J. Rademann and R. Schmidt, *Carbohydr. Res.,* 1995, **269**, 217.
22. G. Lich, H. Paulsen, B. Meyer, M. Meldal, K. Bock, *Carbohydr. Res.,* 1997, **299**, 33.
23. M. Meldal, M.A. Juliano, A.M. Jansson, *Tetrahedron Lett.,* 1997, **38**, 2531.
24. L. Lay, M. Meldal, F. Nicotra, L. Panza, G. Russo *J. Chem. Soc., Chem Commun,* 1997, 1469.
25. E. Meinjohanns, M. Meldal, H. Paulsen, K. Bock, *J. Chem. Soc., Perkin Trans. 1,* 1995, 401.
26. E. Meinjohanns, A. Vargas-Berenguel, M. Meldal, K. Bock, *J. Chem. Soc., Perkin Trans. 1,* 1995, 2165.
27. K. Jensen, P. Hansen, D. Venugopal, G. Barany, *J. Am. Chem. Soc.,* 1996, **118**, 3148.
28. J. S. Debenham, S. D. Debenham, B. Fraser-Reid, *Bioorg. Med. Chem.,* 1997, **4**, 1909.
29. E. Meinjohanns, M. Meldal, K. Bock, *Tetrahedron Lett.,* 1995, **36**, 9205.
30. U. K. Saha and R. R. Schmidt, *J. Chem. Soc., Perkin Trans. 1,* 1997, 1855.
31. M. Elofsson, S. Roy, L.A. Salvador, J. Kihlberg, *Tetrahedron Lett.,* 1996, **37**, 7645.
32. C. Unverzagt, *Angew. Chem. Int. Ed.,* 1996, **35**, 2350.
33. Z. Györgydeak, L. Szilágyi, H. Paulsen, *J. Carbohydr. Chem.,* 1997, **12**, 139.
34. E. Meinjohanns, M. Meldal, H. Paulsen, R.A. Dwek, K. Bock, *J. Chem. Soc., Perkin Trans. 1,* 1997, 549.
35. J. Y. Roberge, X. Beebe, S. J. Danishefsky, *J. Am. Chem. Soc.,* 1998, **120**, 3915.
36. H. Paulsen, A. Schleyer, N. Mathieux, M. Meldal, K. Bock, *J. Chem. Soc., Perkin Trans. 1,* 1997, 281.
37. M. Renil and M. Meldal, *Tetrahedron Lett.,* 1996, **37**, 6185.
38. A. Schleyer, M. Meldal, M. Renil, H. Paulsen, K. Bock, *Angew. Chem. Int. Ed.,* 1997, **109**, 2064.
39. M. Meldal, M. Renil, M.A. Juliano, A.M. Jansson, E. Meinjohanns, J. Buchardt, A. Schleyer. "Peptides 1996. Proceedings of the 24th European Peptide Symposium", R. Epton. and R. Ramage, Eds.; Mayflower Scientific Ltd., Kingswinford, 1998, p.141.
40. K. Witte, P. Sears, R. Martin, C.-H. Wong, *J. Am. Chem. Soc.,* 1997, **119**, 2114.
41. K. Haneda, T. Inazu, M. Mizuno, K. Yamamoto, K. Fujimori, H. Kumagai, *Pept. Sci.* 1996, **34** 113.
42. K. Haneda, T. Inazu, K. Yamamoto, H. Kumagai, Y. Nakahara, A. Kobata, *Carbohydr. Res.,* 1996, **292**, 61.
43. a) L.X. Lee, J.Q. Fan, Y.C. Lee, *Tetrahedron Lett.,* 1996, **37**, 1975. b) I.L. Deras, K. Takegawa, A. Kondo, I. Kato, Y.C. Lee, *Bioorg. Med. Chem. Lett.,* 1998, **8**, 1763.
44. K. Witte, O. Seitz, C.-H. Wong, *J. Am. Chem. Soc.,* 1998, **120**, 1979.
45. P.M. St. Hilaire, T. Lowary, M. Meldal, K. Bock. "Peptides 1996. Proceedings of the 24th European Peptide Symposium", R. Epton. and R. Ramage, Eds.; Mayflower Scientific Ltd., Kingswinford, 1998, p. 817.
46. P.M. St. Hilaire, M. Meldal, K. Bock, "Peptides 1997. Proceedings of the 15th American Peptide Symposium" *In press.*
47. R.G. Lorenz, P.M. Allen, *Immunol. Reviews,* 1988, **106**, 115.
48. T. Jensen, L. Galli-Stampino, S. Mouritsen, K. Frische, S. Peters, M. Meldal, O. Werdelin, *Eur. J. Immunol.,* 1996, **26**, 1342.

49. K. Frische, M. Meldal, O. Werdelin, S. Mouritsen, T. Jensen, L. Galli-Stampino, K. Bock, *J. Pept. Sci.*, 1996, **2**, 212.
50. L. Galli-Stampino, E. Meinjohanns, K. Frische, M. Meldal, T. Jensen, O. Werdelin, S. Mouritsen, *J. Cancer. Res.*, 1997, **57**, 3214.
51. T. Jensen, P. Hansen, L. Galli-Stampino, S. Mouritsen, K. Frische, E. Meinjohanns, M. Meldal, O. Werdelin, *J. Immunol.*, 1997, 3769.

Nitric Oxide Synthase Inhibition

SELECTIVE INHIBITION OF NITRIC OXIDE SYNTHASES

Hui Huang, Younghee Lee, Henry Q. Zhang, Walter Fast, Brigit Riley, and Richard B. Silverman*

Department of Chemistry, Department of Biochemistry, Molecular Biology, and Cell Biology, and the Drug Discovery Program, Northwestern University, Evanston, Illinois 60208-3113, USA

1 INTRODUCTION

Nitric oxide (NO) has been the object of study for more than 200 years since its discovery by Joseph Priestley in 1772 (two years before his discovery of molecular oxygen).[1] Despite many years of research on NO as an intermediate in commercial nitrate production, a component of photochemical smog, and its implication in the depletion of the ozone layer, the relatively recent interest in NO has arisen not from ecological concerns, but from the surprising discovery that NO serves as a biological messenger and is synthesized *in vivo* from oxygen and L-arginine by the enzyme nitric oxide synthase[2] (NOS, E.C.1.14.13.39; Scheme 1).

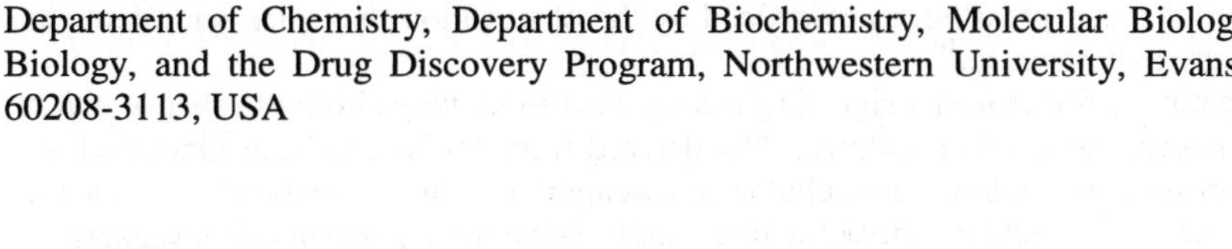

L-Arginine → (O₂, NADPH, NOS) → L-Citrulline + :N=O:

Scheme 1

1.1 Nitric Oxide Synthase Isozymes

NO has been implicated in numerous biological functions,[3] however, most systems can be grouped into one of three paradigms for NO function which correlate with three known isoforms of NO synthase: the immune response (inducible macrophage NOS, iNOS), smooth muscle relaxation (endothelial NOS, eNOS), and neuronal signaling (neuronal NOS, nNOS). In the immune response, cytokines induce the transcription of iNOS in macrophages, leading to high levels of NO production. NO can freely diffuse into an adjacent tumor cell and recombine with the tyrosyl radical of ribonucleotide reductase, stopping DNA synthesis.[4] NO (or its metabolites[5]) can also react with iron sulfur clusters in aconitase and complexes I and II of the mitochondrial electron transport chain, an accumulative effect that blocks new ATP formation.[6] Homozygotic iNOS knockout mice show drastically increased susceptibility to infection by the parasite *Leishmania major*, as well as other immune response abnormalities.[7]   The antimicrobial properties of NO may even account for some of the healing properties of dog saliva: nitrite, secreted in canine saliva, can be acidified on contact with skin, forming NO non-enzymatically.[8]   While beneficial for defense against invading microorganisms, production of NO by iNOS also can have deleterious effects, namely, in animal models of septic shock, the induction of iNOS has been shown to be responsible for the increased vasodilation and resulting hypotension.

Smooth muscle relaxation is regulated by NO produced by the endothelial isoform (eNOS).   NO interacts with the heme of guanylate cyclase forming a ferrous-nitrosyl, six coordinate complex which activates the cyclase activity.[9]   This pathway is responsible for the regulation of blood pressure, as evidenced by the observation that mice with disrupted eNOS genes are hypertensive.[10]

The NO pathway for neuronal signaling is suspected to be very similar to that of eNOS, yet nNOS has some distinctive features.   NO derived from nNOS has been identified as a retrograde messenger and/or intracellular messenger in the formation of olfactory memories in sheep.[11]   nNOS knockout mice show behavioral abnormalities (aggressive behavior and excess, inappropriate sexual behavior).[12]   In pathophysiology, nNOS derived NO exacerbates acute ischemic injury during stroke while vascular NO seems to have the opposite effect.[13]   The effects of NO on ischemic brain injury appear to be either protective or destructive depending on the stage of the ischemic injury or the source of NO.[14]

The three isoforms of NO synthase have all been cloned and overexpressed.[15] Interspecies sequence similarity between NOS is quite high (81-93% identical), but similarity between each of the three isoforms within a species is much lower (51-57% identical), suggesting specific functions, interactions, or specificity for each isoform.

1.2 Cofactor Requirements and Enzyme Domains

NOS has been shown to bind a number of cofactors, including one FAD, one FMN, one protoporphyrin IX, and one tetrahydrobiopterin per monomer.   Each isoform is also calmodulin dependent, although iNOS has been shown to bind calmodulin tightly regardless of Ca^{2+} concentration fluctuations.[16]   Inspection of the sequence allows the identification of two major domains of NOS, the C-terminal and the N-terminal domain. The C-terminal domain of NOS shows a high sequence similarity to cytochrome P450 reductase and contains sequences for FAD, FMN, and NADPH binding sites (Figure 1).[17] This domain serves as an adapter from the obligate two electron donor, NADPH, through the one or two electron acceptor / donors, FAD and FMN, to the one electron acceptor, heme.   In the resting state, an air stable flavin semiquinone (presumably FMNH•) has been detected by EPR.[18]

Unlike the C-terminal domain, the N-terminal domain, which binds the heme and tetrahydrobiopterin,[19] does not show any sequence similarity to known cytochrome P450 related proteins.   In the resting state, NOS contains a ferric, five coordinate heme with a cysteine thiolate as the proximal axial ligand.   The N-terminal and C-terminal domains are joined by a short sequence encoding a calmodulin binding site which serves as a sort of switch. Ca^{2+}/calmodulin are required to allow electron transfer between the C-terminal and N-terminal domains.   Lack of calmodulin binding abolishes electron transfer to the N-terminal domain.[20]   However, if the two domains are held in the proper orientation / distance by bound Ca^{2+} / calmodulin,[20,21] reducing equivalents are allowed to pass from the C-terminal flavin domain to the N-terminal heme / oxygenase domain.

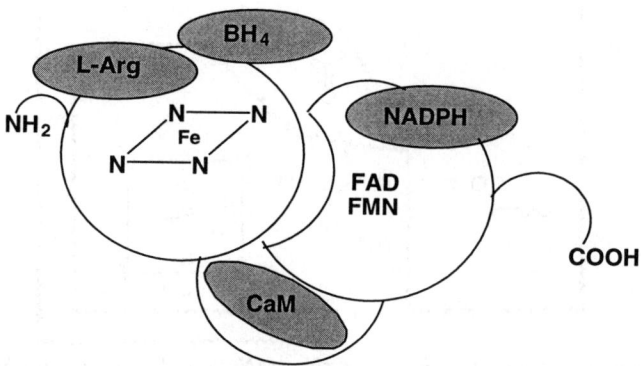

Figure 1 *Schematic diagram of the structure of nitric oxide synthase*

1.3 Crystal Structures

Recently, two published X-ray crystal structures have been important in elucidating the structural basis behind NOS catalysis. Wang et al.[22] have reported the structure of NADPH-cytochrome P450 reductase and Crane et al. [23] have reported the structure of a portion of the N-terminal domain of murine iNOS. The heme environment is particularly interesting with regard to determining the chemical mechanism and designing potent and/or selective inhibitors. A cysteine residue serves as the distal iron ligand, however, the charge density of this thiolate is presumably reduced from that of typical P450s, due to additional hydrogen bonds to this cysteine from a peptide nitrogen, an indole nitrogen from a tryptophan, and and a guanidino nitrogen from an arginine. The authors suggest that the reduced electronegativity of the distal heme ligand may favor formation of the peroxo-iron species like the peroxidases rather than the oxo-iron like the P450s. The distal heme environment consists of isoform conserved hydrophobic residues, with some more polar residues located at the pocket's edge. No sidechains that could serve as a general acid during the presumed formation of an Fe^V=O species are located at an appropriate distance from the heme iron.

Two crystal structures of iNOS with complexed ligands are reported, one containing two imidazoles, and one containing one imidazole and one aminoguanidine, a mechanism based inactivator[24] of iNOS.[23] In both structures, one imidazole provides a sixth iron ligand, tilted 10⁰ from the heme normal toward the second ligand binding site, and making no hydrogen bonds with any protein residues. The second ligand binding site contains either the second imidazole or the aminoguanidine. The guanidino group is tilted 45⁰ from the heme plane with both of the terminal nitrogens forming a hydrogen bond to a backbone carbonyl and one of the terminal nitrogens hydrogen bonding to a glutamate residue (Figure 2).

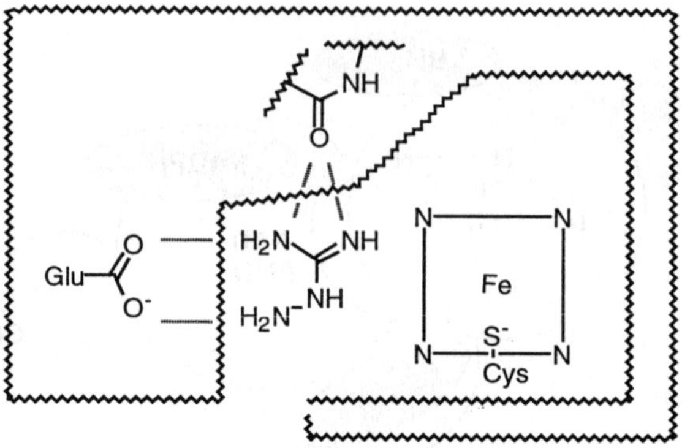

Figure 2 *Schematic diagram of the active site of iNOS with aminoguanidine bound*

The other terminal nitrogen atom is pointed toward the distal heme pocket and is positioned over the meso carbon between pyrrole rings A and B. The third guanidino nitrogen is pointed toward the pocket opening and the amino nitrogen is hydrogen bonding to the same glutamate. Modeling suggests that the arginine backbone also protrudes from the pocket in this direction. Considering the extensive hydrogen bonding of one of the terminal guanidino nitrogens versus that of the other, this structure is consistent with substrate studies on N^ω-substituted arginines that show that oxidation occurs exclusively at the functionalized terminal guanidino nitrogen.[25]

1.4 Relevance of NOS to Disease States

Overproduction of NO has been implicated in a wide variety of diseases.[26] NO overproduction by nNOS has been associated with strokes,[27] septic shock,[28] seizures,[29] schizophrenia,[30] migraine headaches,[31] and Alzheimer's disease.[32] iNOS overproduction of NO has been associated with tolerance to and dependence on morphine,[33] development of colitis,[34] tissue damage and inflammation,[35] overproduction of osteoclasts, leading to osteoporosis, Paget's disease, and rheumatoid arthritis,[36] and destruction of photoreceptors in the retina.[37]

1.5 Importance of NOS Inhibition to Treatment for Diseases

The fact that NO overproduction can be involved in the etiologies of various diseases suggests that inhibition of NOS could be a significant beneficial approach to the treatment of these disease states. A wide variety of compounds have been shown to inhibit NOS in general.[38] However, because of the importance of NO to human health as well as its potential involvement in disease states, *selective* inhibition of the isoforms of NOS is essential. Selectivity for the three isoforms already has been reported to some degree. The early inhibitors were analogues of L-arginine. N^ω-Methyl-L-arginine (**1**, R = Me) and N^ω-ethyl-L-arginine (**1**, R = Et), however, show only about a factor of 2 selectivity; N^ω-methyl-L-arginine is selective for eNOS and nNOS over iNOS and N^ω-ethyl-L-arginine is selective for iNOS over nNOS and eNOS.[39] N^ω-Nitro-L-arginine (**1**, R = NO2) is about 300-fold selective for nNOS over iNOS.[40] 2-Amino-5,6-dihydro-6-methyl-4*H*-1,3-thiazine (**2**) and

S-ethylisothiourea (**3**, R = Et) have 10-40 fold selectivity in favor of iNOS,[41] and a series of

2-iminoazaheterocycles (**4**) show selectivities of 1-10 in favor of iNOS over nNOS.[42] Various indazole analogues (**5**) have selectivities of 5-10 for either nNOS or iNOS.[43] Imidazole analogues (**6**) exhibit selectivities in the range of 3-6 fold in favor of iNOS.[44] Aminoguanidine (**7**, R = NH$_2$) shows a 50-fold selectivity for iNOS over nNOS and 500-fold over eNOS.[45] Several series of isothiourea analogues (**3**, R = various) favor inhibition of iNOS over eNOS by factors of between 2- and 6-fold with one analogue being 19-fold selective; bisisothioureas (**8**) are more selective, with one analogue showing a selectivity of 190-fold in preference for iNOS.[46] *S*-Methyl- and *S*-ethyl-L-thiocitrulline (**9**, R = Me or Et) are 10- and 50-fold, respectively, more selective for nNOS than eNOS.[47] L-*N*^{6}-(1-Iminoethyl)lysine (**10**), another inhibitor of NOS, favors the inhibition of iNOS by a factor of 30 over eNOS and by 13 over nNOS.[48]

2 DESIGN OF NEW INHIBITORS OF NOS

Since only a structure of iNOS has been published, and the coordinates are not yet available, this structure does not help much in terms of selective inhibitor design, particularly when our interest is more in nNOS inhibition. Consequently, we thought that a conformationally-restricted analogue approach might be useful to define differences in active-site binding modes.

2.1 Conformationally-Restricted Analogues of L-Arginine

By fixing the orientation of various binding determinants within the active site, it may be possible to determine which arrangement of atoms favors the different NOS isozymes. If some selectivity arises, then this would be a key component in the design of selective inhibitors of the NOS isozymes. On that basis we designed four conformationally-restricted arginine analogues, (*E*)-(**11**)- and (*Z*)-3,4-didehydro-*D,L*-arginine (**12**), *m*-guanidino-*D,L*-phenylglycine (**13**), and 5-keto-*D,L*-arginine (**14**), as potential alternative substrates or inhibitors.

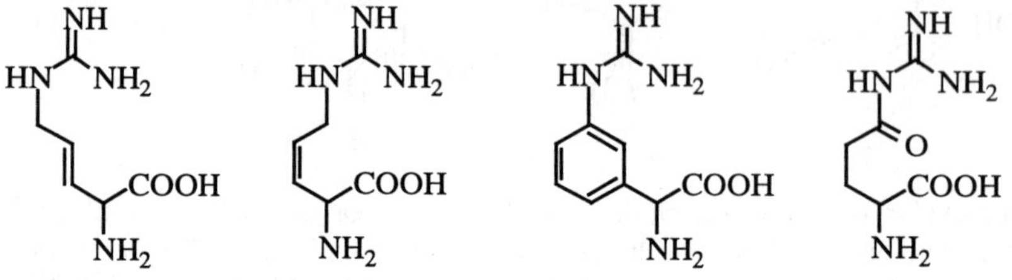

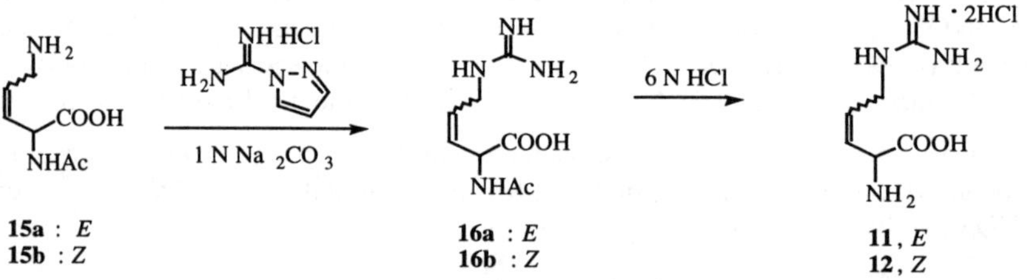

Scheme 2

The syntheses of **11** and **12** are shown in Scheme 2. (*E*)- and (*Z*)-5-Amino-2-acetamido-3-pentenoic acids (**15**)[49] are condensed with 1*H*-pyrazole-1-carboxamidine hydrochloride[50] to give **16**; acid hydrolysis of **16** gives **11** and **12**. The synthesis of **13** is shown in Scheme 3.

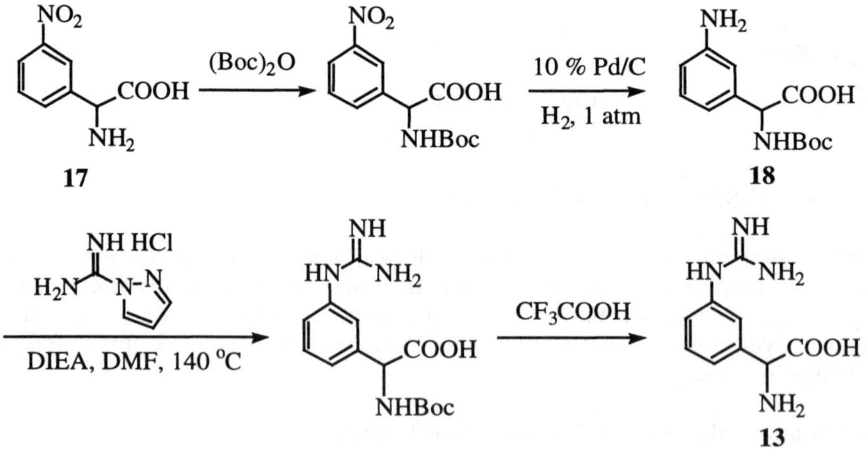

Scheme 3

Boc protection of (3-nitrophenyl)-D,L-glycine (**17**, Aldrich) followed by hydrogenolysis gives N-Boc-(3-aminophenyl)-D,L-glycine (**18**). Treatment of **18** with with 1*H*-pyrazole-1-carboxamidine hydrochloride and diisopropyl ethyl amine in DMF at 140 °C followed by acid deprotection gives **13**. Condensation of 5-methyl L-glutamate (**19**, Aldrich) with guanidine at room temperature produces **14** (Scheme 4).

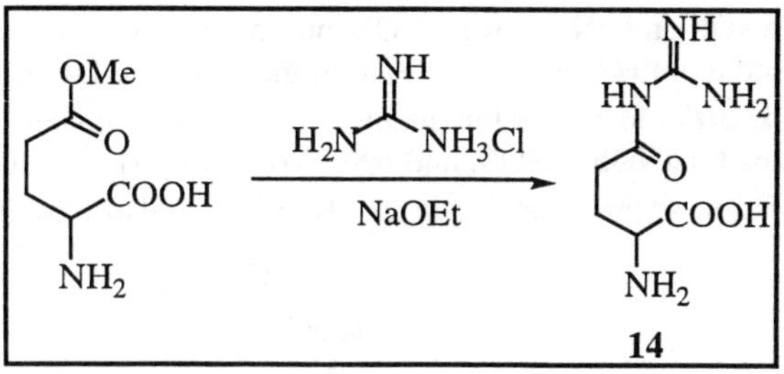

Scheme 4

Compounds **11-13** are both substrates and inhibitors of all three isozymes of NOS, and **14** is only an inhibitor (Table 1); little selectivity ($\leq$ 25-fold) is displayed by any of these compounds.

On the basis of K_m alone, the **11** selectivities for nNOS and eNOS over iNOS are 23- and 35-fold, respectively; however, on the basis of k_{cat}/K_m values the selectivities are a much more modest 8- and 4-fold, respectively. The Z-isomer (**12**) was found to be a much poorer substrate for all three isoforms of NOS than was **11**, having K_m values of 3-4 orders of magnitude greater than **11** and k_{cat}/K_m values 3-5 orders of magnitude lower than those for **11**. This indicates that the binding orientation of arginine in the active site is better mimicked by an extended sidechain than a folded one. To further rigidify **11** in an attempt to increase the binding effectiveness, analogue **13** was synthesized. However, this compound is a much poorer substrate than **11**; its K_m and k_{cat}/K_m values are 3-4 orders of magnitude worse than those for **11** (Table 1), suggesting a possible steric hindrance to its binding at the arginine binding site by the aromatic ring.

Table 1 *Kinetic and Binding Data to the NOS Isozymes for 11-14*

Compound	nNOS			iNOS			eNOS		
	K_m	IC$_{50}$	k_{cat}/K_m	K_m	IC$_{50}$	k_{cat}/K_m	K_m	IC$_{50}$	k_{cat}/K_m
		(s^{-1}mM^{-1})			(s^{-1}mM^{-1})			(s^{-1}mM^{-1})	
L-Arginine	2.7 μM		180	9.5 μM		110	1.1 μM		73
11	1.5 μM	100 μM	110	35 μM	100 μM	13	1.0 μM	250 μM	26
12	5 mM	25 mM	1.8 x 10^{-3}	20 mM	30 mM	6.9 x 10^{-3}	15 mM	50 mM	8.0 x 10^{-4}
13	40 mM	2.0 μM^a	1.8 x 10^{-2}	20 mM	8 μM^a	4.4 x 10^{-3}	25 mM	50 μM^a	7.9 x 10^{-2}
14		3.0 mM			2 mM			14 mM	

^{a} K_i value

5-Ketoarginine (**14**) also was investigated as a conformationally-restricted analogue of arginine because of the potential intramolecular hydrogen bonding properties of the ketone carbonyl with the terminal NH$_2$ group. This compound was not a substrate at all for any of the isozymes of NOS, and showed only weak inhibitory properties (Table 1), consistent with the finding that the Z-isomer is a poor substrate. In addition to its possible cyclic structure, as a result of hydrogen bonding, the carbonyl group will strongly decrease the pK_a of the guanidino group. One or both of these effects may be detrimental to the binding.

While this work was in progress, a different set of conformationally-restricted arginine analogues (**20** and **21**, R = NH_2, $NHCH_3$, or CH_3) was reported that had no better than a 12-fold selectivity of eNOS and nNOS over iNOS, but the potencies were better than those for **11-14**.[51] They found that the folded (Z-like) arginine analogue **20** (R = NH_2), in which the amino acid group is *ortho* to that of the guanidine group, inhibits all three human NOS isoforms with K_i values for iNOS, eNOS, and nNOS of 2.60 µM, 0.25 µM, and 0.37 µM, respectively. The extended arginine analogue **21** (R = NH_2) also inhibits all isoforms of

20   21

NOS, but the inhibitory potencies are lower (100, 46, and 44 µM, respectively) than those of **20**. On the basis of these results, it was proposed that arginine based inhibitors preferentially bind to the active sites of all of the NOS isoforms in a folded conformation. This hypothesis is ostensibly different from what we proposed above using **11-13**, in which the extended E-arginine analogue is a better substrate than the folded Z-isomer. However, this is not an appropriate comparison, because **21** is an analogue of homoarginine, not arginine (there is an extra carbon in **21** relative to arginine), whereas **11-13** are all analogues of arginine. The additional carbon should impart very different binding orientations of the amino acid group and the guanidino group than what is observed in **11-13**. There also should be a difference between **13** and **21** in their selectivities for the NOS isoforms. In fact, there is. Whereas the iNOS/eNOS selectivity for **13** is in favor of iNOS by a factor of 6, the selectivity in the case of **21** is reversed (eNOS/iNOS) by a factor of 2. The nNOS/eNOS selectivity for **13** favors nNOS inhibition by a factor of 25, but for **21** there is no selectivity.

2.2 Imidazolylalkylglycine Derivatives

Although it is apparent from the results in Table 1 that the E-isomer binds more effectively than the Z-isomer, the selectivity between the different isozymes is not very great. Since the crystal structure of iNOS with an imidazole bound[23] showed that the imidazole interacts with the heme iron, the next inhibitor design was to determine if there is a distance difference between the amino acid binding site and the heme in the three isozymes. To accomplish that goal, a series of imidazolylalkylglycine derivatives (**22**) having different distances between the amino acid moiety and the imidazole was synthesized by the route in Scheme 5. These compounds are extended histidines, having one to five more methylenes than has histidine. The analogue with two methylenes could not be prepared by the route used for the other analogues because elimination of HBr from the dibromoethane, rather than substitution, occurred exclusively. However, ethylene oxide worked well in the substitution of the imidazole ring.

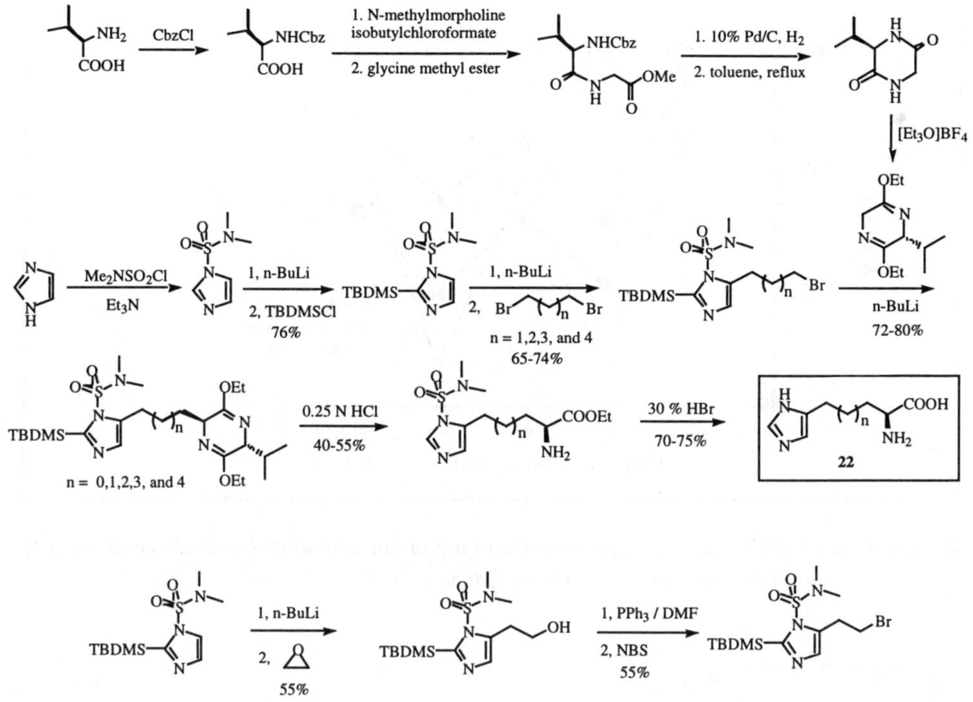

Scheme 5

Table 2 *Binding Data to the NOS Isozymes for 22 (n = 0-4)*

Compound 22	K_i (μM)			Selectivity	
	nNOS	iNOS	eNOS	nNOS/iNOS	nNOS/eNOS
n = 0	170	950	500	5.6	2.9
n = 1	2	10	33	5.0	16.5
n = 2	65	35	150	0.5	2.3
n = 3	2	8	50	4	25
n = 4	150	40	250	0.3	1.7

All of these compounds were inhibitors of the three isozymes of NOS (Table 2). The best selectivity was a factor of 5 in favor of nNOS over iNOS which both **22** (n = 0) and **22** (n = 1) had. However, the potency of **22** (n =1) was almost two orders of magnitude greater than that for **22** (n = 0).

The graph in Figure 3 makes it clear that the potencies of these inhibitors is a function of the chain length; the chain lengths with an odd number of methylene groups are much more potent than the ones with an even number of methylenes. This may be related to the geometry of the amino acid group relative to the imidazole group as a function of the number of methylenes. As demonstrated in Figure 4, each homologue increase changes the orientation of these two functionalities.

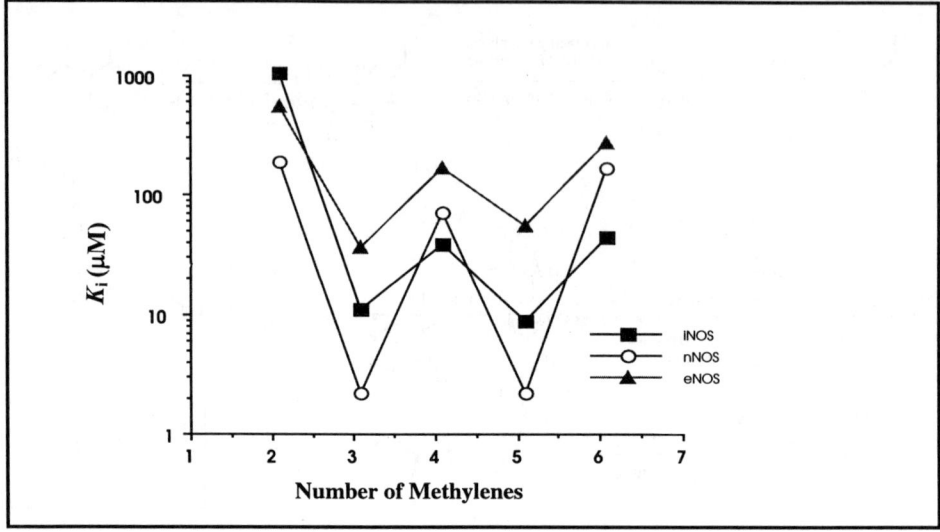

Figure 3 *Effect of the number of methylenes in the imidazolylalkylglycine derivatives (22) on binding potency to the three isozymes of NOS*

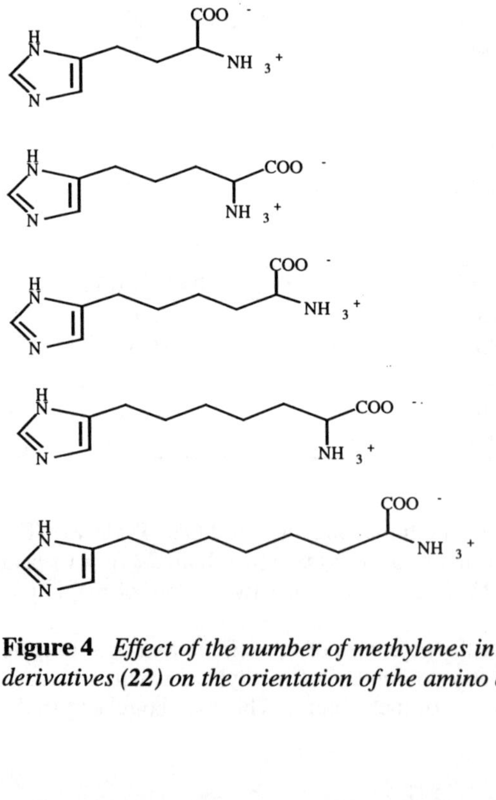

Figure 4 *Effect of the number of methylenes in the imidazolylalkylglycine derivatives (22) on the orientation of the amino acid functionality*

2.3 (1-Phenylimidazolylalkyl)glycine Derivatives

Wolff and Gribin[52] reported that, although imidazole was a good inhibitor of iNOS, 1-phenylimidazole was almost 7 times more potent. Therefore, a series of (1-phenylimidazolylalkyl)glycine derivatives (**23**) was synthesized (Scheme 6) to determine if the potency increases by the addition of the phenyl group. Surprisingly, as shown in Table 3, these compounds were both less potent and less selective than the imidazole analogues (**22**).

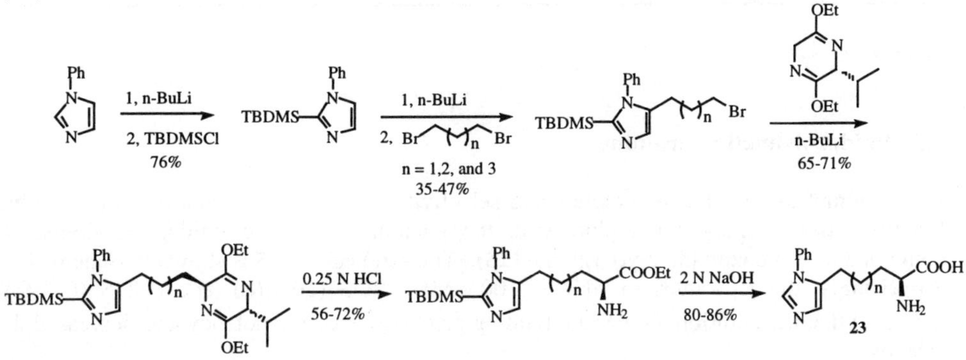

Scheme 6

Table 3 *Binding Data to the NOS Isozymes for 23 (n = 1-3)*

	K_i (μM)			Selectivity	
Compound 23	nNOS	iNOS	eNOS	nNOS/iNOS	nNOS/eNOS
n = 1	80	100	50	1.25	0.63
n = 2	100	50	17	0.5	0.17
n = 3	70	120	350	1.7	5

2.4 (Imidazolylmethyl)guanidine

In sections 2.2 and 2.3 above the compounds were tested to determine the effect of amino acid binding and heme binding. Next, we were interested in determining the effect of a molecule that binds to the guanidine binding site and the heme, so we synthesized (imidazolylmethyl)guanidine (**24**, Scheme 7).

The K_i values for iNOS, nNOS, and eNOS are 375, 120, and 1500 μM, respectively, indicating that the potency and selectivity for nNOS/iNOS were poor and the selectivity of nNOS/eNOS was weak.

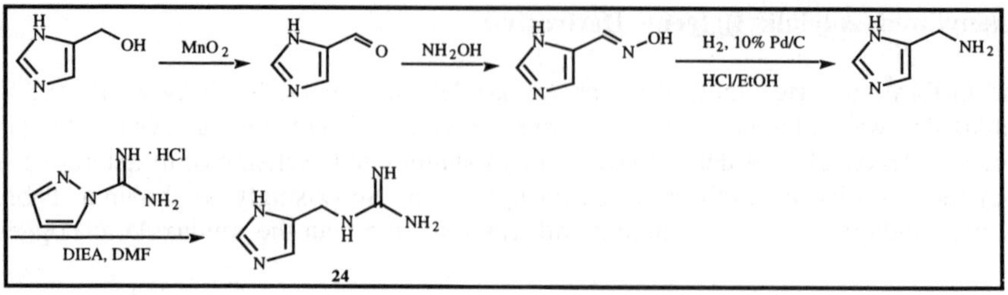

Scheme 7

2.5 (Imidazolylmethyl)arginine

In an attempt to increase the potency and selectivity, it was thought that all three binding functionalities, namely, the amino acid, the guanidine, and the imidazole, should be combined as (imidazolylmethyl)arginine (**25**). The synthesis of **25** is shown in Scheme 8. Surprisingly, the K_i values for iNOS (300 μM), nNOS (50 μM), and eNOS (65 μM) indicated that the additional potential binding group lowered the potency and decreased the selectivity.

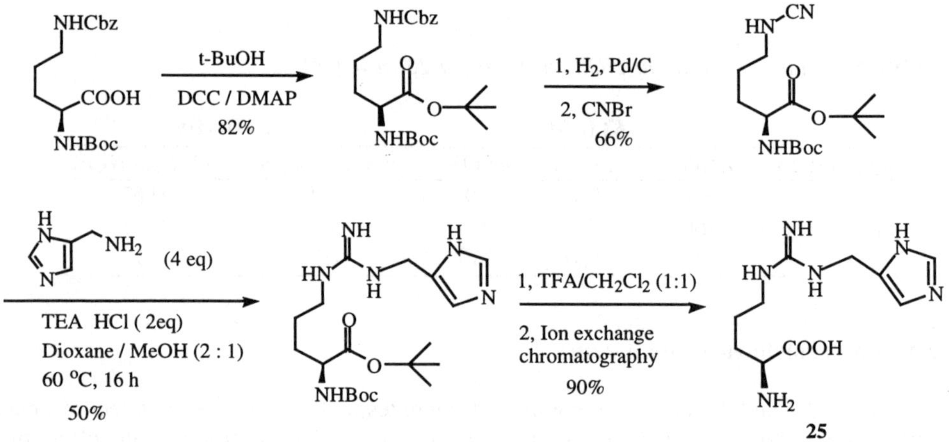

Scheme 8

3 DESIGN OF NEW SELECTIVE INHIBITORS OF NOS

3.1 N^{ω}-Alkyl-L-arginine Derivatives

The next approach was to determine if there was a difference in the active site environment about the terminal guanidine amino group. This was monitored by simple alkyl substitution at the N^{ω}-position. Three different syntheses of a series of N^{ω}-alkyl-L-arginines were carried out (Schemes 9-11), giving the series shown in Figure 5.

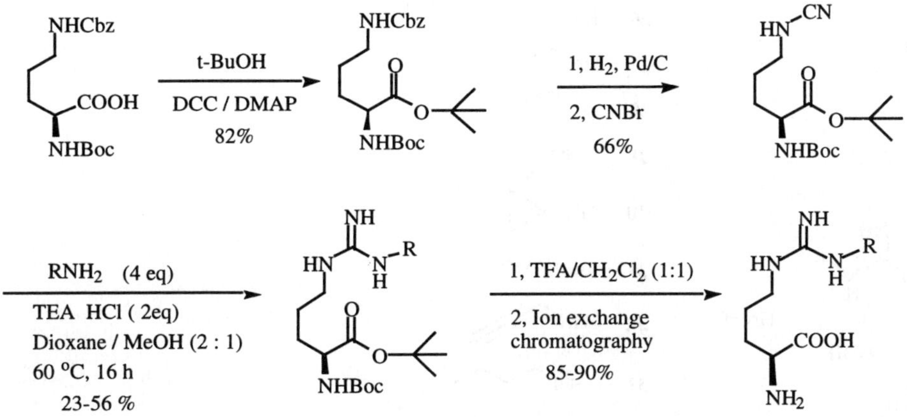

Scheme 9

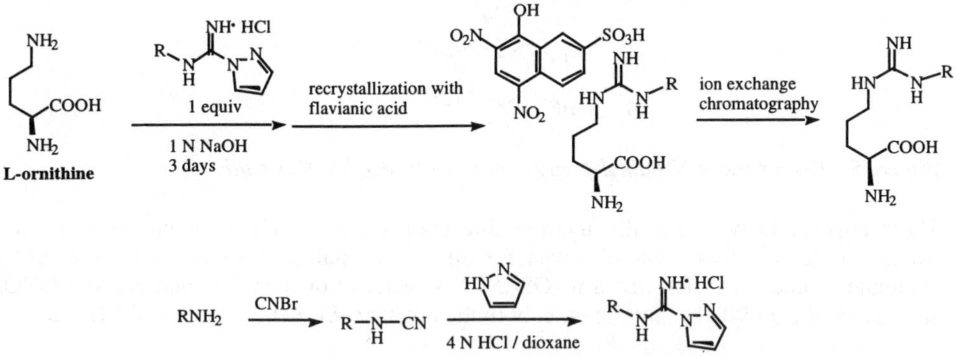

Scheme 10

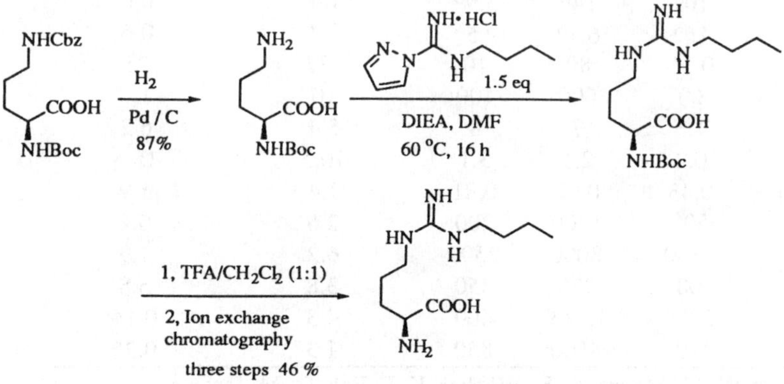

Scheme 11

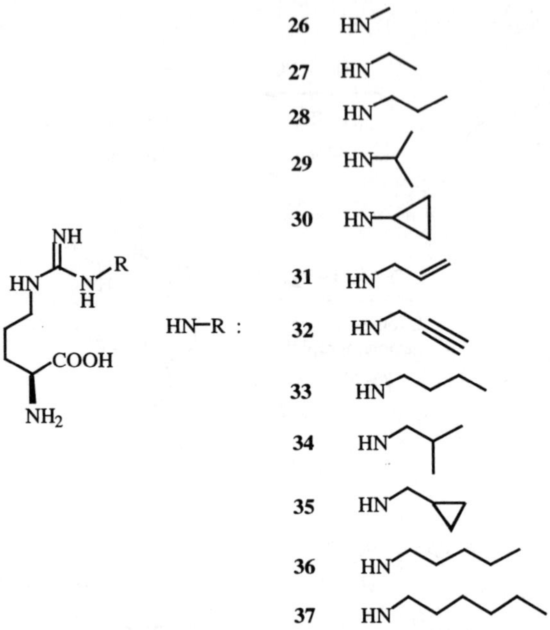

Figure 5 *The series of N^{ω}-alkyl-L-arginines synthesized in this study*

There appears to be a specific hydrophobic pocket into which these compounds bind, because little selectivity was observed for all of the analogues except for the *n*-propyl analogue (Table 4), which had a nNOS/iNOS selectivity of over 700 and a nNOS/eNOS selectivity of over 90. In this one case, both the potency ($K_i = 107$ nM for nNOS) and

Table 4 *Binding Data to the NOS Isozymes for 26 Analogues*

	K_i (μM)			Selectivity	
Compound	nNOS	iNOS	eNOS	nNOS/iNOS	nNOS/eNOS
26[a] (Me)	10[b]	14[b]	5.9[b]	1.4	0.6
27[a] (Et)	16[b]	6.1[b]	9.5[b]	0.4	0.6
28 (n-Pr)	0.11	80	10	727	91
29 (i-Pr)	60	600	100	10	1.7
30 (cyc-Pr)	1.3	7	8	5.4	6.2
31 (allyl)	0.2	2.1	3.1	10.5	15.5
32 (propargyl)	0.43	0.62	0.81	1.4	1.9
33 (n-Bu)	500	1300	200	2.6	0.4
34 (i-Bu)	1300	8000	2500	6.2	1.9
35 (cycMe)	60	230	350	3.8	5.8
36 (n-Pen)	2800	12000	400	4.3	0.14
37 (n-Hex)	3400	4500	850	1.3	0.25

[a]Data taken from W. M. Moore, R. K. Webber, K. F. Fok, G. M. Jerome,
C. M. Kornmeier, F. S. Tjoeng, M. G. Currie, Bioorg. Med. Chem., 1996, **4**, 1559.
[b]IC_{50} values, not K_i values

selectivity for nNOS are outstanding. It should be noted that the K_i values shown for the *n*-propyl analogue vary somewhat from those reported previously for that compound;[53] however, the selectivities are only about a factor of four different for nNOS/iNOS and a factor of two different for nNOS/eNOS.

3.2 N^w-Nitroarginine- and Phenylalanine-Containing Dipeptides and Dipeptide Esters

Nitro-L-arginine was reported to have about a 300-fold selectivity in favor of nNOS over iNOS.[54] Furthermore, whereas nitro-L-arginine showed typical irreversible inhibition patterns with nNOS and eNOS, it was only a reversible inhibitor of iNOS.[55] Because of the differences in inhibitory activities and high selectivity for these two isozymes, we decided to determine if N^ω-nitroarginine-containing dipeptides and dipeptide esters would have even greater selectivity for nNOS than the parent amino acid. Since L-arginine methyl ester, L-argininamide, and L-arginine-containing dipeptides are substrates for NOS,[56,57] it was thought that the dipeptide esters may be functional as well. Furthermore, the fact that N^ω-nitro-L-arginine inhibits nNOS in vivo following intraperitoneal injection[55a] suggests that it crosses the blood-brain barrier.

To test this question, peptides with phenylalanine as the partner amino acid were selected for initial evaluation. The peptides were synthesized manually in solution by the route shown in Scheme 12. Chiral HPLC, which can separate all four epimers of each dipeptide, indicated the presence of only one epimer for each compound, and all dipeptides were pure by NMR and elemental analysis and were optically active. Thirteen different dipeptides and dipeptide esters containing N^ω-nitroarginine and phenylalanine were synthesized; the data for the inhibition of the three isozymes of NOS are compiled in Table 5.

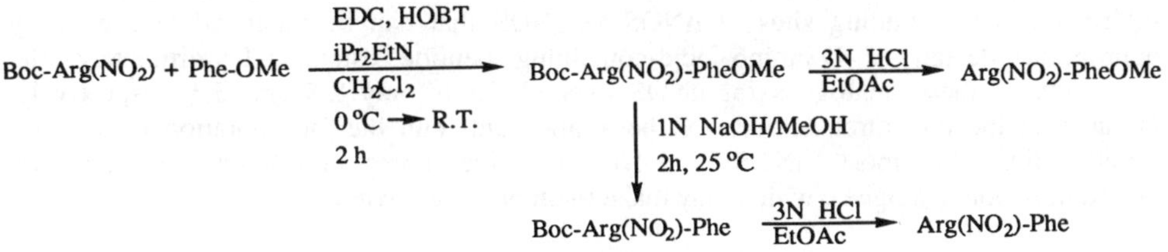

Boc-Arg(NO$_2$) + Phe-OMe $\xrightarrow[\substack{CH_2Cl_2 \\ 0\,°C \rightarrow R.T. \\ 2\,h}]{\substack{EDC,\ HOBT \\ iPr_2EtN}}$ Boc-Arg(NO$_2$)-PheOMe $\xrightarrow[EtOAc]{3N\ HCl}$ Arg(NO$_2$)-PheOMe

Boc-Arg(NO$_2$)-PheOMe $\downarrow$ 1N NaOH/MeOH, 2h, 25 °C

Boc-Arg(NO$_2$)-Phe $\xrightarrow[EtOAc]{3N\ HCl}$ Arg(NO$_2$)-Phe

Scheme 12

All of the dipeptides and dipeptide esters are competitive inhibitors of the three isoforms of NOS, except for the ones that contain D-Arg^{NO2} (**45-47, 49, 50**), which are uncompetitive inhibitors of iNOS, but competitive inhibitors of nNOS and eNOS. None of the dipeptides or dipeptide esters tested (**38, 39, 49, 50**) exhibit time-dependent inhibition of any of the isoforms, unlike N^w-nitro-L-arginine itself, which we found had K_I and k_{inact} values of 9.3 mM and 0.132 min^{-1}, respectively with nNOS. None of the kinetic plots show curvature, suggesting that there is no time dependence to the inhibition.

Table 5 *Binding Data for N^w-Nitroarginine- and Phenylalanine-Containing Dipeptides and Dipeptide Esters*

Inhibitor	K_i (mM)			Selectivity	
	nNOS	iNOS	eNOS	nNOS/iNOS	nNOS/eNOS
38 L-Arg^{NO2}-L-Phe	18	160	395	8.9	21.9
39 L-Phe-L-Arg^{NO2}	140	93	490	0.66	3.5
40 L-Arg^{NO2}-L-Phe-OMe	14	45	400	3.2	28.6
41 L-Phe-L-Arg^{NO2}-OMe	370	204	1350	0.55	3.6
42 L-Arg^{NO2}-L-Phe-OBn	55	18	125	0.33	2.3
43 L-Phe-L-Arg^{NO2}-OBn	110	45	250	0.41	2.3
44 L-Arg^{NO2}-D-Phe-OMe	92	100	525	1.1	5.7
45 D-Arg^{NO2}-L-Phe-OMe	290	6400[a]	1400	22.1	4.8
46 D-Arg^{NO2}-D-Phe-OMe	150	6500[a]	375	43.3	2.5
47 L-Phe-D-Arg^{NO2}-OMe	90	7500[a]	150	83.3	1.7
48 D-Phe-L-Arg^{NO2}-OMe	400	1200	8500	3	21.3
49 D-Phe-D-Arg^{NO2}-OMe	2	3600[a]	5	1800	2.5
50 D-Phe-D-Arg^{NO2}	17	13600[a]	90	800	5.3

[a] Uncompetitive inhibition

The order of the amino acids in the dipeptide is important to selectivity, but it depends on the chirality of the amino acids. Thus, for the dipeptide and dipeptide methyl ester pairs **38** vs **39** and **40** vs **41**, having L-Arg^{NO2} at the N-terminus favors nNOS over iNOS and eNOS inhibition , but when the L-Phe is at the N-terminus, it favors iNOS over nNOS and eNOS inhibition. In the case of the corresponding benzyl ester (**42** vs **43**), both dipeptides favor iNOS over nNOS and eNOS inhibition. All of the dipeptide methyl esters containing a D-amino acid, however, exhibit an inhibitory preference for nNOS over iNOS and eNOS. The most impressive selectivities observed are 1800-fold and 800-fold for **49** and **50**, respectively, in favor of nNOS over iNOS. These results indicate that there is a difference in the binding sites of nNOS vs iNOS that can be taken advantage of by appropriate design of D-amino acid-containing peptide esters. Unfortunately, the selectivity of these analogues for nNOS over eNOS is only 2.5 and 5.3, respectively. Because of the unnatural chirality of the amino acids and the incorporation of an ester functionality, the most nNOS vs iNOS-selective compounds (**49** and **50**) are peptidomimetic analogues, which may make them orally bioavailable.

3.3 N^w-Nitroarginine-Containing Dipeptide Amides

Because of the initial success with N^w-nitroarginine- and phenylalanine-containing dipeptides and dipeptide esters, dipeptides containing N^w-nitroarginine- and other amino acids (except arginine) were synthesized. All four isomers of each dipeptide combination (L,L; L,D; D,L; D,D) were synthesized with N^w-nitroarginine at either the N-terminus or the C-terminus. In addition to the standard 19 natural amino acids (excluding arginine), N^w-nitroarginine-containing dipeptide amides with L- and D-ornithine or with L-2,3-diaminopropionic acid (Dpr) or with L-2,4-diaminobutyric acid (Dbu) also were synthesized. The amides were made because these 150 compounds were synthesized with an automated peptide synthesizer, which cleaves the dipeptide from the resin with ammonia to give the dipeptide amide. In preliminary tests we found that the dipeptide amides were comparable to the corresponding dipeptides in their inhibitory properties with NOS. These compounds were initially screened by measuring IC_{50} values to get an estimate of their potencies with the three isozymes of NOS. Those results are shown in Figures 6-11. From those data the 11 most selective compounds (the ones with an asterisk over the bar graph) were chosen for more accurate K_i data measurements. Those results are given in Table 6. The following comments refer only to these most selective compounds. It is apparent that an amino group on the sidechain of the amino acid (Lys, Orn, Dbu, Dpr) is the most effective selectivity determinant. With regard to these analogues having L-Arg^{NO2} at the N-terminus (**51-55**), as the number of methylenes in the sidechain decreases from L-Lys to L-Dbu (**51-53**), the binding to nNOS and iNOS increases, but that to eNOS decreases. This shows up in the selectivities; nNOS/iNOS remains relatively constant, but the nNOS/eNOS ratio increases dramatically. The smaller sidechain, Dpr (**54**), leads to decreased binding to nNOS, and a slight decrease to iNOS and eNOS. When the ornithine is changed to the D-isomer (**55**), binding to nNOS and eNOS decreases, but to iNOS remains the same. In all cases, however, selectivities for nNOS/iNOS and nNOS/eNOS are excellent.

When the nitroarginine is at the N-terminus, it is more selective as the L-isomer, but when it is at the C-terminus, it is more selective as D-Arg^{NO2}. That would imply that the best isomer for nitroarginine-nitroarginine should be L-Arg^{NO2}-D-Arg^{NO2}, but it is the L-L isomer (**58**) that is the most selective for nNOS (the L-D isomer was next best). D-Lys-D-Arg^{NO2} (**60**) binds better to all three isozymes relative to L-Lys-D-Arg^{NO2} (**59**), but the selectivities are smaller for **60** than for **59**. Whereas decreasing the sidechain length of **51** to **52** leads to increased nNOS binding, decreased eNOS binding, and no change in iNOS binding, doing the same when these amino acids are at the N-terminus and the nitroarginine is the D-isomer, decreases the nNOS binding, increases the iNOS binding, and leaves the eNOS binding the same. When one shorter methylene sidechain is incorporated (**61**) relative to **59**, binding to nNOS decreases, binding to iNOS increases, and binding to eNOS stays the same. Overall, then, the selectivities of **61** for nNOS relative to iNOS and eNOS decrease relative to those for **59**.

Table 6 *Binding Data for N^{ω}-Nitroarginine- and Another Amino Acid-Containing Dipeptide Amides*

	K_i (μM)			Selectivity	
Dipeptide amides	**nNOS**	**iNOS**	**eNOS**	**nNOS/iNOS**	**nNOS/eNOS**
51 L-Arg^{NO2}-L-Lys-NH$_2$	0.45	104	141	231	313
52 L-Arg^{NO2}-L-Orn-NH$_2$	0.33	97	245	294	742
53 L-Arg^{NO2}-L-Dbu-NH$_2$	0.13	25	200	192	1538
54 L-Arg^{NO2}-L-Dpr-NH$_2$	1.1	61	261	55	237
55 L-Arg^{NO2}-D-Orn-NH$_2$	2.0	103	1290	51	638
56 L-Arg^{NO2}-D-Asn-NH$_2$	0.32	8.9	412	28	1287
57 D-Arg^{NO2}-L-Ser-NH$_2$	1.25	1180	218	944	174
58 L-Arg^{NO2}-L-Arg^{NO2}-NH$_2$	0.77	62	33	80	43
59 L-Lys-D-Arg^{NO2}-NH$_2$	1.7	4700	229	2765	135
60 D-Lys-D-Arg^{NO2}-NH$_2$	0.89	910	30	1022	34
61 L-Orn-D-Arg^{NO2}-NH$_2$	5.0	2041	218	408	44

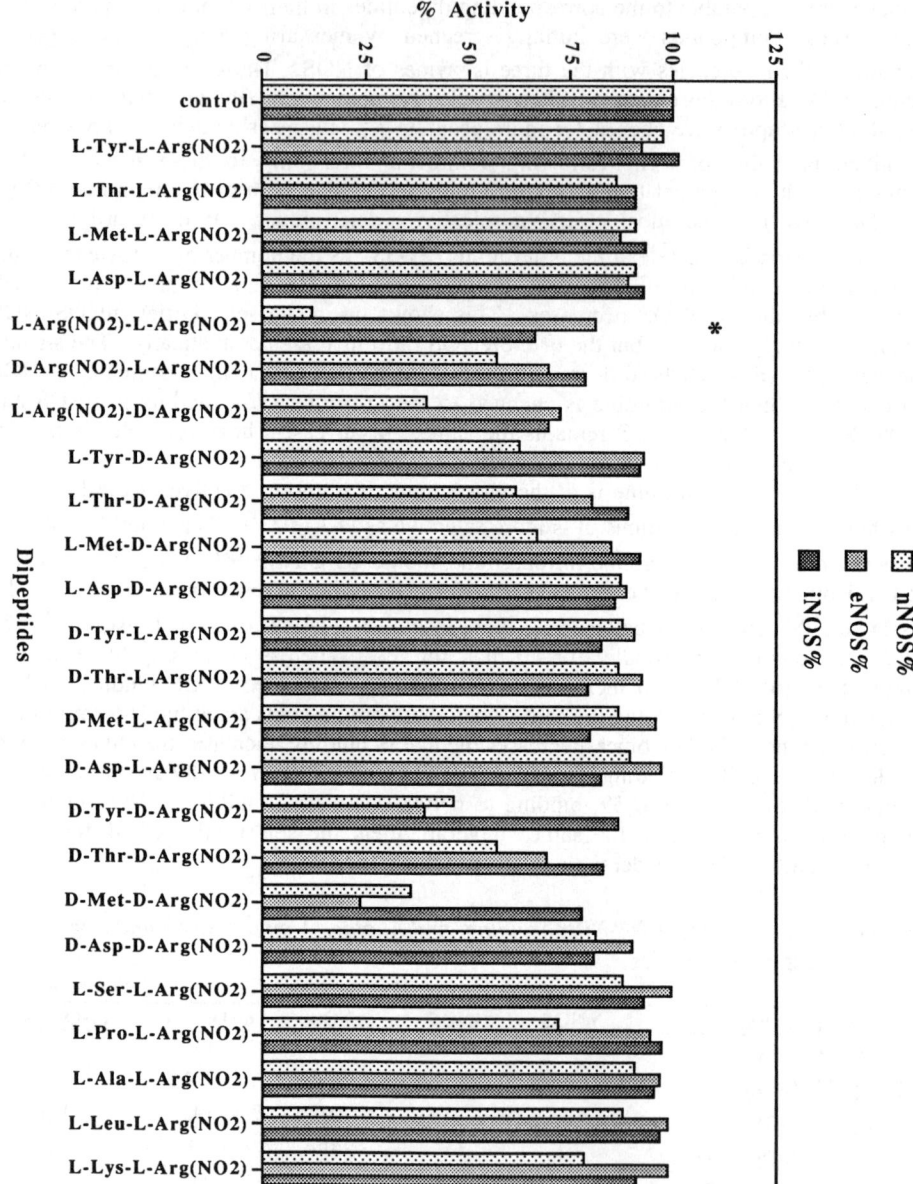

Figure 6 *Percent loss of each NOS isozyme activity by a family of Nω-*
nitroarginine-containing dipeptide amides

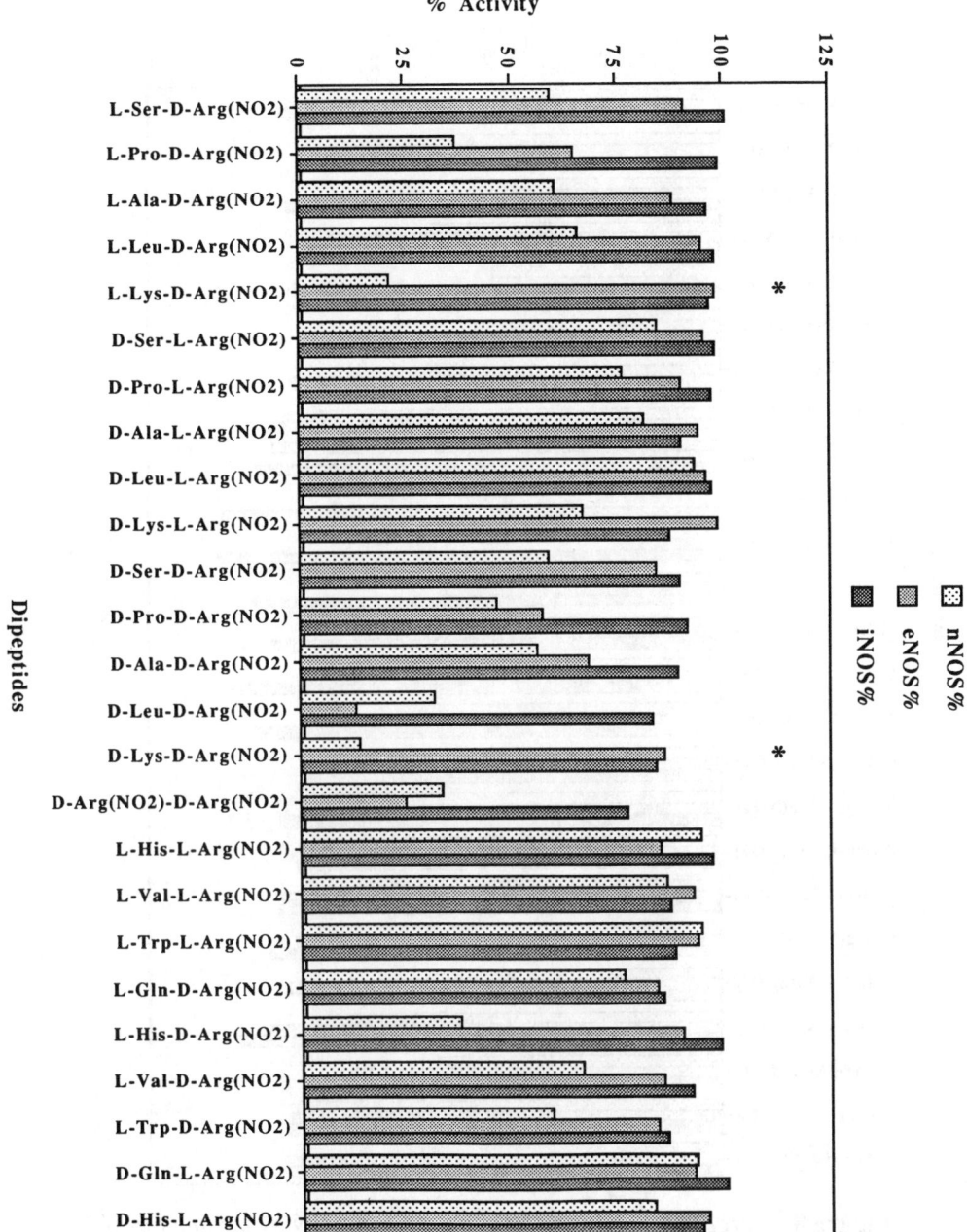

Figure 7 *Percent loss of each NOS isozyme activity by a family of N^ω-nitroarginine-containing dipeptide amides*

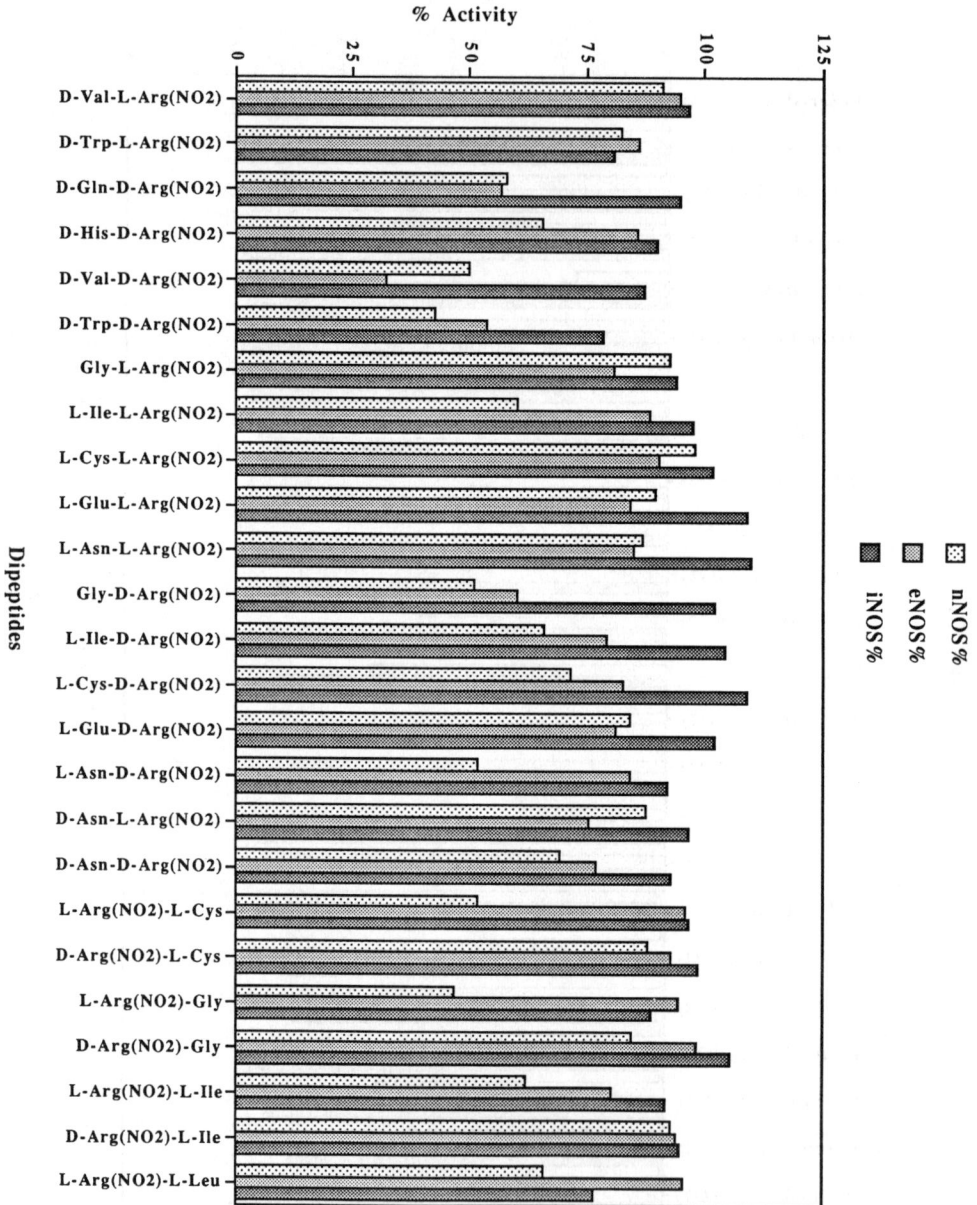

Figure 8 *Percent loss of each NOS isozyme activity by a family of N$^{\omega}$-*
nitroarginine-containing dipeptide amides

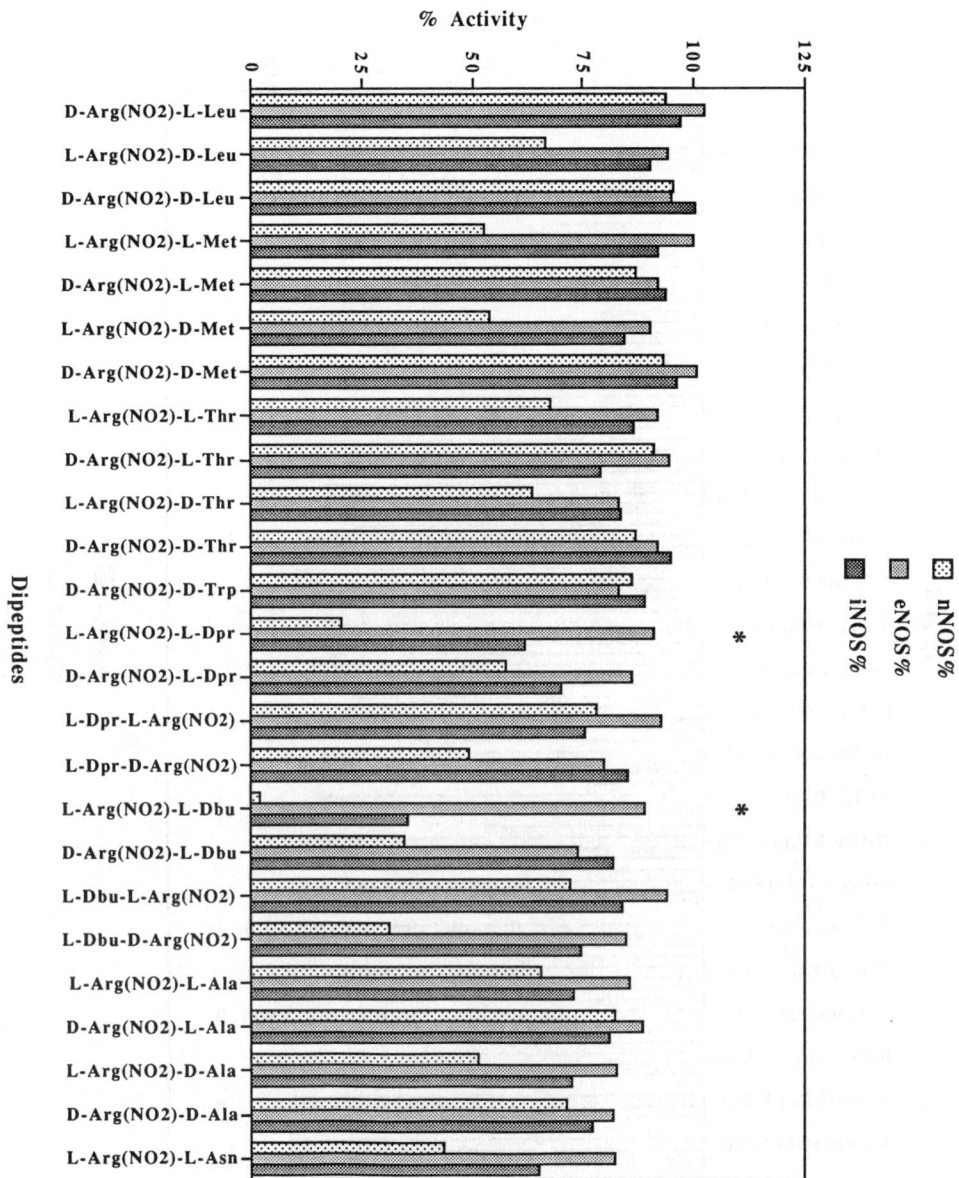

Figure 9 *Percent loss of each NOS isozyme activity by a family of Nω-nitroarginine-containing dipeptide amides*

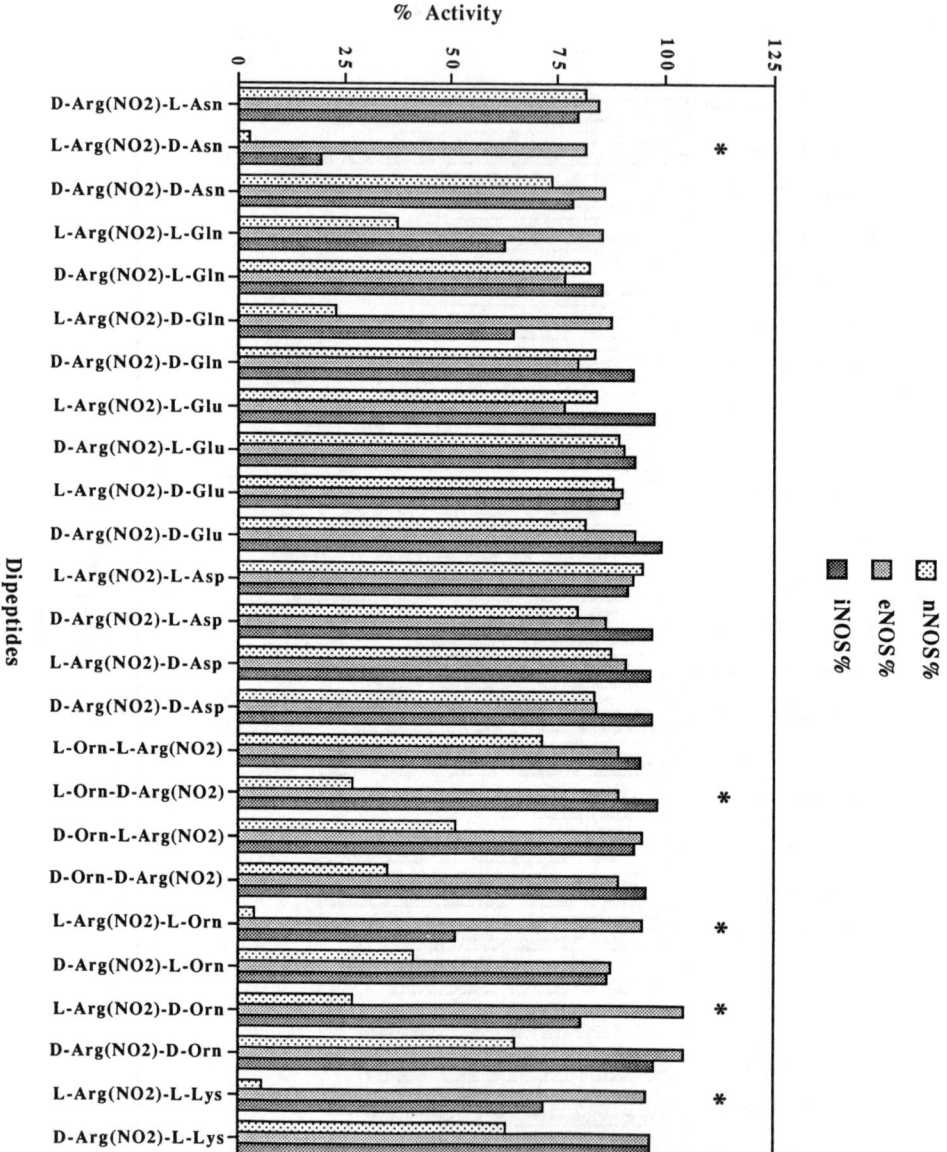

Figure 10 *Percent loss of each NOS isozyme activity by a family of N^ω-nitroarginine-containing dipeptide amides*

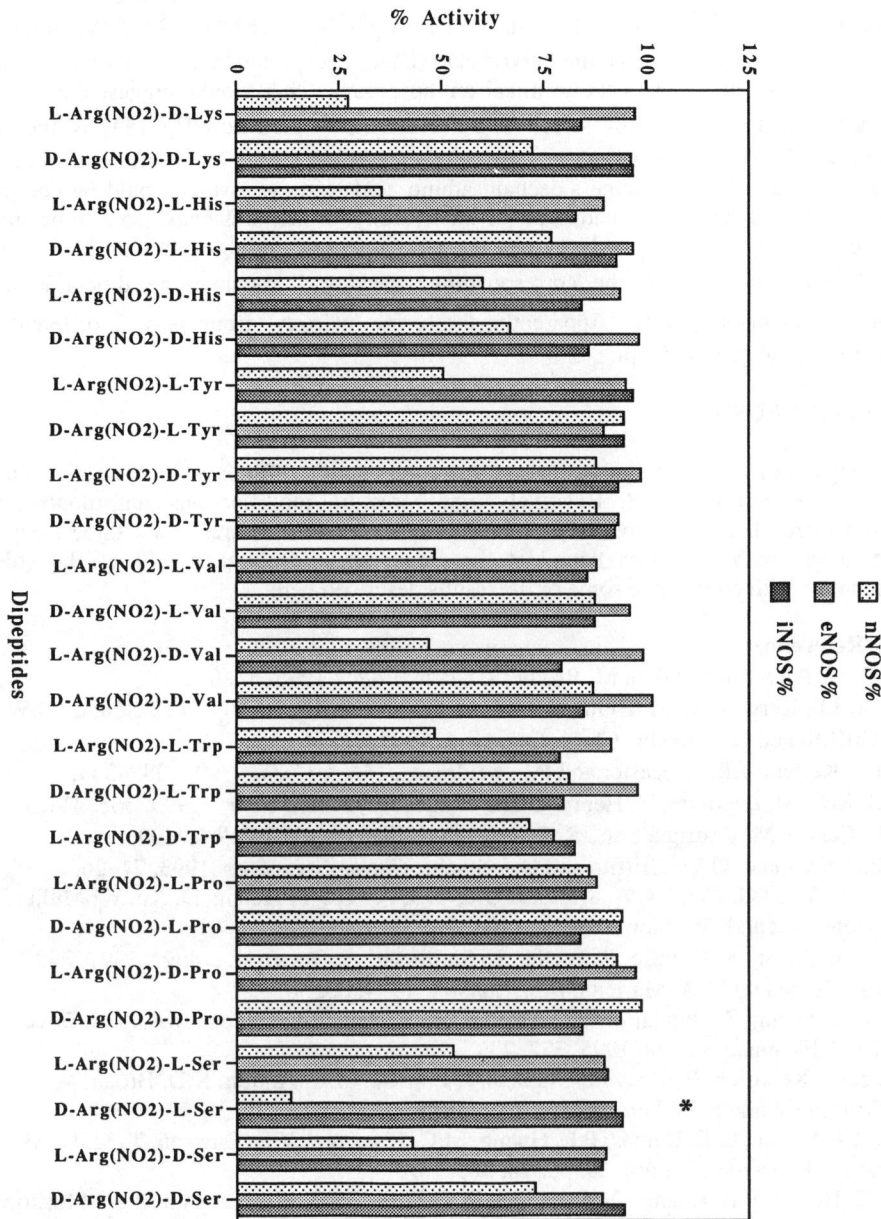

Figure 11 *Percent loss of each NOS isozyme activity by a family of N*$^\omega$*-nitroarginine-containing dipeptide amides*

Only two nonamine sidechain amino acid-containing dipeptide amides are highly selective: L-Arg^{NO2}-D-Asn-NH$_2$ (**56**) and D-Arg^{NO2}-L-Ser-NH$_2$ (**57**). Asparagine is similar in structure to 2,4-diaminobutyric acid (Dbu), except that there is a carbonyl instead of a methylene adjacent to the terminal amino group. That would suggest that the L-L isomer should be the most selective, since L-Arg^{NO2}-L-Dbu-NH$_2$ (**53**) is the most selective. However, that is not the case, so the carbonyl must play a different role in the binding. The other nonamine sidechain amino acid is serine, which could be compared with 2,3-diaminopropionic acid (Dpr), if the hydroxyl of serine is considered to be similar to the amino group of Dpr. However, the most selective serine-containing isomer is D-Arg^{NO2}-L-Ser-NH$_2$ and the corresponding selective Dpr-containing dipeptide is L-Arg^{NO2}-L-Dpr-NH$_2$ (**54**). Apparently, the serine hydroxyl group plays a different role than the amino group of Dpr.

4 CONCLUSION

It is apparent from these studies that there are very subtle differences in the active sites of the three isozymes of NOS. Relatively minor structural modifications can produce highly divergent results. It will be very exciting to see crystal structures of all three isozymes, particularly with inhibitors bound at the active site. Only then will we be able to rationalize with confidence some of the results described here.

References
1. E. W. Ainscough and A. M. Brodie, *J. Chem. Educ.*, 1995, **72**, 686.
2. (a) J.F. Kerwin and M. Heller, *Med. Res. Rev.* 1994, **14**, 23. (b) P.L. Feldman, O.W. Griffith and D.J. Stuehr, *Chem. Eng. News* 1993, **71**, 26.
3. J.F. Kerwin, J.R. Lancaster and P.L. Feldman, *J. Med. Chem.* 1995, **38**, 4343.
4. B. Roy, M. Lepoivre, Y. Henry and M. Fontecave, *Biochemistry* 1995, **345**, 411.
5. L. Castro, M. Rodriguez and R. Radi, *J. Biol. Chem .*, 1994, **269**, 29409.
6. P.L. Feldman, O.W. Griffith and D.J. Stuehr, *Chem. Eng. News* 1993, **71**, 26.
7. X.Q. Wei, I.G. Charles, A. Smith, J. Ure, G.J. Feng, F.P. Huang, D. Xu, W. Muller, S. Moncada and F.W. Liew, *Nature* 1995, **375**, 408.
8. N. Benjamin, S. Pattullo, R. Weller, L. Smith and A. Ormerod, *Lancet* 1997, **349**, 776.
9. J.R. Stone and M.A. Marletta, *Biochemistry* 1994, **33**, 5636.
10. P.L. Huang, Z. Huang, Z., H. Mashimo, K.D. Bloch, M.A. Moskowitz, J.A. Bevan and M.C. Fishman, *Nature* 1995, **377**, 239.
11. K.M. Kendrick, R. Guevara-Guzman, J. Zorrilla, M.R. Hinton, K.D. Broad, M. Mimmack and S. Ohkura, *Nature* 1997, **388**, 670.
12. R.J. Nelson, G.E. Demas, P.L. Huang, M.C. Fishman, V.L. Dawson, T. M. Dawson and S.H. Snyder, *Nature* 1995, **378**, 383.
13. Z. Huang, P.L. Huang, N. Panahian, T. Dalkara, M.C. Fishman and M.A. Moskowitz, *Science* 1994, **265**, 1883.
14. C. Iadecola, *Trends. Neurosci.* 1997, **20**, 132.
15. R.G. Knowles and S. Moncada, *Biochem J.* 1994, **198**, 249.
16. H.J. Cho, Q. Xie, J. Calaycay, R.A. Mumford, K.M. Swiderek, T.D. Lee and C. Nathan, *J. Exp. Med.* 1992, **176**, 599.
17. D.S. Bredt, P.M. Hwang, C.E. Glatt, C. Lowenstein, R.R. Reed, and S.S. Snyder, *Nature* 1991, **351**, 714.

18. D.J. Stuehr and M. Ikeda-Saito, *J. Biol. Chem.* 1992, **267**, 20547.
19. L.E. Lambert, J.F. French, J.P. Whitten, B.M. Baron and I.A. McDonald, *Eur. J. Pharmacol.* 1992, **316**, 131.
20. H.M. Abu-Soud, L.L. Yoho and D.J. Stuehr, *J. Biol. Chem.* 1994, **269**, 32047.
21. E.A. Sheta, K. McMillan and B.S. Siler-Masters, *J. Biol. Chem.* 1994, **169**, 15147.
22. M. Wang, D.L. Roberts, R. Paschke, T.M. Shea and B.S.S. Masters, *Proc. Natl. Acad. Sci. U.S.A.* 1997, **94**, 8411.
23. B.R. Crane, A.S. Arvai, R. Gachhui, C. Wu, D.K. Ghosh, E.D. Getzoff, D.J. Stuehr and J.A. Tainer, *Science* 1997, **378**, 425.
24. (a) R. B. Silverman, "Mechanism-Based Enzyme Inactivation: Chemistry and Enzymology," CRC Press, Boca Raton, FL, 1988, Vols. I and II. (b) R. B. Silverman, Methods Enzymol. 1995, **249**, 240.
25. N.M. Olken, Y. Osawa and M.A. Marletta, *Biochemistry* 1994, **33**, 14784.
26. J.F. Kerwin and M. Heller, *Med. Res. Rev.*, 1994, **14**, 23.
27. D.W. Choi and S.M. Rothman, *Annu. Rev. Neurosci.*, 1990, **13**, 171 (b) J. Garthwaite, In "The NMDA Receptor", J. C. Watkins, G. L. Collingridge, Eds., Oxford University Press, Oxford, England, 1989, p. 187.
28. K.L. Crossin, *Trends Biochem. Sci*, 1991, **16**, 81.
29. J.A. Ferrendelli, A.C. Blank and R.A. Gross, *Brain Res.* 1980, **200**, 93.
30. I. Das, N.S. Khan, B.K. Puri, S.R. Sooranna, J. de Belleroche, S.R. Hirsch, *Biochem. Biophys. Res. Commun.* 1995, **212**, 375.
31. L.L. Thomsen, H.K. Iversen, L.H. Lassen and J. Olesen, *CNS Drugs* 1994, **2**, 417.
32. M.-A. Dorheim, W.R. Tracey, J.S. Pollock, and P. Grammas, *Biochem. Biophys. Res. Commun.* 1994, **205**, 659.
33. H.N. Bhargava, *Gen. Pharmacol.* 1995, **26**, 1049
34. H.G. Seo, I. Takata, M. Nakamura, H. Tatsumi, K. Suzuki, J. Fujii, and N. Taniguchi, *Arch. Biochem. Biophys.* 1995, **324**, 41.
35. P. Kubes, M. Suzuki and D.N. Granger, *Proc. Natl. Acad. Sci. USA* 1991, **88**, 4651.
36. I. MacIntyre, M. Zaidi, A.S.M. Towhidul Alam, H.K. Datta, B.S. Moonga, P.S. Lidbury, M. Hecker and J.R. Vane, *Proc. Natl. Acad. Sci. USA* 1991, **88**, 2936.
37. C.A. Ross, A.Bredt and S.H. Snyder, *Trends Neurosci.* 1990, **13**, 216.
38. M.A. Marletta, *J. Med. Chem.* 1994, **37**, 1899. (b) M. Nakane, V. Klinghofer, J. E. Kuk, J. L. Donnelly, G. P. Budzik, J. S. Pollock, F. Basha, and G. W. Carter, *Mol. Pharmacol.*, 1995, **47**, 831.
39. W.M. Moore, R.K. Webber, K.F. Fok, G.M. Jerone, G.M.; C.M.Kornmeier, F.S. Tjoeng and M.G. Currie, *Bioorg. Med. Chem.* 1996, **4**, 1599.
40. E.S. Furfine, M.F. Harmon, J.E. Paith and E.P. Garvey, *Biochemistry*, 1993, **32**, 8512.
41. M. Nakane, V. Klinghofer, J.E. Kuk, J.L. Donnelly, G.P. Budzik, J.S. Pollock, F. Basha and G.W. Carter, *Mol. Pharmacol.* 1995, **47**, 831.
42. (a) W.M. Moore, R.K. Webber, K.F. Fok, G.M. Jerone, J.R. Connor, P.T. Manning, P.S. Wyatt, R.P. Misko, F.S. Tjoeng and M.G. Currie, *J. Med. Chem.* 1996, **39**, 669. (b) R.K.Webber, S. Metz, W.M.Moore, J.R. Connor, M.G. Currie, K.F. Fok, T.J. Hagen, D.W. Hansen, Jr., G.M. Jerome, P.T. Manning, B.S. Pitzele, M.V. Toth, M. Trivedi, M.E. Zupec and F.S. Tjoeng, *J. Med. Chem.* 1998, **41**, 96.(c) D.W. Hansen, Jr. K.B. Peterson, M. Trivedi, S.W. Kramer, R.K. Webber, F.S. Tjoeng, W.M. Moore,

G.M. Herone, C.M. Kornmeier, P.T. Manning, J.R. Connor, T.P. Misko, M.G. Currie and B.S. Pitzele, *J. Med. Chem.* 1998, **41**, 1361.

43. D.J. Wolff and B.J. Gribin, *Arch. Biochem. Biophys.* 1994, **311**, 300.

44. D.J. Wolff and B.J. Gribin, *Arch. Biochem. Biophys.* 1994, **311**, 293.

45. D.J. Wolff and A. Lubeskie, *Arch, Biochem. Biophys.* 1995, **316**, 290.

46. E.P. Garvey, J.A. Oplinger, G.J. Tanoury, P.A. Sherman, M. Fowler, S. Marshall, M.F. Harmon, J.E. Paith and E.S. Furfine, J. Biol. Chem. 1994, **269**, 69.

47. E.S. Furfine, M.F. Harmon, J.E. Paith, R.G. Knowles, M. Salter, R.J. Kiff, C. Duffy, R. Hazelwood, J.A. Oplinger and E.P. Garvey, J. Biol. Chem. 1994, **269**, 26677.

48. W.M. Moore, .R.K. Webber, G.M. Jerone, F.S. Tjoeng, T.P. Misko and M.G. Currie, J. Med. Chem. 1994, **37**, 3886.

49. L. Havlicek, J. Hanus, P. Sedmera and J. Nemecek, J. Collect. Czech. Chem. Commun. 1990, **55**, 2074.

50. M.S. Bernatowicz, Y. Wu, and G.R. Matsueda, J. Org. Chem. 1992, **57**, 2497.

51. B.G. Shearer, S. Lee, K.W. Franzmann, H.A.R. White, D.C.J. Sanders, R.J. Kiff, E.P. Garvey, and E.S. Furfine, Bioorg. Med. Chem. Lett. 1997, **7**, 1763.

52. D. J. Wolff and B. J. Gribin, Arch. Biochem. Biophys., 1994, **311**, 293.

53. H. Q. Zhang, W. Fast, M.A. Marletta, P. Martasek, R. B. Silverman, J. Med. Chem., 1997, **40**, 3869.

54. E. S. Furfine, M. F. Harmon, J. E. Paith, and E. P. Garvey, Biochemistry, 1993, **32**, 8512.

55. (a) M. A. Dwyer, D. S. Bredt, and S. H. Snyder, Biochem. Biophys. Res. Commun., 1991, **176**, 1136. (b) D. W. Reif and S. A. McCreedy, Arch. Biochem. Biophys., 1995, **320**, 170.

56. R. Iyengar, D. J. Stuehr, and M. A. Marletta, Proc. Natl. Acad. Sci. USA 1987, **84**, 6369.

57. M. Hecker, D. T. Walsh, and J. R. Vane, FEBS Lett., 1991, **294**, 221.

Predicting DMPK

ABSORPTION AND DISTRIBUTION AS FACTORS IN DRUG DESIGN

Dennis Smith

Drug Metabolism, Pfizer Central Research, Ramsgate Road, Sandwich, Kent, CT13 9NJ

1 INTRODUCTION

The oral absorption of a drug is dependant on the compound dissolving in the aqueous contents of the gastro-intestinal tract (dissolution) and then traversing the actual barrier of the gastro-intestinal tract to reach the blood.

$$\text{Solid drug} \xrightarrow{\text{dissolution}} \text{Drug in solution} \xrightarrow[\text{transfer}]{\text{membrane}} \text{Absorbed drug}$$

2 DISSOLUTION

Dissolution can be looked at from the perspective of rate of the solid drug dissolving and also the extent. The process depends on the surface area of the dissolving solid and the solubility of the drug at the surface of the dissolving solid. Considering these factors separately surface area is manipulated by the processing and formulation of the compound. Solubility itself, is manipulated mainly by the structure of the drug. In general, solubility is inversely proportional to the number and type of lipophilic functions within the molecule and the tightness of the crystal packing of the molecule. Yalkowski [1] has produced a general solubility equation, for organic non electrolytes incorporating the entropy of melting and melting point as a measure of crystal packing and logP as a measure of lipophilicity.

The rate of dissolution is effected by surface area and solubility. The extent of dissolution as represented by the actual concentration of drug in the bulk of the solution (aqueous contents of gastrointestinal tract) depends only on solubility. Concentration of drug in solution is the driving force of the membrane transfer of drug into the body and low aqueous solubility can continue to present itself as a problem even after formulation improvements. Examples of low solubility, dissolution limited drugs include danazole, griseofulvin, halofantrine, ketoconzaole, nitrofurantoin, phenytoin and triamterene.

3 MEMBRANE TRANSFER

The barrier of the gastro-intestinal tract is similar to any other that involves the crossing of biological membranes. Biomembranes are composed of a lipid-bilayer. The bilayer results from the orientation of the lipids (phospholipids, glycolipids and cholesterol) in the aqueous medium. Phospholipids are ampipathic with polar head groups and lipid "tails" and align so that the polar head groups orientate to the aqueous medium and the lipid tails form an inner hydrophobic core. Tight junctions are formed by the interaction of membrane proteins at the contact surfaces between single cells. Tight junctions can be viewed as small aqueous filled pores. The number of the tight junctions depends on the cell or tissue. A useful guide to the total number of aqueous channels available is provided by comparing the total surface area of membrane available for absorption compared to the area of aqueous pores. For the small intestine the area of aqueous pores amounts to about 0.01% of the whole surface.

Compounds can cross biological membranes by two passive processes, transcellular and paracellular. For transcellular diffusion the compound can distribute into the lipid core of the membrane and diffuse within the membrane to the basolateral side or alternatively, the solute may diffuse across the apical cell membrane cross, the cytoplasm of the cell before exiting across the basolateral membrane. Both processes involve diffusion through the lipid core of the membrane and therefore the ability of the compound to 'dissolve' in lipid (lipophilicity) is critical. Paracellular absorption involves the passage of the compound through the aqueous filled pores. Clearly all compounds can be absorbed by this route but the process is invariably slower than transcellular (surface area of pores *versus* surface area of the membrane) and is very dependant on molecular size due to the finite dimensions of the aqueous pores.

Compounds crossing the gastrointestinal tract *via* the transcellular route can usually be absorbed throughout the length of the tract. In contrast the paracellular route is only, readily, available in the small intestine and the term "absorption window" is often applied. The calculated human pore sizes (radii) are jejenum 6-8Å, ileum 2.9-3.8 Å and colon less than 2.3 Å . In practice the small intestine transit time is around six hours whilst transit of the whole tract is approximately 24 hours. For lipophilic compounds (providing dissolution is adequate) transcellular passage across membranes is rapid. Moreover, since the drug is absorbed throughout the g.i. tract the proportion of a dose absorbed is high (complete). For hydrophilic compounds, transfer between cells by the paracellular pathway is slow and broadly inverseley proportional to molecular size. Moreover, the "absorption window" referred to above means that the drug rapidly moves away from the absorption site. Consequently most paracellularly absorbed compounds show incomplete absorption with the proportion decreasing with molecular size. Examples of compounds absorbed by the paracellular route include atenolol. nadolol, ranitidine, cimetidine and sumatriptan.

For simple molecules, like β-adrenoceptor antagonists octanol log $D_{7.4}$ values are remarkably predictive of absorption potential. Compounds with $logD_{7.4}$ values below 0 are absorbed by the paracellular route and compounds with $logD_{7.4}$ values above 0 are absorbed by the transcellular route. For instance atenolol ($logD_{7.4}$ -1.5, mwt 266), sotalol ($logD_{7.4}$ - 1.7, mwt 272) nadolol ($logD_{7.4}$ - 2.1, mwt 309) and xamoterol ($logD7.4$), - 1.0, mwt 339) are absorbed by the paracellular route to the extent of 50,100, 13 and 9%. In contrast propranol ($logD_{7.4}$, 0.9, mwt 295) is 100% absorbed by the transcellular route. Note, however, that lipophilicity correlates with increased metabolic lability and such compounds may have their apparent systemic availabilities decreased by metabolism as they pass through the gut and the liver. Thus the systemic bioavailability of propranolol is reduced to 5-50% [2].

More complex structures result in a more complex interplay of physicochemical parameters. As the number of H-bonding functions rise so octanol $logD_{7.4}$, in isolation, becomes a progressively less valuable predictor. For such compounds desolvation and breaking of H-bonds becomes the rate limiting step in transfer across the membrane. Methods to calculate H-bonding potential range from simple H-bond counts (number of donors and acceptors), through systems that assign a value of 1 for donors and 0.5 for acceptors to sophisticated scoring systems such as the Raevsky H-bond score [3].

None of these methods gives a perfect prediction, particularly because H-bonding potential needs to be overlayed over intrinsic lipophilicity. For this reason Lipinski's rule of five become valuable as defining the outer limits in which chemists can work in [4]. Lipinski defined the boundaries of good absorption potential by demonstrating that poor permeabilty was produced by:

a) More than five H-bond donors (sum of OH's and NH's).

b) More than ten H-bond acceptors (sum of N's and O's).

c) Molecular weight over 500

and poor dissolution by:

a) Log P over 5.

Pro-drug approaches have been tried extensively as a way to overcome the intrinsic properties of a molecule that attenuate absorption, that of poor solubility [5] and low lipophilicity / high H-bonding potential [6].

To overcome poor solubility assuming the molecule has good properties for membrane transfer the resultant pro-drug must:

(i) still retain the properties (lipophilicity) required for membrane transfer and be hydroysed rapidly in the systemic circulation or liver .

(ii) or be hydrolysed at the aqueous / membrane barrier of the gastrointestinal tract.

Benzimadazole compounds provide [5] examples of the former (i). N-alkoxycarbonyl prodrugs of drugs, such as mebendazole , decrease the melting point and increase aqueous solubility, whilst maintaining the lipophilicity to values favourable for membrane transfer (figure 1).

Phenytoin provides an example that may relate more to the latter case (ii) than the former (i)[5].

These prodrugs rely on a hydroxymethyl spacer (figure 1) that will spontaneously decompose to phenytoin upon the hydrolysis of the pro-moiety. The phosphate ester derivative relies largely on hydrolysis at the brush border membrane and increases oral bioavailabilty 3.5 fold.

To overcome poor membrane transfer the pro-drug must lower H-bonding potential and raise liophilicity without adverse effects on solubility. Lipinski's rules also offer a framework for drug design since the addition of the pro moiety must not raise the molecular weight much over 500. In addition the pro-drug must be rapidly converted back to the parent compound once entry to the systemic circulation has been achieved.

The esterification of a carboxylic acid function in a molecule has the immediate effect of a reduction in H-bonding potential and an increase in lipophilicity. Such parameters are important in the oral absorption of compounds as described before. Candoxatrilat (figure 2), an inhibitor of neutral endopeptidase, has poor oral bioavailability[7] .

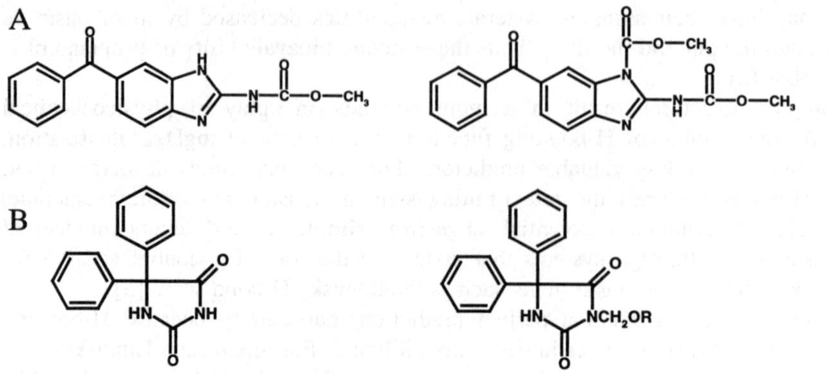

A

B

Figure 1 *Approaches to overcome poor aqueous solubility: (A) mebendazole (right) and its N-alkoxy derivative (left) and (B) phenytoin (right) and the unstable hydroxymethyl linking unit used in a prodrug series (left)*

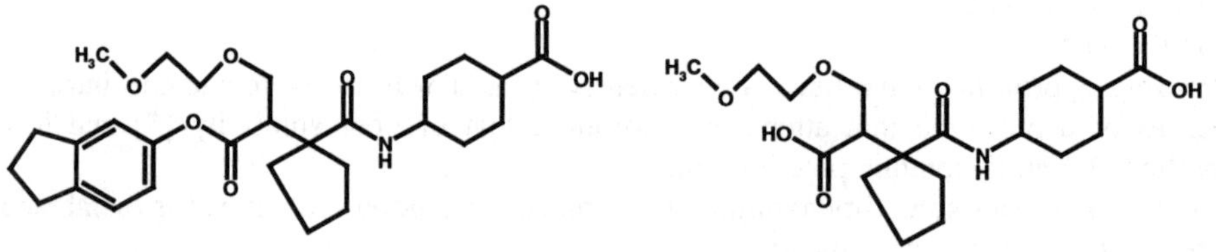

Figure 2 *Structures of candoxatril (right) and candoxatrilat (left)*

The compound has a log D of -2 and a molecular weight of 399. The indanyl ester analogue[7] candoxatril (figure) has a log D of 1.5, and a Raevsky score reduced by 2.3 units and a molecular weight of 515. As such the compound is within or close to most of the properties expected for an oral agent. The prodrug is well absorbed, rapidly hydrolysed but complete conversion is not achieved. The proportion of candoxatrilat liberated depends on competing clearance processes for candoxatril clearance e.g. hepatic uptake/ biliary clearance. For candoxatrilat the values of systemic availability after oral administration to mouse, rat, dog and man are 88, 53, 17 and 32% and depend on the esterase activity (of which rodents have the highest) and the competing processes (of which man is probably the lowest).

The pro-drug route is often problematic due to the complex interplay of the many factors referred to above and later. Fluconazole is an example of a drug design programme aimed at producing an antifungal agent [8] with superior intrinsic pharmacokinetics to the first orally active azole antifungal drug ketoconazole (Fig 3). Ketoconazole is cleared primarily by hepatic metabolism and is difficult to formulate to obtain consistent absorption, due to poor aqueous solubility (and resultant poor dissolution) due to its high lipophilicity log P>5. Fluconazole was designed to have sufficient lipophilicity for good membrane transfer, but also high water solubility. The compound is the least lipophilic (log P 0.5) member of a series of metabolically stable bis-triazoles. This physicochemical profile gives fluconazole rapid dissolution and complete transcellular absorption. This complete absorption is not affected by food, achlorhydria, etc.

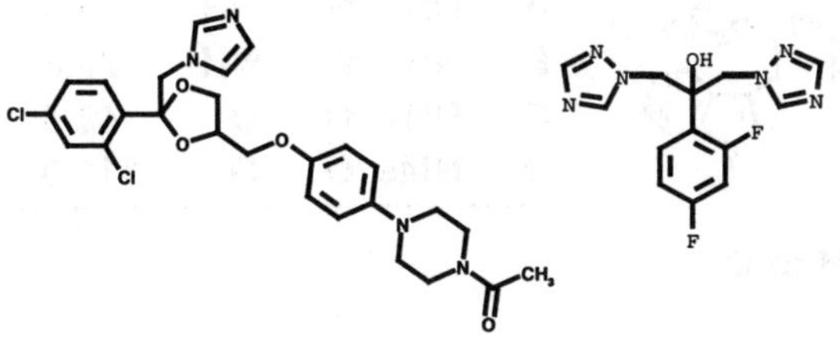

Figure 3 *Structures of ketoconazole (left) and fluconazole (right), fluconazole was designed for optimum pharmacokinetics, including oral absorption*

The incorporation of an ionisable centre, such as an amine or similar function, into a template can bring a number of benefits including water solubility. A key step [9] in the discovery of indinavir was the incorporation of a basic amine (and a pyridine) into the backbone of hydroxyethylene transition state mimic compounds (figure 4) to enhance solubility (and potency).

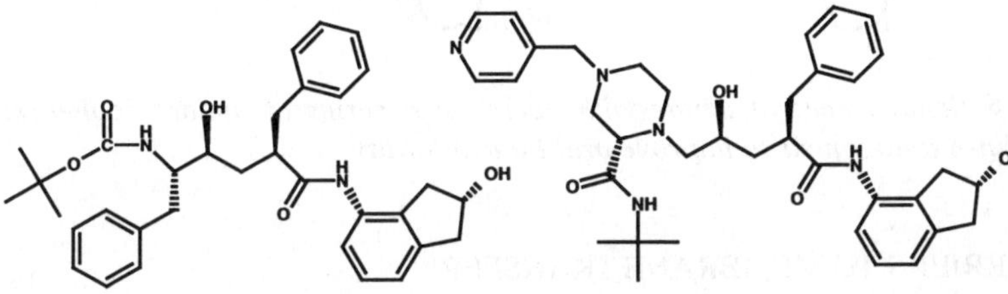

Figure 4 *Structures of lead compound L-685,434 (right) and indinavir (left) which incorporates basic functions aiding water solubility*

The medicinal chemist can use the above rules, understand the boundaries and work to lowering these values in more complex structures. Figure 5 shows a synthetic strategy aimed at removing H-bond donors from a series of endothelin antagonists and a resultant increase in apparent bioavailability as determined by intradueodenal AUC [10]. Noticeably cLogP values vary only marginally with the changes in structure, values being 4.8, 5.0, 4.8 and 5.5 for compounds A, B, C and D respectively. In contrast the number of H-bond donors is reduced by 3 and the Raevsky score from 28.9 (A) to 21.4 (D).

A similar example, also from endothelin antagonists, is the replacement of the amide group (figure 6) in a series of amidothiophenesulfonamides with acetyl [11]. This move retained in vitro potency, but markedly improved oral bioavailability.

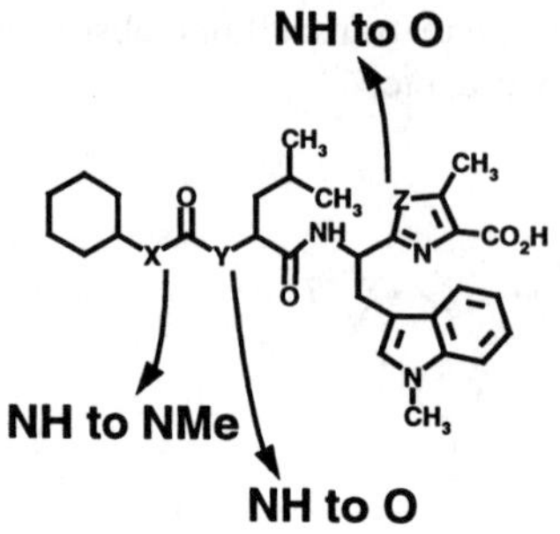

	X	Y	Z	i.d. AUC (µg.min/ml)
A	NH	NH	NH	0.3
B	NH	O	NH	20.8
C	NH	O	O	48.9
D	NMe	O	O	110.3

Figure 5 *Synthetic strategy for a series of azole containing endothelin antagonists aimed at improving bioavailability by lowering H-bonding potential*

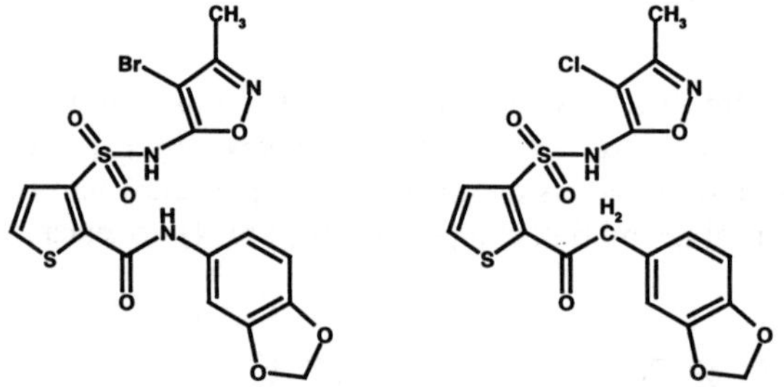

Figure 6 *Replacement of amide with acetyl in a series of amidothiophenesulfonamide endothelin-A antagonists to improve oral bioavailabilty*

4 BARRIERS TO MEMBRANE TRANSFER

The cells of gastrointestinal tract contain a number of enzymes of drug metabolism and also various transport proteins. Of particular importance in the attenuation of absorption/ bioavailability are the glucuronyl and sulphotransferases which metabolise phenol containing drugs sufficiently rapidly to attenuate the passage of intact drug across the gastrointestinal tract. Cytochrome P450 enzymes are also present, in particular CYP3A4 (see later) and again certain substrates for the drug may be metabolised during passage across the tract. This effect may be greatly enhanced by the action of the efflux pumps, in particular p-glycoprotein P [12]. P-glycoprotein's range of substrates is large but includes a number of relatively large molecular weight drugs which are also CYP3A4 substrates. Cyclosporin is one example. The drug shows significant attenuation of absorption across the gastrointestinal tract due to efflux and metabolism [12]. It can be postulated that in effect absorption of the drug is followed by secretion back into the lumen of the gut by p-glycoprotein. This cyclical process effectively exposes cyclosporin to "multi-pass" metabolism by CYP3A4 and a resultant lowered appearance of intact cyclosporin in the circulation.

There is not yet a concise structure-activity relationship for p-glycoprotein. What is available suggests that the best substrates (for efflux) are lipophilic, often basic and contain a number of H-bond donor and acceptor functions. Most substrates have molecular weights approaching 500 or greater. The physiochemical properties leading to poor permeability characteristics outlined earlier, probably include a component of efflux alongside the factors limiting membrane transfer *per se*: ability to shed the surrounding water sheaf and traverse the lipid core of the membrane.

These considerations are important in pro-drug design and add to the complexity referred to earlier. Many active principles in pro-drug programmes are non lipophilic compounds, possessing a number of H-bond donor and acceptor functions (amide or peptide linkages). Addition of a pro-moiety will raise the lipophilicity and molecular weight. In doing so the final molecule may have the characteristics to traverse the lipid core of a membrane, but this advantage is lost by it becoming a substrate for efflux. An example (figure 7) of this is the fibrinogen receptor antagonist L-767,679, a low lipophilicity compound (log P<-3) with resultant low membrane flux. The benzyl ester (L-775,318) analogue (log P 0.7) also showed limited absorption, and studies in Caco-2 cells showed the compound to be effluxed by P-glycoprotein [13].

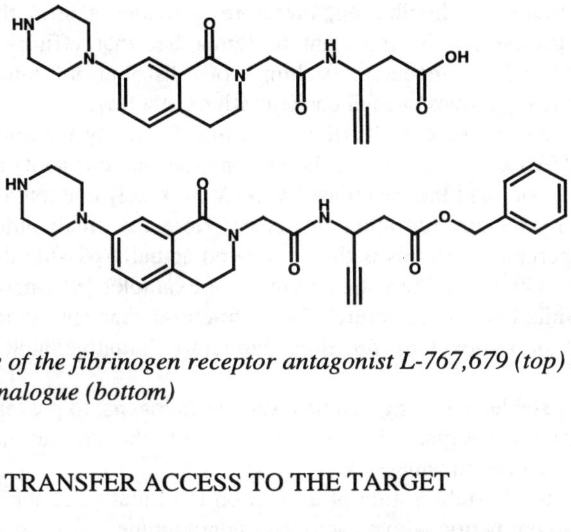

Figure 7 *Structures of the fibrinogen receptor antagonist L-767,679 (top) and its benzyl ester (L-775,318) analogue (bottom)*

5 MEMBRANE TRANSFER ACCESS TO THE TARGET

Drug targets can be broadly classified as cell surface, cell surface within the CNS and intracellular. Penetration from the circulation into the interstitial fluid is rapid for all drugs since the aqueous pores present in capillary membranes have a mean diameter of between 50-100 angstroms. Thus there is ready access to targets located at the surface of cells such as G-protein coupled receptors. The exception to this is the cerebral capillary network, since here there is a markedly reduced number of pores due to the continuous tight intercellular junctions. Thus to readily access the CNS drugs have to cross the lipid bilayer. This is particularly so since there is a net outflow of ECF and CSF effectively clearing water soluble drugs from the brain. For intracellular targets, if only the free drug is considered then at steady state the concentrations present inside the cell and in the circulation should be similar for a drug that readily crosses the cell membrane.

Key factors in crossing the membranes of cells to access targets parallel those for oral absorption: lipophilicity, as defined by partition coefficient, hydrogen bonding capacity [14], and molecular size [15]. For simple small molecules with a minimum of N or O containing groupings a positive log D value is a good indicator of ability to cross the membrane. For more complex molecules size and H-bonding count become important. The work of Young et al. [14] provides a good specific example of the role of increased H-bonding potential in preventing access to the CNS (crossing capillary and astrocyte cell membranes. Lipinski's rules provide a general framework, however a positive (>0) log P or Log D value is needed to dissolve in the lipid bilayer. For a small drug molecule, penetration to the target may often be easier to achieve than duration of action. Assuming duration of action is linked to drug half life then distribution can be an important factor.

6 VOLUME OF DISTRIBUTION AND DURATION

The volume of distribution of a drug molecule is, as described previously, a theoretical number that assumes the drug is at equal concentration in the tissue to that in the circulation and represents what volume (or mass) of tissue is required to give that concentration. Volume of distribution, therefore, provides a term that partially reflects tissue affinity. However, it is important to remember that affinity may vary between different tissues and a moderate volume of distribution may reflect moderate concentrations in many tissues or high concentrations in a few.

The importance of volume of distribution is in influencing the duration of drug effect. Since half-life ($0.693/kel$, where kel is the elimination rate constant) is determined by the volume of distribution and the clearance ($Cl_{(f)}=Vd_{(f)} \times kel$) manipulation of volume is an important tool for changing duration of action. Here the small amount of drug in the circulation is important, since this is the compound actually passing through the organs of clearance (liver and kidney). There are a number of examples [8] where clearance has been modulated by manipulation of structure. These structural changes (figure 8) include:

(i) Replacement or removal of functions labile to oxidative attack such as benzylic positions.

(ii) Introduction of stable "blocking" groups, such as halogens, to prevent oxidative attack.

(iii) Structural changes that place the compound outside the structure-activity relationships of particular oxidative enzymes.

(iv) Steric and electronic stabilisation of a function to oxidative attack such as replacement of a teriary amine function with a cyclic secondary amine.

(v) Use of bioisosteres (such as indole) to replace phenolic functions labile to conjugation processes.

Taking the simplest case of neutral drugs, where no additional interactions exist, log $D_{7.4}$ reflects increased binding to lipophilic sites present in cellsand their membranes, there is a trend for increasing volume of distribution with increasing lipophilicity (Fig. 10).

Simple changes in lipophilicity (per se) are not normally sufficient to increase duration. This is because as a general rule increases in lipophilicity are reflected in increases in hepatic (metabolic) clearance. For example nifedipine (figure 9) has a log P value of 2.86, a free unbound plasma clearance of 175 ml/min/kg and a free unbound volume of distribution of 19.5L/kg, resulting in an elimination half life of 1.8 hours [20]. Nimodipine (figure 9), in contrast, has a log P of 4.18 resulting in increased distribution volume (85L/kg) but also increased clearance (950 ml/min/kg) compared to nifedipine [20]. These changes, effectively, cancel each other out and the elimination half life is similar to nifedipine (1.1 hours).

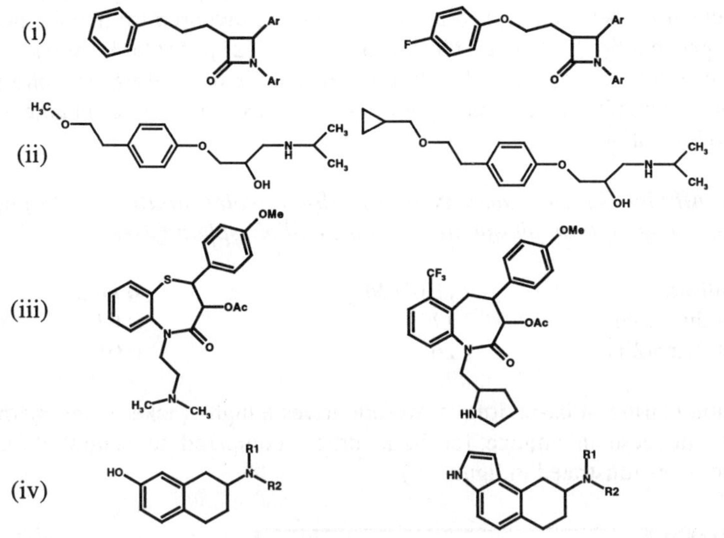

Figure 8 *Examples of (i), removal of a labile benzylic function and halogen substitution to prevent oxidation by P450 [16], (ii),Structural changes that move the molecule outside the SAR of a particular P450 [17] , (iii), Stabilisation of an amine function by substitution of a tertiary amine for a cyclic secondary amine [18] and (iv) use of a hydrogen bond donor isostere to allow receptor interaction but prevent glucuronide and sulphate conjugation [19] by the gut and liver. The susceptible compound is shown on the left, the "stable" derivative on the left.*

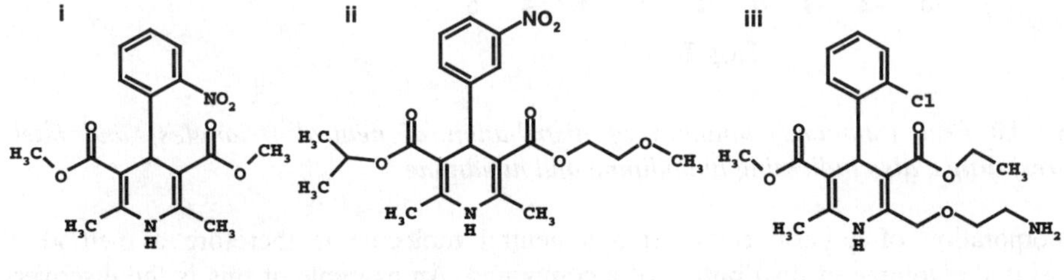

Figure 9 *Structures of (i) nifedipine, (ii) nimodipine, (iii) amlodipine*

In most cases, the volume of distribution is highest for basic drugs ionized at physiological pH (figure 10). As a model of distribution in the body the partitioning of molecules has been studied between unilamellar vesicles of dimyristoylphosphatidylcholine and aqueous buffers. These systems allow the interaction of molecules to be studied with the whole membrane which includes the charged polar head group area (hydrated) and the highly lipophilic carbon chain region. Such studies indicate that partitioning for basic compounds which are ionised at physiological pH, partioning into the membrane is highly favoured, compared to neutral or acidic compounds. This is due to electrostatic interactions with the charged phospholipid head group[21] combined with lipophilic interactions, in contrast to simple lipophilic interactions.. Such ionic interactions between basic drugs are even more favoured for membranes containing " acidic" phospholipids such as phosphatidylserine [22]. Table 1 shows the preferential binding in terms of capacity of chlorphentermine, a basic drug, for phosphatidylserine containing membranes, the phospholipid with overall acidic charge, compared to phosphatidylcholine.

Table 1 *Affinity (k) and capacity (moles drug/ moles lipid) of chlorphentermine for liposomes prepared from phosphatidylcholine and phosphatidylserine*

Phospholipid	k [10^{-4}] M	n *max*
phosphatidylserine	2.17	0.67
phosphatidylcholine	1.26	0.05

This ion-pairing of basic drugs therefore drives a high uptake by the membrane and the resultant   increase in volume for basic drugs, compared to neutral drugs, at a given lipophilicity (as illustrated in figure 10).

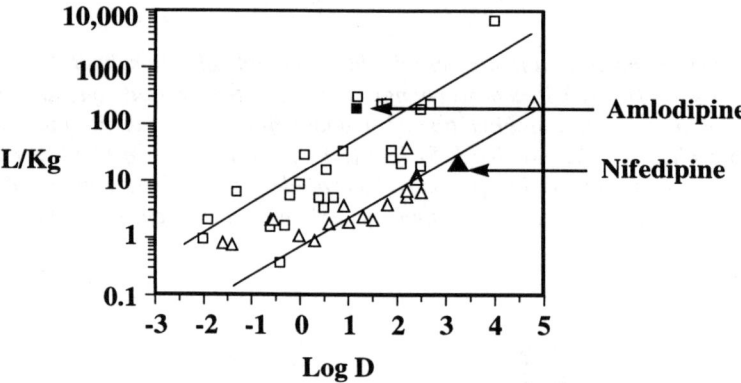

Figure 10 *Free (unbound) volumes of distribution of neutral (triangles) and basic (squares) drugs, also indicating amlodipine and nifedipine*

Incorporation of a basic centre into a neutral molecule is therefore a method of increasing the volume of distribution of a compound. An example of this is the discovery of the series of drugs based on Rifamycin SV (figure 12). This compound was one of the first drugs with high activity against Mycobacterium tuberculosis. Its clinical performance [23], however, was disappointing due to poor oral absorption (dissolution) and very short duration ascribed at the time to rapid biliary elimination (clearance).

Many different analogues were produced, including introduction of basic functions with a goal of increased potency, solubility, and reduction in clearance. Rifampicin is a methyl-piperazinyl amino methyl derivative [20] with much better duration and a successful drug. The basic functionality however does not alter clearance but increases volume substantially (figure 11).

A

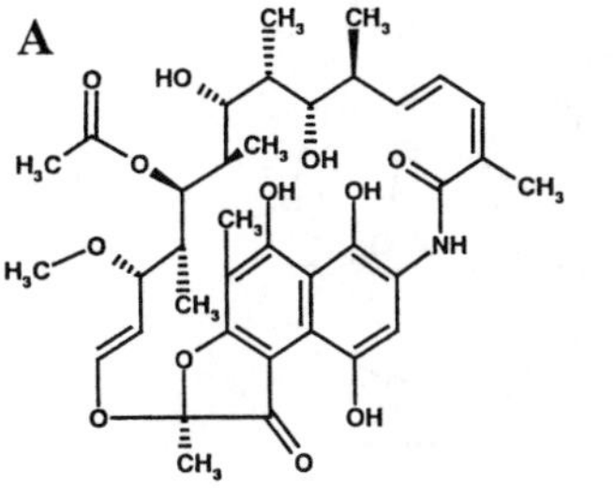

B

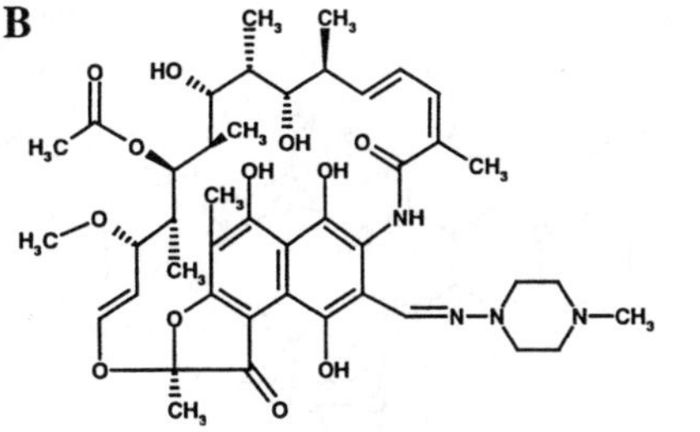

C

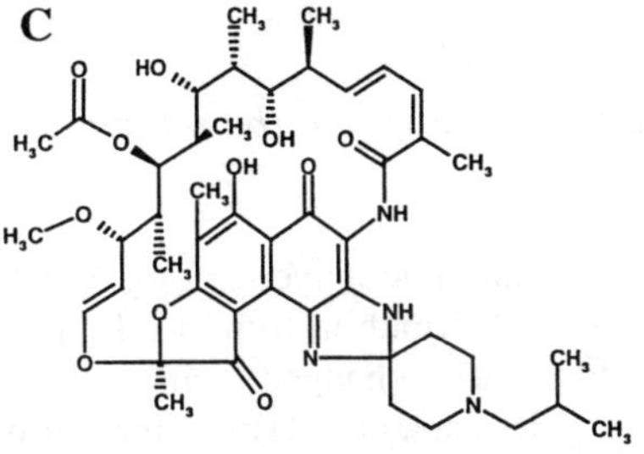

	Vd l/Kg	Clp ml/min/Kg	T1/2 hours
A	1.2	80	0.33
B	8.8	32	3.5
C	266	80	45

Figure 11 *Structures of (A) rifamycin, (B) rifampicin and (C) rifabutin, together with their pharmacokinetic properties. Volume of distribution (Vd) and plasma clearance (Clp) are for free unbound drug*

Duration is enhanced further [20] by the more basic spiropiperidyl analogue, rifabutin (figure 11). Again the desirable pharmacokinetic (and pharmacodynamic) properties are due to effects on volume rather than effects on clearance.

This strategy of modification of a neutral molecule by addition of basic functionality was employed in the discovery of the dihydropyridine calcium channel blocker amlodipine. The long plasma elimination half-life (35 hours) of the dihydropyridine calcium channel blocker, amlodipine (figure 9), is due, in large part, to its basicity and resultant high volume of distribution (free unbound volume: 228 L/kg) [8].

These pharmacokinetic parameters are unique among dihydropyridine calcium channel blockers and allow once-a-day dosing of amlodipine, without the need for sustained release technology. The large volume of distribution is achieved despite the moderate lipophilicity of amlodipine and can be compared to the prototype dihydropyridine drug nifedipine which is of similar lipophilicity but neutral (figure 9). Notably, these changes in structure do not trigger a large change in clearance from that of nifedipine (free unbound drug clearance of amlodipine: 85 ml/min/kg). Detailed investigations have confirmed a specific ionic interaction between the protonated amino function and the charged anionic oxygen of the phosphate headgroups, present in the phospholipid membranes, is responsible for amlodipine's high volume of distribution.

Another basic drug where minor structural modification results in a dramatic increase in volume of distribution is the macrolide antibiotic, azithromycin [8]. The traditional agent in this class is erythromycin, which contains one basic nitrogen in the sugar sidechain (figure 12)

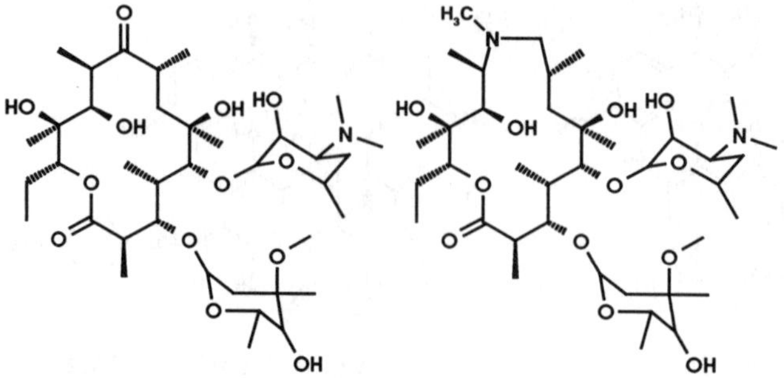

Figure 12 *Stuctures of the macrolide antibiotics erythromycin (basic) and azithromycin (di-basic)*

Introduction of a second basic center into the macrolide aglycone ring in azithromycin (figure 12) increases the free (unbound) volume of distribution from 4.8 L/kg to 62 L/kg. Free (unbound) clearance of the two compounds is also changed from 55 ml/min/kg for erythromycin to 18 ml/min/kg for azithromycin . The apparent plasma elimination half-life is, therefore, increased from 3 to 48 hs.   Overall, the pharmacokinetic properties of azithromycin provide adequate concentrations for efficacy on a once-daily dosing regimen. The high and prolonged concentrations of azithromycin achieved provide a long duration of action and only 3-5 day courses of treatment are required (8).

A similar example to azithromycin, but in a small molecule series is pholcodine (figure 13), where a basic morpholino side chain replaces the methyl group of codeine. Unbound clearance is essentially similar (10 ml/min/Kg) but the free unbound volume is increased approximately 10 fold (4 to 40 l/Kg) with a corresponding increase in half life (3 to 37 hours) [18].

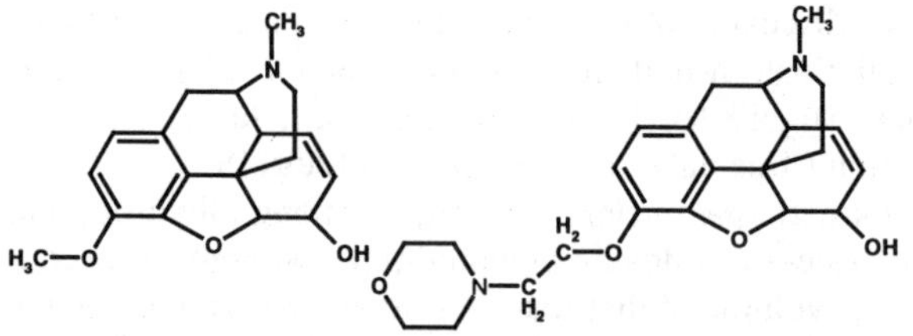

Figure 13 *Structures of the monobasic codeine (right) and the dibasic pholcodine (left)*

A high volume of distribution andaccumulation of drug in tissues explains the 7-times longer elimination half-life of the dibasic antiarrhythmic, disobutamide (figure 14) compared to the monobasic agents, disopyramide.   The elimination half-life of disobutamide is 54 hours compared to approximately 7 for disopyramide [25]

Disobutamide has been shown to accumulate extensively in tissues in contrast to disopyramide [25]. It is important to note, in these examples, that the high volume of distribution and tissue affinity of the well tolerated, anti-infective, azithromycin, is viewed as a pharmacokinetic and therapeutic advantage. Whilst the same properties are viewed as disadvantageous for the low safety margin, antiarrhythmic, disobutamide.

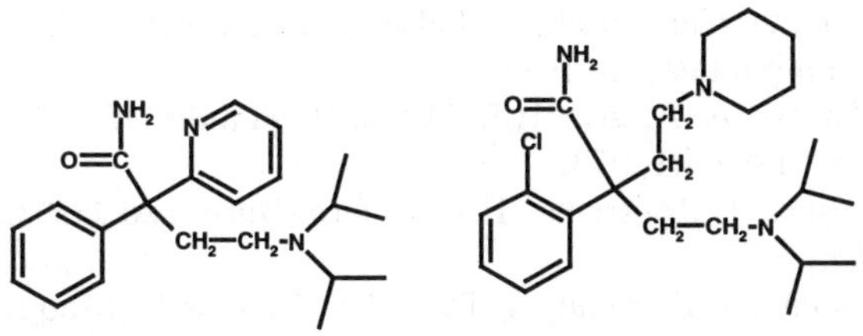

Figure 14 *Structures of the monobasic antiarrythmic disopyramide (right) and the dibasic analogue disobutamide (left)*

Obviously, different therapeutic areas impose different restrictions on the ideal pharmacokinetic profile for management of each condition, hence careful consideration should be paid to this at an early stage in drug discovery programs.

In summary, features that provide good dissolution properties, such as the incorporation of ionised (basic) functionality can markedly improve the duration of a compound in the body. In working with neutral lead compound, part of the synthetic strategy should be directed at achieving this objective.

References

1. S.H. Yalkowski and S.C. Valvani,, *J. Pharm. Sci.*, 1980 **69**, 912.
2. A. Hayes, and R.G. Cooper, *J. Pharm. Exper. Ther.*, 1971 **176**, 302..
3. O.A. Raevsky, V.Y. Grifor'er,. D.B. Kireev, and N.S. Zefirov, *Q.S.A.R.* 1992, **14**, 433.
4. C.A. Lipinski,, F. Lombardo, B.W. Dominy and P.J. Feeney, *Adv. Drug Delivery Rev.* 1997, **23**, 3.
5. D. Fleisher., R. Bong and B.H. Stewart,*Adv. Drug Delivery Rev.*, 1996, **19**, 115
6. M.D. Taylor, 1996, *Adv. Drug Delivery Rev.*, **19**, 131.
7. B. Kaye, C.J. Brearley, N.J. Cussans, M. Herron, M.J. Humphrey, M.J. and A.R. Mollatt, Xenobiotica, 1997, **27**, 1091.
8. D.A. Smith, B.C. Jones and D.K. Walker, *Medicinal, Research Reviews*,1996, **16**, 243.
9. J.P. Vacca, B.D. Dorsey,, W.A. Schleif, R.B. Levin., S.L. McDaniel, P.L. Darke, J. Zugay, J.C. Quintero, O.M. Blahy, E. Roth, V.V. Sardana, A.J. Schlabach, P.L. Graham,, J.H. Condra, L. Gotlib, M.K. Holloway, J. Lin, I.W. Chen, K. Vastag, D. Ostovic, P.S. Anderson, E.A. Emini and J.R. Huff, 1994, *Proc. Natl. Acad. Sci. 1994*, **91**, 4096
10. T.W.Von Geldern, D.J. Hoffman, J.A. Kester, H.N. Nellans, B.D. Dayton, S.V. Calzadilla, K.C.Marsh, L. Hernandez, and W. Chiou, , *J. Med. Chem.*, 1996, **39**, 982
11. C. Wu,, M.F. Chan, F. Stavros, B. Raju, I. Okun,, S. Mong, K.M. Keller, T. BROCK, T.P. Kogan, and R.A.F. Dixon, *J. Med. Chem.*, 1997, **40** , 1690
12. P.B.Watkins, *Adv. Drug Del., 1997, Rev* **27**, 161.
13. T. Prueksaritanont, P. Deluna, L.M. Gorham, M.A. Bennett, D. Cohn, J. Pang, X. Xu, K. Leung and J.H. Lin, *Drug Met. and Disp,.1998,* **26**, 520..
14. R.C. Young, R.C. Mitchell, T.H. Brown, C.R. Ganellin, R. Griffiths, M. Jones, K.K. Rana, D. Saunders, I.R. Smith, N.N.E. Sore, T.J. Wilks, 1988, *J. Med. Chem.*, 1988, **31**, 656.

15. H. Van de Waterbeemd, G. Camenisch, G. Folkers, J.R. Chrétien, Q.A. Raevsky, , J. Pharm. Sci., accepted for publication. 1988.
16. S.B. Rosenblum, T. Huynh, A. Afonso, H.R. Davis, N. Yumibe, J.W. Clader. And D.A. Burnett, *J. Med. Chem.*, 1998, **41**, 973.
17. P.M. Manoury, J.L. Binet, J. Rousseau, F.M. Leferre-Borg, and I.G. Cavero, *J. Med. Chem*, 1987, **30**, 1003.
18. D.M. Floyd, S.D. Dimball, J. Drapcho, J. Das, C.F. Turk, R.V. Moquin, M.W. Lago, K.J. Duff, V.G. Lee, R.E. White, R.E. Ridgewell, S. Moreland, R.J. Brittain, D.E. Normandin, S.A. Hedberg, and G.C. Cucinotta, *J. Med. Chem.*,1992, **35**, 756.
19. P. Stjernlof, M. Gullme, T. Elebring, B. Andersson,, H. Widstrom, S. Lagerquist, K. Svensson, K. Ekman., A. Carlsson and S. Sundell, *J. Med. Chem.* 1993, **36**, 2059.
20. L.Z. Benet, S. Oie, S. and SCHWARTZ, J.B., 1995, 'Design and optimization of doseage regimens; pharmacokinetic data, in The Pharmacological Basis for Therapeutics', McGraw-Hill, New York.pp. 1707.
21. R.P. Austin, A.M. Davis, C.N. Manners, . *J. Pharm. Sci.*,1995, **84**, 1180
22. .H. Lullman and M. Wehling, *Biochem. Pharmacol.1979,* **28**, 3409.
23. N. Bergamini and G. Fowst, 1965, Rifamycin SV, *Arznl-Forsch.1965,* **15**, 951.
24. W.A. Fowle, A.S.E. Butz, E.C. Jones, B.C. Weatherly, R.M. Welch, J. Posner, *Brit. J. Clin. Pharmol, 1986,* **22**, 61.
25. C.S. Cook, S.J. McDonald, A. Karim, *Xenobiotica*, 1993, **23**, 1299.

DRUG METABOLISM AND THE MEDICINAL CHEMIST : SO WHAT ?

Bernard Testa

Institut de Chimie thérapeutique, Ecole de Pharmacie, Université de Lausanne, CH-1015 Lausanne, Switzerland

ABSTRACT

Drug design as practised today is mainly a ligand design aimed at discovering compounds with high affinity towards predefined biological targets. Modern high-throughput techniques have rendered this strategy immensely successful, but they have done nothing, to say the least, to shorten the path leading from a high-affinity ligand to a pharmacokinetically and toxicologically well-behaved drug candidate. To decrease the costly and time-consuming development of active compounds ultimately doomed by hidden pharmacokinetic and toxicological defects, medicinal chemists are coming to integrate metabolic considerations into drug design strategies. Various aspects of drug metabolism of interest to medicinal chemists are:

- The chemistry and biochemistry of reactions of toxication and detoxication;
- Predictions of drug metabolism based on expert systems, quantitative structure-metabolism relationships, and molecular modelling of enzymatic sites.
- Prodrug and soft drug design;
- Changes in physicochemical properties (acidity, basicity, lipophilicity, etc) resulting from biotransformation;

1 INTRODUCTION

Drug design as practised today is mainly a ligand design aimed at discovering compounds with high affinity towards predefined biological targets. Modern high-throughput techniques have rendered this strategy immensely successful, but they have done nothing, to say the least, to shorten the path leading from a high-affinity ligand to a pharmacokinetically and toxicologically well-behaved drug candidate. To decrease the costly and time-consuming development of active compounds ultimately doomed by hidden pharmacokinetic and toxicological defects, medicinal chemists are coming to integrate relevant considerations into drug design strategies.

Reactions of biotransformation play a major role in influencing the nature, intensity and duration of wanted and unwanted effects of drugs and drug candidates[1-4]. Such influences can schematically be classified into two categories, as shown in Table 1.

Table 1   *Major   pharmacodynamic   and   pharmacokinetic   consequences   of drug metabolism*

Major pharmacodynamic consequences of drug metabolism
The drug yields one or more metabolites which contribute to its therapeutic effects; The drug is inactive per se (prodrug) but is transformed into an active metabolite responsible for the therapeutic effects; The drug yields one or more metabolites which are responsible for unwanted or downright toxic effects (toxication).
Major pharmacokinetic consequences of drug metabolism
Whether active and/or inactive metabolites are formed, the rate of metabolism affects the duration and intensity of action of the drug. The drug induces one or several enzymes mediating its metabolism (auto-induction), resulting in a therapeutic response that changes over days or weeks. A metabolite acts as inhibitor of one of the metabolic pathways, resulting in complex kinetics. One or more metabolites have physicochemical properties vastly different from those of the parent compound, for example a very high lipophilicity resulting in tissue accumulation and residue retention.

A good understanding of drug metabolism is thus an essential component of the medicinal chemist's intellectual assets. In more practical terms, Table 2 summarises four aspects of drug metabolism of interest to medicinal chemists who, starting from hits or lead compounds, have it as their objective to design promising drug candidates with reduced risks of pharmacokinetic or toxicological failure.

Table 2   *Aspects   of   drug   metabolism   of   major   interest   to   medicinal chemists*

The chemistry and biochemistry of reactions of toxication and detoxication (see section 2); Predictions of drug metabolism based on expert systems, quantitative structure-metabolism relationships, and molecular modelling of enzymatic sites[5]; Prodrug and soft drug design[2,6,7]; Changes in physicochemical properties (acidity, basicity, lipophilicity, etc) resulting from biotransformation[8-12] (see section 3).

Among the four aspects listed in Table 2, two are discussed in the present review, namely toxication and detoxication, and the chamges in physicochemical properties resulting from biotransformation. This is not to say, of course, that prediction of drug metabolism or prodrug design are not of major interest to medicinal chemists. The few references given in Table 2 may serve to guide the interested reader.

2 TOXICATION RESULTING FROM METABOLIC REACTIONS

Toxication resulting from biotransformation is a major issue in the molecular toxicology of xenobiotics[13-15]. It is also of great significance in drug design[16], and as such constitutes one of the two main arguments of this mini-review.

2.1 Biochemical Aspects of Toxication Reactions

The relation between metabolism and toxicity must first be seen in a broad context comprising both partners (the drug and its metabolites) and the two types of pharmacodynamic effects they can produce, namely wanted (therapeutic) and unwanted (unfavourable and toxic) effects. As shown in Figure 1, both the drug and its metabolites can contribute to wanted and unwanted effects.

Metabolism and Pharmacodynamic Activity

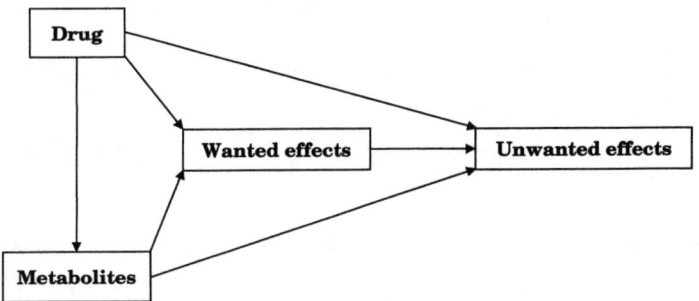

Figure 1 *A priori, both the drug and its metabolites can produce wanted and unwanted effects (reproduced with permission).*

Biotransformation can conveniently be subdivided into reactions of functionalisation (phase I reactions, which create or modify a functional group), and reactions of conjugation (phase II reactions) which covalently attach an endogenous moiety to a xenobiotic or a metabolite thereof[1-4]. The two classes of reactions can produce toxic metabolites which may elicit various types of cellular damage, but by essentially different mechanisms. Major molecular mechanisms of toxicity include:

- covalent binding to biological macromolecules (reactive intermediates produced by phase I and II reactions);
- oxidative stress produced by oxygen-activating phase I metabolites;
- interference with physiological pathways;
- accumulation of lipophilic residues formed by conjugation reactions.

   A number of functional groups (the so-called toxophoric groups) are well known for their potential to undergo metabolic toxication. Important toxophoric groups activated by a reaction of oxidation or reduction are listed in Table 3.
   Thus, the cytochrome P450 catalyzed activation of ethynyl groups (Figure 2; R' = H; reaction **a**) leads to reactive products such as oxirenes (reaction **b**), ketenes (reaction **c**), or intermediates that immediately bind covalently to the catalytic site of the enzyme (reaction

d). The oxidation of aromatic amines and the reduction of aromatic nitro compounds yields a variety of reactive metabolites, for example arylhydroxylamines that may interact in various ways with oxyhemoglobin (Figure 3).

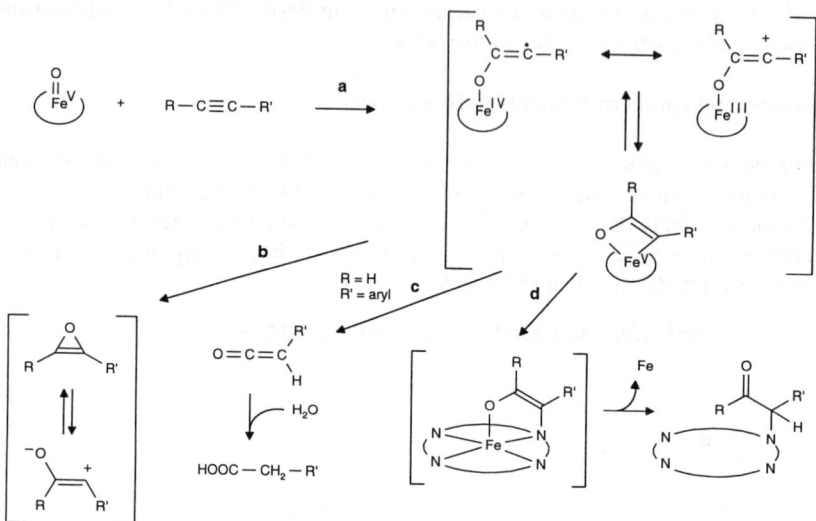

Figure 2 *Cytochrome P450 catalyzed activation of ethynyl groups (R' = H) (modified from ref. 1)*

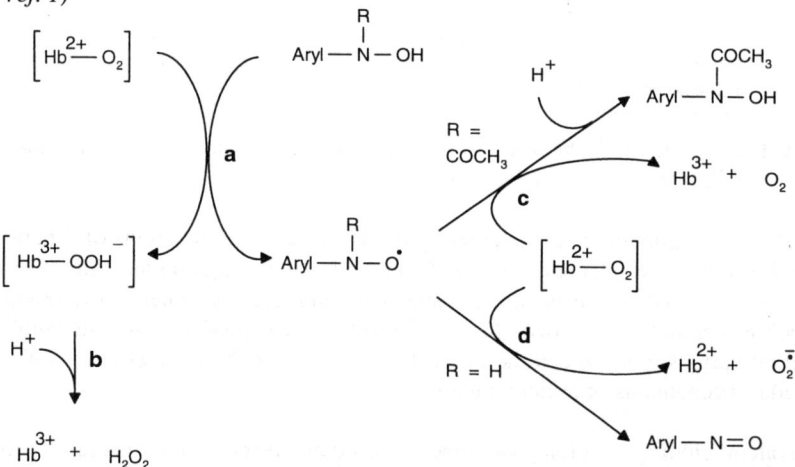

Figure 3 *Mechanisms of interaction of arylhydroxylamines (R = H) and N-hydroxy-N arylacetamides (R = acetyl) with oxyhemoglobin (modified from ref. 1).*

Such examples could be multiplied, and indeed they fill an entire book[1]. Of significance in the present context is the common characteristics of such toxication reactions. As represented in Figure 4, nucleophilic radicals formed by reduction, and electrophilic metabolites formed by oxidation can bind covalently to biomacromolecules. They can also reduce molecular oxygen, the first step in oxidative stress.

Table 3 *Some potential toxophoric groups activated by functionalisation (A) or conjugation (B)*

A	• Aromatic rings oxidised to epoxides and quinones • Aromatic ethynyl groups oxidised to ketenes • Polyhalogenated alkyl groups reduced to radicals • Primary aromatic amines oxidised to reactive N-oxygenated species and nitrenium ions • Aromatic nitro compounds reduced to reactive N-oxygenated species and nitrenium ions • Thiocarbonyl compounds oxidised to sulfenes • Thiols oxidised to mixed disulfides
B	• Carboxylic acids forming reactive acylglucuronides • Carboxylic acids yielding Coenzyme A conjugates that can interfere with lipid biochemistry and/or form lipophilic, retained residues

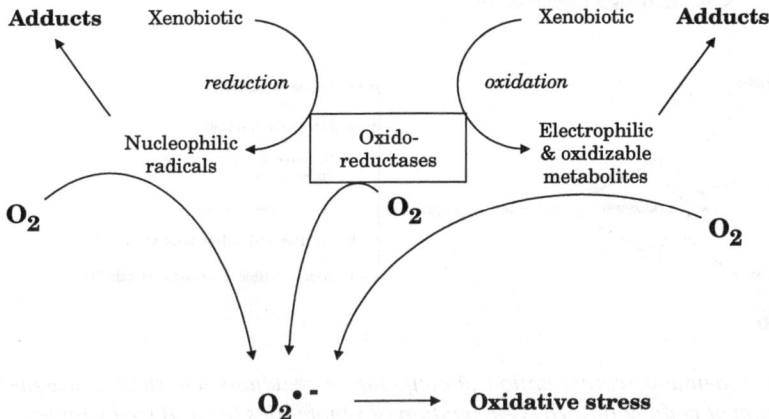

Figure 4 *A general scheme showing the interconnected roles of oxidoreductases in the initiation of oxidative stress and the toxication of xenobiotics (modified from ref. 1)*

As a rule, reactions of conjugation involve the coupling of an endogenous moiety (the so-called endocon) to a substrate, as mediated by an enzyme known as transferase. A few non-enzymatic conjugations are known (e.g., formation of Schiff's bases and some glutathione conjugations)[2,3]. Reactions of conjugation are all important in the metabolism of endogenous compounds, waste products and nutrients, with a variety of possible consequences as shown in Figure 5a. They are also of major significance in the metabolism of xenobiotics and their phase I metabolites, with comparable but not identical consequences (Figure 5b).

Xenobiotic carboxylic acids have received much interest since some of these compounds may yield reactive acylglucuronides that bind convalently to proteins by an Amadori reaction (Table 3). Also of interest is the fact that a few reactions of conjugation produce metabolites of high lipophilicity which tend to accumulate as tissular residues. The substrates of such reactions are xenobiotics that undergo esterification by fatty acids, or xenobiotic carboxylic acids which form hybrid triglycerides or cholesteryl esters. Some anti-inflammatory arylpropionic acids (e.g. ibuprofen) are a case in point[13,15].

Conjugation of endobiotics

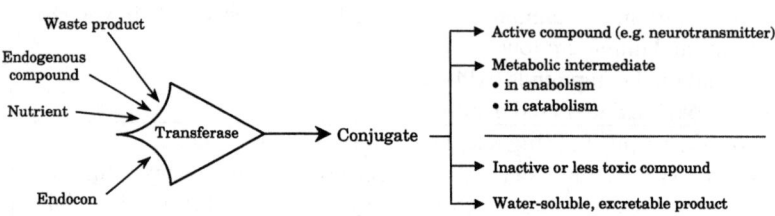

Figure 5a

Conjugation of xenobiotics

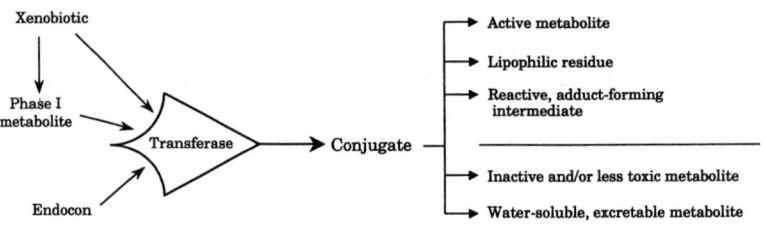

Figure 5b

Figure 5 *Schematic representation of conjugation reactions and their consequences. (A) Conjugation of endobiotics; (B) Conjugation of xenobiotics (reproduced with permission)*

2.2   Beyond the Biochemical Level

From the above, it is clear that the formation of macromolecular adducts from reactive metabolites or metabolic intermediates plays a key role in reactions of toxication. Some characteristics of macromolecular adducts, and how and why they may elicit toxicity, are summarised in Table 4.

At this stage of the argument, it might be concluded that potential toxophoric groups should be avoided at all costs by medicinal chemists. This is an unsustainable position for the reasons discussed below, and one just needs to count the number of marketed drugs bearing a carboxylic group to suspect that there is much more to molecular toxicology than the mere presence or absence of a toxophoric group.

Table 4   *Macromolecular adducts as a class of toxic metabolites*

• Their <u>formation</u> may be:
         • non-enzymatic
         • enzymatic
         • post-enzymatic
• The <u>covalent bond</u> may be:
         • strong (C–N, C–O, C–S)
         • of medium energy (S–S)
• The <u>target macromolecule</u> may be:
         • soluble
         • membrane-bound
• The <u>fate of macromolecular adducts</u> is largely unknown
• Macromolecular adducts may be <u>toxic due to</u>, e.g.:
         • loss of the macromolecule's original functions
         • antigenic activity
         • toxic breakdown products.

There are several reasons for this state of affairs. Indeed, given the presence of a toxophoric group in a compound, a number of factors will operate to render the latter either toxic or non-toxic, e.g.:

a)   The molecule in its entirety must be a substrate of the metabolic reaction and must thus fulfil a number of structural conditions such as size, shape, lipophilicity and electronic properties, as revealed by structure-metabolism relationships. In other words, the molecular properties of the substrate will increase or decrease its affinity and reactivity towards toxication or detoxication pathways.

b)   The second reason is found in the many biological factors that can markedly influence the substrate specificity, rate and capacity of a metabolic reaction. Such factors include the animal species, genotypes and phenotypes, sex, tissue, age, and external influences (diet, inducers or inhibitors).

c)   Metabolic reactions of toxication are always accompanied by competitive and/or sequential reactions of detoxication which compete with the formation of the toxic metabolite and/or inactivate it. The biological factors listed above also control the relative effectiveness of these competitive and sequential pathways.

d)   The reactivity and half-life of a reactive metabolite condition its sites of action and determine whether it will reach sensitive sites.

e)   Dose, rate and route of entry into the organism are all factors of known significance.

f)   Above all, one must consider metabolic reactions in their biological context (Figure 6). Indeed, the toxicological impact at the level of the organism will be influenced qualitatively and quantitatively by many known and unknown events that may be detrimental (e.g. damage amplification) or protective (e.g. repair mechanisms, immunological removal). Organisms could not survive had they not evolved essential mechanisms which operate to repair molecular lesions, remove them immunologically, and/or regenerate the lesioned sites.

Thus, it would be wrong to conclude from the above that the presence of a toxophoric group necessarily implies toxicity. Reality is far less gloomy, as only potential toxicity is indicated. In fact, the presence of a toxophoric group is by far not a sufficient condition for observable toxicity. Nor is it a necessary condition since other mechanisms of toxicity exist.

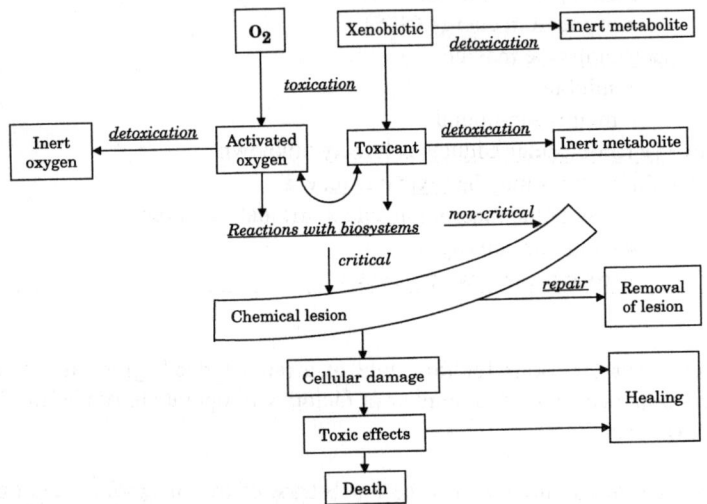

Figure 6 *General scheme placing reactions of toxication and detoxication in a broader biological context. Damaging and protective events are represented by vertical and horizontal arrows, respectively (reproduced with permission)*

3 CHANGES IN PHYSICOCHEMICAL PROPERTIES RESULTING FROM METABOLIC REACTIONS

The changes in physicochemical properties resulting from metabolic reactions represent a poorly explored field. It is often assumed that metabolism markedly increases water solubility and facilitates urinary excretion, but quantitative data are very scarce indeed. Certainly information on the ionization and lipophilicity of metabolites would help to make educated guesses on their distribution and excretion.

3.1 Some Reactions of Oxidation and Reduction

Systematic studies are in progress to assess and rationalize the changes in lipophilicity caused by some metabolic reactions. Thus, N-oxygenation is an important metabolic reaction for many medicinal tertiary amines containing a pyridine-type motif or an (N,N-dimethyl)alkylamino moiety. The resulting N-oxides are usually inactive, but they may be reduced back to the parent compound, in effect creating a futile cycle. To understand better and rationalize the properties of aromatic N-oxides, we compared the octanol/water partition coefficients of a number of 4-substituted pyridines and their N-oxides, also including quinoline and isoquinoline in the study[9]. The results (Table 5) show that the N-oxides are indeed more hydrophilic than their parent compounds, with a decrement ranging from −0.87 to −2.24.

Table 5   *Compared octanol/water partition coefficients of pyridines and their N-oxides[9]*

4-Substituted pyridine, R =	logP of parent compound	logP of N-oxide	$\log P^{N(O)}$ minus $\log P^{parent}$
H	0.65	−1.26	−1.91
COCH$_3$	0.48	−0.93	−1.41
NO$_2$	0.33	−0.54	−0.87
CN	0.46	−0.91	−1.37
CH$_3$	1.22	−0.88	−2.10
N(CH$_3$)$_2$	1.34	−0.90	−2.24
Cl	1.28	−0.44	−1.72
OCH$_2$CH$_3$	1.57	−0.53	−2.10
C$_6$H$_5$	2.59	0.93	−1.66
Quinoline	2.03	0.41	−1.62
Isoquinoline	2.08	0.26	−1.82

Such a broad range and strong influence of the *para*-substituent were rather unexpected, but could be explained by the electronic density on the oxygen atom. Indeed, the decrement in lipophilicity ($\log P^{N(O)}$ minus $\log P^{parent}$) was highly correlated with the partial charge on the oxygen atom, such that the greater the partial charge, the greater the decrement:

$$(\log P^{N(O)} \text{ minus } \log P^{parent}) = 16.5(\pm 3.7) \bullet \text{Charge}_{oxygen} + 5.4(\pm 1.6)$$
$$n = 11; \ r^2 = 0.92; \ s = 0.12; \ F = 100 \qquad \text{(Eq. 1)}$$

This was interpreted in terms of intermolecular forces to mean that the greater the hydrogen-bond acceptor basicity of the N-oxide group, the greater the decrement in lipophilicity.

In another study, the influence on lipophilicity of SO-to-SO$_2$ oxygenation and SO-to-S reduction was investigated using model compounds and the sulfoxide drugs sulindac and sulfinpyrazone[10]. In model compounds, the lipophilicity decrement from −S− to −SO− was −2.4 ± 0.15, and that from −S− to −SO$_2$− was −2.3 ± 0.2. The sulfide and sulfone metabolites of sulindac showing a normal partitioning behaviour, whereas the lipophilicity of sulfinpyrazone and its metabolites was markedly affected by tautomeric and conformational equilibria.

3.2 Glucuronidation

We also compared the lipophilicity of O-glucuronides and their aglycones in *n*-octanol/buffer systems[12]. The experimentally determined global influence of glucuronidation on lipophilicity, obtained as the difference (decrement) log $P_{(glucuronide)}$ minus log $P_{(aglycone)}$, was found to be -1.30 ± 0.16 (n = 4) for glucuronides of alcohols (methyl, menthyl, neomenthyl and chloramphenicol O-glucuronide). The mean decrement was -2.06 ± 0.31 (n = 9) for glucuronides of phenols (phenyl, *p*-nitrophenyl, 1-naphthyl, 6-bromo-2-naphthyl, 4-methylumbelliferyl, 3-coumarinyl, phenolphthalein, 4'-benzophenonyl O-glucuronide, and diflunisal phenolic glucuronide). For the acylglucuronide of diflunisal and its rearrangement isomers, the mean decrement was -1.80 ± 0.08 (n = 4; range -1.7 to -1.9).

Differences in through-bond proximity effects as parametrized in the CLOGP algorithm seem to account for much of this difference. The influence of conformational factors, which is considered significant in the lipophilicity of morphine O-glucuronides[11], appears modest and unassessable for the glucuronides investigated here.

Considering only the neutral forms, glucuronidation was shown to decrease the lipophilicity of phenols and carboxylic acids more efficiently than that of alcohols. The results presented in this study appear of interest in a pharmacokinetic perspective. Given their pK_a value of about 3, glucuronides exist mainly as anions in physiological fluids. The log P of carboxylic acids is known to decrease by about 3 units upon deprotonation, indicating that the log P of anionic glucuronides is too low to be measurable. In the body, both the anion and the neutral form of glucuronides may be involved in distribution and excretion. Our results imply that *in vivo* glucuronidation could facilitate the excretion of phenols more than that of alcohols. For carboxylic acids, glucuronidation also produced an important decrease in log P values, but in contrast to phenols and alcohols this is not done with an additional ionizable group being introduced, and thus the overall facilitation of excretion may be less important.

4 CONCLUSION

Table 1 lists the major pharmacodynamic and pharmacokinetic consequences of drug metabolism. These are of such significance that they simply cannot be ignored in any rational drug design, leading medicinal chemists to become increasingly aware of the role of metabolism in drug research and development and in modulating the clinical response. Those aspects that are most relevant to drug design are compiled in Table 2.

The main conclusion to be drawn from Table 1 is that the traditional view of an organism reacting to a drug like a machine reacts to a switch is simply false. Not only does an organism react, it also acts to eliminate xenobiotics by excretion and biotransformation. This is turn affects the nature, intensity and kinetics of a drug's effects, the resulting interplay of actions and reactions being characteristic of complex interactive systems.

This mini-review has focused on two important consequences of drug metabolism, namely a pharmacodynamic one (the production of toxic metabolites) and a physicochemical/pharmacokinetic one. An enormous body of evidence now exists to appreciate the variety of reactions of toxication and detoxication. However, the diversity of underlying factors (chemical and biological) is such that it precludes any quantitative prediction on the metabolism of a new compound. And even some qualitative predictions may prove erroneous, leading many medicinal chemists to await with great expectation the development of truly comprehensive and reliable expert systems to guide their steps. But expert systems are not instruments of discovery, they are simply experts at handling existing information, and they cannot be better than the existing body of knowledge in a given field.

As for the changes in physicochemical properties resulting from biotransformation, little knowledge is currently available, but systematic studies are in progress. Unfortunately, even less is known on the influence of such changes on the pharmacokinetic behaviour of the resulting metabolites. This is a field where a much better understanding is needed.

Drug researchers are eager to learn all there is to know about the infinitely complex interactions between living organisms and xenobiotics. Paradoxically, what we miss most at present are not expert systems, but information and knowledge, from which new concepts and understanding will emerge. The time is ripe for teams of medicinal chemists, biologists, pharmacologists and other experts to contribute significant advances in our knowledge of xenobiotic metabolism and its broad biological context.

REFERENCES

1. B. Testa. The Metabolism of Drugs and Other Xenobiotics - Biochemistry of Redox Reactions, Academic Press, London, 1995.
2. G.J. Mulder, ed., Conjugation Reactions in Drug Metabolism. Taylor & Francis, London, 1990.
3. B. Testa. "Drug Metabolism". In Burger's Medicinal Chemistry and Drug Discovery, 5th Edition (M.E. Wolff, Ed.). Wiley-Interscience, New York, 1995, pp. 129-180.
4. C.G. Wermuth and B. Testa. "Biotransformation Reactions". In The Practice of Medicinal Chemistry (C.G. Wermuth, Ed.). Academic Press, London, 1996, pp. 615-641.
5. D.A. Smith, M.J. Ackland and B.C. Jones, "Properties of cytochrome P450 isoenzymes and their substrates. Part 1: active site characteristics. Part 2: properties of cytochrome P450 substrates", Drug Discovery Today 2, 406-414, 479-486 (1997).
6. B. Testa and J. Caldwell. "Prodrugs revisited: The ad hoc approach as a complement to ligand design", Med. Res. Rev. 16, 233-241 (1996).
7. S. Gangwar, G.M. Pauletti, B. Wang, T.J. Siahaan, V.J. Stella and R.T. Borchardt. "Prodrug strategies to enhance the intestinal absorption of peptides", Drug Discovery Today 2, 148-155 (1997).
8. B. Walther, P. Vis and A. Taylor. "Lipophilicity of metabolites and its role in biotransformation." In Lipophilicity in Drug Action and Toxicology (V. Pliska, B. Testa and H. van de Waterbeemd, Eds). VCH, Weinheim, 1996, pp. 253-261.
9. G. Caron, P.A. Carrupt, B. Testa, G. Ermondi and A. Gasco. "Insight into the lipophilicity of the aromatic N-oxide moiety", Pharm. Res. 13, 1186-1190 (1996).
10. G. Caron, P. Gaillard, P.A. Carrupt and B. Testa. "Lipophilicity behavior of model and medicinal compounds containing a sulfide, sulfoxide or sulfone moiety", Helv. Chim. Acta 80, 449-462 (1997).

11. P. Gaillard, P.A. Carrupt and B. Testa. "The conformation-dependent lipophilicity of morphine glucuronides as calculated from their molecular lipophilicity potential". Bioorg. Medic. Chem. Lett. 4, 737-742 (1994).
12. Y. Giroud, P.A. Carrupt, P. Gaillard, B. Testa and R.G. Dickinson. "Intrinsic and intramolecular lipophilicity effects in O-glucuronides". Helv. Chim. Acta 81, 330-341 (1998).
13. J. Mayer and B. Testa. "Stereoselectivity in metabolic reactions of toxication and detoxication". In Pharmacokinetics of Drugs (P.G. Welling and L.P. Balant, Eds). Springer Verlag, Berlin, 1994, pp. 209-231.
14. B. Testa and J. Mayer. "Molecular toxicology and the medicinal chemist". Farmaco 53, 287-291 (1998).
15. J.M. Mayer, B. Testa, M. Roy-de Vos, C. Audergon and B. Testa. "Interactions between the in vitro metabolism of xenobiotics and fatty acids. The case of ibuprofen and other chiral profens", Arch. Toxicol. Suppl. 17, 499-513 (1995).
16. W.A. Korfmacher, K.A. Cox, M.S. Bryant, J. Veals, K. Ng, R Watkins and C.C. Lin, "HPLC-API/MS/MS: a powerful tool for integrating drug metabolism into the drug discovery process", Drug Discovery Today 2, 532-537 (1997).

RAPID ASSESSMENT OF DRUG METABOLISM IN THE PHARMACEUTICAL DISCOVERY PROCESS

Marc Bertrand, Peter Jackson and Bernard Walther

Technologie Servier, Orléans, FRANCE.

1 INTRODUCTION

The aim of the pharmaceutical discovery and development process is to identify new pharmacological active chemical entities that, once released on the market, can be simply and safely prescribed by physicians.

For the past few decades, new chemical entities were only screened for pharmacological activities following which pre-clinical and clinical development was performed in order that some of them would be marketed with success. Today, the increasing cost of pre-clinical and especially clinical development, together with the impossibility to increase in parallel the marketed price, have pushed companies to select new chemical entities to be developed in a different manner. This has been achieved by combining chemistry and pharmacology together with biopharmacy during earlier discovery stages in order to improve candidate selection for pre-clinical and clinical development.

In vitro tools, introduced several years ago into the drug metabolism field, are today mastered. When pharmacokinetic parameters are combined to physicochemical descriptors in the discovery process, a prediction of the major biopharmaceutical characteristics of new drugs is possible. Integration of these tools into the drug discovery process has been possible because of the recent advancement of powerful analytical tools such as LC-MS-MS together with the increased use of laboratory automation.

Today, the emphasis is more on the appropriate use of these tools to solve the right problem at the right moment, or in other words to master the balance between their systematic or selective use. Now that data can be generated very quickly, different issues such as validation and data analysis have arisen.

This paper will focus, with specific examples, on the real predictive value of certain metabolic tools at the various stages of the drug discovery process.

2 BIOPHARMACEUTICAL SCREENING

A critical step in any quantitative drug design approach has always been to describe the molecules with the appropriate physicochemical parameters. This also applies to the drug metabolism field. Because the biological processes involved in drug metabolism, described by Testa et al [1] as the interaction level with biological environments, are of multidimensional nature, it is not possible to explore all these processes at a screening stage. The tools used in drug metabolism screening must correspond to each important individual limiting factor for the development of a new chemical entity, i.e. absorption characteristics including solubility and permeability, rate and extent of metabolism, but also enzyme inhibition and induction potential as well as the identity of liver enzyme implicated in primary metabolic pathways (Figure 1).

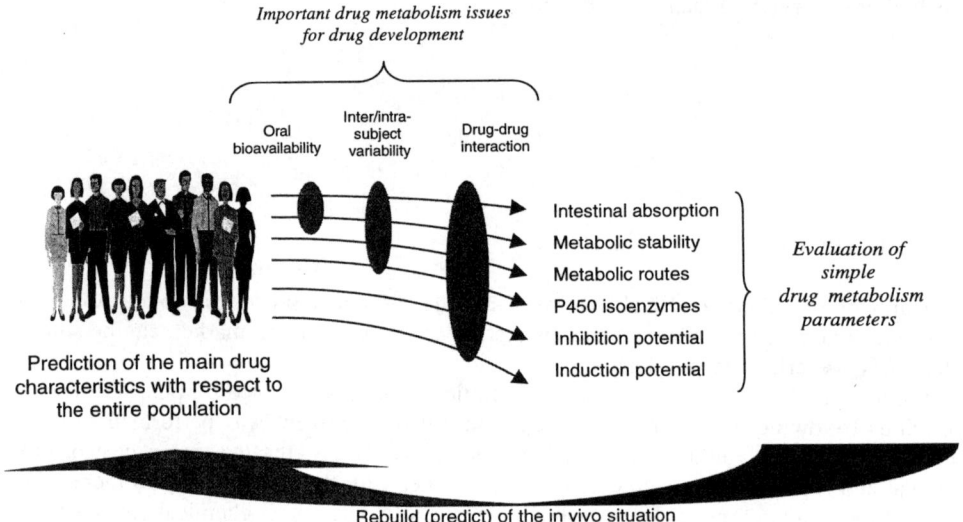

Figure 1 *Objective of a rapid assessment of drug metabolism parameters*

The different types of screening tools to be used in a screening process may depend on many factors. These may differ depending on individual company strategies on the integration of biopharmaceutics into the screening process, but in all cases, the screening tools must be adapted to the number of molecule to be tested. Rodrigues [2] is emphasising the need of a rational approach in drug screening in applying the right model to the right problem or to the right number of molecules.

If the objective is to sort a large number of molecules in the High Throughput Screening (HTS) process, in order to discover a lead (Figure 2), there is no doubt that the only way is to obtain the most simple parameters by the simplest methods (e.g. a single data point for the evaluation of the rate of biotransformation).

In contrast, when working on a smaller chemical series, in a more Selective Drug Screening (SDS) process, one can generate more informative parameters (e.g. Km and Vmax values in metabolic stability tests) which can be used to directly predict the corresponding *in vivo* parameters. This stage can follow a preceding HTS stage, with the aim of transforming leads into drugs of potential clinical use, or can correspond to a backup for drugs which have failed during the development process (Figure 2). In the latter case, some *in vivo* data may be available to aid the choice of the screening tool to be used.

In many of the screening strategies described, SDS results (or even HTS ones) need to be confirmed (validated) for molecules that have been selected for pre-clinical development using, for example, animal n-in-one dosing. This serves to increase the confidence in a particular model that is used in the screening stages. When more informative parameters are obtained during the SDS stages, accurate predictions of drug plasma levels, variability in clearance and risk of drug-drug interactions maybe made (Figure 2).

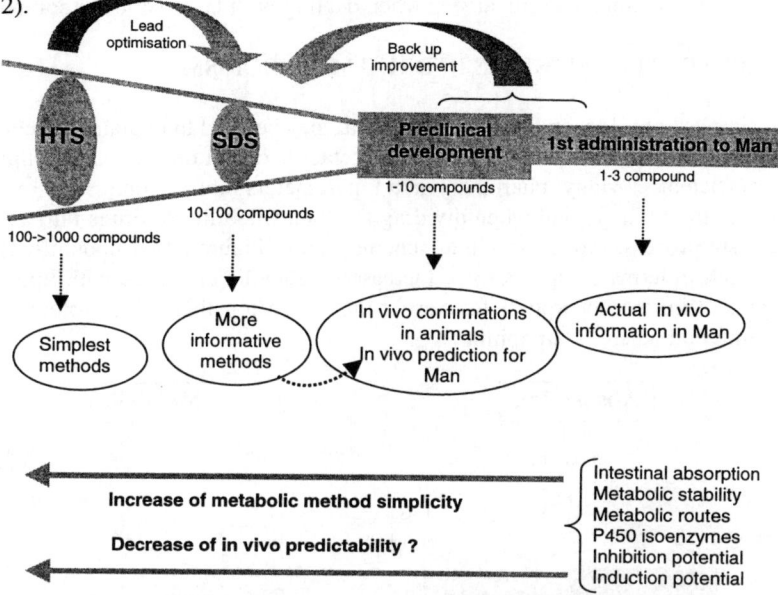

Figure 2 *Scheme of the screening process respective to the type of screening tools*

However, the screening process does not stop there and has been stretched out to the first administration to man with administration of several compounds having a good *in vitro* profile (e.g. good metabolic stability) but having very limited *in vivo* animal data.

The major question to be addressed is whether we are really able to maintain *in vivo* predictability when *in vitro* methods are simplified so they become more applicable to HTS (Figure 2), or is it that when we deal with large series of compounds we are just generating simple metabolic parameters that have no real predictive use?

3 EVOLUTION IN ASSAY TECHNOLOGIES

Rapid assessment of drug metabolism parameters has been possible with the combined use of:

- Robotics, using 96 well (or more) plates, for manipulations including preparation of compound solutions, *in vitro* incubations, and the preparation of the samples before analysis [3]. These robotic systems increase the throughput of all simple *in vitro* tools, allowing scientists to focus more on the interpretation of data rather than on the routine aspects of sample manipulation.

- LC/MS/MS as an analytical tool, the sole able to achieve the sensitivity as well as the specificity in the presence of complex mixtures of chemicals to be tested [4].

Using these two techniques, analytical method development has been shortened and is therefore generally no longer a critical step when dealing with large series of compounds.

4 SCREENING PARAMETERS IN DRUG METABOLISM

One of the simplest physiochemical parameters that may be used to explain the relationship between chemical structure and certain drug metabolism parameters is the lipid/water partition coefficient.   Many pharmacokinetic parameters are in one way or another correlated with lipophilicity and when dividing the bioavailability of drugs into absorption and metabolism processes, one can see as schematised in Figure 3 that lipophilicity has an importance both in terms of metabolism (increased metabolic clearance with lipophilicity) and absorption (limited absorption for very polar drugs together with protein efflux or solubility limitation when too lipophilic).

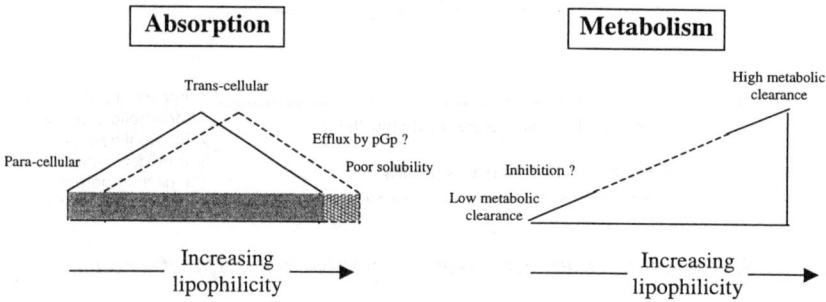

Figure 3 *Schematic representation of drug absorption and metabolism evolution with lipophilicity*

But this is probably too much of a simplification and classical lipophilicity calculations or measurements are not completely satisfactory as physicochemical descriptors for drug metabolism studies at the screening stage and are applicable only within structurally related homologous series. Their real value lies in the fact that they can be assessed very early, on a strict theoretical base, and when used as alerts for synthetic chemists , provide boundaries helping ultimately to improve the pharmacokinetic profile of a drug.

4.1 Pre-systemic Metabolism

Oral absorption of a drug is a function of the <u>pharmaceutical phase</u> corresponding to the *in vivo* dissolution of the drug as well as physiological conditions such as gastric emptying etc. and the drugs <u>passage</u> through the intestinal membrane.

Drug solubility and intestinal permeability are therefore the two major factors to be evaluated in this biological process [5]. The less a drug is absorbed, the more pharmacokinetic variation will occur from one subject to another (and sometimes within the same subject), and the more it may be influenced by exogenous factors such as diet co-absorption and regimen differences, leading to a more difficult clinical development.

4.1.1 Solubility. Solubility in physiological conditions is probably the first physicochemical parameter one should evaluate before starting any *in vitro* drug metabolism parameter measurements (permeability, metabolic stability, etc.). This probably should also apply to *in vitro* pharmacological testing, where solubility can often be a problem.

This parameter can either be estimated by calculation [6], or experimentally by turbimetry (96 well plates) in a HTS mode [5] or by pHmetry that gives a pH related solubility profile [7] when fewer compounds are to be tested (SDS mode).

4.1.2 Intestinal Passage. Once solubility is thought no longer a limiting factor, *in vitro* intestinal passage can be assessed to predict the *in vivo* absorption. Intestinal passage is a function of intestinal permeability and pre-systemic metabolism. Different *in-situ* animal models have been described to estimate these two parameters [8] but are, in general, more adapted to the study of complex drug absorption problems than to screening. There are several cellular models available which may be adapted to screening but the Caco2 model, is today, the standard method to study the passage of a given drug. The success of this human cell line is mainly due to the simplicity of cell culture which is able to spontaneously differentiate as a monolayer of enterocytes with well developed tight junctions after only several days of culture. This allows measurement of both trans-cellular passive transport through the membrane and para-cellular passive transport through the tight junctions. Active transport including pGp efflux may also be studied relatively simply. While allowing permeability comparisons within a chemical series (in SDS context), with use of calibration curves based on compounds with a known *in vivo* absorption, this model can be used to predict *in vivo* absorption for large numbers of compounds [9]. The combined use of n-in-one incubations with LC/MS/MS analyses can increase throughput to such an extent where its use in HTS may be envisaged.

One of the drawbacks of this cell line, is the difference in metabolic competence of these cells compared to normal intestinal enterocytes with an over-expression of P450 1A1 and a down regulation of the constitutive P450 3A4. This can be overcome by restoring P450 3A4 (Figure 4) with a simple modification of the culture medium [10], or by using CYP3A4 expressing Caco 2 cell lines [11]. When validated against *in vivo* intestinal metabolism, the intestinal passage, with both its permeability and metabolism components, could then be assessed using the same *in vitro* model.

Intestinal permeability may also be estimated using other simpler techniques which may be better adapted to the HTS context. LogP as a descriptor of lipophilicity is often used but more recently, interactions of compounds between an immobilised artificial membrane (IAM) have been shown to provide a better correlation with intestinal permeability than LogP alone [12]. This method comprises of a LC columns with a stationary phase consisting of phospholipids covalently bound to silica particles through long chain diacylated hydrocarbons. LC capacity factors (log (k'$_{IAM}$)) with this column therefore describe the interactions of drugs between an aqueous medium and phospholipids similar to those found in biological membranes. This method has been tested in our

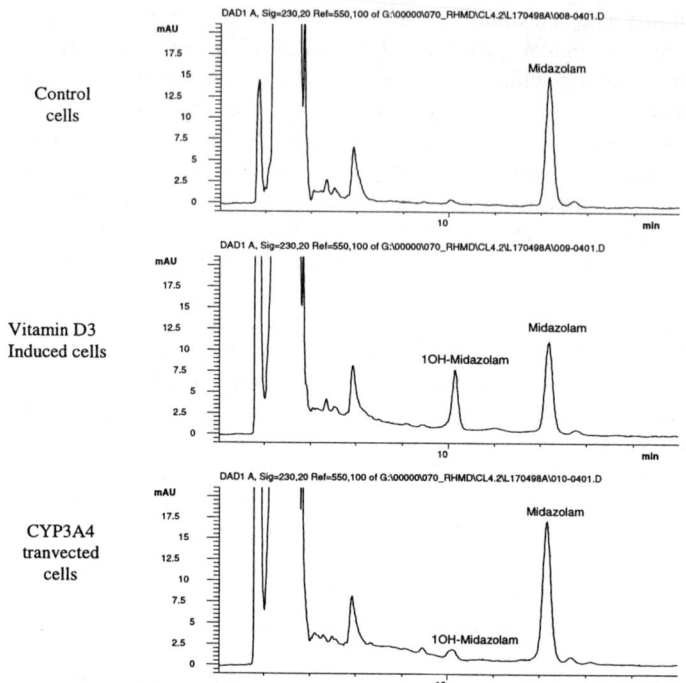

Figure 4 *Activity of CYP3A4 in either vitamin D3 induced or CYP3A4 transfected CaCO2 cells. Method derived from Schmiedlin-Ren et al, 1997*

laboratories for the prediction of intestinal permeability of drugs or for the possible link with other parameters (e.g. rate of drug metabolism).

As shown in Figure 5, the log (k'_{IAM}) values obtained from a chemically related series (which shows better correlations with the rate of metabolism than with Caco2 permeability) are very close to those obtained from a second chemically distinct series (correlated more with Caco2 permeability than with the rate of metabolism).

First chemical series

Metabolic bioavailability prediction in Man %	Caco2 absorption prediction %	Caco2 Papp	Log K' IAM
100	100	24.7	0.66
100	100	19.4	0.81
100	100	16.8	1.00
74	100	61.6	1.95
10	100	26.3	2.14
100	100	36.5	2.18
88	100	33.6	2.54
8	100	8.6	3.20

Second chemical series

Log K' IAM	Caco2 Papp	Caco2 absorption prediction %	Metabolic bioavailability prediction in Man %
1.8	0.12	10	100
1.9	0.09	10	100
2.1	0.07	10	100
2.6	0.20	10	100
3.4	1.03	70	100

Figure 5 *Log K' with IAM column between two different chemical series*

However, even with similar log (k'_{IAM}) values, absorption prediction using the Caco2 models is excellent (100 %) for the first series but rather poor (10 %) for the second one. These recent in-house results illustrate the possible use of this descriptor for screening within a series, once a correlation with a measurable parameter (e.g. absorption, metabolism or pharmacological activity) has been described with a few molecules of the series. However, a high risk exists if correlations are attempted with chemically diverse series.

4.2 Metabolism

In addition to the important connection with oral bioavailability, metabolic parameters are also important in pharmaceutical development as they may explain inter-subject variability, drug-drug interactions, non-linear pharmacokinetics, toxic effects etc. If we define metabolism as the chemistry of enzymatic and non-enzymatic processes [13] this covers the rate (metabolic stability) and extent (metabolic routes) of metabolism as well as encompassing enzyme inhibition and induction potential of the drug concerned as well as the identity of the liver enzymes implicated in its primary metabolic pathways. *In vitro* models for the evaluation of drug metabolic stability have already been extensively used within the pharmaceutical industry and have easily been adapted to screening stages. Other tools in drug metabolism (i.e. those used to study metabolic routes, inhibition or induction potential, etc.) are currently being introduced progressively into the drug screening processes.

4.2.1 Metabolic Stability. The hepatic microsomal model, when compared to more complete cellular models such as liver slice cultures (expressing all the liver enzymes together with intact cell-cell communications) or hepatocyte cultures (expressing all the metabolic liver enzymes), is the simplest and the most adapted tool for early drug screening stages.

Hepatic microsomal fractions are available for all animal species including man, and express all the P450 superfamily of enzymes together with other oxido-reduction and conjugation enzymes. This model may be used to obtain simple or more complex metabolic parameters which may be used to predict the *in vivo* situation in man. It is also easily used in an automated system and is therefore totally adapted to HTS.

However, because of it's simplicity of use, one tends to forget the importance of the experimental conditions used in these *in vitro* incubations. Many of the enzymes implicated in the metabolism of xenobiotic compounds are highly saturable and therefore methods of calculating the rates of metabolism which use a single substrate concentration may, if this substrate concentration is not low enough, produce results that drastically underestimate the *in vivo* situation. For this reason, we use a simple method which provides apparent Km and Vm determinations for the overall metabolism of a drug, based on the computer fitting of *in vitro* drug disappearance curves using the Michaelis-Menten integrated equation (Figure 6). When used with certain physiological parameters (hepatic blood flow, liver weight, microsomal yields, etc.), these simple *in vitro* data can then be scaled up to the *in vivo* situation to obtained an estimate of the metabolic oral bioavailability. Simulations may be performed for different oral dose levels in order to model saturation of first pass metabolism and therefore non-linear pharmacokinetic effects and good predictions were obtained as shown in Figure 7 with an interspecies example.

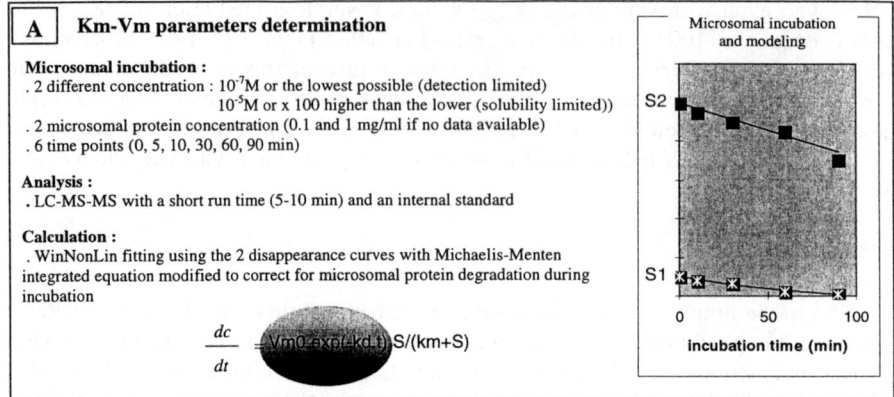

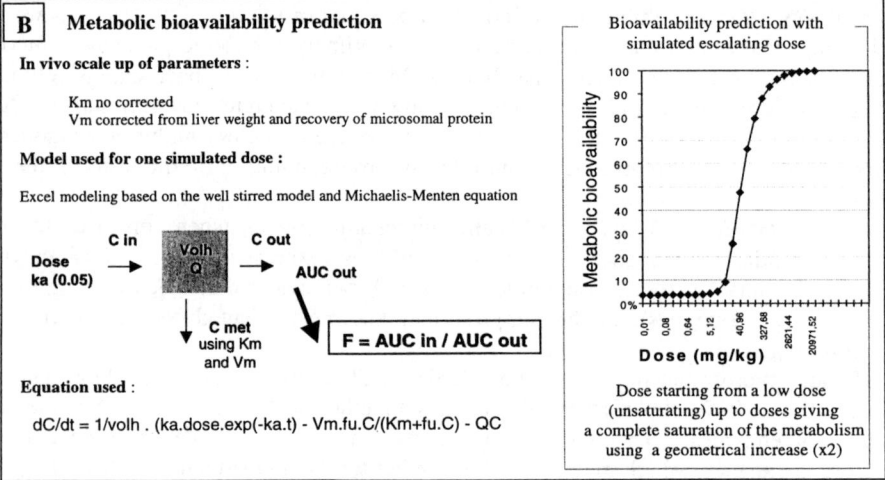

Figure 6 *Non-linear regression fitting of the Michaelis-Menten equation for calculation of apparent Vm and Km metabolic parameters (A) and their use for metabolic bioavailability prediction for a simulated administered dose (B)*

When predicting non saturated metabolic bioavailability as shown in Figure 8, this method has given a good *in vitro - in vivo* correlation for about 20 structurally diverse compounds in the rat. In man with a fewer number of compounds, good correlations has also been observed.

This approach, which has been developed primarily in order to predict the metabolic clearance in man before the first administration of the drug, is also well adapted to SDS when automated and used with cassette LC-MS-MS analyses. Its throughput may be increased by using the same hepatic microsomal model but with only one very low substrate (drug) concentration in order to minimise any risk of metabolic saturation, allowing a determination of intrinsic clearance (CLint).

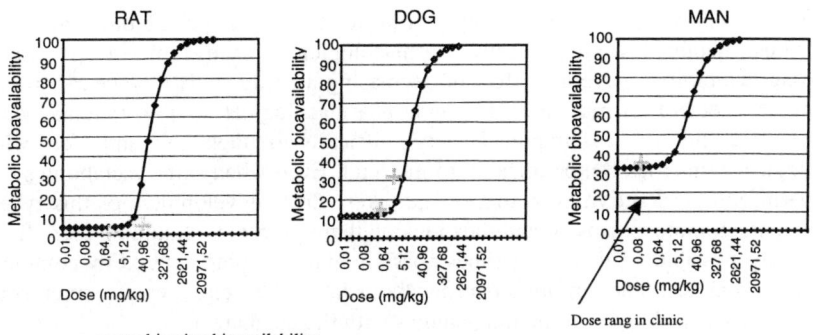

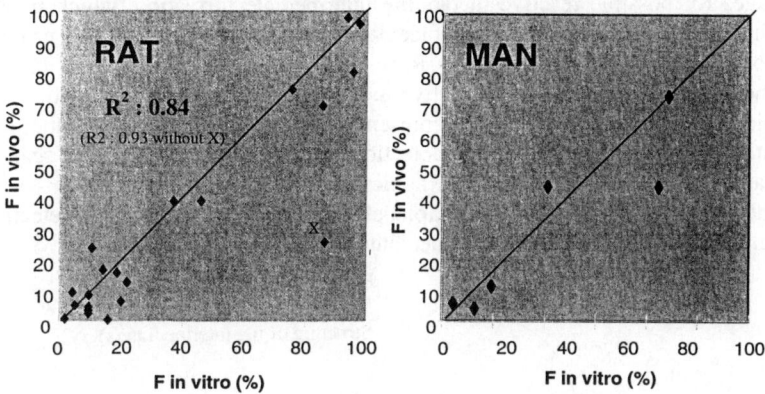

Figure 7 *Prediction of metabolic bioavailability and of its evolution with dose compared with actual bioavailability in vivo in Rat, Dog and Man*

A good correlation between these two methods were observed on a set of 37 compounds for which the low concentration used was 10^{-7} M, even if a risk of enzymes saturation (and thus of underestimated Clint) must be taken into account when using this simpler method. Further simplification (e.g. one substrate concentration and one time point) allows an estimation of metabolic rate which is extremely well adapted to HTS, allowing sorting of molecules on their *in vitro* rate of metabolism. However, a higher risk exists for the prediction of *in vivo* clearances using these simplified methods.

Figure 8 *Correlation between metabolic bioavailability predicted in vitro from Km/Vm values and actual measured in vivo for rat and man*

Assessing Km and Vmax for certain compounds at a SDS level with animal and human material in parallel together with *in vivo* animal data on selected molecules reinforces *in vivo* predictions for man if the *in vitro/in vivo* correlation for rat within a series is good. This may secure future predictions with new compounds from different series and can be seen as a useful validation step.

Non P450 dependent pathways, can be assessed either with hepatic microsomes or with hepatocytes or liver slices. For peptides or esters, blood plasma or cell culture medium incubations can be carried out for large series of compounds.

4.2.2 Metabolic routes. Metabolism can be regarded as a biological process which, when extensive, may make a very chemically pure compound (bioavailability or purity of

100 %) become very non pure (bioavailability or purity of 1 %) within a biological system. The impurities (metabolites) are generally pharmacologically inactive and in certain cases may be toxic, which may have the effect of decreasing a compounds therapeutic margin. In some cases, metabolites may have a reverse pharmacological activity (agonist versus antagonist) resulting in a compound very difficult to develop, and for which pharmacological properties can be difficult to interpret. Even when some metabolites have similar pharmacological activities to that of the parent drug, development of this type of drug is, in general, more expensive and thus ideally, these problems should be identified at the earliest step possible. This is sometimes possible in a screening context by combining a pharmacological test and an *in vitro* metabolism model (e.g. hepatic microsomal incubations together with receptor binding studies with the incubates).

Higher-throughput methods for *in vitro* formation and mass spectrometric characterisation of microsomal drug metabolites have been developed [14]. However, it is doubtful that the determination of the extent of metabolism at the HTS and SDS stages has any other interest than to fill databases even if this approach can be valuable if performed on a few selected molecules of a series which can help the chemist apply metabolic stabilising modifications. Because the numerous examples of metabolism switching, where a different route of metabolism can be favoured by blocking of a previous route, the real impact of any chemical modifications on metabolic stability must be verified by further *in vitro* experiments.

More interesting is the possible use of structural tools such as LC-MS-MS at the HTS or SDS stages for possible reactive metabolite intermediate formation, which may cause problems in terms of binding to macromolecules such as proteins and DNA providing a risk of hepatic toxicity or genotoxicity. Recent in-house experiments have provided a sensitive method for the detection of reactive metabolites by using a combination of drug hepatic microsomal incubations in the presence of GSH (used to trap the reactive intermediate) together with LC-MS-MS detection of GSH adducts using the scan of the possible parent ions of a GSH specific fragment. As shown in Figure 9, this sensitive method that may be performed on non-radiolabelled compounds allows both detection and structural elucidation of the reactive intermediates.

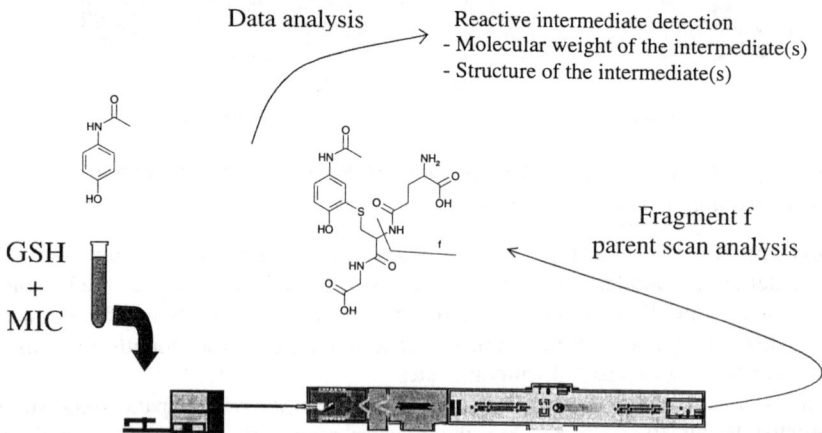

Figure 9 *Reactive intermediate detection by parent scan analysis and structural elucidation*

4.3 P450 Isoenzymes: Identity, Inhibition and Induction potential

P450 enzymes are involved in many of the primary metabolic routes of xenobiotics, and as such, they are determinant in the metabolic clearance of many drugs. P450 enzymes may also be inhibited or induced by drugs. When combined, these three characteristics (the type of isoenzyme involved in the primary metabolic routes, the inhibition and the induction potential of the drug tested) give a good understanding of the risk of metabolism based drug-drug interactions between drugs that may be co-administered.

These parameters are generally assessed during the pre-clinical stage, and we have developed an *in vitro* strategy for the rational assessment of the risks of interaction, mainly to rationalise the use of these pre-clinical data into the clinical context (Figure 10). This strategy, which defines the type of drug-drug interaction risk and the type of clinical interaction study requirements, is based on three different *in vitro* data types:

- the number and the nature of P450 isoenzymes involved in the drug metabolism (one major P450 implicated producing a risk of interaction of the studied drug with all drugs that can inhibit the P450 implicated),

- the inhibition potential based on the nature of the P450 inhibited and on the *in vivo* extrapolation of the inhibition constant (Ki) (a Ki close to the *in vivo* hepatic concentration estimated from the plasmatic one producing a risk of interaction of the drug on all other co-administered drugs metabolised by the P450 inhibited),

- the induction potential based also on the nature and the dose require to produce an induction (*in vivo* for animal data or *in vitro* for hepatocytes culture data) compared to the *in vivo* dose regimen (an induction potential at a low dose being a risk of interaction of the drug on all other co-administered drugs metabolised by the P450 induced).

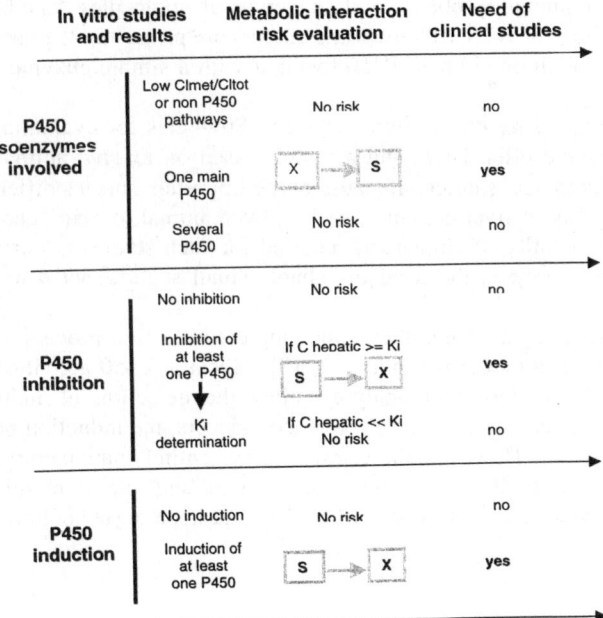

Figure 10 *Rational strategy for drug–drug interaction risk assessment*

This strategy, adapted to the FDA or European Medical Agency guidelines, give a better understanding of how to use some *in vitro* data (sometimes generated in the earlier stages of development) but is there a need for integrating these parameters into the screening process ?

4.3.1 Inhibition of Drug Metabolism Enzymes. The tools available for inhibition studies are the same as those previously described for metabolic stability determinations. Hepatic microsomes are used in combination with specific substrates for the P450 isoenzymes. It allows us to assess the inhibition potential of new drugs as well as the nature of the P450 isoenzymes inhibited.

The inhibition potential can be assessed on large series of molecules and is a valuable drug metabolism parameter as regards to drug-drug interaction predictions, the most hazardous being irreversible inhibition (e.g. suicide substrates). Inhibition potential can therefore be considered as a useful parameter to be measured at earlier screening stages. Particular attention must be paid to compounds with a very good metabolic stability to control that this stability is not linked with an inhibition potential.

However, if some reversible inhibition is observed, the inhibition constant (Ki) must be measured in order to scale up the effect of inhibition on plasma levels by comparison with predicted or actual *in vivo* concentrations in Man. This approach is therefore limited to a small number of chemicals.

The inhibition measurement can be simplified by only using an IC50 estimation, making it adapted to SDS. Some methods based on the use of a non-specific P 450 substrate (e.g. testosterone for rat P450 inhibition) or better still a mixture of specific substrate and n-in-one LC-MS-MS analysis (as could be possible by the *in vitro* use of the Pittsburg cocktail proposed for *in vivo* studies [15]) could allow HTS inhibition measurements if a known problem with a series of molecules has been previously identified. In parallel, for HTS screens, some authors are proposing the use of microsomes of cells transvected with one human P450 together with a simple enzyme assay detection method [16].

4.3.2 Induction of drug metabolism enzymes. Strategies for evaluating the induction potential of drugs have often been, and are still, based on *ex-vivo* animal liver analyses obtained from toxicological studies. Because of the important species differences observed in terms of inducibility of liver enzymes, scaling from animal to man becomes hazardous. In addition, as the quantity of compound required for such studies is normally a limiting factor in a screening project, the need of reliable small scale *in vitro* tools has become necessary.

Today, long term hepatocyte cultures allowing the induction process to be completely established, are the most predictive tools [17, 18]. Specific P450 activities measurements complemented with Western blot analyses allow the detection of induction potential directly with human hepatocytes. Using reference inducers, the induction potency of tested drugs can be evaluated. However, these tests are not entirely satisfactory because of the limitations in the availability of human hepatocytes and the inter-subject variability observed in response to inducers. For SDS screens, rat hepatocytes cultures are a possible alternative and can be used in order to sort molecules with respect to their induction potential.

The use of cell lines with inductive properties such as HepG2 [19] or FAO [20] have been shown many times to be suitable models for in vitro induction studies but, as regards to the complete panel of P450's potentially inducible in human, these models has not been convincing either for prediction of in vivo induction potential in Man. Tools that will become really adapted to large scale screens will probably be genetically engineered cells expressing all or part of the P450 regulation machinery coupled with a reporter gene protein which may be easily quantified. These tools could be envisaged as becoming part of the HTS process.

4.3.3 Identification of liver enzymes. Identifying the number and the nature of isoenzymes implicated in the metabolism of drugs allows us to predict for the potential variability in *in vivo* clearance as a result of the involvement of drug metabolism enzymes. P450 enzymes are known to be particularly variable from one subject to another for certain enzymes (e.g. 1A1, 1A2, 3A4) and polymorphic for others (e.g. 2D6, 2C19).

Classically, for the identification of the isoenzymes implicated in the metabolism of a drug, three approaches are run in parallel:

- the use of specific inhibitors of the different isoenzymes in a pool of human microsomes gives information on the nature but also the relative importance of the enzymes implicated,
- the use of heterologous expressed drug metabolising enzymes expressing a single P450, allowing to assess the nature of the P450's but also their implication in the metabolic routes.
- the correlation with specific P450 activities obtained with a bank of human microsomes previously characterised with specific P450 substrates.

Most of these techniques are potentially applicable to HTS, but their systematic use to screen for a particular isoenzyme at the HTS level cannot be motivated in order to only explain variability in pharmacokinetics. Effectively, most of the major P 450's are variable and there is no reason to point towards a particular one. These screens should therefore be more oriented to a back-up approach to avoid cases where only one P450 is involved in the metabolism of the drug in order to avoid the risk of drug-drug interaction.

5 HANDLING THE DATA

With the introduction of automated systems at most of the stages of the screening program we have seen an incredible leap in the data acquisition process. Attention is now focusing on their storage in structured databases, data retrieval together with appropriate analysis.

Because the biological processes involved in drug metabolism are of multidimensional nature, one has, in fact, in the screening process isolated each pharmacokinetic parameter or even used surrogates (e.g. log (k'_{IAM})) as drug metabolism descriptors. The analysis step is effectively an attempt to rebuild partially or completely this biological process.

One can see in Figure 11 that data are collected at all stages of the screening process and that the type of information collected can vary according to the model used. From primary parameters such as a single data point allowing the evaluation of the rate of metabolism within a series of compounds, to secondary parameters combining a number of primary parameters as bioavailability or drug-drug interaction prediction, up to collection of *in vivo* data from n-in one dosing experiments in animals.

At a very early step, when dealing with large series of compounds, integration of warnings for the chemists, is a simple way to use the data. This has already been applied for the solubility of compounds [5] and can be easily applied to sort molecules in large series. Another classical way to analyse these data is to use quantitative methods already applied in the QSAR field in a quantitative structure- pharmacokinetic relationship (QSPR) approach [13]. These data can then be combined with pharmacological activity data and the information can be used within but also across chemical series in a more predictive way.

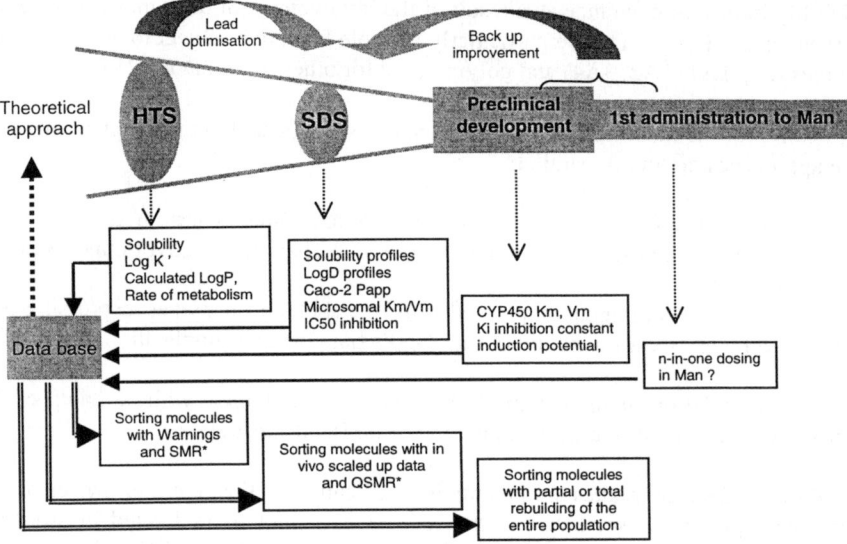

* SMR : Structure Metabolism Relationships – QSMR : Quantitative Structure Metabolism Relationships

Figure 11 *Flow of metabolic data information during the drug screening and development processes*

However, ideally, in drug metabolism, all the individual data collected at each stage should enter an appropriate model helping us to rebuild the biological process and hence validate the approach by appropriate *in vivo* experiments performed with a certain number of compounds. This has already been shown with bioavailability predictions from the Km and Vm enzymatic parameters. We can however, go further and one way of illustrating these approaches is the prediction of drug plasma levels using a physiologically-based pharmacokinetic (PBPK) model based solely on *in vitro* data [21].

This model is based on the determination of the enzymic kinetic parameters (Km and Vmax) for the main metabolic pathways using enzyme preparations containing single P450 enzymes. Together with scaling factors based on the levels of individual P450 enzymes present in a bank of microsomes, expected rates of metabolism of a drug are calculated with Michaelis-Menten or first order equations and compared with actual rates within the same microsomal bank. After the incorporation of kinetic parameters and scaling factors into the PBPK model, prediction of ranges of plasma levels can be made, representing the inter-individual variability of the bank used. Figure 12 shows the predicted range of plasma concentrations compared to the actual range found in clinical studies. This model allows the quantitative description of the potential impact of metabolism based inter-individual variability on *in vivo* pharmacokinetic parameters. This model can be further completed by adding Ki values and predict even more precisely the drug-drug interaction risks. This tool can also be simplified by using apparent Km and Vmax values for the total drug disappearance helping us again to predict drug plasma levels in Man.

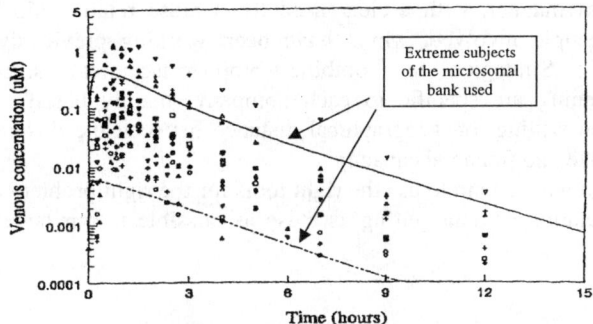

Brief description of the method used :
. Km-Vm determination for the main P450 implicated in primary pathways
. In vitro validation by comparison of actual/calculated rates of metabolism for the different subjects within a bank of microsomes
. In vivo scale up of individual Vm and Km parameters for each subject of the bank using interindividual variability in P450 expression
. Estimated or measured secondary *in vitro* pharmacokinetic parameters are needed (blood/plasma and blood/tissues ratios, protein binding)

Figure 12 *Prediction of in vivo drug plasma concentration using PB/PK modeling of in vitro data*

Even if this approach is not useful at the HTS level, where the objective is more to sort molecules on primary parameters, all the tools used in this example are adapted to series up to 50 to 100 compounds, and these types of modelling methods give a better understanding on how to use more simple *in vitro* data generated during earlier stages of development.

6 CONCLUSION

Things have moved fast in the field of drug metabolism over the past 5 years, providing a large range of metabolism tools (some of these described in this article) which are perfectly adapted to screening, allowing a rapid understanding of the *in vivo* fate of a potential drug, long before clinical studies.

In one sense, things have moved too fast as most of the automated *in vitro* tests are already in place but we have not yet completely resolved a way to handle the enormous amount of data being produced.

Rapid metabolism assessment in all the discovery process is a continuing process. The simpler the metabolism tests are, the more adapted to sorting large series of compounds

(HTS) they are, producing warnings or *in vitro* SARs. The more sophisticated tests used on smaller series (SDS) produce more informative metabolism parameters, confirming possible previous results and leading to an *in vivo* prediction for animal and Man using simple or more sophisticated modelling techniques, up to the *in vivo* study in animal with n-in-one dosing confirming some in vitro results. Inversely, the more sophisticated tools and/or modelling give a better understanding to how to use more simple HTS data and allow the rapid development of new tests. This bi-directional continuum is the sole way to combine both validation and use of the new tools whilst the domain is still moving fast; and yet we are facing already another possible leap with the advent of genomics.

These new possibilities and/or necessities have pushed companies to handle the R and D interface in a different manner, with a clear need for a close relationship between the different groups of people involved, which have been working previously in a more sequential manner. Strategies to combine biopharmaceutical screening with pharmacological screening are specific to each company, mainly based on their past experiences, on human willing, on geographical distance between the different expertise and on the company size and financial capacity.

The difficulty is and will remain to use the right tools for the right problem and the right model for the interpretation of data getting as close as possible to our target, the whole human organism.

REFERENCES

1. B. Testa, L. B. Kier and P-A. Carrupt, *Med. Res. Rev.*, 1997, **17**, <u>4</u>, 303

2. D. Rodrigues, *Pharm. Res.*, 1997, **14**, <u>11</u>, 1504

3. H. Simpson, A. Berthemy, D. Buhrman, R. Burton, J. Newton, M. Kealy, D. Wells and D. Wu, *Rapid Communications In Mass Spectrometry*, 1998, **12**, 75

4. J. Berman, K. Halm, K. Adkison and J. Shaffer, *J. Med. Chem.*, 1997, **40**, <u>6</u>, 827

5. C. A. Lipinski, F. Lombardo, B. W. Dominy and P. J. Feeney, *Adv. Drug Deliv. Rev.*, 1997, **23**, 3

6. Advanced Chemistry Development, 133 Richmond St West, Suite 605, Toronto, Ontario, M5H 2L3, USA.

7. Sirius Analytical Instrument Ltd, Riverside, Forest Row Business Park, Forest Row, East Sussex RH18 5DW, England.

8. N. Schurgers, J. Bijdendijk, J.J. Tukker and D.J.A. Crommolin, *J. Pharm. Sci.*, 1986, **75** 117

9. S. Yee, *Pharmaceutical Research*, 1997, **14**, <u>6</u>, 763

10. P. Schmiedlin-Ren, K. E. Thummel, J. M. Fisher, M. F. Paine, K. S. Lown and P. B. Watkins, *Mol. Pharmacol.*, 1997, **51**, 741

11. C. L. Crespi, B. W. Penman and M. Hu, *Pharm. Res.*, 1996, **13**, <u>11</u>, 1635

12. S. Ong, H. Liu and C. Pidgeon, *J. Chromatogr. A*, 1996, **728**, 113

13. J. M. Mayer and H. van de Waterbeemd, *Environ. Health Perspect.*, 1985, **61**, 295

14. R. B. van Breemen, D. Nikolic and J. L. Bolton, *Drug Metab. and Dispos.*, 1998, **26**, 2, 85

15. R. F. Frye, G. R. Matzke, A. Adedoyin, J. A. Porter and R. A. Branch, *Pharmacokinet. Drug Dispos.*, 1997, **62**, 4, 365

16. C. L. Crespi, V. P. Miller and B. W. Penman, *Anal. Biochem.*, 1997, **248**, 190

17. J-B. Ferrini, L. Pichard, J. Domergue and P. Maurel, *Chem.-Biol. Interact.*, 1997, **107**, 31

18. A. Guillouzo, *In vitro methods in pharmaceutical research*, 1997, 17, ISBN 0-12-163390-X, Academic Press Ltd

19. H. Doostdar, M.H. Grant, W.T. Melvin, C.R. Wolf and M.D. Burke, *Biochemical Pharmacology*, 1993, **46**, 4, 629

20. I. de Waziers, J. Bouguet, P. H. Beaune, F. J. Gonzalez, B. Ketterer and R. Barouki, *Pharmacogenetics*, 1992, **2**, 12

21. J. J.P. Bogaards, A. M. Hissink, M. Briggs, R. Weaver, R. Jochemsen, P. Jackson, M. Bertrand, and P. J. van Bladeren, submitted *to J. Pharmacol. Exp. Ther.*

MODELING THE ACTIVE SITES OF CYTOCHROME P450s, ONE OF THE MOST IMPORTANT BIOTRANSFORMATION ENZYMES

Nico P.E. Vermeulen, Marcel J. de Groot, John H.N. Meerman and Jennifer Venhorst*

Vrije Universiteit – LACDR, Pharmacochemistry – Div. Molecular Toxicology, De Boelelaan 1083 – 1081 HV, Amsterdam, The Netherlands

* Part of this paper was published in a Ph.D.-thesis (M.J. de Groot) and in *Drug. Metab. Rev.* (modified)

ABSTRACT

In the absence of (bio)chemical information concerning the active sites of biotransformation enzymes, both indirect and direct computer modeling techniques can be used to get more insight into the active sites and the mechanisms of action of these enzymes. Indirect modeling or small molecule modeling uses a variety of substrates, inhibitors or metabolic products in order to derive a mold of the active site, while direct modeling or homology modeling uses crystal structures of similar proteins to derive a more or less complete model of the protein including the active site. Knowledge about the shape and properties of the active sites may indicate amino acids important for binding of substrates and/or the activation of substrates or intermediates. In this paper, the requirements and the assumptions for the generation of small molecule models and homology models, and the drawbacks and limitations of these models are discussed. Also some methods for the experimental validation of small molecule models and protein models are mentioned. Examples concerning one of the most important biotransformation enzyme families, viz. cytochrome P450, is indicated and discussed.

1 INTRODUCTION

Biotransformation enzymes catalyze various metabolic reactions in xenobiotic and endogenous compounds, e.g., oxidation, reduction, and conjugation reactions. The binding of competitive inhibitors and the conversion of substrates by enzymes takes place in the active sites of the enzymes. In order to get more insight into the mechanism of action of an enzyme, the substrate selectivity of an enzyme, and into the factors determining whether or not a compound will be metabolized by a certain enzyme, a detailed description of the shape and the physico-chemical properties of the active site is a prerequisite. For some (iso)enzymes crystal structures are available. However, the structures of the active sites of most of the important enzymes (both mammalian and non-mammalian) are not known yet. In recent years, this lack of knowledge has resulted in the prediction of various enzyme structures using computer aided molecular modeling techniques. The primary aim of this review is to summarize and discuss the requirements and the assumptions for the various computer modeling techniques used, the drawbacks and limitations of these modeling techniques, and furthermore to indicate some of the possible experimental methods to validate the modeled structures of the proteins and, specifically, the active sites. One important class of enzymes, metabolizing both xenobiotic and endogenous substrates, will be used to illustrate these aspects: cytochromes P450 (P450s).

1.1 Cytochromes P450

P450s constitute a large superfamily of heme-containing enzymes, capable of oxidizing and reducing a variety of substrates, both of endogenous and exogenous origin. P450 isoenzymes have been classified into families and subfamilies. P450s belong to a separate family when the primary sequence homology with any other family is $\leq 40\%$ (1). For mammalian P450 amino acid sequences within the same subfamily the identity is usually $> 55\%$ (1).

Characteristics which allow a large number of structurally different compounds to be metabolized by a limited number of enzymes include a broad substrate specificity and a broad regio- and stereoselectivity. Cytochromes P450 generally detoxify potentially dangerous compounds, but in a number of cases non-toxic compounds are bioactivated to toxic reactive intermediates, and procarcinogens are activated into their ultimate carcinogens (2). P450s also catalyze key reactions in steroidogenesis in animals, resistance in insects and plants, and flower coloring (3). The metabolic activities of P450s can be divided in (a) monooxygenase activity, usually resulting in incorporation of an oxygen atom into the substrate, (b) oxydase activity, resulting in formation of superoxide anion radicals and hydrogen peroxide (uncoupling of the catalytic cycle (4)), and (c) reductase activity, usually producing free radical intermediates under anaerobic conditions (4, 5).

P450s can also be classified according to the electron transfer chain delivering the electrons required for the one-electron reductions from NAD(P)H to the P450: class I P450s are found in the mitochondrial membranes of eukaryotes and in bacteria and require an FAD (flavin adenine dinucleotide) containing reductase and an iron-sulfur protein (putidaredoxin) (3, 6, 7), while class II P450s are bound to the endoplasmic reticulum and interact directly with a cytochrome P450 reductase (containing FAD and FMN (flavin mononucleotide)) (3, 7, 8).

2 SMALL MOLECULE MODELS

One possibility to derive a model for the active site of an enzyme is the creation of a small molecule model or pharmacophore model (Scheme 1). With this technique information on the active site is derived (indirectly) from the shape, electronic properties and conformations of substrates, inhibitors or metabolic products. Various substrates, inhibitors or metabolic products (which have been characterized by a variety of experimental data) are fitted onto each other by superimposing chemically similar groups. By using this approach a mold is created which describes the size of the active site and the electrostatic distribution therein. The general procedure followed to construct a small molecule model is depicted in Scheme 1. In the text a small molecule model for substrates will be used as an example. Small molecule models for inhibitors or metabolites can be obtained in a similar manner.

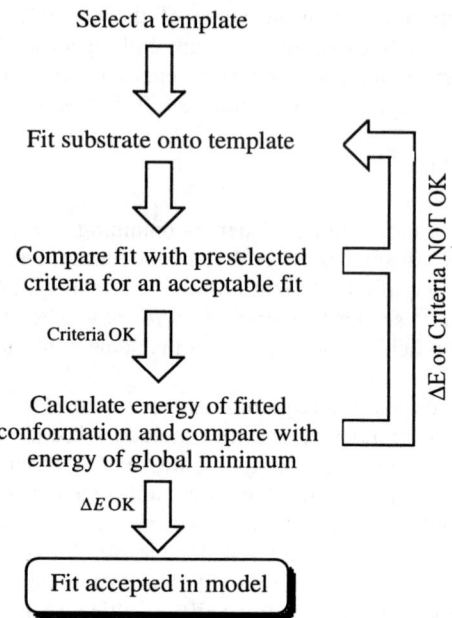

Scheme 1 *General Procedure for the Construction of a Small Molecule (Pharmacophore) Model*

2.1 Requirements

In order to build a small molecule model for substrates of a specific enzyme, a first prerequisite is a template molecule upon which the model will be built. The template is usually a substrate which ideally: (a) is specifically metabolized by the (iso)enzyme under investigation, (b) is large, in order to describe as much as possible of the active site area, (c) is relatively rigid, as flexible molecules will have too much conformational freedom, which complicates the selection of the "active" conformation, (d) contains essential functional groups, and (e) is regio- and/or stereoselectively metabolized. A second prerequisite is the availability of appropriate enzyme kinetic and metabolic data concerning a variety of additional substrates, which are specifically metabolized by the (iso)enzyme under investigation. Thirdly, a computer program is needed which contains the molecular forcefield parameters required for modeling the substrates under investigation.

After selection of a template molecule, additional substrates are superimposed onto the template molecule. Some pre-defined fit-criteria have to be met, otherwise the fit onto the template is rejected. When the fit is accepted, an energy calculation is performed in order to determine the energy difference between the global minimum energy conformation and the fitted conformation. If this energy difference (ΔE) is within a pre-defined range, the fit of the compound on the template molecule is accepted in the model.

2.2 Assumptions

The first assumption concerns the geometry of the substrates. Substrates are usually energy minimized, using either semi-empirical or *ab initio* methods, or their geometry derived from the Cambridge structural database (CSD(9)). In case of energy minimization using semi-empirical or *ab initio* methods, the *in vacuo* geometry of the substrate is calculated. This may give a reasonably correct geometry for the biologically active geometry only when (a) the active site of the enzyme is mainly hydrophobic in nature, (b) charge stabilization by the apoprotein is not significant during the reaction, and (c) the metabolic reaction of the substrate is "chemical like" and does not strictly require specific interactions between the substrate and the apoprotein thereby altering the geometry of the substrate *(10,11)*. Geometries derived from crystal structures, on the other hand, are usually influenced by crystal packing effects. The geometries used, either derived from calculations or from crystal structures, do *not necessarily* correspond to the biologically active conformation. In some studies, geometries obtained from the CSD (9) and calculated geometries are both used, despite the fact that this may lead to erroneous conclusions. A second assumption for small molecule modeling is that all substrates will be oriented in a similar manner (both electronically and sterically) in the active site of the enzyme. Without these assumptions the construction of a small molecule model would be impossible.

2.3 Drawbacks of Small Molecule Models

In small molecule models steric, electronic and other interactions with the protein are neglected. However, if a substrate can be accommodated in a small molecule model but experimentally the formation of a certain metabolite does not occur, this does not necessarily imply a steric (or electronic) restriction which is neglected by the model. Possibly other metabolic pathways and/or (iso)enzymes may compete with the specific metabolic reaction of the (iso)enzyme for the substrate under the experimental conditions applied. The absence of a certain (predicted) metabolite may also be due to kinetic rather than thermodynamic effects (10) for example when the specific metabolic conversion is very slow compared to other metabolic reactions.

When using substrate geometrics directly from the CSD (9), the geometries are usually influenced by packing effects, which are absent in the biological environment. In a similar way, the geometries of the substrates/inhibitors derived from optimizations using semi-empirical or *ab initio* methods are not necessarily identical to the geometries in a biological environment (as indicated above).

Small molecule models for inhibitors are generally more difficult to construct compared to small molecule models for substrates or metabolic products. The specific site of reaction (e.g., oxidation or conjugation) is lacking in inhibitor models and can therefore not be used as an easily identifiable site to be superimposed.

2.4 Experimental Validation

After building a small molecule model, in principle metabolic predictions can be made using the model. In order to validate the model and the predictions based on the model, experiments can be designed to test the hypotheses. Generally, the predicted metabolites of a substrate have to be identified in incubations using the purified or heterologously expressed (iso)enzyme. When using microsomes for such experiments, other (iso)enzymes can be responsible for the metabolites as well. The metabolite pattern found in the metabolism experiments can subsequently be compared with the predicted metabolites. In case the predicted metabolite is not detected experimentally, this does not unequivocally indicate that the small molecule model is erroneous (as indicated under Drawbacks).

Several parameters that can be easily determined, such as Michaelis-Menten constants (K_m), inhibition constants (K_i) and binding constants (K_s) appeared not very useful for the validation of small molecule models. The most useful constant is most likely K_s, as a small molecule model can only give information on the binding of substrates, and not on overall reaction rates. However, for a series of compounds used in a small molecule model for rat glutathione S-transferase (see below (12), recent experiments indicated that although the K_m appeared to correlate well with differences observed in the small molecule model, the differences in K_s (and to a lesser extent the differences in K_i) were almost indistinguishable for the various substrates.

In the following paragraphs, small molecule models for P450s will be reviewed.

2.5 Small Molecule Models for Cytochrome P450 Isoenzymes

Small molecule models have been derived for only a limited number of P450 isoenzymes. In recent years, more elaborate computational techniques were used, compared to the relative simple calculations performed in the 1980's.

    2.5.1 P450 1A1. A very simple small molecule model for rat P450 1A1 was first derived by Jerina and coworkers (13) using benzo[*a*]pyrene and a variety of other polycyclic aromatic hydrocarbons (PAHs) (Figure 1a). Benzo[*a*]pyrene is converted stereoselectively *via* 7,8-epoxidation by P450 1A1, hydration by epoxide hydrolase, and 9,10-epoxidation by P450 1A1 to the ultimate carcinogen benzo[*a*]pyrene 7(*R*),8(*S*)-diol 9(*S*),10(*R*)-epoxide. Based on the PAH-substrates used, this model described the active site of P450 1A1 as a hydrophobic cleft, asymmetrically oriented relative to the heme. This original model of P450 1A1 substrates was later extended to accommodate several other PAHs (14, 15). The original model had to be extended considerably (Figure 1b) or a certain degree of flexibility in the position of the substrates had to be incorporated (Figure 1c).

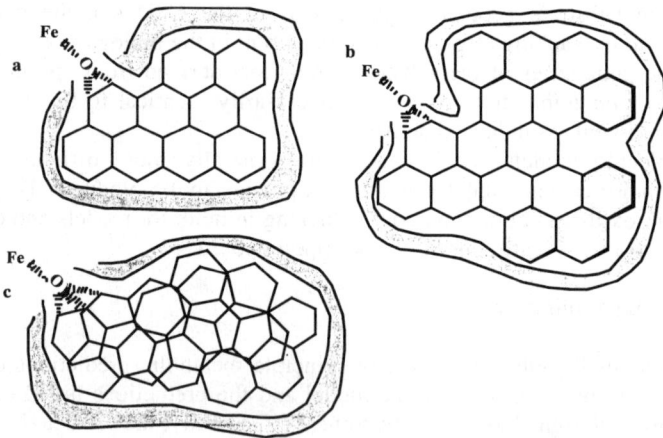

Figure 1   (a) *Steric model of the active site of P450 1A1 based on the metabolism of benzo[a]pyrene; the binding site is asymmetrically positioned toward the activated oxygen species bound at the iron atom (13). (b) Expansion of model (a) in order to accommodate several other PAHs (14, 15). (c) Proposed model in which some flexibility in the angle of the oxygen addition to the substrate is allowed (14). Taken from reference (16)*

Rat P450 1A1 is also known to metabolize, in a regio- and stereoselective manner, a variety of small non-PAH substrates, like 7-ethoxycoumarine and zoxazolamine (16). The binding and orientation of these small substrates in the active site of P450 1A1 was suggested to be the result of a hydrogen bonding interaction and aromatic interactions between these small substrates and the protein (16).The sites of oxidation and the heteroatoms responsible for the hydrogen bonding interaction with the protein were superimposed, as indicated in Figure 2 (16). The combination of the model for PAHs (13, 15) and the model for small molecules (16).presents a rough small molecule model which can accommodate many substrates of rat P450 1A1.

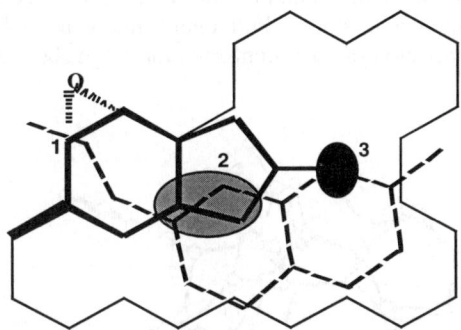

Figure 2 *Small molecule model for P450 1A1 for small non-PAH substrates (16). 7-Ethoxycoumarine (solid line) and zoxazolamine (dashed line) are shown when superimposed onto the steric model for PAHs as shown in Figure 1a (13) with (1) site of oxidation, (2) region of presumed π–π interactions between substrates and protein, and (3) location of heteroatoms in the substrates proposed to form a hydrogen bonding interaction with the protein. Taken from reference (16)*

2.5.2 P450 1A1/1A2. Computational analysis of compounds oxidized by rat P450 1A1 and 1A2 [1] indicated that these isoenzymes preferentially catalyze the hydroxylation of essentially flat molecules, further characterized by a small depth and a large area/depth ratio (17, 19). These studies used substrate geometries from crystal structures and from MINDO/3 semi-empirical calculations (17). As crystal structures may be influenced by crystal packing effects, a direct comparison is, however, not necessarily warranted (see Assumptions). The substrates were fitted onto each other, based only on size and shape, no specific groups within the substrates were superimposed (17).

2.5.3 P450 2B1/2B2. A simple computational analysis of compounds metabolized by rat P450 2B1 and 2B2 [1], suggested these substrates to be rather bulky, non-planar molecules characterized by small area/depth ratios and a larger flexibility in molecular conformation, when compared to substrates of the rat P450 1A1 and 1A2 (17, 19). Again crystal structures and MINDO/3 optimized geometries were used interchangeably. The substrates were not superimposed in this study and only sizes and shapes were compared (17).

2.5.4 P450 2C9. Human P450 2C9 is an isoenzyme, which is involved in the metabolism of a large number of antiinflamatory drugs, which exist as anions at physiological pH (20). Based on 12 substrates a small molecule model for P450 2C9 was

---

[1] No distinction was made between these isoenzymes in this study.

recently derived, using two rigid substrates, namely phenytoin and (S)-warfarin, as templates (20). The geometries of the substrates were partially derived from crystal structures, and partially from molecular mechanical calculations using the "consistent valence forcefield" (20). It was possible to superimpose the substrates with their sites of hydroxylation, and to bring all anionic heteroatoms in the various substrates at a distance between 3.5 Å and 4.8 Å from a common (hypothetical) cationic interaction site within the P450 2C9 protein (20) (Figure 3). Since the positions of the anionic heteroatoms were rather different for the various substrates, a hydrogen bond to the protein was suggested not to be possible with all the substrates. Instead, a purely cation-anion interaction was presumed (20). As indicated by the authors, more calculations are needed to substantiate this preliminary small molecule model. Full conformational analysis of the substrates including geometry optimization of each conformation might be a useful approach for all substrates.

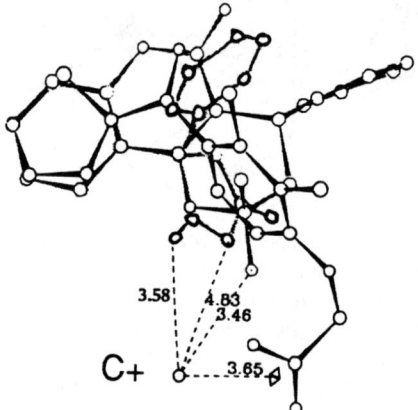

Figure 3   *Superposition of the hydroxylation sites and hydroxylated aromatic rings of warfarin, phenytoin, and tienilic acid. Possible interaction of their anionic sites with a cationic site of P450 2C9 (C^+) is shown. Taken from reference (20)*

2.5.5   P450 2D6.   Human P450 2D6 is a polymorphic member of the P450 superfamily and is absent in 5-9% of the Caucasian population as a result of a recessive inheritance of gene mutations (21, 23). This results in deficiencies in drug oxidations known as the debrisoquine/sparteine polymorphism which affect the metabolism of numerous drugs. A decreased metabolism of these drugs is found in poor metabolizers which have two non-functional P450 2D6 alleles, compared to extensive metabolizers with at least one functional allele. Small molecule models predicting the involvement of P450 2D6 may identify potential problems for poor metabolizers when either a drug is not metabolized or a prodrug is not activated due to the dependence on the lacking P450 2D6. A relatively large number of small molecule models has been derived for this particular human P450 isoenzyme, using a variety of substrates or inhibitors (24, 28).

The first substrate models were based on substrates containing a basic nitrogen atom at a distance of either 5Å (Figure 4a (45)) or 7Å (Figure 4b (25)) from the site of oxidation, and an aromatic ring system which was coplanar in both models (24,25). In the 5Å model

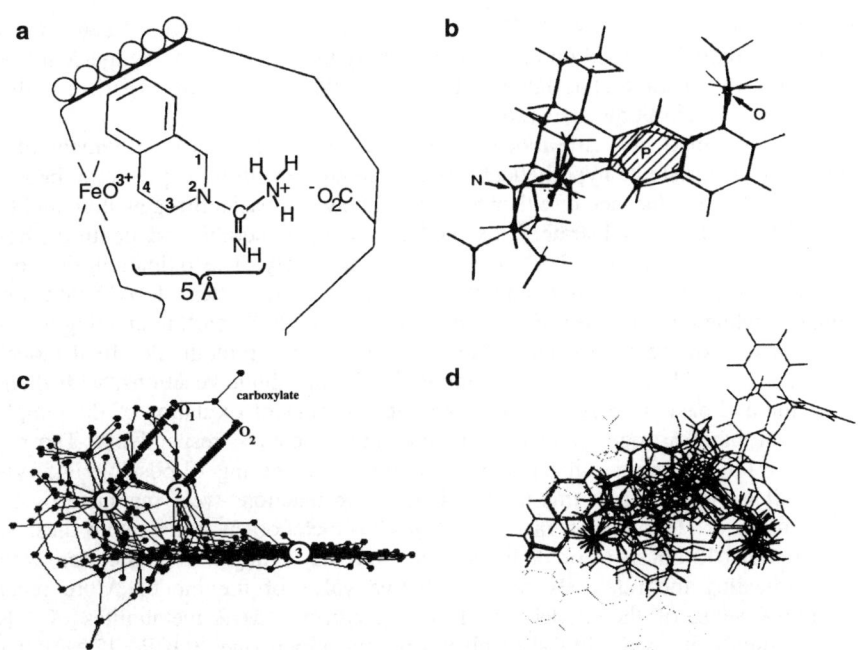

Figure 4 (a) *Initial 5 Å small molecule model for P450 2D6. Debrisoquine is shown at the active center with the basic nitrogen atom attached to a carboxyl group and the site of oxidation adjacent to the iron-oxo complex. The heavy line and six circles denote a hydrophobic region. Taken from reference (32). (b) Initial 7 Å small molecule model for P450 2D6. Juxtaposition of dextromethorphan and bufuralol, with N: basic nitrogen atom, P: lipophilic plane, and O: oxidation site. Taken from reference (26). (c) Combined 5 Å - 7 Å small molecule model for P450 2D6 (27). Oxidation sites (3) of all molecules are superimposed. Basic nitrogen atoms are fitted either on the basic nitrogen atom of debrisoquine (2), or onto that of dextromethorphan (1) and interact with one of the carboxylic oxygen atoms (O_1 or O_2). Taken from reference (16). (d) Refined small molecule model for P450 2D6 (33), containing the heme moiety (gray) and aspartic acid residue 301 (2) derived from a protein model for P450 2D6 (34). The site of oxidation is indicated (1). Adapted from reference (33)*

no substrates were actually fitted onto each other (24). The main problem of these initial models was that neither of the two models could explain the other group of substrates.

An extended model was derived by Islam *et al.* (26) which indicated a distance between a basic nitrogen atom and the site of oxidation between 5 and 7 Å. This small molecule model also contained the heme moiety from the crystal structure of P450 101 (P450$_{cam}$ (29) above which the template molecule of this small molecule model (debrisoquine) was positioned arbitrarily (26), in a manner resembling the orientation of camphor in the P450 101 crystal structure (30) The model also included an oxygen atom bound to the iron of the heme moiety, which is involved in the P450 hydroxylation reaction (26). A set of 15 compounds was fitted onto the template debrisoquine onto which some of the known substrates of P450 2D6 (e.g., sparteine and amitriptyline) could not be fitted (26). One prediction based on this model, namely that NNK (4-(*N*-methyl-*N*-nitrosamino)-1-(3-pyridyl)-1-butanone) is not a substrate for P450 2D6, was experimentally verified using human liver microsomes (26).

Another small molecule model for P450 2D6 was derived by Koymans *et al.* (27). This model suggested a hypothetical carboxylate group within the protein to be responsible for a well defined distance of either 5 Å or 7 Å between basic nitrogen atom and the site of oxidation within the substrate. This model used debrisoquine and dextromethorphan as templates for the 5 Å and 7 Å compounds, respectively. The oxidation sites of the two templates were superimposed and the areas next to the sites of oxidation were fitted coplanar, while the basic nitrogen atoms were placed 2.5 Å apart, interacting with different oxygen atoms of the postulated carboxylate group in the protein. The final model (Figure 4c) consisted of 16 substrates, accounting for 23 metabolic reactions, with their sites of oxidation and basic nitrogen atoms fitted onto the sites of oxidation of the templates, and one of the basic nitrogen atoms of the template molecules, respectively. The model was verified by predicting the metabolism of 4 compounds giving 14 possible P450-dependent metabolites. According to this model, 4 oxidative reactions were mediated by P450 2D6 while the other 10 were not. *In vivo* and *in vitro* metabolism studies with these substrates indicated that 13 out of 14 predictions (3 positive and 10 negative predictions) were correct (27) indicating the relatively high predictive value of the model. More recently, the predictive value of the model was further confirmed as 2 metabolites of 1-[2-[bis(4-fluorophenyl)methoxy]-ethyl]-4-(3-phenyl-propyl)-piperazine (GBR 12909) were also correctly predicted and shown to be formed by heterologous expressed P450 2D6 (31). The relatively large GBR 12909 extended considerably from the region described by the small molecule model, indicating an extension of the model in certain directions to be necessary (31).

Recently, the actual positions of the heme moiety and the I-helix, containing Asp[301] (derived from a protein model of P450 2D6, see below (34) have been added to this small molecule model, thereby incorporating some steric restrictions and orientational preferences into the small molecule model (33). Involvement of Asp[301] in substrate binding was initially predicted using homology modeling techniques (35) (see below) and recently confirmed with site specific mutation and expression experiments, to be important for the activity of P450 2D6 (36). In this refined small molecule model, an aspartic acid residue was coupled to the basic nitrogen atom of the substrates, thus enhancing the small molecule model with the direction of the hydrogen bond between the aspartic acid in the protein and the (protonated) basic nitrogen atom (33). Debrisoquine and dextromethorphan were still used as template molecules. The site of oxidation above the heme moiety was one of the two possible sites of oxidation as suggested by the recently derived protein model for P450 2D6 (see below) (34) and is located above pyrrole ring B of the heme moiety. In this model the sites of oxidation in the substrates were fitted onto the defined oxidation site above pyrrole ring B of the heme moiety, while the C_α and C_β atoms of the attached aspartic acid moiety were fitted onto the C_α and C_β atoms of Asp[301], respectively (33). A schematic representation of the refined small molecule model of P450 2D6 is given in Figure 4d. A variety of substrates fitted in the original substrate model for P450 2D6 (27, 31, 37) were successfully fitted into the refined substrate model (for example GBR 12909), indicating that the refined substrate model for P450 2D6 (with extra sterical and directional restraints) can accommodate the same variety in molecular structures as the original substrate model. The refined small molecule model also gives a more accurate description of the active site of P450 2D6.

Parallel to the substrate models for P450 2D6, an inhibitor model has been derived. As no suitable template inhibitor (see Requirements) was available, the template of this model was derived by fitting 6 strong reversible inhibitors of P450 2D6 onto each other (28). The basic nitrogen atoms were superimposed and the aromatic planes of these inhibitors were fitted coplanar. All inhibitors used were relatively flexible, resulting in various low-energy conformations. The final template consisted of those conformations of ajmalicine, quinidine, chlorpromazine, trifluperidol, prodipine, and lobeline, that could be fitted relatively well onto each other (49). Consecutively, other inhibitors, such as derivatives of ajmalicine and quinidine, were fitted onto the derived template. The derived preliminary pharmacophore model consisted of a tertiary nitrogen atom (protonated at physiological pH) and a flat hydrophobic region (A in Figure 5 (49)). Furthermore, there appeared to be a region (B in Figure 5) in which functional groups with lone pairs seemed to cause enhanced inhibitory potency, while in another region (C in Figure 5) hydrophobic groups seemed to be allowed but caused no enhanced inhibitory effect (28). The inhibition data were obtained from experiments using human liver microsomes and bufuralol as substrate (28). The uncertainties in both the template used and the inhibition experiments used to verify this model were relatively large (27, 31). The features derived for this inhibitor based small molecule model (28) were very similar to the features of the proposed substrate models of P450 2D6 (26, 27, 33). For this reason it is not unlikely that the substrate based and inhibitor based small molecule models can be combined.

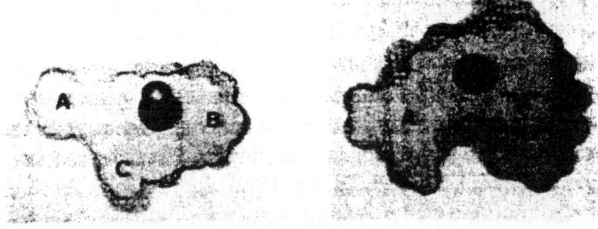

Figure 5   *View of a 2D6 inhibitor model represented by the overall surface of strong inhibitors. (a, right) Top view of the model. (b, left) Model rotated 90° around the x-axis. The protonated nitrogen atom is depicted in dark gray, with the proton in light gray. Three regions are indicated: a flat hydrophobic region (A), a region in which functional groups with lone pairs seemed to enhance the inhibitory effect (B), and a region in which hydrophobic groups were allowed but did not seem to cause an enhanced inhibitory potency (C) (28). Taken from reference (28)*

2.6 Summary

A wide variety of substrates specifically metabolized by a certain P450 isoenzyme is generally available. This usually enables the selection of a suitable template for the small molecule model. In case no suitable template molecule is available, a combination of several structurally different compounds may also be successfully used as a template. The earliest reported substrate models for P450s are relatively crude small molecule models, while the more recently derived models are much more advanced and are constructed using more sophisticated computational modeling techniques. The latter models (e.g., for P450 1A1 (16), P450 2C9 (20) and P450 2D6 (26-28, 33)) demonstrate a clear potential to predict the possible involvement of specific P450 isoenzymes in the metabolism of selected substrates and the nature of hypothetical interaction sites in the active site of the protein. For the polymorphic isoenzyme P450 2D6, small molecule models have already been used for the prediction of the involvement of P450 2D6, in order to identify potentially large inter-individual differences between extensive and poor metabolizers. This may pose risks to poor metabolizers in case either a drug is not metabolized or a prodrug is not activated due to dependence on a lacking P450 2D6. Furthermore, these models might be used to rationalize inhibitory properties of various compounds.

3 PROTEIN MODELS

Another computer assisted approach to obtain structural information on the active site of a protein (e.g., an enzyme) is the construction of a protein or homology model (direct modeling). Homology modeling yields information on the active site by constructing a three-dimensional model of the protein based on the amino acid sequence and the crystal structure of one or more similar proteins. By using this method a three-dimensional representation of the protein, and more specifically, the active site may be obtained, as well as information on amino acids (potentially) involved in binding of substrates and inhibitors, and in the catalytic process (10). The general procedure for constructing a protein model is schematically depicted in Scheme 2.

Initially, an alignment is made between the amino acid sequences of the unknown structure and (a) crystallized template protein(s), ideally supplemented with structural or biochemical data. Subsequently, homologous regions present in both the crystal structure(s) and the unknown structure are directly copied from the crystal structure(s) to the homology model, while the non-identical parts are calculated for the model. In the next step the homology model is energy minimized (using molecular mechanical methods). In the minimized homology model, substrates, inhibitors or metabolites can be docked. The constructed homology model can also be validated and refined with experimental data (e.g., from site-directed mutagenesis and/or site-specific modification experiments).

One of the aims of this review is to summarize and discuss various aspects of homology modeling techniques. Although a good amino acid alignment between similar proteins is a prerequisite for the construction of a protein model, it is not our aim to discuss the various methods and software programmes used for obtaining (automatic) alignments.

3.1 Requirements

In order to build a homology model of a protein, at least one crystal structure of a similar protein is required, as well as an alignment describing corresponding amino acids in the protein under investigation and the crystallized protein(s). The crystal structure(s) should preferably have a high resolution (1.5-2.5 Å) and a high (primary sequence) homology with the protein under investigation. Ideally the crystallized protein(s) belong(s) to the same family of proteins ((iso)enzymes). The reliability of the alignment depends on the homology between the crystal structure(s) used and the protein under investigation. When the homology is relatively low, the alignment will contain parts of questionable reliability and consequently various alignments will be possible for such regions. In case of low homology, the algorithm used to derive the alignment also has an important influence on the final homology model, as different algorithms give rise to different alignments, and consequently different protein models. Generally, an automatically generated alignment needs to be adjusted manually based on available additional information, such as site-directed mutagenesis data (38). Use of multi-alignment techniques and secondary structure predictions can also help aligning specific regions with a very low homology (39).

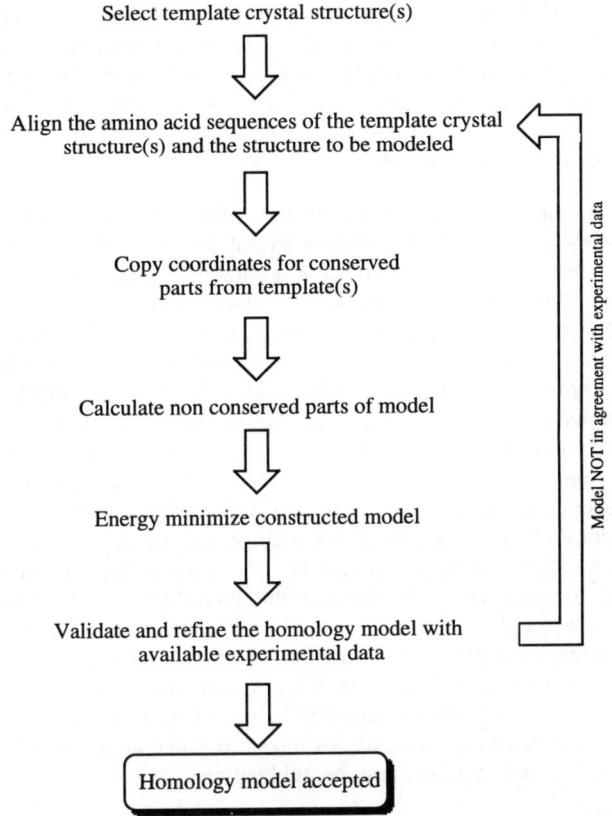

Scheme 2 *General procedure for the construction of a homology (protein) model*

3.2 Assumptions

The most important assumption inherent to homology modeling is that the three-dimensional structure of the protein constructed is similar to that of the crystallized protein used as a "template". The validity of this assumption, of course, depends on the specific protein under investigation and the availability of homologous crystal structures.

An important factor determining the quality of a homology model is the forcefield used for the molecular mechanical calculations. Various programs, employing a variety of forcefields, can be used to build a homology model, to energy minimize the model, and to dock the substrates, inhibitors or metabolites into the model. As the energy terms and parameters in the different forcefields are not identical, generally no direct comparison can be made of total energies obtained for homology models of the same protein by different programmes. Structural comparisons can be made to a certain extent. However, differences in the forcefields employed will usually have consequences for the final geometry of the protein model. In order to select a forcefield, one should first determine whether that specific forcefield gives an appropriate description of all aspects of the protein model under construction, e.g., that it contains the correct parameters, in case of P450s for example for the description of a heme moiety. To describe this heme moiety, a specific set of parameters has been derived (40, 41). These parameters give an appropriate description of this moiety, but are not available in all homology modeling programs. *Ab initio* calculations would circumvent the dependency of homology models on forcefields, but protein/enzyme structures are generally far too large for *ab initio* approaches.

3.3 Drawbacks of Homology Models

The drawbacks of homology models are closely related to the assumptions mentioned above. A homology model will to a certain extent resemble the crystal structure from which it has been derived. This resemblance might be real or merely a consequence of the methodology used. When the homology between the available crystal structure(s) and the protein/enzyme for which the model is constructed is relatively low, the alignment of the respective sequences is not straightforward. In the modeling studies mentioned in the P450 section, several alignment programs have been used. Most of the automated alignments have been manually adjusted to incorporate additional information (e.g., site-directed mutagenesis data) and to remove errors (e.g., insertions or deletions in α-helices). Although these manual adjustments introduce uncertainties, different authors have nevertheless independently derived almost identical alignments (8, 34, 38).

The dependency of the geometry of the final protein model on the forcefield used is another drawback. It is therefore advisable to perform the geometry optimization calculations used to construct and optimize the homology model also on the crystal structure(s) used as a template, and to determine first the changes occurring in the template structure(s) due to this procedure. As several homology modeling programs and forcefields have been used to geometry optimize the resulting protein models, a comparison of the various models has to be considered carefully. Even when identical software is used, the forcefield parameters used in the various optimization procedures are not always identical and unfortunately often not mentioned in the publications.

3.4 Experimental Validation

The validation of protein models has to come from crystallization experiments or from other methods, such as three-dimensional NMR. Often, however, homology modeling techniques are used when protein structure determinations using three-dimensional NMR or crystallization were not successful. Predictions as to e.g., the possible role of specific amino acids in binding of substrates and/or inhibitors and in the mechanism of catalysis can often be verified experimentally using site-directed mutagenesis experiments or site-specific modification experiments. Predictions concerning available space in the active site above different pyrrole rings can be assessed using reactions between arylhydrazines or aryldiazenes with heme proteins, leading to different iron N-arylporphyrins (42). Sucg information can be derived from NMR spin relaxation studies as well, as performed recently for a number of P450 2D6 substrates (62)

In the following paragraphs, homology models derived for P450s will be reviewed.

3.5 Homology Models for Cytochrome P450 Isoenzymes

Crystal structures have been resolved for several soluble bacterial P450s: P450 101 (P450$_{cam}$, schematically shown in Figure 6a) without substrate (29,45) with camphor as bound substrate (30), with adamantanone, adamantane, camphane, norcamphor or thiocamphor as bound substrate analogs (44, 45) with metyrapone or 1-, 2- or 4-phenylimidazole as bound inhibitors (46) with both enantiomers of a chiral, multifunctional inhibitor bound (47), and with 5-*exo*-hydroxycamphor as bound catalytic product (78), P450 102 (P450$_{BM3}$, schematically shown in Figure 6b) without substrate (6, 49-51), P450 107A (P450$_{eryF}$) with 6-deoxyerythronolide B as bound substrate (52, 53) and P450 108 (P450$_{terp}$) without substrate (49, 54). Furthermore, crystals have been reported for a soluble eukaryotic *Fusarium oxysporum* P450, P450 55 (P450nor) without substrate, although no coordinates have been reported as yet (57). In contrast to other P450s, P450 55 does not possess monooxygenase activity, but reduces nitric oxide instead (7). Despite extensive efforts, no eukaryotic, membrane-bound P450s have been crystallized so far. The core region of P450s, containing the D-, E-, I- and L-helices and the heme coordination region, of all available crystal structures is very similar (3, 8, 53, 58, 59) indicating that the three-dimensional structure of these regions is well conserved despite a low sequence homology, while other regions (e.g., the active site region containing the B'-helix (3, 53, 59), the loops between the C- and D-helices, the region spanning the F- and G-helices and some parts of the β-sheets) are less similar 3, 8, 53, 58, 59) For this reason, the core region of homology models of P450s based on these crystal structures will likely be a reliable representation, while other parts will remain speculative.

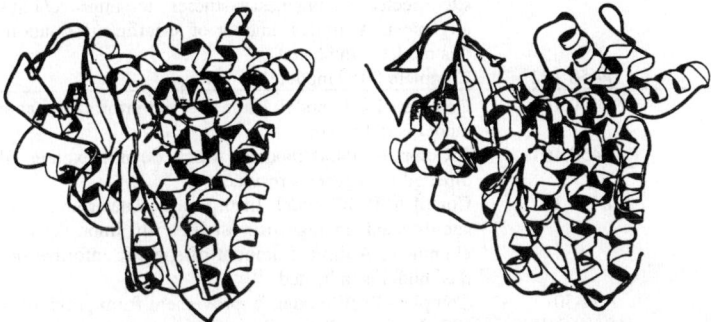

Figure 6   *Schematic diagram of* (a, left) *P450 101 and* (b, right) *P450 102. Helices are represented as rods and β-sheets as flat arrows. Taken from reference (7)*

Table 1 summarizes homology models built so far based on the available soluble bacterial P450 crystal structures. In principle, a crystal structure of a membrane bound P450 would be the best starting point for a homology modeling study on a membrane bound P450. In the absence of such a crystal structure, however, P450 102 (a class II P450, to which many eukaryotic P450s belong) might be a better template P450 for homology building studies (6, 50, 51), than to P450 101 and P450 108 (class I P450s). Due to its larger endogenous substrates (long-chain fatty acids, alcohols and amides), P450 102 is expected to have an active site that more closely resembles the active sites of other P450s than P450 101 (60). This was recently confirmed by a homology modeling study on human thromboxane A$_2$ synthetase (TXAS, P450 5) using both P450 101 and P450 102 (separately) as templates (61). The authors further suggested that models based solely on P450 101 should be reexamined closely, using the new crystal structures (61) in order to improve these models.

The most reliable homology models so far have been constructed based on multiple alignments and use site-directed mutagenesis data to enhance the reliability of the alignment. Some models have been experimentally validated by site-directed mutagenesis experiments, while in other cases the site-directed mutagenesis experiments were based on initial predictions from the homology models. Recently, a set of protein models for P450 2D6 was reported which incorporated distance restraints derived from NMR data in order to enhance the quality of these models (56, 62, 93). Although several homology models are based on multiple alignments methods, the use of site-directed mutagenesis data is less widely used (10, 34, 62-67).

Table 1 *Overview of Homology Models Built for P450s, the Crystal Structure(s) Used as a Template, and Some Specifications of the Homology Models*

Enzyme model (1)	Template P450(s)	Specification [a]	Ref.
P450 1[b]	P450 101	Complete P450 model. Little specific information about this model is indicated.	(68)
P450 1A1	P450 101	Complete P450 model. Alignment in conflict with experimental data for P450 2A4/2A5 (69)	(70)
P450 1A1	P450 101	Complete P450 model.	(71)
P450 1A1/A2	P450 102	Complete P450 model	(94)
P450 2A1	P450 102	Complete P450 model.	(67)
P450 2A4	P450 102	Complete P450 model.	(67)
P450 2A5	P450 102	Complete P450 model.	(67)
P450 2A6	P450 102	Complete P450 model. Incorporates data from a variety of site-directed mutagenesis studies to improve/adjust the alignment. A limited amount of specific information about this model is indicated.	(66)
P450 2A6	P450 102	Complete P450 model.	(67)
P450 2B[b]	P450 101	Complete P450 model. Little specific information about this model is indicated.	(68)
P450 2B1	P450 101	1 Complete P450 models, which do not explain all site-directed mutagenesis results.	(65)
P450 2B1	P450 102	Complete P450 model. Incorporates data from a variety of site-directed mutagenesis studies to improve/adjust the alignment. A limited amount of specific information about this model is indicated.	(66)
P450 2B1	P450s 101/102/108	Complete P450 model. Improvement from previous models (95). Explains all site-directed mutagenesis results.	(68)
2B1/2B4/2B6	P450 102	Incorporates data: consistent with site directed mutagenesis antibody recognition sites residues associated with binding redox partner interactions.	(95)
P450 2B4	P450 102	Complete P450 model. Incorporates data from a variety of site-directed mutagenesis studies to improve/adjust the alignment. A limited amount of specific information about this model is indicated.	(66)
P450 2C3	P450 102	Complete P450 model. Incorporates data from a variety of site-directed mutagenesis studies to improve/adjust the alignment. A limited amount of specific information about this model is indicated.	(66)
P450 2C3v	P450 102	Complete P450 model. Incorporates data from a variety of site-directed mutagenesis studies to improve/adjust the alignment. A limited amount of specific information about this model is indicated.	(66)
P450 2C9	P450 101	Complete P450 model. Site-directed mutagenesis data used to improve the multi-alignment of the 2-family.	(10)
P450 2C9	P450 102	Complete P450 model. Incorporates data from a variety of site-directed mutagenesis studies to improve/adjust the alignment. A limited amount of specific information about this model is indicated.	(66)

Table 1   *Overview of Homology Models Built for P450s, the Crystal Structure(s) Used
as a Template, and Some Specifications of the Homology Models (continued)*

Enzyme model (1)	Template P450(s)	Specification ^{a}	Ref.
P450 2D1	P450 102	Complete P450 model. Incorporates data from a variety of site-directed mutagenesis studies to improve/adjust the alignment. A limited amount of specific information about this model is indicated.	(66)
P450 2D6	P450 101	Preliminary P450 model, only containing active site regions of the protein (11 segments). Indicated Asp^{301} as an important amino acid for catalytic activity.	(35)
P450 2D6	P450 102	Complete P450 model. Incorporates data from a variety of site-directed mutagenesis studies to improve/adjust the alignment. A limited amount of specific information about this model is indicated.	(66)
P450 2D6	P450s 101/102/108	A set of 13 complete P450 models. Uses structural alignment method, multiple alignment (16 P450 sequences) and NMR derived distance restraints. In close agreement with (34)	(62)
P450 2D6	P450s 101/102/108	Semi-complete P450 model containing active site region and well conserved regions (3 segments, only highly variable loops omitted). Uses structural alignment method and multiple alignment (66 P450 sequences). Incorporates data from site-directed mutagenesis results concerning the 2-family to improve/adjust the alignment. Improvement from preliminary homology model (35) In close agreement with (62).	(34)
P450 2D6	P450 102	Complete P450 model. Incorporates data of allelic variants and site directed mutagenesis studies.	(97)
P450 2D6	P450s 101/102/107A/108	Two sets of P450 models. Same approach, comparative modeling, used as (92) now only also inclusing P450 107A as remplate protein.	(56)
P450 2D6	P450 101/102/107A/108	A set of 9 complete P450 models, same approach as (92) and (93) was used.	(93)
P450 2E1	P450 102	Complete P450 model. Includes data of species differences between rat, mouse and man .	(96)
P450 3A4 (P450$_{NF}$)	P450 101	Complete P450 model. Only partially geometry optimized.	(73)
P450 3A4	P450 101/102/107A/108	Complete P450 model. Uses structure based alignment and consensus strategy.	(98)
P450 4A4	P450 102	Complete P450 model. Incorporates data from a variety of site-directed mutagenesis studies to improve/adjust the alignment. A limited amount of specific information about this model is indicated.	(66)
P450 4A11	P450 102	Complete P450 model. Incorporates data from a variety of site-directed mutagenesis studies to improve/adjust the alignment. A limited amount of specific information about this model is indicated.	(66)
P450 5 (TXAS)	P450 101	Complete P450 model.	(61)
P450 5 (TXAS)	P450 102	Complete P450 model. Comparison with model derived from P450 101 (see directly above) indicated P450 102 to be a better template for microsomal P450 and that it might be necessary to reexamine microsomal P450 structures predicted based on P450 101.	(61)

Table 1   *Overview of Homology Models Built for P450s, the Crystal Structure(s) Used as a Template, and Some Specifications of the Homology Models (continued)*

Enzyme model (1)	Template P450(s)	Specification [a]	Ref.
P450 11A (P450$_{scc}$)	P450 101	Complete P450 model.	(74)
P450 17 (P450$_{17\alpha}$)	P450 101	Complete P450 model.	(75)
P450 17 (P450$_{17\alpha}$)	P450 101	Complete P450 model.	(64)
P450 19 (P450$_{arom}$)	P450 101	Partial P450 model only containing heme region and I-helix.	(63)
P450 19 (P450$_{arom}$)	P450 101	Complete P450 model. Little specific information about this model is indicated.	(68)
P450 19 (P450$_{arom}$)	P450 101	Complete P450 model.	(76)
P450 19 (P450$_{arom}$)	P450s 101/102/108	Semi-complete P450 model. Uses structural alignment. Improvement from previous model (63). Lacks membrane spanning region.	(58)
P450 19 (P450$_{arom}$)	P450s 101/102/108	Partial P450 model containing heme moiety, I-helix and C-terminus.	(77)
P450 51 (P450$_{14\alpha}$)	P450 101	Complete P450 model.	(78)
P450 19 (P450$_{arom}$)	P450s 101/102/108	Partial P450 model containing heme moiety, I-helix and C-terminus.	(77)
P450 51 (P450$_{14\alpha}$)	P450 101	Complete P450 model.	(78)
P450 51 (P450$_{14\alpha}$)	P450 101	Complete P450 model	(79)
P450 105A1 (P450SU1)	P450 101	2 Complete P450 models (different alignments)	(80)
P450 105B1 (P450SU2)	P450 101	3 Complete P450 models (different alignments)	(80)

[a] Complete P450 model = model constructed for complete enzyme, including regions with (very) low homology; Partial P450 model = regions with low homology have been omitted; Semi-complete/Preliminary P450- model = regions with low homology, non-essential for catalytic activity have been omitted. [b] Specific isoenzyme not given.

Three recently constructed homology models using all three available P450 crystal structures (P450 101, P450 102, and P450 108) and a variety of site-directed mutagenesis data (Table 1) will be discussed below: P450 2B1 (38), P450 2D6 (55) and P450 19 (58).

3.5.1   P450 2B1.   P450 2B1 is one of the most active and versatile cytochromes P450 in the rat, which catalyzes the androstenedione 16β-hydroxylation with a high degree of specificity (81). A homology model for P450 2B1 was constructed using a consensus modeling method in which the coordinates of the model are weighted averages of the coordinates of the three crystal structures (38). The alignment of the sequences of the three crystal structures was done using a structure-based alignment (8), in which positions of secondary structure elements were aligned based on a structural superposition, rather than on an alignment based on primary amino acid sequences. Molecular mechanical and molecular dynamical techniques were used to optimize the protein model (38). The substrates androstenedione and progesterone were docked into the active site area of the protein model and all site-directed mutagenesis data available for P450 2B1 could be explained by this model, in contrast to previous homology models constructed based on P450 101 alone (65). This indicates the superiority of homology models which use all available crystal structures and combine these with site-directed mutagenesis experiments or other protein biochemistry data, relative to models solely constructed from the crystal structure of P450 101. A stereo view of androstenedione docked into the active site of the homology model for P450 2B1 (38) is shown in Figure 7.

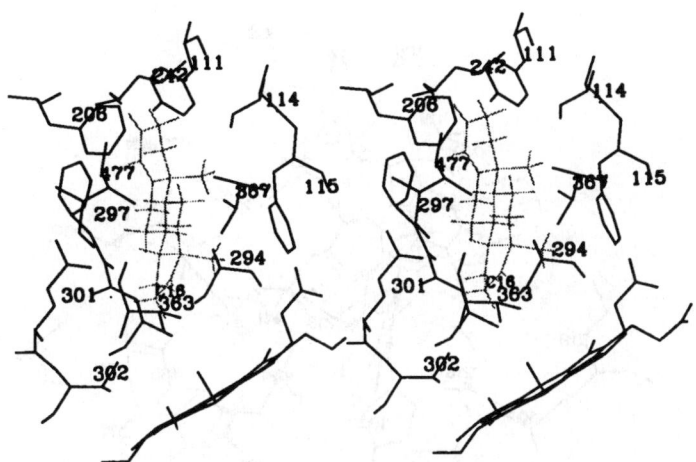

Figure 7 *Androstenedione docked into the upper part of the binding pocket of the P450 2B1 model in a 16β-binding orientation. The substrate is shown in gray, with all hydrogen atoms displayed. Taken from reference (38)*

The active site could be distinguished in an upper part containing residues Ile[114] and Ile[290] (not shown in Figure 7), and a lower part with residues Gly[478] and Ile[480], which were shown to be important for activity (81-87). These two groups of residues could not interact with the substrate androstenedione simultaneously when it is docked in a 16α- or 16β-binding orientation. The key amino acids indicated by site-directed mutagenesis experiments were changed in the model after which androstenedione was docked into the mutant protein model in 16α-, 16β- and 15α-binding orientations, thereby confirming key roles of residues Ile[114], Phe[206], Ile[290], Thr[302], Val[363], and Gly[478], in agreement with site-directed mutagenesis data (81-87) and with the previously derived homology model for P450 2B1 (65).

3.5.2 P450 2D6. A homology model was recently constructed for human P450 2D6, a polymorphic member of the P450 superfamily which is absent in 5-9% of the Caucasian population (21-23). First, the sequences of the crystal structures of the bacterial P450 101, P450 102, and P450 108 isoenzymes were structurally aligned (34) using a method similar to that described by Hasemann *et al.* (8). Then a multi-alignment for 66 members of the P450 2-family was constructed (34), which facilitated the alignment of P450 2D6 with the structural alignment of the three crystal structures. This multi-alignment also enabled the use of site-directed mutagenesis data of other members of the P450 2-family to improve the alignment between P450 2D6 and the structural alignment of the sequences of the three crystal structures (34). Molecular mechanical calculations were used to optimize the constructed homology model (34). The active site consisted of the heme moiety, the F-, I- and K-helices, the loop between helices B and B', the loop between the B' and the C-helix, and β-sheets 3 and 5 (34). Three known substrates (debrisoquine, dextromethorphan and GBR 12909 (Figure 8)) and one inhibitor (ajmalicine) were docked into the active site of the P450 2D6 model (34) indicating the protein model to be able to accommodate large substrates, which extended considerably the boundaries of the previously derived small molecule model for P450 2D6 (27, 31) described in a previous section.

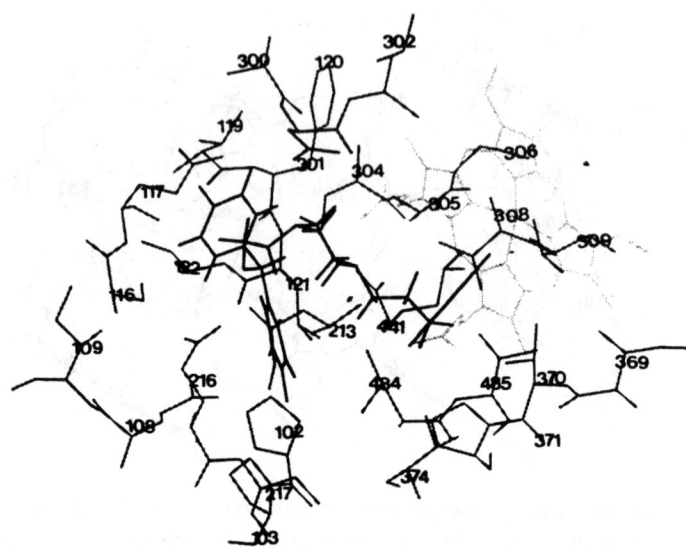

Figure 8   *Orientation of GBR 12909 leading to benzylic hydroxylation docked into the*
active site of the homology model for P450 2D6 (34). The heme moiety is
shown in light gray. The protein is depicted in gray with Asp[301] highlighted in
black. GBR 12909 is shown in black, with all hydrogen atoms displayed.
Adapted from reference (34)

The orientation of the substrates relative to each other when docked into the active
site, the position of the heme moiety and the position of the I-helix containing Asp[301] (an
amino acid proposed (35) and shown (36)to be crucial for the catalytic activity of P450
2D6), were used to improve the previously described small molecule model for P450 2D6
substrates (see Figures 4c (27)and 4d (33). The two amino acids in P450 2D6 for which
site-directed mutagenesis data are available, namely Asp[301] (36) and Val[374] (88, 89) were
indeed part of the active site of the derived protein model (34). Especially Asp[301] is an
important residue for catalytic activity as it forms a hydrogen bond with the basic nitrogen
atom present in the substrates of P450 2D6 (as indicated above). As no further site-directed
mutagenesis data are available for P450 2D6 as yet, no validation could be given of the
importance of other amino acids in the active site of P450 2D6 indicated by the model. The
homology model indicated a region of the active site to be a hydrophobic envelope in
which only planar substrates could be accommodated, in close agreement with previously
derived small molecule models for P450 2D6 (25, 27, 29, 32, 33). Furthermore, this protein
model for P450 2D6 was in close agreement with a recently described set of protein
models for P450 2D6 based on a similar structural alignment and NMR derived distance
restraints (62).

3.5.3 P450 19. P450 19 (P450 aromatase) catalyzes the conversion of C19 steroids
to estrogens, which is one of the most complex and least understood P450 catalyzed
reactions (58). A recently built model for P450 19 (58)was based on the core structure of
the three crystallized P450s using a structure-based alignment (8) based on a combination
of previously reported alignments from Hasemann *et al.* (54) and Ravishandran *et al.* (6).
Molecular mechanics and molecular dynamics were used to optimize the homology model
(58). The active site was formed by the heme moiety, the loop between helices B' and C,

the I-helix and β-sheets 1 and 4 (58). The loop between helices B and B' was not in the active site of this homology model (58), in contrast with an earlier homology model for P450 19 (76) based solely on P450 101 and in contrast with the homology model for P450 2D6 (34) based on the crystal structures of P450 101, P450 102 and P450 108, as described above. Two enantiomers of vorozole, a known inhibitor of P450 19, were docked into the active site of the protein model explaining experimentally observed results (58), like the necessity for a kink in the I-helix which can be accomplished by either a proline residue or two glycine residues. Residues indicated by site-directed mutagenesis experiments to be important for catalytic activity, i.e., Glu^{302} (63), Asp^{309} (90, 91), Thr^{310} (90, 91), and Ile^{474} (92), were indeed part of the active site (58), Regions important for binding of P450 19 and its redoxpartner were also predicted, indicating that P450 19 cannot be classified as a class I or a class II P450, but is an intermediate P450 type (58).

3.6 Summary

All homology models of mammalian P450s based on four (bacterial) crystal structures presently available indicate certain regions in the P450 isoenzymes which can be modeled with relative ease and high accuracy (e.g., the oxygen binding domain near the heme, the helices D, E, I and L and some β-sheets (34, 38, 58)) and certain regions in which the models are less reliable due to large differences between the available bacterial P450 crystal structures in these four regions (e.g., the B'-helix, the loops between the C- and D-helices, the region spanning the F- and G-helices and some parts of the β-sheets) (34, 38, 58). Generally, however, very useful information concerning amino acids important for substrate and/or inhibitor binding can be obtained using homology models, although due to the relative low homology in the substrate/inhibitor binding site region between the various P450s, these predictions should always be considered carefully and verified experimentally. Homology models can therefore be very useful to guide site-directed mutagenesis or site-specific modification experiments, but they cannot completely replace them. Concerning amino acids responsible for the catalytic activity of a certain P450, homology models can merely be used to verify whether the observed differences can be rationalized using the modeled structure, as kinetic information on catalytic activities cannot be obtained from theoretical interaction studies.

4 GENERAL CONCLUSIONS

For P450s a number of small molecule models have been derived, either based on suitable template molecules or on a variety of substrates or inhibitors when a single compound was inappropriate as a template molecule. Several of these small molecule models were shown to have a good predictive value concerning metabolism and substrate/inhibitor selectivity, a property especially relevant for isoenzymes which are subject to genetic polymorphisms (e.g., P450 2D6 (27, 28, 33)). Despite the potential benefits (especially for the chemical and pharmaceutical industry) the development of small molecule (pharmacophore) models for biotransformation enzymes has received relatively little attention as yet, in contrast to pharmacophore models for receptors in medicinal chemistry.

The homology models for P450s indicate that certain regions in the proteins can be modeled with relative ease and high accuracy (e.g., the oxygen binding domain near the heme, the helices D, E, I, and L and some β-sheets (34, 38, 58)) while in certain other regions the homology models are less reliable due to large differences between available crystal structures and the modeled P450s (e.g., the B'-helix, the loops between the C- and D-helices, the region spanning the F- and G-helices and some parts of the β-sheets (34, 38, 58). The topology of homology models are generally prejudiced by the template crystal structure. However, due to crystal packing effects, the crystal structure conformation might differ from the conformation of the protein in solvent. For this reason additional information from three-dimensional NMR techniques would be useful to supplement the crystal structures.

Generally, useful information concerning amino acids important for substrate and/or inhibitor binding can be obtained using homology models, although due to the relatively low homology in the case of P450s in the substrate binding site region these predictions have to be considered carefully and should be verified experimentally. Homology models can be used to guide site-directed mutagenesis and site-specific modification experiments, but cannot completely replace them. Concerning the role of amino acids in the catalytic activity of a certain P450, homology models can merely be used to verify whether the observed differences can be rationalized using the modeled structure as information on catalytic activities cannot be obtained from these theoretical interaction studies. Cautious indications of substrate selectivity can be give in specific cases, although these predictions also have to be considered carefully and verified experimentally.

Combination of small molecule models and homology models enhances the value of computational chemistry techniques for biotransformation enzymes as specific sites in the ligand (substrate, inhibitor or metabolic product) are combined with specific sites in the protein.

References

(1) Nelson, D. R., Kamataki, T., Waxman, D. J., Guengerich, F. P., Estabrook, R. W., Feyereisen, R., Gonzalez, F. J., Coon, M. J., Gunsalus, I. C., Gotoh, O., Okuda, K., and Nebert, D. W. (1993) *DNA Cell Biol.* **12**, 1-51.
(2) Vermeulen, N. P. E. (1996) In *Cytochromes P450: metabolic and toxicological aspects.* (Ioannides, C., Ed.) CRC Press Inc., Boca Raton, Florida, USA, pg. 29-53.
(3) Graham-Lorence, S., and Peterson, J. A. (1996) *FASEB J.* **10**, 206-214.
(4) Guengerich, F. P. (1991) *J. Biol. Chem.* **266**, 10019-10022.
(5) Goeptar, A. R., Scheerens, H., and Vermeulen, N. P. E. (1995) *Crit. Rev. Toxicol.* **25**, 25-65.
(6) Ravichandran, K. G., Boddupalli, S. S., Hasemann, C. A., Peterson, J. A., and Deisenhofer, J. (1993) *Science* **261**, 731-736.
(7) Degtyarenko, K. N. (1995) *Protein Eng.* **8**, 737-747.
(8) Hasemann, C. A., Kurumbail, R. G., Boddupalli, S. S., Peterson, J. A., and Deisenhofer, J. (1995) *Structure* **2**, 41-62.
(9) Allan, F. H., and Kennard, O. (1993) *Chemical Design Automation News* **8**, 31-37.
(10) Korzekwa, K. R., and Jones, J. P. (1993) *Pharmacogenetics* **3**, 1-18.
(11) de Groot, M. J., Donné-Op den Kelder, G. M., Commandeur, J. N. M., van Lenthe, J. H., and Vermeulen, N. P. E. (1995) *Chem. Res. Toxicol.* **8**, 437-443.
(12) de Groot, M. J., van der Aar, E. M., Nieuwenhuizen, P. J., van der Plas, R. M., Donné-Op den Kelder, G. M., Commandeur, J. N. M., and Vermeulen, N. P. E. (1995) *Chem. Res. Toxicol.* **8**, 649-658.
(13) Jerina, D. M., Michaud, D. P., Feldman, R. J., Armstrong, R. N., Vyas, K. P., Thakker, D. R., Yagi, H., Thomas, P. E., Ryan, D. E., and Levin, W. (1982) In *Microsomes, drug oxidations, and drug toxicity.* (Sato, R., and Kato, R., Eds.) pp 195-201, Japan Scientific Societies Press, Tokyo.

(14) Kadlubar, F. F., and Hammons, G. J. (1987) In *Mammalian cytochromes P450*. (Guengerich, F. P., Ed.) pp 81-130, CRC Press, Boca Raton.

(15) Yang, S. K. (1988) *Biochem. Pharmacol.* **37**, 61-70.

(16) Koymans, L. M. H., Donné-Op den Kelder, G. M., te Koppele, J. M., and Vermeulen, N. P. E. (1993) *Drug Metab. Rev.* **25**, 325-387.

(17) Lewis, D. F. V., Ioannides, C., and Parke, D. V. (1986) *Biochem. Pharmacol.* **35**, 2179-2185.

(18) Ioannides, C., and Parke, D. V. (1987) *Biochem. Pharmacol.* **36**, 4197-4207.

(19) Lewis, D. F. V., Ioannides, C., and Parke, D. V. (1989) *Chem.-Biol. Interact.* **70**, 263-280.

(20) Mancy, A., Brotto, P., Dijols, S., Dansette, P. M., and Mansuy, D. (1995) *Biochemistry* **34**, 10365-10375.

(21) Mahgoub, A., Idle, J. R., Dring, L. G., Lancaster, R., and Smith, R. L. (1977) *Lancet* **11**, 584-586.

(22) Eichelbaum, M., Spannbrucker, N., Steineke, B., and Dengler, H. J. (1979) *Eur. J. Clin. Pharmacol.* **16**, 183-187.

(23) Armstrong, M., Fairbrother, K., Idle, J. R., and Daly, A. K. (1994) *Pharmacogenetics* **4**, 73-81.

(24) Wolff, T., Distlerath, L. M., Worthington, M. T., Groopman, J. D., Hammons, G. J., Kadlubar, F. F., Prough, R. A., Martin, M. M., and Guengerich, F. P. (1985) *Cancer Res.* **45**, 2116-2122.

(25) Meyer, U. A., Gut, J., Kronbach, T., Skoda, C., Meier, U. T., Catin, T., and Dayer, P. (1986) *Xenobiotica* **16**, 449-464.

(26) Islam, S. A., Wolf, C. R., Lennard, M. S., and Sternberg, M. J. E. (1991) *Carcinogenesis* **12**, 2211-2219.

(27) Koymans, L. M. H., Vermeulen, N. P. E., van Acker, S. A. B. E., te Koppele, J. M., Heykants, J. J. P., Lavrijsen, K., Meuldermans, W., and Donné-Op den Kelder, G. M. (1992) *Chem. Res. Toxicol.* **5**, 211-219.

(28) Strobl, G. R., von Kreudener, S., Stöckigt, J., Guengerich, F. P., and Wolff, T. (1993) *J. Med. Chem.* **36**, 1136-1145.

(29) Poulos, T. L., Finzel, B. C., and Howard, A. J. (1986) *Biochemistry* **25**, 5314-5322.

(30) Poulos, T. L., Finzel, B. C., Gunsalus, I. C., Wagner, G. C., and Kraut, J. (1985) *J. Biol. Chem.* **260**, 16122-16130.

(31) de Groot, M. J., Bijloo, G. J., Hansen, K. T., and Vermeulen, N. P. E. (1995) *Drug Metab. Dispos.* **23**, 667-669.

(32) Wolff, T., Distlerath, L. M., Worthington, M. T., and Guengerich, F. P. (1987) *Arch. Toxicol.* **60**, 89-90.

(33) de Groot, M. J., Bijloo, G. J., Martens, B. J., van Acker, F. A. A., and Vermeulen, N. P. E. (1997) *Chem. Res. Toxicol.*, **10**(1): 41-48 .

(34) de Groot, M. J., Vermeulen, N. P. E., Kramer, J. D., van Acker, F. A. A., and Donné-Op den Kelder, G. M. (1996) *Chem. Res. Toxicol.*, **9** (7), 1079-1091.

(35) Koymans, L. M. H., Vermeulen, N. P. E., Baarslag, A., and Donné-Op den Kelder, G. M. (1993) *J. Comp.-Aided. Mol. Design* **7**, 281-289.

(36) Ellis, S. W., Hayhurst, G. P., Smith, G., Lightfoot, T., Wong, M. M. S., Simula, A. P., Ackland, M. J., Sternberg, M. J. E., Lennard, M. S., Tucker, G. T., and Wolf, C. R. (1995) *J. Biol. Chem.* **270**, 29055-29058.

(37) de Groot, M. J., Bijloo, G. J., van Acker, F. A. A., Fonseca Guerra, C., Snijders, J. G., and Vermeulen, N. P. E. (1996) *Xenobiotica*, **27**(4), 357-368.

(38) Szklarz, G. D., He, Y. A., and Halpert, J. R. (1995) *Biochemistry* **34**, 14312-14322.

(39) Ouzounis, C. A., and Melvin, W. T. (1991) *Eur. J. Biochem.* **198**, 307-315.

(40) Paulsen, M. D., and Ornstein, R. L. (1991) *Proteins: Structure, Function and Genetics.* **11**, 184-204.

(41) Paulsen, M. D., and Ornstein, R. L. (1992) *J. Comp.-Aided Mol. Design* **6**, 449-460.

(42) Ortiz de Montellano, P. R. (1995) *Biochimie* **77**, 581-593.

(43) Poulos, T. L., Finzel, B. C., and Howard, A. J. (1987) *J. Mol. Biol.* **195**, 687-700.

(44) Raag, R., and Poulos, T. L. (1989) *Biochemistry* **28**, 917-922.

(45) Raag, R., and Poulos, T. L. (1991) *Biochemistry* **30**, 2674-2684.

(46) Poulos, T. L., and Howard, A. J. (1987) *Biochemistry* **26**, 8165-8174.

(47) Raag, R., Li, H., Jones, B. C., and Poulos, T. L. (1993) *Biochemistry* **32**, 4571-4578.

(48) Li, H., Narasimhulu, S., Havran, L. M., Winkler, J. D., and Poulos, T. L. (1995) *J. Am. Chem. Soc.* **117**, 6297-6299.

(49) Boddupalli, S. S., Hasemann, C. A., Ravichandran, K. G., Lu, J. Y., Goldsmith, E. J., Deisenhofer, J., and Peterson, J. A. (1992) *Proc. Natl. Acad. Sci. U.S.A.* **89**, 5567-5571.

(50) Li, H., and Poulos, T. L. (1994) *Structure* **2**, 461-464.

(51) Li, H., and Poulos, T. L. (1995) *Acta Crystallogr. D-Biol. Cryst.* **51**, 21-32.

(52) Cupp-Vickery, J. R., Li, H. Y., and Poulos, T. L. (1994) *Protein-Struct. Funct. Genet.* **20**, 197-201.

(53) Cupp-Vickery, J. R., and Poulos, T. L. (1995) *Nature Struct. Biol.* **2**, 144-153.

(54) Hasemann, C. A., Ravichandran, K. G., Peterson, J. A., and Deisenhofer, J. (1994) *J. Mol. Biol.* **236**, 1169-1185.

(55) Abola, E. E., Bernstein, F. C., Bryant, S. H., Koetzle, T. F., and Weng, J. (1987) In *Crystallographic databases- information contents, software systems, scientific applications.* (Allen, F. H., Bergerhoff, G., and Sievers, R., Eds.) pp 107-132, Data Commision of the International Union of Crystallography, Bonn/Cambridge/Chester.

(56) Modi, S., Gilham, D.E., Sutcliffe, L.-Y., Primrose, W.U., Wolf, C.R. and Roberts, G.C.K. (1997) *Biochemistry*, **36**, 3361-4470.

(57) Nakahara, K., Shoun, H., Adachi, S., Iizuka, T., and Shiro, Y. (1994) *J. Mol. Biol.* **239**, 158-159.

(58) Graham-Lorence, S., Amarneh, B., White, R. E., Peterson, J. A., and Simpson, E. R. (1995) *Protein Sci.* **4**, 1065-1080.

(59) Poulos, T. L. (1995) *Curr. Opin. Struct. Biol.* **5**, 767-774.

(60) Jones, J. P., Shou, M., and Korzekwa, K. R. (1995) *Biochemistry* **34**, 6956-6961.

(61) Ruan, K. H., Milfeld, K., Kulmacz, R. J., and Wu, K. K. (1994) *Protein Eng.* **7**, 1345-1351.

(62) Modi, S., Paine, M. J., Sutcliffe, M. J., Lian, L. Y., Primrose, W. U., Wolf, C. R., and Roberts, G. C. K. (1996) *Biochemistry* **35**, 4540-4550.

(63) Graham-Lorence, S., Khalil, M. W., Florence, M. C., Mendelson, C. R., and Simpson, E. R. (1991) *J. Biol. Chem.* **266**, 11939-11946.

(64) Lin, D., Zhang, L. H., Chiao, E., and Miller, W. L. (1994) *Mol. Endocrinol.* **8**, 392-402.

(65) Szklarz, G. D., Ornstein, R. L., and Halpert, J. P. (1994) *J. Biomolec. Struct. Dynamics.* **12**, 61-78.

(66) Lewis, D. F. V. (1995) *Xenobiotica* **25**, 333-366.

(67) Lewis, D. F. V., and Lake, B. G. (1995) *Xenobiotica* **25**, 585-598.

(68) Lewis, D. F. V., and Moereels, H. (1992) *J. Comp.-Aided Design* **6**, 235-252.

(69) Lindberg, R. L., and Negishi, M. (1989) *Nature* **399**, 632-634.

(70) Zvelebil, M. J. J. M., Wolf, C. R., and Sternberg, M. J. E. (1991) *Protein Eng.* **4**, 271-282.

(71) Lewis, D. F. V., Ioannides, C., and Parke, D. V. (1994) *Toxicol. Lett.* **71**, 235-243.

(72) Skrzypczak-Jankun, E., and Kurumbail, R. G. (1996) *Acta Crystallogr. C-Cryst. Str.* **C52**, 189-191.

(73) Ferenczy, G. G., and Morris, G. M. (1989) *J. Mol. Graphics* **7**, 206-211.

(74) Vijayakumar, S., and Salerno, J. C. (1992) *Biochim. Biophys. Acta* **1160**, 281-286.

(75) Laughton, C. A., Neidle, S., Zvelebil, M. J. J. M., and Sternberg, M. J. E. (1990) *Biochem. Biophys. Res. Commun.* **171**, 1160-1167.

(76) Laughton, C. A., Zvelebil, M. J. J. M., and Neidle, S. (1993) *J. Steroid Biochem. Molec. Biol.* **44**, 399-407.

(77) Koymans, L. M. H., Moereels, H., and Bossche, H. V. (1995) *J. Steroid Biochem. Mol. Biol.* **53**, 191-197.

(78) Ishida, N., Aoyama, Y., Hatanaka, R., Oyama, Y., Imajo, S., Ishiguro, M., Oshime, T., Nakazato, H., Noguchi, T., Maitra, U. S., Mohan, V. P., Sprinson, D. B., and Yoshida, Y. (1988) *Biochem. Biophys. Res. Commun.* **155**, 317-323.

(79) Boscott, P. E., and Grant, G. H. (1994) *J. Mol. Graphics* **12**, 185-192.

(80) Braatz, J. A., Bass, M. B., and Ornstein, R. L. (1994) *J. Comp.-Aided Mol. Design* **8**, 607-622.

(81) Halpert, J. R., and He, Y. (1993) *J. Biol. Chem.* **268**, 4453-4457.

(82) Aoyama, T., Korzekwa, K., Nagata, K., Adesnik, M., Reiss, A., Lapenson, D. P., Gillette, J., Gelboin, H. V., Waxman, D. J., and Gonzalez, F. J. (1989) *J. Biol. Chem.* **264**, 21327-21333.

(83) Kedzie, K. M., Balfour, C. A., Escobar, G. Y., Grimm, S. W., He, Y., Pepperl, D. J., Regan, J. W., Stevens, J. C., and Halpert, J. R. (1991) *J. Biol. Chem.* **266**, 22515-22521.

(84) He, Y. A., Balfour, C. A., Kedzie, K. M., and Halpert, J. R. (1992) *Biochemistry* **31**, 9220-9226.

(85) He, Y., Luo, Z., Klekotka, P. A., Burnett, V. L., and Halpert, J. R. (1994) *Biochemistry* **33**, 4419-4424.

(86) Hasler, J. A., Harlow, G. R., Szklarz, G. D., John, G. H., Kedzie, K. M., Burnett, V. L., He, Y. A., Kaminsky, L. S., and Halpert, J. R. (1994) *Mol. Pharmacol.* **46**, 338-345.

(87) He, Y. Q., He, Y. A., and Halpert, H. R. (1995) *Chem. Res. Toxicol.* **8**, 574-579.

(88) Ellis, S. W., Rowland, K., Harlow, J. R., Simula, A. P., Lennard, M. S., Woods, H. F., Tucker, G. T., and Wolf, C. R. (1994) *Br. J. Pharmacol.* **112**, 244P.

(89) Ellis, S. W., Rowland, K., Ackland, M. J., Rekka, E., Simula, A. P., Lennard, M. S., Wolf, C. R., and Tucker, G. T. (1996) *Biochem. J.* **316**, 647-654.

(90) Chen, S., and Zhou, D. (1992) *J. Biol. Chem.* **267**, 22587-22594.

(91) Amarneh, B., Corbin, C. J., Peterson, J. A., Simpson, E. R., and Graham-Lorence, S. (1993) *Mol. Endocrinol.* **7**, 1617-1624.

(92) Zhou, D., Cain, L. L., Laughton, C. A., Korzekwa, K. R., and Chen, S. (1994) *J. Biol. Chem.* **269**, 19501-19508.

(93) Smith, G., Modi, S., Pilla, I., Lian, L-Y., Sutcliffe, M.J., Pritchard, M.P., Friedberg, T., Roberts, C.K. and Wolf, C.R. (1998) *Biochrem.J.*, **331,** 783-792.

(94) Lewis, D.F.V. and Lake, B.G. (1996), *Xenobiotica*, **26**(7), 723-753.

(95) Lewis, D.F.V. and Lake, B.G. (1997), *Xenobiotica*, **27**(5), 443-478.

(96) Lewis, D.F.V., Bird, N.G. and Parke, D.V. (1997), *Toxicology*, **118**, 93-113.

(97) Lewis, D.F.V., Eddershaw, P.J., Goldfarb, P.S. and Tarbit, M.H. (1997), *Xenobiotica*, **27**(4), 319-340.

(98) Szklarz, G.D. and Halpert, J.R. (1997), *J. of Computer-Aided Mol. Design,* **11,** 265-272.